Lawrie's meat science

Related titles:

Improving the safety of fresh meat
(ISBN-13: 978-1-85573-955-0; ISBN-10: 1-85573-955-0)
It is widely recognised that food safety depends on effective intervention at all stages in the food chain, including the production of raw materials. Contaminated raw materials from agricultural production increase the hazards that subsequent processing operations must deal with, together with the risk that such contamination may survive through to the point of consumption. This book provides an authoritative reference summarising the wealth of research on reducing microbial and other hazards in raw and fresh red meat.

Meat processing: improving quality
(ISBN-13: 978-1-85573-583-5; ISBN-10: 1-85573-583-0)
This major collection summarises key developments in research, from improving raw meat quality and safety issues to developments in meat processing and specific aspects of meat product quality such as colour, flavour and texture.

HACCP in the meat industry
(ISBN-13: 978-1-85573-448-7; ISBN-10: 1-85573-448-6)
Following the crises involving BSE and *E coli*, the meat industry has been left with an enormous consumer confidence problem. In order to regain the trust of the general public the industry must establish and adhere to strict hygiene and hazard control systems. HACCP is a systematic approach to the identification, evaluation and control of food safety hazards. It is being applied across the world, with countries such as the USA, Australia, New Zealand and the UK leading the way. However, effective implementation in the meat industry remains difficult and controversial. This book is a survey of key principles and best practice, providing an authoritative guide to making HACCP systems work successfully in the meat industry.

Details of these books and a complete list of Woodhead titles can be obtained by:

* visiting our web site at www.woodheadpublishing.com
* contacting Customer Services (e-mail: sales@woodhead-publishing.com; fax: +44 (0) 1223 893694; tel.: +44 (0) 1223 891358 ext.30; address: Woodhead Publishing Ltd, Abington Hall, Abington, Cambridge CB1 6AH, England)

Lawrie's meat science

SEVENTH EDITION

R. A. Lawrie
Emeritus Professor of Food Science,
University of Nottingham,

in collaboration with

D. A. Ledward
Emeritus Professor of Food Science,
University of Reading

CRC Press
Boca Raton Boston New York Washington, DC

WOODHEAD PUBLISHING LIMITED
Cambridge England

Published by Woodhead Publishing Limited, Abington Hall, Abington
Cambridge CB1 6AH, England
www.woodheadpublishing.com

Published in North America by CRC Press LLC, 6000 Broken Sound Parkway, NW,
Suite 300, Boca Raton, FL 33487, USA

First English edition 1966 Pergamon Press, reprinted 1968
Spanish edition 1967
German edition 1969
Japanese edition 1971
Russian edition 1973
Second English edition 1974, reprinted 1975
Second Spanish edition 1977
Third English edition 1979
Italian edition 1983
Fourth English edition 1985, reprinted 1988
Fifth English edition 1991
Sixth English edition 1998 Woodhead Publishing Limited, reprinted 2002
Third Spanish edition 1998
Brazilian edition 2005
Seventh English edition 2006 Woodhead Publishing Limited and CRC Press LLC

© 2006, Woodhead Publishing Limited
The authors have asserted their moral rights.

British Library Cataloguing in Publication Data
A catalogue record for this book is available from the British Library.

Library of Congress Cataloging in Publication Data
A catalog record for this book is available from the Library of Congress.

Woodhead Publishing ISBN-13: 978-1-84569-159-2 (book)
Woodhead Publishing ISBN-10: 1-84569-159-8 (book)
Woodhead Publishing ISBN-13: 978-1-84569-161-5 (e-book)
Woodhead Publishing ISBN-10: 1-84569-161-X (e-book)
CRC Press ISBN-13: 978-0-8493-8726-5
CRC Press ISBN-10: 0-8493-8726-4
CRC Press order number: WP8726

The publishers' policy is to use permanent paper from mills that operate a sustainable forestry policy,
and which has been manufactured from pulp which is processed using acid-free and elementary
chlorine-free practices. Furthermore, the publishers ensure that the text paper and cover board used
have met acceptable environmental accreditation standards.

Typeset by SNP Best-set Typesetter Ltd., Hong Kong
Printed by TJ International, Padstow, Cornwall, England

Contents

Preface to seventh edition

Although 40 years have passed since this book was first published and, in the interim, there have been many developments in meat science, I have seen no reason to alter the general plan in which the subject is presented.

Since the publication of the sixth edition, the science of bioinformatics has emerged, whereby complex computer techniques have made it possible to simultaneously identify, in a cell or tissue, all the possible modes of transcription of nuclear DNA by RNA (transcriptomics), the entirety of the protein species present (proteomics) and all the metabolites produced during functioning (metabolomics). Nanotechnology has made it possible to identify – and usefully manipulate – biological structures at the molecular level where the properties may vary in important respects from those exhibited at more conventional dimensions. These developments provide a new approach to the understanding and potential control of eating quality and nutritive value in meat.

The different characteristics of the individual muscles in a carcass – long recognized by biochemists – are now being related to new methods of slaughter and carcass dressing whereby specific cuts or individual muscles can be economically produced; and consumers may anticipate, before long, being able to demand and obtain meat of the precise colour, juiciness, tenderness and flavour which they personally desire. Such 'muscle profiling' is already being developed in the USA.

More detailed information is becoming available on the complexity of the protein of muscle, on the proteolysis responsible for tenderizing during ageing, on the central role of Ca^{++} ions in contraction, proteolysis, water-holding capacity and the action of many enzymes on both the membranes and interiors of cells.

New techniques (e.g. 'nose space' analysis) are elucidating the mechanism of olefaction and revealing the concomitant involvement of factors, such as viscosity, in modifying their expression.

Vastly increased understanding of the mode of action of genes and of the nature of DNA has provided reliable means for the identification of species (even in severely processed meat products), revealed the mechanism of such defects as pale, soft, exudative pork, and afforded the means of analysing the multiplicity of toxins produced by pathogenic micro-organisms.

A new concept, 'quorum sensing', has shown how micro-organisms communicate, thereby influencing their potential for growth and survival, in various environments.

Further advances have been made in the processing of meat by high pressure, thermal treatment, ionizing radiation and storage below its freezing point.

There is continuing interest in the significance of meat eating for the health of consumers. Thus, insofar as saturated fatty acids are less beneficial than those that are polyunsaturated, a number of stratagems have been developed whereby poly-unsaturated acids from feed can be incorporated even into the flesh of ruminants. Again, it is now known that meat is an important source of selenium and zinc – micronutrients whose nutritional importance has recently been recognized.

Respecting potential hazards of meat consumption, there is still no proof that the consumption of flesh from animals suffering from bovine spongioform encephalitis induces mental degeneration in human beings. Increasingly sophisticated studies appear to show a relationship between meat eating and the induction of cancer; but the biochemical basis for such a relationship has not been established. New strains of antibiotic-resistant micro-organisms associated with meat continue to emerge; and cause ephemeral concerns.

Whatever the merits and demerits – real or inferred – of meat, its true signifi-cance for the consumer must await the means of specific biochemical identification of each individual's metabolism. In the interim there is no reason to doubt that meat should be included in a balanced diet both for its content of essential nutrients and for its widely appreciated organoleptic characteristics.

R. A. LAWRIE
Sutton Bonington

Preface to first edition

The scientific study of food has emerged as a discipline in its own right since the end of the 1939–45 war. This development reflects an increasing awareness of the fact that the eating quality of food commodities is determined by a logical sequence of circumstances starting at conception of the animal, or at germination of the seed, and culminating in consumption. From this point of view, the food scientist is inevitably involved in various aspects of chemistry and biochemistry, genetics and microbiology, botany and zoology, physiology and anatomy, agriculture and horticulture, nutrition and medicine, public health and psychology.

Apart from the problems of preserving the attributes of eating quality and of nutritive value, it seems likely that food science will become increasingly concerned with enhancing the biological value of traditional foods and with elaborating entirely new sources of nourishment, as the pressure of world population grows. Moreover, a closer association of food science and medicine can be anticipated as another development. This will arise not only in relation to the cause or remedy of already accepted diseases, but also in relation to many subclinical syndromes which are as yet unappreciated. Such may well prevent us as individuals and as a species from attaining the efficiency and length of life of which our present evolutionary form may be capable.

Meat is one of the major commodities with which food science is concerned and is the subject of the present volume. It would not be feasible to consider all aspects of this vast topic. Instead, an attempt has been made to outline the essential basis of meat in a sequence of phases. These comprise, in turn, the origin and development of meat animals, the structural and chemical elaboration of muscular tissue, the conversion of muscle to meat, the nature of the adverse changes to which meat is susceptible before consumption, the discouragement of such spoilage by various means and, finally, the eating quality. The central theme of this approach is the fact that, because muscles have been diversified in the course of evolution to effect specific types of movement, all meat cannot be alike. It follows that the variability, in its keeping and eating qualities, which has become more apparent to the consumer with the growth of prepackaging methods of display and sale, is not capricious. On the contrary, it is predictable and increasingly controllable.

Those aspects of meat which have not been introduced in the present volume have mainly economic implications and do not involve any concept which is incompatible with the basic approach adopted. They have been thoroughly considered by other authors.

In addition to acknowledging my specific indebtedness to various individuals and organizations, as indicated in the following paragraphs, I should like to express my appreciation of the co-operation of many colleagues in Cambridge and Brisbane during the 15 years when I was associated with them in meat research activities.

I am especially grateful to Mr D. P. Gatherum and Mr C. A. Voyle for their considerable help in the preparation of the illustrations. I should also like to thank Prof. J. Hawthorn, F.R.S.E., of the Department of Food Science, University of Strathclyde, for useful criticism.

R. A. LAWRIE
Sutton Bonington

Acknowledgements

I wish to thank the following individuals for their kindness in permitting me to reproduce the illustrations and tables indicated:

Prof. M. E. Bailey, Department Food Science & Nutrition, University of Missouri, Columbia, USA (Table 5.5); Mr J. Barlow, M.B.E., formerly of A.F.R.C. Food Research Institute, Bristol (Fig. 6.1); Dr E. M. Barnes, formerly of A.F.R.C. Food Research Institute, Norwich (Fig. 6.6); Dr J. A. Beltran, Univ. of Zarogoza (Table 4.25); the late Dr J. R. Bendall, Histon, Cambridge (Fig. 4.2); Dr E. Bendixen, Danish Institute of Agricultural Service, Tjele (Fig. 4.1); Mr C. Brown, Meat & Livestock Commission, Milton Keynes (Table 1.3); Mr D. Croston, Meat & Livestock Commission, Milton Keynes (Table 1.2); Mr A. Cuthbertson, formerly Head, Meat Quality Unit, Meat & Livestock Commission, Milton Keynes (Fig. 3.2); Dr C. E. Devine, Meat Industry Research Institute of New Zealand Inc. (Fig. 10.4); Dr M. R. Dickson, Meat Industry Research Institute of New Zealand, Inc. (Fig. 5.2); Dr J. B. Fox, Jr., US Department of Agriculture, Philadelphia, USA (Fig. 10.1); Prof. Marion Greaser, Muscle Biology Laboratory University of Wisconsin, USA (Fig. 3.9); Prof. J. Gross, Massachusetts General Hospital, Boston, USA (Fig. 3.5); the late K. C. Hales, Shipowners Refrigerated Cargo Research Council, Cambridge (Fig. 7.2); Prof. R. Hamm, former Director, Bundesanstalt für Fleischforschung, Kulmbach, Germany (Figs. 8.1, 8.2, 8.4 and 10.2); the late Sir John Hammond, FRS, Emeritus Reader in Animal Physiology, University of Cambridge (Figs. 1.1, 1.2 and 1.3); Dr H. E. Huxley, FRS, M.R.C. Unit for Molecular Biology, Cambridge (Figs. 3.7(f), 3.7(g), 3.8(a) and 3.8(c)); Prof. H. Iwamoto, Kyushu Univ., Japan (Fig. 4.8); Mr N. King, A.F.R.C. Food Research Institute, Norwich (Fig. 3.7(e)); Prof. G. G. Knappeis, Institute of Neurophysiology, University of Copenhagen, Denmark (Fig. 3.8(e)); Dr Susan Lowey, Harvard Medical School, USA (Fig. 3.8(d)); Prof. B. B. Marsh, former Director, Muscle Biology Laboratory, University of Wisconsin, USA (Fig. 4.6); Dr M. N. Martino, La Plata, Argentina (Fig. 7.6); the late Dr H. Pálsson, Reykjavik, Iceland (Fig. 2.1); Dr I. F. Penny (Fig. 8.7), the late Dr R. W. Pomeroy (Fig. 3.2) and Mr D. J. Restall (Figs. 3.7(d) and (e)), all formerly of A.F.R.C. Food Research Institute, Bristol; Dr R. W. D. Rowe, formerly of C.S.I.R.O. Meat

Investigations Laboratory, Brisbane, Queensland (Figs. 3.4 and 3.6); Dr R. K. Scopes, University of New England, Australia (Fig. 5.4); Dr Darl Scwartz, Indiana University Medical School, USA (Fig. 3.9); the late Dr W. J. Scott, formerly of C.S.I.R.O. Meat Investigations Laboratory, Brisbane, Queensland, Australia (Fig. 6.4); the late Dr J. G. Sharp, formerly of Low Temperature Research Station, Cambridge (Figs. 3.7(a), 5.5 and 8.3); Prof. K. Takahashi, Hokkaido University, Sapporo, Japan (Figs. 3.3 and 5.4); Dr M. C. Urbin, Swedish Convenant Hospital, Chicago, USA (Fig. 10.3); Mr C. A. Voyle, formerly of A.F.R.C. Food Research Institute, Bristol (Figs. 3.7(c), 3.7(d), 3.10 and 3.11); Mr G. E. Welsh, British Pig Association (Table 1.4); and Drs O. Young and S. R. Payne, Meat Industry Research Institute of New Zealand Inc. (Fig. 7.5).

I am similarly indebted to the following publishers and organizations.

Academic Press, Inc., New York (Figs. 6.6, 8.1, 8.2, 8.4 and 10.2); American Meat Science Association, Chicago (Fig. 3.9); Butterworths Scientific Publications, London (Figs. 2.1 and 5.1; Table 4.1); Cambridge University Press (Fig. 3.2); Commonwealth Scientific and Industrial Research Organization, Melbourne, Australia (Figs. 6.4, 7.1, 7.6, 10.4 and 10.5); Elsevier Applied Science Publishers Ltd., Oxford (Figs. 3.3, 3.4, 3.6, 4.1, 4.8, 5.4, 7.5 and 7.6); Food Processing and Packaging, London (Fig. 8.7); Garrard Press, Champaign, Illinois, USA (Fig. 10.1); Heinemann Educational Books Ltd., London (Fig. 4.2); Controller of Her Majesty's Stationery Office, London (Figs. 6.2, 6.3 and 8.3); Journal of Agricultural Science, Cambridge (Fig. 3.2 and Table 4.32); Journal of Animal Science, Albany, NY, USA (Fig. 10.3); Journal of Cell Biology, New York (Fig. 3.8(e)); Journal of Molecular Biology, Cambridge (Fig. 3.8(d)); Journal of Physiology, Oxford (Fig. 4.7); Journal of Refrigeration, London (Fig. 7.8); Royal Society, London (Fig. 7.3); Meat Industry Research Institute of New Zealand Inc. (Fig. 5.2); Science and the American Association for the Advancement of Science, Washington, USA (Figs. 3.8(a) and 3.8(c)); Scientific American Inc., New York (Fig. 3.5); Society of Chemical Industry, London (Figs. 4.6, 5.5, 8.5 and 8.6); and the Novosti Press Agency, London (Fig. 7.4).

Chapter 1

Introduction

1.1 Meat and muscle

Meat is defined as the flesh of animals used as food. In practice this definition is restricted to a few dozen of the 3000 mammalian species; but it is often widened to include, as well as the musculature, organs such as liver and kidney, brains and other edible tissues. The bulk of the meat consumed in the United Kingdom is derived from sheep, cattle and pigs: rabbit and hare are, generally, considered separately along with poultry. In some European countries (and elsewhere), however, the flesh of the horse, goat and deer is also regularly consumed; and various other mammalian species are eaten in different parts of the world according to their availability or because of local custom. Thus, for example, the seal and polar bear are important in the diet of the Inuit, and the giraffe, rhinoceros, hippopotamus and elephant in that of certain tribes of Central Africa: the kangaroo is eaten by the Australian aborigines: dogs and cats are included in the meats eaten in Southeast Asia: the camel provides food in the desert areas where it is prevalent and the whale has done so in Norway and Japan. Indeed human flesh was still being consumed by cannibals in remote areas until only recently past decades; (Bjerre, 1956).

Very considerable variability in the eating and keeping quality of meat has always been apparent to the consumer; it has been further emphasized in the last few years by the development of prepackaging methods of display and sale. The view that the variability in the properties of meat might, rationally, reflect systematic differences in the composition and condition of the muscular tissue of which it is the post-mortem aspect is recognized. An understanding of meat should be based on an appreciation of the fact that muscles are developed and differentiated for definite physiological purposes in response to various intrinsic and extrinsic stimuli.

1.2 The origin of meat animals

The ancestors of sheep, cattle and pigs were undifferentiated from those of human beings prior to 60 million years ago, when the first mammals appeared on Earth. By

2–3 million years ago the species of human beings to which we belong (*Homo sapiens*) and the wild ancestors of our domesticated species of sheep, cattle and pigs were probably recognizable. Palaeontological evidence suggests that there was a substantial proportion of meat in the diet of early *Homo sapiens*. To tear flesh apart, sharp stones – and later fashioned stone tools – would have been necessary. Stone tools were found, with the fossils of hominids, in East Africa (Leakey, 1981).* Our ape-like ancestors gradually changed to present day human beings as they began the planned hunting of animals. There are archaeological indications of such hunting from at least 500,000 BC. The red deer (*Cervus elaphus*) and the bison (referred to as the buffalo in North America) were of prime importance as suppliers of hide, sinew and bone, as well as meat, to the hunter-gatherers in the areas which are now Europe and North America, respectively (Clutton-Brock, 1981). It is *possible* that reindeer have been herded by dogs from the middle of the last Ice Age (about 18,000 BC), but it is not until the climatic changes arising from the end of this period (i.e. 10,000–12,000 years ago) that conditions favoured domestication by man. It is from about this time that there is definite evidence for it, as in the cave paintings of Lascaux.

According to Zeuner (1963) the stages of domestication of animals by man involved first loose contacts, with free breeding. This phase was followed by the confinement of animals, with breeding in captivity. Finally, there came selected breeding organized by man, planned development of breeds having certain desired properties and extermination of wild ancestors. Domestication was closely linked with the development of agriculture and although sheep were in fact domesticated before 7000 BC, control of cattle and pigs did not come until there was a settled agriculture, i.e. about 5000 BC.

Domestication alters many of the physical characteristics of animals and some generalization can be made. Thus, the size of domesticated animals is, usually, smaller than of their wild ancestors.** Their colouring alters and there is a tendency for the facial part of the skull to be shortened relative to the cranial portion; and the bones of the limbs tend to be shorter and thicker. This latter feature has been explained as a reflection of the higher plane of nutrition which domestication permits; however, the effect of gravity may also be important, since Tulloh and Romberg (1963) have shown that, on the same plane of nutrition, lambs to whose back a heavy weight has been strapped, develop thicker bones than controls. (As is now well documented, exposure to prolonged periods of weightlessness causes loss of bone and muscle mass.) Many domesticated characteristics are, in reality, juvenile ones persisting to the adult stage. Several of these features of domestication are apparent in Fig. 1.1 (Hammond, 1933–4). It will be noted that the domestic Middle White pig is smaller (45 kg; 100 lb) than the wild boar (135 kg; 300 lb), that its skull is more juvenile, lacking the pointed features of the wild boar, that its legs are shorter and thicker and that its skin lacks hair and pigment.

Apart from changing the form of animals, domestication encouraged an increase in their numbers for various reasons. Thus, for example, sheep, cattle and pigs came

---

* Rixson (2000) presented convincing arguments showing how the development of butchery skills, deriving from the use of stone tools, promoted a settled communal life; and, thereafter, led to civilized societies.
** It appears, however, that the sizes of domestic cattle, sheep and pigs in Anglo-Saxon times were much smaller than those of their modern counterparts (Rixson, 2000).

Fig. 1.1 Middle White Pig (aged 15 weeks, weighting 45 kg; 100 lb) and Wild Boar (adult, weighting about 135 kg; 300 lb), showing difference in physical characteristics. Both to same head size (Hammond, 1933–4). (Courtesy of the late Sir John Hammond.)

to be protected against predatory carnivores (other than man), to have access to regular supplies of nourishing food and to suffer less from neonatal losses. Some idea of the present numbers and distribution of domestic sheep, cattle and pigs is given in Table 1.1 (Anon., 2003).

1.2.1 Sheep

Domesticated sheep belong to the species *Ovis aries* and appear to have originated in western Asia. The sheep was domesticated with the aid of dogs before a settled

Table 1.1 Numbers of sheep, cattle and pigs in various countries, 2003

Country	Approx. million head		
	Sheep	Cattle	Pigs
Argentina	12.5	51	4
Australia	98	27	3
Brazil	4	189.5	33
China	144	103.5	47
Denmark	negligible	2	13
Eire	5	7	2
France	9	19.5	15
Germany	3	14	26
Italy	11	6	9
Japan	negligible	4.5	10
Kazakhstan	10	4.5	1
Netherlands	1	4	11
New Zealand	39	10	0.5
Poland	negligible	5.5	19
Russian Federation	14	26.5	17
Turkey	27	10.5	negligible
Ukraine	1	9	9
UK	36	10.5	5
USA	6	96	59.5

agriculture was established. The bones of sheep found at Neolithic levels at Jericho, have been dated as being from 8000–7000 BC (Clutton-Brock, 1981). Four main types of wild sheep still survive – the Moufflon in Europe and Persia, the Urial in western Asia and Afghanistan, the Argali in central Asia and the Big Horn in northern Asia and North America. In the United Kingdom, the Soay and Shetland breeds represent remnants of wild types.

By 3500–3000 BC several breeds of domestic sheep were well established in Mesopotamia and in Egypt: these are depicted in archaeological friezes. Domestication in the sheep is often associated with a long or fat tail and with the weakening of the horn base so that the horns tend to rise much less steeply. The wool colour tends to be less highly pigmented than that of wild sheep.

Nowadays about 55 different breeds of sheep exist in the United Kingdom. Some of these are shown in Table 1.2. Further information on numbers of sheep in each breed, the size of crossbred ewe populations and the general structure of the sheep industry can be found in 'Sheep in Britain' (Meat & Livestock Commission, 1988).

The improved breeds, such as the Suffolk, tend to give greater carcass yield than semi-wild breeds such as the Soay or Shetland sheep, largely because of their increased level of fatness (Hammond, 1932a). Again, of the improved breeds, those which are early maturing, such as the Southdown and Suffolk, have a higher percentage of fat in the carcass than later maturing breeds, such as the Lincoln and Welsh; moreover, the subcutaneous fat appears to increase, particularly in the former. The English mutton breeds (e.g. Southdown and Cotswold) have a greater development of subcutaneous connective tissue than wool breeds, e.g. Merino.

Table 1.2   Some breeds of sheep found in the United Kingdom (courtesy D. Croston, Meat & Livestock Commission)

Hill breeds
Scottish Blackface, Swaledale, Welsh Mountain, North Country Cheviot, Dalesbred, Hardy Speckled Face, South Country Cheviot, Derbyshire Gritstone, Beulah, Shetland, Roughfell, Radnor

Longwool crossing breeds
Bluefaced Leicester, Border Leicester, Bleu de Maine, Rouge de l'Ouest, Cambridge

Longwool ewe breeds
Romney Marsh, Devon and Cornwall Longwool, Devon Closewool

Terminal sire breeds
Suffolk, Southdown, Texel, Oxford Down, Shropshire, Hampshire Down, Ile de France, Charollais, Berrichon du Cher, Vendeen

Shortwool ewe breeds
Clun Forest, Poll Dorset, Lleyn, Kerryhill, Jacob

The coarseness of grain of the meat from the various breeds tends to be directly related to overall size, being severe in the Large Suffolk sheep: the grain of the meat from the smaller sheep is fine. Breed differences manifest themselves in a large number of carcass features – in the actual and relative weights of the different portions of the skeleton, in the length, shape and weight of individual bones, in the relative and actual weights of muscles, in muscle measurements, colour, fibre size and grain and in the relative and actual weights and distribution of fat (Pállson, 1939, 1940).

The shape of the *l. dorsi** muscle (back fillet) in relation to fat deposition is shown for several breeds of sheep in Fig. 1.2: the relative leanness of the hill sheep (Blackface) will be immediately apparent.

1.2.2   Cattle

The two main groups of domesticated cattle, *Bos taurus* (European) and *B. indicus* (India and Africa), are descended from *B. primigenius*, the original wild cattle or aurochs. The last representative of the aurochs died in Poland in 1627 (Zeuner, 1963). Although variation in type was high amongst the aurochs, the bulls frequently had large horns and a dark coat with a white stripe along the back. These characteristics are found in the cave paintings of Lascaux. Certain wild characteristics survive more markedly in some domestic breeds than in others, for example, in West Highland cattle and in the White Park cattle. Some of the latter may be seen at Woburn Abbey in England: similar animals are also represented pictorially at Lascaux.

Domestication of cattle followed the establishment of settled agriculture about 5000 BC. Domesticated hump-backed cattle (*B. indicus*, 'Zebu') existed in Mesopotamia by 4500 BC and domesticated long-horned cattle in Egypt by about

---

* In this text the term '*longissimus dorsi*' (abbrev. '*l. dorsi*') signifies '*M. longissimus thoracis et lumborum*' (or parts thereof).

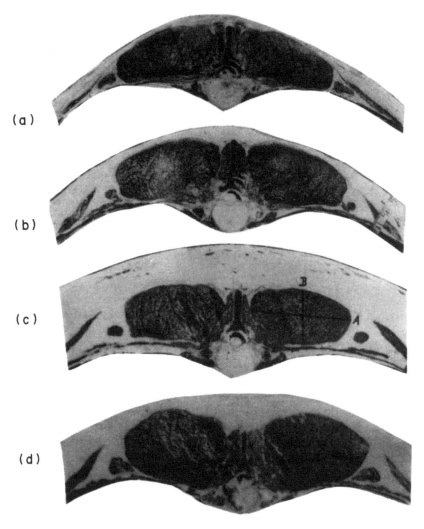

Fig. 1.2   The effect of breed on the shape and fat cover of the *L. dorsi* muscle of sheep (Hammond, 1936).

	A (mm)	B (mm)	Shape index
(a) Blackface	52	24	46
(b) Suffolk	65	35	54
(c) Hampshire	58	33	57
(d) Southdown	62	43	69

All the photographs have been reduced to the same muscle width (A) in order to show the proportions. (Courtesy of the late Sir John Hammond.)

4000 BC: both of these appear on pottery and friezes of the period (Zeuner, 1963). Several breeds of domesticated cattle were known by 2500 BC. An interesting frieze from Ur, dating from 3000 BC, shows that cows were then milked from the rear. According to Zeuner, this is further evidence that the domestication of sheep preceded that of cattle. About this same time the fattening of cattle by forced feeding was practised in Egypt.

According to Garner (1944) the more immediate wild predecessor of most breeds of British cattle was *B. longifrons*, which was of relatively small frame, rather than *B. primigenius*, which is said to have been a massive animal. Indirectly, the development of many present British breeds was due to the early improvements initiated by Bakewell in the middle of the eighteenth century, who introduced in-breeding, the use of proven sires, selection and culling. In the United Kingdom prior to that time cattle had been developed, primarily, for draught or dairy purposes. A deliberate attempt was now made to produce cattle, primarily for meat, which would fatten quickly when skeletal growth was complete. During the last 200 years the trend has been towards smaller, younger and leaner animals; and there has been growing realization that breed potential will not be fully manifested without adequate food given at the right time in the growth pattern of the animal (Hammond, 1932a; Garner, 1944). Some of the present breeds of British cattle are listed in Table 1.3; they are grouped according to whether they are of beef, dairy or dual-purpose types.

A beef animal should be well covered with flesh, blocky and compact – thus reducing the proportion of bone. Muscle development should be marked over the hind, along the back and down the legs. In a dairy animal, on the other hand, the frame should be angular with relatively little flesh cover, the body should be cylindrical (thus accommodating the large digestive tract necessary for efficient conversion of food into milk) and mammary tissue should be markedly developed.

Aberdeen Angus has been regarded as the premier breed for good-quality meat (Gerrard, 1951). The carcass gives a high proportion of the cuts which are most in demand; there is, usually, a substantial quantity of intramuscular (marbling) fat and the eating quality of the flesh is excellent; on the other hand, the carcass is relatively light. One of the reasons for the good eating quality of the Aberdeen Angus is its tenderness, which is believed to be partly due to the small size of the muscle bundles, smaller animals having smaller bundles. Because of the small carcass, however, such meat is relatively expensive. One way of making available large quantities of the relatively tender meat would be to use large-framed animals at an early age when the muscle bundles would still be relatively small (Hammond, 1963a). This may be done by feeding concentrates such as barley to Friesians (Preston *et al.*, 1963). Aberdeen Angus, Herefords and Shorthorns (beef-types) have been extensively used to build up beef herds overseas, as in Argentina and Queensland.

Table 1.3 Some breeds of cattle found in the United Kingdom (courtesy G. Brown, Meat & Livestock Commission)

(a) **Principal beef breeds**
     Charollais, Limousin, Simmental, Hereford, Aberdeen Angus, Belgian Blue, Blonde d'Aquitaine, South Devon, Beef Shorthorn, Welsh Black, Devon, Lincoln Red, Murray Grey, Sussex, Galloway
(b) **Dairy breeds**
     Holstein/Friesian, Jersey, Ayrshire, Guernsey, Dairy Shorthorn
(c) **Dual-purpose breeds**
     Meuse Rhine Issel, Dexter, Red Poll

In terms of numbers. Holstein/Friesian are predominant and the Hereford is now about the fifth most popular beef breed, following the Charollais, Limousin, Simmental and Aberdeen Angus. In the United Kingdom about 64 per cent of home killed beef is derived from dairy breeds.

Callow (1961) suggested that selection for beef qualities has brought about various differences between beef and dairy breeds. Thus, Friesians (a milk breed) have a high proportion of fat in the body cavity, and low proportion in the subcutaneous fatty tissue. In Herefords (a beef breed), on the other hand, the situation is reversed. The distribution of fat in Shorthorns (a dual-purpose breed) is intermediate between that of Herefords and Friesians. In the United Kingdom about 65 per cent of home-killed beef is derived from dairy herds.

There are, of course, many other modern breeds representative of *B. taurus,* for example the Simmental in Switzerland, the 'Wagyu' in Japan, the Charollais in France; and, in warmer areas, *B. indicus* is widely represented. Attempts have been made to cross various breeds of *B. indicus* (Indian Hissar – 'Zebu' – cattle have been frequently involved) with British breeds, to combine the heat-resisting properties of the former with the meat-producing characteristics of the latter. Such experiments have been carried out for example in Texas and Queensland. A fairly successful hybrid, the Santa Gertrudis, consists of three-eighths 'Zebu' and five-eighths Shorthorn stock.

Unusual types of cattle are occasionally found within a normal breed. Thus, dwarf 'Snorter' cattle occur within various breeds in the USA; and pronounced muscular hypertrophy, which is often more noticeable in the hind quarters and explains the name 'doppelender' given to the condition, arises in several breeds – e.g. Charollais and South Devon (McKeller, 1960). Recessive genes are thought to be responsible in both cases.

1.2.3   Pigs

The present species of domesticated pigs are descendants of a species-group of wild pigs, of which the European representative is *Sus scrofa* and the eastern Asiatic representative *S. vittatus*, the banded pig (Zeuner, 1963). As in the case of cattle, pigs were not domesticated before the permanent settlements of Neolithic agriculture. There is definite evidence for their domesticity by about 2500 BC in what is now Hungary, and in Troy. Although pigs are represented on pottery found in Jericho and Egypt, dating from earlier periods, these were wild varieties. The animal had become of considerable importance for meat by Greco-Roman times, when hams were salted and smoked and sausages manufactured.

About 180 years ago European pigs began to change as they were crossed with imported Chinese animals derived from the *S. vittatus* species.

These pigs had short, fine-boned legs and a drooping back. Then in 1830, Neapolitan pigs, which had better backs and hams, were introduced. According to McConnell (1902) it was customary in the past to classify British pigs by their colour – white, brown and black – and the older writers mention 30 breeds. Few of these are now represented.

The improvement of pigs has not been continuous in one direction, but has been related to changing requirements at different periods. Of the improved breeds of pig now in use in the world the majority originated in British stock (Davidson, 1953). The first breed to be brought to a high standard was the Berkshire: it is said to produce more desirably shaped and sized *l. dorsi* muscles than any other breed. Berkshire pigs, crossed with the Warren County breed of the USA, helped to establish the Poland China in that country a century ago. The change of type which can be swiftly effected within a breed is well exemplified by the Poland China, which altered over only 12 years from a heavy, lard type to a bacon pig (Fig. 1.3: Hammond, 1932b). Berkshire pigs have also been employed to upgrade local breeds in Germany, Poland and Japan.

In Britain about 70 per cent of the pigs slaughtered are produced from F1 hybrids of Large White x Landrace. The predominant sire type used is the Large White, with an increasing use of 'meat type' sires produced by the major pig breeding companies. When considering pedigree breeds, the Large White is the most numerous in the United Kingdom (Table 1.4).

In recent years Landrace pigs from Scandinavia have strongly competed with them as bacon producers. The Landrace was the first breed to be improved

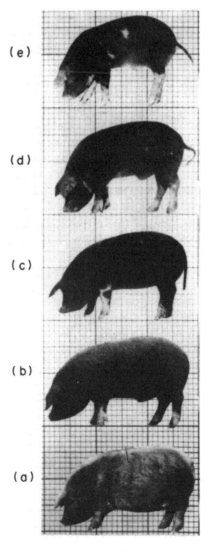

Fig. 1.3   The effect of intensive selection over 12 years on the conformation of the Poland China pig in changing from a lard to a bacon type (Hammond, 1932b): (a) 1895–1912, (b) 1913, (c) 1915, (d) 1917, (e) 1923. (Courtesy of the late Sir John Hammond.)

Table 1.4   Relative numbers of pigs of various breeds in
the United Kingdom (based on 1995 data supplied by
G. E. Welsh, Chief Executive, British Pig Association)

Breed	Per cent
Large White	49.5
Landrace	35.5
Welsh	4.5
British Saddleback	2
Gloucester Old Spot	2.5
Berkshire	2
Tamworth	1.5
Middle White	1
Large Black	1
Others	0.5

scientifically. In Denmark, these animals have been intensively selected for leanness, carcass length and food-conversion efficiency with a view to the production of Wiltshire bacon. Pigs of 200 lb (100 kg) live weight, irrespective of breed, have been used for pork, bacon or manufacturing purposes in Denmark, according to the conformation and level of fatness (Hammond, 1963b). In Hungary, there is a meat pig (the Mangalitsa) which is particularly useful for making salami, partly because it has a rather highly pigmented flesh.

1.3   Current trends and developments

The increasing pressure of world population, and the need to raise living standards, has made the production of more and better meat, and its more effective preservation, an important issue. Thus, progeny testing, based on carcass measurement, is being increasingly recognized as an efficient way of hastening the evolution of animals having those body proportions which are most desirable for the meat consumer. It has been applied especially to pigs (Harrington, 1962); but progeny testing of both cattle and sheep is developing. Artificial insemination has afforded a means of vastly increasing the number of progeny which can be sired by a given animal having desired characteristics. In the future, it may well be that young bulls of under 15 months will increasingly replace steers of this age since they produce the lean flesh which is now in demand in greater quantities – and more economically. The somewhat higher incidence of 'dark-cutting' beef in bulls is probably a reflection of their stress susceptibility (cf. §5.1.2) and can be overcome by careful handling. During recent decades, and especially since the report on the relationship between diet and cardiovascular disease by the Committee on Medical Aspects of Food Policy (1984), there has been a marked reduction in the percentage of saturated fat derived from meat. The fat content of beef, pork and lamb has fallen from 20–26 per cent to 4–8 per cent (Higgs, 2000). This has been achieved not only by selective breeding for leanness (aided by the development of carcass classification schemes by the Meat & Livestock Commission (UK)), but also by changed methods of butchery applied to the hot carcass, whereby not only is backfat removed, but also intermuscular fat by 'seaming out' the muscles (cf. §§ 5.2.2 and 7.1.1.3). This trend has been strengthened by the increasing sale of meat as consumer-portion, prepackaged cuts. For this purpose the larger continental breeds have certain advantages over traditional British beef animals. Such breeds as Limousin, Charollais and Chianina produce leaner carcasses at traditional slaughter weights; and attain these weights faster. There are occasionally reproductive problems; but these can be controlled by improved management (Allen, 1974). There has been a tendency towards the consumption of lamb in recent years, since it is more tender than mutton and produces the small joints now in demand. To some extent the increased costs which this trend entails have been offset by increasing the fertility of the ewe and thus the number of lambs born. The Dorset Horn ewe breeds throughout the year; but ewes of other breeds are being made to breed with increased frequency by hormone injections which make them more responsive to mating with the rams (Hammond, 1963b). The goat, being able to thrive in poor country, may well be developed more intensively. Public pressure to reduce the use of pesticides in crops has led to the development of so-called 'organic' farming, in which no 'artificial' additives are employed to assist the growth of plants and animals. Nevertheless, this approach is not ideal. Thus, 'organically' reared pigs show no organoleptic benefits over those reared conventionally, and, indeed, in some respects, compare unfavourably with the latter (Ollson et al., 2003).

Increasing attention is being directed to the potential of hitherto unexploited animals for meat production. Berg and Butterfield (1975), in studying the muscle/weight distribution in a number of novel species, noted that those which were more *agile* had greater muscle development in the fore limbs: in *mobile* species the musculature of all limbs was highly developed. In the elephant seal, the abdominal muscles are especially involved in locomotion, and their relative development is about threefold that of corresponding muscles in cattle, sheep or pigs.

In large areas, such as Central Africa, where the more familiar European types of domestic animal do not thrive well, there are a number of indigenous species in game reserves, well adapted to the environment, which could be readily used for meat production, e.g. the giraffe, roan antelope and springbok (Bigalke, 1964). Satisfactory canned meats can be prepared from the wildebeest antelope, if it is processed on the day of slaughter (Wismer Pedersen, 1969a). The meat may become pale and watery if the animals are not killed by the first shot. Of the East African ungulates the meat quality of wildebeest, buffalo and zebra is probably the most acceptable organoleptically. Onyango et al. (1998), in a comparative study of game as meat in Kenya, found that the lipids of zebra were markedly more unsaturated than those of beef. Combined with its high content of myoglobin, this causes zebra meat to undergo rapid oxidative deterioration under aerobic conditions.

As game farming has developed in South Africa, there has been increasing interest in the impala as a meat animal. They feed well on the bushveld and are able to consume the foliage of both trees and bushes. Their flesh has low levels of intermuscular and intramuscular fat and has a high titre of polyunsaturated fatty acids (Hoffman et al., 2005).

The water buffalo is a species which shows considerable promise. The world population of buffalo is already one-ninth of that of cattle; in the Amazon basin they are increasing at 10 per cent per year (Ross Cockrill, 1975). The eating quality of the meat is similar to that of beef (Jocsimovic, 1969); and, indeed, may be preferred in some areas. Having less fat, the flesh of the water buffalo conforms to current trends. On the other hand the flesh has more connective tissue, and is darker, features which tend to make it compare less favourably with beef (Robertson et al., 1986). It thrives in the wet tropics – an extensive area which European cattle find distressing. The eland antelope shows particular promise for development in Africa. For example, it has behavioural and physiological characteristics which enables it to survive even when no drinking water is available and temperatures are high. It feeds mainly at night when the bushes and shrubs have a tenfold higher water content than in day-time (Tayler, 1968).

Such species as oryx can withstand body temperatures of 45 °C for short periods by a specialized blood flow whereby the brain is kept relatively cool (Tayler, 1969). The meat of the oryx has a lower myoglobin content than that of beef, but it is more susceptible to the formation of metmyoglobin (Onyango et al., 1998).

In those parts of Africa where drought conditions prevail, the one-humped camel (dromedary) thrives much better than cattle: it constitutes an important source of meat in arid regions. The proportion of edible meat on the camel carcass is comparable with that of cattle, red muscles contributing ca. 60 per cent of the overall yield (Babiker, 1984). Most of the joints are devoid of fat: the exception is the sirloin because it includes the hump. Most of the camel's fat is deposited in the hump rather than being distributed throughout the carcass (Yousif and Babiker, 1989). The meat of young camels is comparable in taste and texture to that of beef (Knoess, 1977), but, not surprisingly, that of those which have been slaughtered after a working life as draught animals is tough.

Since cattle eat grasses wherein the proportion of lignin in the stem is below a certain maximum and eland prefer to eat the leaves of bushes, there are advantages in mixed stocking (Kyle, 1972). Indeed a surprising number of species can subsist in the same area, without encroaching upon one another's feed requirements, by eating different species of plant, or different parts of the same species of plant, and by feeding at different heights above the ground (Lamprey, 1963).

In Scotland there is interest in the development of the red deer as an alternative meat producer to sheep in areas where cattle rearing or agriculture is not feasible. It has been shown that, when fed on concentrates after weaning, stags can achieve feed conversion efficiencies better than 3 lb (1.4 kg) feed dry matter per pound (kilogram) of gain (Blaxter, 1971–2). This conversion rate is better than that achieved with cattle or intensive lamb production.

In New Zealand, the introduction of deer for sport led to serious denudation of plant species; and culling was thus undertaken, using helicopters to reach otherwise inaccessible areas. Thereafter the development of an export trade in venison, and an even more profitable one in velvet from the antlers of stags, has stimulated interest in the controlled production of deer. Half of the world's farmed deer population is now found in New Zealand (Wiklund et al., 2001), and this has greatly increased interest in the red deer as meat. Live deer are now being captured from the air, immobilization (prior to aerial transport) being effected by firing tranquillizing darts, or pairs of electrodes (for anaesthetization), into the animals. Because deer and goats are naturally lean species, procedures are being sought to reduce their fat content even further by selection since there is currently a demand for lean meat. For both species, a wide range of breed sizes are available, making this objective relatively easy (Yerex and Spiers, 1987). In Scandinavia the meat of the reindeer is eaten. It is a relatively small animal and its reputed tenderness may well be a function of the correspondingly small diameter of the muscle fibres (Keissling and Keissling, 1984).

In the period 1965–85 world goat numbers increased by 30 per cent, particularly in developing countries such as Africa. Because of their early sexual maturity and the relative shortness of their gestation period, goats are a valuable species in situations where herd numbers require to be rapidly built up after drought (Norman, 1991). Moreover, because goats have low per head feed requirements, they are able to utilize marginal grazing land and small plots on which larger ruminants could not thrive. Yet goat meat accounts for only *ca.* 1.5 per cent of total world meat production. It is true, of course, that goat meat tends to be less desirable in flavour and tenderness than beef, lamb and pork when samples of comparable maturity and fatness are considered (Smith *et al.*, 1974); but the acceptability of the meat of any species is often determined by local custom. At a time when populations are increasingly moving from rural areas into cities in developing countries, further use of a species which can quickly respond to intensification and to fluctuations in demand would seem desirable (Norman, 1991).

A more general interest in the exploitation of non-mammalian species for meat is reflected by the increasing availability of flesh from the crocodile, the emu and the ostrich. Meat from the ostrich is derived mainly from the muscles of the well-developed legs. It has a relatively high myoglobin content, resembling beef or mutton rather than pork or poultry. Since it has relatively less cholesterol and total lipid, and a higher content of polyunsaturated fatty acids, than beef (Paleari *et al.*, 1998), whilst its tenderness is greater than that of the latter, its consumption could well become more popular. Although the ostrich has been farmed for many years in South Africa, primarily for its hide and plumage, the species has been introduced into other countries wherein the meat of the ostrich is now available to the public.

Currently there is increasing concern – whether soundly based or unfounded – expressed by consumers respecting the safety of meat (e.g. chemical residues, allergens, microbial and parasitic hazards) and increasing selectivity in the demand for palatability (e.g. guaranteed and reproducible levels of eating quality attributes) (Tarrant, 1998). Improved methods of preservation (e.g. refrigeration, high pressure) are being devised and authoritative assurances on the safety of meat subjected to low levels of ionizing radiation, in combination with chilling, predict its renewed importance.

Techniques for identifying the molecular morphologies that are essential for generating the attributes of eating quality in meat (and knowledge of the means of controlling their expression, once identified) are developing rapidly. Genetic manipulation of the live animal, to eliminate undesirable features in its meat and to incorporate those which are desirable, is now a reality (de Vries *et al.*, 1998).

In studying biological systems it has hitherto been necessary to isolate their components and, therefrom, to deduce the nature of the systems from which they were derived; but it has long been appreciated that these systems are exceedingly complex and highly organized and, that from their components in isolation, only limited information can be obtained about their interactions *in vivo*. Recently, however, techniques such as two-dimensional electrophoresis have made it possible to obtain patterns that show all the representatives of groups such as genes, nucleic acids, proteins and functional metabolites simultaneously. Concomitantly, the rapid growth of computing science has afforded the means of distinguishing and classifying the patterns obtained whereby they can be related to specific tissues and, in the case of muscle, to organoleptic properties of the meat postmortem. (Eggen and Hocquette, 2003) The potential of proteomics ('panoramic protein characterization') has been reviewed by Bendixen (2005) and its value in accurately understanding and controlling organoleptic properties has already been established.

Such developments demonstrate that meat continues to be a significant commodity for the human consumer.

Chapter 2

Factors influencing the growth and development of meat animals

2.1 General

'As an animal grows up two things happen: (i) it increases in weight until mature size is reached; this we call Growth and (ii) it changes in its body conformation and shapes, and its various functions and faculties come into full being; this we call Development' (Hammond, 1940). The curve relating live weight to age has an S-shape and is similar in sheep, cattle and pigs (Brody, 1927). There is a short initial phase when live weight increases little with increasing age: this is followed by a phase of explosive growth; then finally, there is a phase when the rate of growth is very low.

When animals are developing, according to Hammond, a principal wave of growth begins at the head and spreads down the trunk: secondary waves start at the extremities of the limbs and pass upwards: all these waves meet at the junction of the loin and the last rib, this being the last region to develop.

The sequence of development of various muscles in the body reflects their relative importance in serving the animal's needs. Thus, the early development of the muscles of the distal limbs confers the mobility required to forage for food; and the development of the jaw muscles promotes effective mastication of the food secured (Berg and Butterfield, 1975).

With the onset of sexual maturity, further differential muscular development occurs, whereby, in the male, the muscles of the neck and thorax grow relatively fast. These assist in fighting for dominance.

In most species of animals, although the female matures earlier, the male is larger and heavier than the female in adult life; and since the different parts of the tissues of the body grow at different rates, the difference in size between the sexes results in a difference in development of body proportions. Castration in either sex tends to reduce sex differences in growth rate and body conformation (Hammond, 1932a). Subjective assessment of the maturity of beef carcasses can be made from the colour of the cartilage at the tips of the dorsal spine of the sacral, lumbar and thoracic vertebrae (Boggs et al., 1998). The accuracy of the prediction can be

Table 2.1 Estimates of heritability of growth characteristics of cattle, sheep and pigs

Character	Species	Average heritability (per cent)
Prenatal growth (birth weight)	Cattle	41
	Sheep	32
Weaning weight	Cattle	30
	Sheep	33
	Pigs	17
Post-weaning weight	Cattle	45
	Sheep	71
	Pigs	29
Feed conversion efficiency	Cattle	46
	Sheep	15
	Pigs	31

increased by objective evaluation of the colour by image processing (Hatem *et al.*, 2003).

Other as yet unidentified influences cause differences in the relative rates of growth of the individual members of the musculature. The pattern is both inherited and extraneously modified.

The establishment of different breeds of sheep, cattle and pigs is partly attributable to artificial selection practised by man under domestication, but the types of pre-existent animals from which such selection could be made have been determined by numerous, long-term extraneous influences, which continue – however much obscured by human intervention. These influences have caused overall alterations in the physiology of the animals concerned, involving the expression, suppression or alteration of physical and chemical characteristics. It must be presumed that such changes have been caused by mutations in the genes in response to the micro- or macro-environment and that they have been subsequently perpetuated by the genes.* In decreasing order of fundamentality, the factors influencing the growth and development of meat animals can be considered in four categories: genetic, physiological, nutritional and manipulation by exogenous agencies.

2.2 Genetic aspects

Genetic influences on the growth of animals are detectable early in embryonic life. Thus Gregory and Castle (1931) found that there were already differences in the rate of cell division between the embryos of large and small races of rabbits 48 h after fertilization. The birth weight of cattle and sheep, but not that of pigs, is influenced to an important extent by the nature of the respective embryos (Table 2.1). More recent data have also emphasized the high heritability of body composition

* The relationship between genes, ribonucleic acid (RNA) and deoxyribonucleic acid (DNA) is considered in § 3.2.1.

traits in comparison with those of reproductive efficiency and meat quality characteristics (Table 2.2; Sellier, 1994).

Among the parameters affected at commercial level is the degree of fatness at comparable carcass weights or animal age. In Tables 2.3 and 2.4 respectively, some relative data for breeds of sheep and cattle are given. The leanness of the carcasses from crosses with the large continental breeds is evident.

At birth the pig is by far the most immature physiologically of the three domestic species. Differences in the physiological age at birth mainly depend on how great a part of the total growing period is spent in the uterus. The birth weight is

Table 2.2 Average heritability of economically important traits in meat-producing mammals

Traits	Range of heritability
Reproductive efficiency (litter size, fertility)	0.02–0.10
Meat quality (colour, pH, tenderness, water-holding capacity)	0.15–0.30
Growth (average daily gain, feed efficiency)	0.20–0.40
Fat quality (fatty acid composition of back fat)	0.30–0.50
Body composition (lean content, fat content, etc.)	0.40–0.60

Table 2.3 Breed differences in percentage fat in sheep carcasses (after Kirton *et al.*, 1974)

Breed/Cross	Fat (% at 20 kg carcass weight)
Suffolk	32.9
Hampshire	33.1
Border Leicester	33.3
Poll Dorset/Dorset Horn	33.8
Romney	34.3
Cheviot	34.4
Southdown	38.5

Table 2.4 Breed differences in percentage fat trim in cattle carcasses (after Koch *et al.*, 1982)

Breed/Cross	Fat trim (% of carcass weight at same age)
Jersey/Hereford	22.1
Red Poll X	21.0
South Devon X	20.0
Simmental X	15.6
Charollais X	15.2
Limousin X	15.2
Chianina X	13.0

influenced by the age, size and nutritional state of the mother, by sex, by the length of the gestation period (5, 9 and 4 months in sheep, cattle and pigs respectively) and by the numbers of young born (Pállson, 1955). An interesting aspect of this latter influence is the finding that embryos next to the top and bottom of each horn of the uterus develop more rapidly than those in intermediate positions (McLaren and Michie, 1960; Widdowson, 1971). The supply of nutrients to these embryos is particularly good since the pressure of blood is high at the top through the proximity of the abdominal aorta and at the bottom through the proximity of the iliac artery. Environmental and genetic factors are closely interrelated: favourable environmental conditions arc necessary for the full expression of the individual's genetic capacity. Irrespective of the birth weight, however, the rate of weight increase in young pigs is largely determined by the establishment of a suckling order: those piglets feeding from the anterior mammary glands grow fastest, probably because the quantity of milk increases in proceeding from the posterior to the anterior glands of the series on each side of the sow (Barber et al., 1955).

In general, the birth weights of the offspring from young mothers are lower than those from mature females and the birth weights of the offspring from large individuals are greater than those from small mothers.

Certain major growth features in cattle are known to be due to recessive genes. One of these is dwarfism (Baker et al., 1951), where the gene concerned (Merat, 1990) primarily affects longitudinal bone growth and vertebral development in the lumbar region, and males rather than females (Bovard and Hazel, 1963). Another is doppelender development (McKellar, 1960; Boccard, 1981), the gene concerned being *mh* (Hanset and Michaux, 1985). Neither has so far proved controllable. The doppelender condition – referred to as 'double muscling' in Britain and the USA, 'a groppa doppia' in Italy and 'culard' in France – has been reviewed by Boccard (1981). The various ways in which the gene responsible for this hereditary hypertrophy has been expressed have generated a corresponding variety of hypotheses on how the condition is transmitted. The higher commercial value of doppelender animals arises from their higher dressing percentage (and higher muscle: bone ratio), the composition of the carcass (which has relatively less fat and offal), and to the distribution of the hypertrophied musculature. The hypertrophy is not uniform: indeed some muscles have relatively less development than the corresponding normal members (Boccard and Dumont, 1974). The most hypertrophied are those with a large surface area; and those which occur near the body surface. This feature has led to the suggestion that a disturbance of collagen metabolism may be implicated (Boccard, 1981; and cf. § 4.3.8).

There is a greatly increased number of muscle fibres in the meat of double muscled cattle and Swatland (1973) suggested that this is not reflected by a corresponding increase in motor nerve units. In such cattle myoblasts appear to have been increased at the expense of fibroblasts. Increase in fibre diameter is less important in contributing to muscle enlargement than the increase in fibre numbers in double muscling. As Deveaux et al. (2000) demonstrated, numerous metabolic functions are altered in doppelender animals, and various genes, other than the myostatin gene, must be involved. The development of oxidative metabolism is delayed in the foetuses of doppelender cattle in comparison with that in normal foetuses (Gagnière et al., 1997).

A recent proteomic study of bovine muscle hypertrophy identified molecular markers which were associated with an 11-base pair deletion in the myostatin gene

– a mutation whereby normal levels of inactive myostatin protein are expressed. It appears that myostatin preferentially controls proliferation of fast twitch glycolytic ('white') muscle fibres – supporting the view that muscle hypertrophy involves increased ratios of glycolytic to oxidative fibres (Deveaux *et al.*, 2003).

Another major gene which has a significant effect on meat animals and on the quality of their flesh, includes the 'Barooroola' gene (F), which affects ovulation rate and litter size in sheep (Piper *et al.*, 1985).

Selection of stock for improved performance seems feasible, however, on the basis of the heritability (or predictability) found for birth weight, growth from birth to weaning, post-weaning growth and feed utilization efficiency (Tables 2.1 and 2.2, after Kunkel, 1961; Sellier, 1994).

There are indications that there exist genetically determined differences in the requirement for essential nutrients by domestic animals, such as vitamin D (Johnson and Palmer, 1939) and pantothenic acid (Gregory and Dickerson, 1952).

A most important aspect of genetic variability is that determining the balance of endocrine control of growth and development. In this context, Baird *et al.* (1952) showed that the pituitary glands of a group of fast-growing pigs contained significantly greater amounts of growth hormone than those of a corresponding group of slow-growing pigs. In the opinion of Ludvigsen (1954, 1957) the intensive selection of pigs for leanness and carcass length thereby increased the numbers of those animals having a high content of growth hormone in the pituitary. In such animals there would appear to be a concomitant deficiency of ACTH (i.e. the hormone which controls the outer part of the adrenal gland) and possibly, therefore, an inability to counteract the initial increase in blood-borne potassium which arises during exposure to stress. Such pigs produce pale soft exudative (PSE) flesh post-mortem, and it appears that this condition reflects another effect of genetic makeup. Two undesirable genes are involved, one being responsible for susceptibility to halothane sensitivity and malignant hyperthermia (Hal^n) and the other for "acid" meat (RN^-).*
Both genes cause watery meat to develop, the first being associated with an abnormally fast rate of pH fall during post-mortem glycolysis, the second with the attainment of an abnormally low ultimate pH (Lawrie *et al.*, 1958; Ollivier *et al.*, 1975; Le Roy *et al.*, 1990) (cf. §§ 3.4.3 and 5.1.2). The differences in water-holding capacity between carriers and non-carriers of the RN^- gene are associated with a more pronounced denaturation of L-myosin and sarcoplasmic proteins in the meat of the former post-mortem (Deng *et al.*, 2002). The RN^- gene codes for an isoform of the γ-unit of AMP-activated protein kinase (Milan *et al.*, 2000), whereby abnormally high levels of glycogen are stored in porcine muscle. The meat from pigs carrying the RN^- gene has been observed to be more tender than that of non-carriers; and this has been attributed to more extensive proteolysis in the former, potentiated by the early attainment of a low pH (Josell *et al.*, 2003).

The level of polyunsaturated n-3 fatty acids (§ 11.1.4) in the polar (but not the neutral) lipids of the muscles of Hampshire pigs is higher in rn^+ animals than in those of the genotype RN^-; whereas that of polyunsaturated n-6 acids is higher in the latter (Högberg *et al.*, 2002). This circumstance may influence the nature of the

* The *RN* gene is named from 'Rendement Napole', since, when in its dominant form, it greatly reduces the yield of cooked, cured ham: Napole is an acronym derived from the names of those who devised a means of measuring the yield. It has been identified in Hampshires, especially in France and Sweden.

cell membranes and thereby help to account for the high glycogen content of the musculature of the Hampshire.

Initially three genotypes of each gene, viz. NN, Nn and nn, and RN^-/RN^-, RN^-/rn^+ and rn^+/rn^+ were recognized. A third allele at the RN locus was identified in Hampshire and Hampshire/Landrace crosses, in 2000 by Milan *et al.*, now designated as rn*, there being, therefore, six different genotypes, viz. RN^-/RN^-, RN^-/rn^+, $RN^-/rn*$, rn^+/rn^+, $rn^+/rn*$ and rn*/rn*.

Their effects on meat quality were studied by Lindahl *et al.* (2004). The RN^- allele was dominant over rn^+ and rn*, and was associated with a low ultimate pH, low water-holding capacity and high cooking loss. The rn* was associated with a higher ultimate pH. All three alleles affected the colour of the porcine musculature.

An inability to prevent excessive release of Ca^{++} ions from the sarcoplasmic reticulum of muscle (§ 3.2.2) is believed to be the immediate prequisite for malignant hyperthermia in pigs and for the development of PSE post-mortem. The plant alkaloid ryanodine is one of the agents which can cause such release and MacLennan *et al.* (1990) implicated the gene which codes for the ryanodine receptor protein. This protein is now believed to be identical to the junctional foot protein which forms a major component of the calcium-releasing complex controlling the contraction of muscle (cf. § 3.2.2). Fujii *et al.* (1991) identified the specific defect responsible, namely, the substitution of thymidine for cytosine at location 1843 on the cDNA, whereby cysteine was coded for at position 615 on the junctional foot protein instead of arginine, with consequent stereochemical changes adverse for the proper functioning of the protein. The genetic defect concerned can be detected in porcine tissues by a DNA–polymerase reaction (Houde and Pommier, 1993).

The condition will be discussed more fully in a later chapter. At this point, however, it is useful to indicate the effects of selection in the UK in recent years. In the 15-year period 1960–75, the feed conversion ratio of both Landrace and Large White pigs steadily improved, the layer of fat above the loin diminished and the cross-section at area of the *l. dorsi* muscle increased (Table 2.5). But, in the period 1972–82, the incidence in the UK of pigs having musculature in which the pH had fallen below 6 within 45 min of death (pH_1) doubled (Chadwick and Kempster, 1983); and although this feature is not an infallible indicator of the subsequent development of pale, soft exudative pork, it is highly prognostic. In a survey of 5500 bacon weight pig carcasses in UK subsequently, Homer and Matthews (1998) found a slight increase in potential PSE meat over the previous 10 years (as indicated by pH values below 6.0 at 45 min, post-mortem), but no evidence for meat of dark-cutting character (as indicated by ultimate pH† values above 6.5): the average ultimate pH was 5.64. This was, however, somewhat higher in the winter, suggesting the utilization of some glycogen reserves for thermoregulation (cf. § 5.1.2).

Singh *et al.* (1956) noted that lambs with intrinsically higher rates of thyroid secretion gained weight more rapidly than those with an intrinsically low rate. There is some evidence that the dwarf gene is associated with an increased sensitivity to insulin (Foley *et al.*, 1960).

Some variations in growth are apparently effected by genetically determined compatibility or incompatibility with the environment. Thus, the resistance of pigs

---

† The pH at which post-mortem glycolysis ceases – usually because the enzymes involved are inactivated – is referred to as the 'ultimate pH' and measured at 24 hours post-mortem (cf. § 5.1.2).

Table 2.5 Average performance of all hogs and gilts tested at MLC test stations (Meat & Livestock Commission, 1974–75)

Year	Feed conversion ratio (lbs (kg) feed/ lbs (kg) liveweight)	Loin fat (mm)	L. dorsi area (sq cm)
Landrace			
1961	3.40	23	28.97
1964	3.03	23	28.97
1967	3.06	23	31.15
1970	2.81	20	32.10
1973	2.76	19	31.70
1975	2.76	19	31.85
Large White			
1961	3.33	25	27.59
1964	2.92	24	29.14
1967	2.92	24	30.35
1970	2.71	21	31.80
1973	2.64	21	32.30
1975	2.63	20	33.30

to brucellosis (Cameron *et al.*, 1943) and of cattle to ticks (Johnson and Bancroft, 1918) can be inherited.

Although careful selection has fostered the expression of genes coding for useful characteristics in animals and in the meat they produce, the procedure has always been lengthy and its outcome has not been invariably successful. Now that it is possible to identify, isolate, multiply and incorporate into animals genes which code for desirable features, and to eliminate those which are associated with problems, the time required to establish new lines has diminished remarkably.

Because of the great potential of recombinant DNA techniques in relation to meat animals, much research is being directed to mapping the entire genome in sheep (Moore *et al.*, 1992), cattle (Bishop *et al.*, 1994) and pigs (Rohner *et al.*, 1994) to identify those genes which code for leanness, muscle morphology and various aspects of the attributes of eating quality. Markers for such genes will be used to make precise selection for desirable traits in breeding programmes. As a most important aspect of the development of genomics, microassay techniques are making possible quantitative assessments of the expression levels of several thousand genes simultaneously, whereby accurate characterization can be made of those components in muscles that determine their eating quality as meat subsequently (Bendixen *et al.*, 2005).

In reviewing transgenic techniques, Pursel *et al.* (1990) concluded that it will be necessary to transfer the nucleic acid sequences which control the expression of the genes coding for the hormones desired, as well as those for the latter *per se* to ensure that overall metabolic balance is maintained.

Recombinant DNA techniques can involve the use of retroviruses. After infecting a cell, the viral RNA is converted into DNA by viral reverse transcriptase. The DNA circularizes and integrates with the genome of the host cell. Clearly retroviruses could be used to transfer a selected foreign gene (Hock and Miller, 1986).

Such techniques can also be effective indirectly. Thus a recombinant bacterium, expressing a powerful cellulase gene, has been developed to improve the digestibility of grasses in the rumen of cattle (Anon., 1994–95).

Recently, embryos which had been reconstructed by transferring nuclei removed from an established somatic cell line into the enucleated cytoplasm of oocytes, have been successfully developed in the uterine horn of ewes and yielded cloned lambs (Campbell et al., 1996). Cattle and pigs are among the other species which have now been cloned. Multiple asexual reproduction of meat animals of desired genotypes can thus be anticipated.

Genetic engineering has made it possible to alter specific features of muscle. Thus, fast-twitch, glycolytic fibres have been induced to develop enhanced mitochondrial oxidative metabolism, reduced fibre diameter and lowered glycolytic potential by transgenic expression of myogenin (Hughes et al., 1999) (cf. § 4.3.5.2). Direct genetic control of the extent to which the attributes of eating quality in meat are expressed may well become a practical procedure.

2.3   Environmental physiology

The subject of heat regulation in farm animals has a wide economic significance. Sheep, cattle and pigs attempt to maintain their body temperature at a constant value which is optimum for biological activity. Of the three domestic species, sheep are most able to achieve this, and pigs least able (Findlay and Beakley, 1954): indeed, newborn pigs are particularly susceptible to succumbing to heat stress. Even so, adult pigs in a cold environment can maintain their skin temperatures at 9 °C when exposed to air at –12 °C (Irving, 1956).

The environmental temperatures normally tolerated by living organisms lie in the range 0–40 °C, but some animals habitually live below the freezing point or above 50 °C. For short periods even more severe conditions are compatible with survival: certain polar animals, for example, will withstand –80 °C (Irving, 1951). Under such conditions, body temperatures are maintained. Nevertheless, the body temperature of certain small mammals can be depressed below the freezing point without causing death (Smith, 1958). In an environment of low temperature the development of many animals is prolonged (Pearse, 1939): under high temperatures it is frequently retarded in unadapted stock. A variable temperature has a greater stimulatory effect on metabolism than those which are uniformly low or high (Ogle and Mills, 1933). Prolonged exposure of an animal to heat or cold involves hormonal changes which are specific to these two stresses, whereas acute exposure of an animal to heat, cold, danger or other aspects of stress elicits a typical complex of reactions from the endocrine system, referred to as the general adaptation syndrome (Selye, 1950; Webster, 1974).

In general it would be expected that in a cold environment a large body would be advantageous since its relatively low surface to volume ratio would oppose heat loss; and that in a warm environment a high surface to volume ratio would help to dissipate heat. This generalization appears to apply to animals of similar conformation (Bergmann, 1847). Some of these principles can be seen to operate with domestic stock (Wright, 1954). Among cattle the yak, which inhabits regions of cold climate and rarefied atmosphere, possesses a heavy, compact body with short legs and neck: it is also covered with thick, long hair. Cattle of more temperate regions have a somewhat less compact frame; while tropical cattle have an angular frame, larger extremities, a large dewlap (i.e. the fold of skin hanging between the throat and brisket of certain cattle) and a coat of very short hair. Thus among dairy cattle in the USA it is found that whereas Holsteins predominate in temperate areas, the smaller Jersey and Guernsey cattle predominate in the hotter, more humid environments. In Australia, the United Kingdom breeds of sheep thrive in the cooler and wetter areas and Spanish Merino sheep in the vast arid regions. Some of the greater heat tolerance of tropical cattle can be attributed to a diminished emphasis of the subcutaneous region as a storage depot for fat (Ledger, 1959). A thick layer of subcutaneous fat would accentuate the stress caused by a hot, humid environment. On average, Hereford cattle possess 20 per cent greater tissue insulation than Friesians of comparable size and condition, although differences in tissue insulation between individuals, due to differences in body composition, can be as much as 60 per cent (Webster and Young, 1970). In sheltered conditions the growth rate of cattle is likely to be unimpaired until the air temperature falls below –20°C. Rapidly growing beef cattle and high-yielding dairy cows produce so much heat that they are unlikely to experience cold severe enough to increase metabolic rate under normal winter conditions in Britain. The zone of thermal neutrality for cattle is that in which metabolic heat production is independent of air temperature (Webster, 1976). Cattle can regulate evaporative heat loss over a wide range at little metabolic cost. Thus, the zone of thermal neutrality for cattle is broad (ca. 20°C). The pig has a much narrower zone of thermal neutrality (Ingram, 1974).

Bearing in mind that radiant energy is absorbed more by dark coloration and reflected more by light, it is not unexpected that many tropical cattle have a lightly coloured coat. Findlay (1950) showed experimentally that cattle having a white, yellow or red coat, especially if the coat is of smooth, glossy texture, will absorb markedly less heat than those of a darker colour. Moreover, under given conditions of heat stress, the temperate breeds will have a higher body temperature than tropical breeds. The mechanisms for heat disposal (evaporation of water from respiratory passages, transudation of moisture through the skin and depression of metabolic rate) are less efficient and temperate stock tend to seek relief by behavioural mechanisms such as voluntary restriction of food intake, inactivity and seeking shade. This must necessarily restrict their development in relation to tropical breeds. There are many records to show that cattle with 'Zebu' blood produce a higher percentage of better-quality carcasses in hot, humid conditions (e.g. Queensland) than do animals of entirely temperate blood (Colditz and Kellaway, 1972). This tendency can be offset to some extent by providing pasture for night grazing.

Among sheep it is found that those in temperate areas are generally of moderate size, of compact conformation and with short legs and a thick wool coat. In tropical areas sheep have long bodies, legs, ears and tails, and a coat of short hair rather than wool. In arid areas sheep frequently develop an enlarged tail, where fat is stored: the metabolism of the latter offsets the environmental scarcity of water and food.

In general it is not the degree of heat alone which causes distress to animals in the tropics but its combination with humidity and the duration of these conditions. 21°C provides a rough division between temperate and tropical stock, the latter functioning efficiently above this temperature (Wright, 1954). A useful measure of the bearability of climate is given by the number of months per year when the mean

24-h wet bulb temperature is above 21 °C (Anon., 1950). Few cattle of the class *Bos taurus* can maintain appetite and growth rate at air temperatures above 25 °C (Colditz and Kellaway, 1972). Normally there are no sudden changes in the form of functions of domestic stock: the characteristics typical of each successive environmental zone will merge with those of its neighbours imperceptibly, this being facilitated by interbreeding at zonal boundaries (Wright, 1954). As witness to this, the number of local breed varieties is very large. Mason (1951) lists 250 breeds of cattle and 200 breeds of sheep.

Intense light has an influence independently of temperature. Radiation in the ultraviolet wavelengths converts the precursor of vitamin D in the skin into the active molecule; and this may explain the rarity of rickets in the tropics. The ultraviolet components of sunlight also elicit the formation of the dark pigment, melanin, in the skin, presumably as a protective mechanism, since they have marked carcinogenic properties when lightly pigmented skin is exposed to them.

The suitability of livestock for introduction into new areas is not limited to physiological reactions: they must meet the economic and social needs of the population. Thus, for example, the excellent draught qualities of Indian cattle should not be sacrificed in an attempt to raise milk yields, and animals introduced into warm areas, because of their heat tolerance, should possess a potential for meat production (Wright, 1954).

An effect on growth, so far unexplained, is the retardation caused by a static magnetic field (Bernothy, 1963).

2.4   Nutritional aspects

'All cold-blooded animals and a large number of warm-blooded ones spend an unexpectedly large proportion of their time doing nothing at all, or at any rate, nothing in particular ... when they do begin, they spend the greater part of their lives eating' (Elton, 1927). Finding enough of the right kind of food is the most important general factor determining the development, dominance and survival of all living organisms.

2.4.1   Plane and quality of nutrition

Differences in the plane of nutrition at any age from the late foetal stage of maturity not only alter growth generally but also affect the different regions, the different tissues and the various organs differentially. Thus animals on different planes of nutrition, even if they are of the same breed and weight, will differ greatly in form and composition (Hammond, 1932a; McMeekan, 1940, 1941; Pomeroy, 1941; Wallace, 1945, 1948). These workers showed that, when an animal is kept on a submaintenance diet, the different tissues and body regions are utilized for the supply of energy and protein for life in the reverse order of their maturity. Under such conditions fat is first utilized, followed by muscle and then by bone; and these tissues are first depleted from those regions of the body which are latest to mature. The relationship between plane of nutrition and development of the different tissues of the body was shown by Hammond (1932a, 1944): the brain and the nervous system have priority over bone, muscle and fat in that order. In general, his findings have been confirmed by later investigators (Butterfield and Berg, 1966; Davies, 1974; Cole

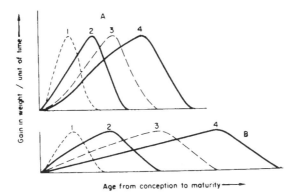

Fig. 2.1   The effect of maturity and plane of nutrition on the rate of increase of different portions of the body (Pállson, 1955). A, early maturity or high plane of nutrition, B, late maturity or low plane of nutrition.

Curves	1	2	3	4
	Head	Neck	Thorax	Loin
	Brain	Bone	Muscle	Femur
	Cannon	Tibia-fibular	Femur	Pelvis
	Kidney fat	Intermuscular fat	Subcutaneous fat	Intramuscular fat

(Courtesy the late Dr H. Pállson.)

et al., 1976). Figure 2.1 shows the order in which the different parts of the body develop on high and low planes of nutrition and in early and late maturing animals (Pállson, 1955). There have been other views, however, on this topic. Thus Tulloh (1964) contended that carcass composition is weight-dependent and largely uninfluenced by age or nutritional regime. In an assessment of the different hypotheses, Lodge (1970) concluded that, while muscle weight appears to increase in proportion to total weight in a simple fashion, complexities arise from the degree of fat, and to a lesser extent that of bone, associated with the muscle. In 1976 Berg and Butterfield proposed an alternative concept of nutritional distribution during growth whereby the effects of a low plane of nutrition are balanced between vital organs, bones, muscle and fat according to the severity of the inanition.

In the view of Hammond (1932a) domestication favours the later developing parts of the body by providing a higher plane of nutrition. Selective breeding, if practised in primitive farming conditions, may fail to improve stock. Such characters as good muscle development in the loin would be incapable of expression to their full genetic potential in underfed animals. Nevertheless, the proportion of muscle in the carcass tends to be higher in wild types than in improved breeds, largely because of the relative absence of fat in the former. The greatest effect of breeding on the chemical composition of the musculature is shown in those muscles which have the greatest rate of postnatal development. That cattle, sheep and pigs differ in the partition of fatty tissues between subcutaneous and intermuscular locations may also be an effect of domestication (Callow, 1948). Subcutaneous fat, which is developed later than intermuscular fat, is favoured more in pigs than in sheep and more in sheep than in cattle. Pigs produce more generations in a given period of time than sheep and sheep more than cattle, so that the effects of domestication would be expected to be most marked in the former and least in the latter. Since the subcutaneous depot fats have a greater proportion of unsaturated fatty acids than do internal ones (Duncan and Garton, 1967), this may be a factor contributing to the greater problems of rancidity development in pork. According to Braude (1967) it will not be long before breeding and feeding policy have produced a desirable carcass so soon after birth that nearly half of the pig's effective life will have been spent *in utero*. This will increase the importance of assessing prenatal influences on growth.

Although fast rates of growth caused by a high plane of nutrition can lead to an earlier onset of the fattening phase of growth, the nature of the diet, not surprisingly, is also an important growth-regulating factor. Thus, when the protein/energy ratio is increased, the fastest-growing animals may become leaner (Campbell and King, 1982). Indeed, when the ratio is very high the growth rate may be diminished (Campbell *et al.*, 1984). Since males have a higher protein/energy requirement than females, this factor can cause differences between the sexes in the composition of the carcasses when the energy intake, at a given ratio, is altered (Campbell and King, 1982).

In 1965 Clausen pointed out that the amino acid requirements for optimum growth of the carcass in pigs can differ from that required for optimum pork quality. It is generally accepted that lysine is the first limiting amino acid for the growing pig: when the level of intake of crude protein is below *ca.* 14 per cent, however, other essential amino acids may become limiting (e.g. threonine) (Taylor *et al.*, 1973, 1974). In this species it may be more useful to formulate the requirements for individual essential amino acids than to provide crude protein (Swan and Cole, 1975).

The relationship of the ruminant to the quality of the diet is different since the intake into the tissues is substantially determined by the rumen microflora, the major source of nitrogen for microbial synthesis being the ammonia derived from the deamination of soluble protein entering the rumen and the activity of the micro-organisms being determined by the amount of energy available to them. Non-incorporated ammonia is excreted and ingested protein may be wasted on high protein diets. On the other hand, non-protein nitrogen (e.g. urea) can be converted into protein when low protein diets are fed. Although the protein-synthesizing activity of the rumen micro-organisms diminishes the effect of differences between the incoming proteins in ruminants, there are differences in the rates at which they are degraded (Butler-Hogg and Cruickshank, 1989). Again, there may be differences in the nature of the proteins synthesized by the micro-organisms and this may require supplementation by amino acids in the feed (Storm and Orskov, 1984).

2.4.2   Interaction with other species

Only humans can catch and destroy animals of any size; and this they have only been able to do in the latter part of their history (Elton, 1927). In general, there are definite upper and lower limits to the size of animal which a carnivore can utilize. It obviously cannot catch and destroy animals above a certain size; and below a certain size it cannot catch enough to get satisfaction. The development and growth of sheep and cattle, which may be conveniently regarded as herbivores, are not, therefore, influenced directly by the availability of food animals; but their numbers may be depleted by the larger carnivores in tropical areas and by smaller carnivores, such as wolves and dogs, in more temperate regions. Such losses need not represent direct attack, but many result from worrying or harassing activities. Moreover, young stock – lambs or calves – may be destroyed by predatory birds. The most serious effects of other animal species on domestic stock, however, arise indirectly when the rate of natural increase of small species, such as mice, rats and rabbits, is strongly enhanced over a short period because of some sudden environmental change or for other as yet obscure reasons.

Plagues of locusts, mice and rats have long troubled humankind by their widespread destruction of growing crops or stored grain – and hence caused substantial losses of stock by depleting their natural food. Such plagues may arise when a species which is normally not troublesome is introduced into a new environment from which its natural enemies are absent. Thus, the rabbit, when introduced into Australia, increased at a phenomenal rate and denuded vast areas of the vegetation upon which stock had previously subsisted. Again, in New Zealand the introduction of starlings, by causing the spread of thorn bushes, brought about much loss of lambs which became entangled in them.

Some species, of course, are beneficial to domestic stock. Ruminants largely depend on the micro-organisms in their digestive tract to break down the cellulose of plant foods to easily assimilable, energy-yielding molecules such as fatty acids; and, in some cases, for the elaboration of vitamins and other accessory food factors. They are thus equipped to make use of poor-quality diets. On the other hand, the rumen micro-organisms tend to degrade high-quality protein if supplied in excess. The feeding of antibiotics to livestock has a conditioning effect upon the contribution of gut micro-organisms to animal nourishment.

Unfortunately, there are a large number of parasitic organisms belonging to various zoological phyla which cause disease involving wasting and even death (Thornton, 1973). The nature of these organisms, and hence of the diseases which they cause, depends on the environment.

2.4.3 Soils and plant growth

Although some animals are carnivores they are ultimately as dependent on plant life for their sustenance as the herbivores. Food represents energy stored in a form assimilable and utilizable by animals and from which it can be released at a rate determined by metabolic needs. On the Earth, the major ultimate source of energy is sunlight; and it is only in the Plant Kingdom that a mechanism exists for its conversion into a stored form – the photosynthesis of carbohydrate from carbon dioxide and water by the green, light-sensitive pigment chlorophyll. In turn the availability of plants depends upon types of soil and the interaction of the latter with climatic conditions. In the last analysis, of course, the remote cosmological events which caused the cooling of the Earth, the primary parting of the elements (Goldschmidt, 1922, 1923), the formation of rocks and the distribution of minerals, and the more recent ones, whereby solar activity and, hence, the Earth's climate, fluctuate cyclically, have been jointly responsible for the types of soils originally formed, eventually produced and currently altered. The fertility of soils depends not only on the chemical nature of the rocks from which they have been formed but also on particle size. The latter determines how much moisture soils will hold and for how long, and the availability of its nutrients to plants. Soil fertility, of course, is also influenced by such extrinsic factors as temperature, rainfall and topography (including the depth of the soil layer above the underlying, unaltered rock). Soil types range from podsols, from which exposure to rainfall of high intensity has leached out the more soluble alkaline minerals and trace elements, to solonized soils, in which alkali and salts have accumulated because of the absence of rainfall and the existence of conditions favouring evaporation. The acid reaction of podsols tends to favour the growth of rank grasses, which have a relatively low nitrogen content, and to discourage the growth of legumes.

The type of rainfall is important: knowledge of the average number of inches of annual precipitation is of limited usefulness in itself. Thus, the *variability* of the rainfall from the mean annual value is especially high in some parts of the world, causing a succession of droughts and floods: its *effectiveness* is important, since much precipitation may be unavailable if conditions of evaporation are high, as would be the case if all the rainfall occurred in the hot season; and its *intensity* (or the amount of rain at any one precipitation) determines the amount of soil erosion and whether moisture is deficiently, optimally or excessively available for plant growth.

Many of the interrelations between soil type, climatic conditions and plant growth, on the one hand, and the breeding and fattening of sheep, cattle and pigs, on the other, are well exemplified by the conditions in Australia (Anon., 1950; Beattie, 1956). During the wet season in tropical Australia, phosphate in the soil is limited. The efficiency of beef production is increased if phosphate is fed to cattle (Meat Research Corporation, 1994–95).

Soils and climate affect plant growth qualitatively as well as quantitatively. As already mentioned, rank grasses of low nitrogen content develop on acid soils: the nitrogen content of grasses and legumes diminishes on prolonged exposure to high temperatures. Again, the digestibility of plants may be altered artificially. The increasing use of plant growth substances, such as gibberellic acid, which increases the internode distance in grasses, tends to lower the nitrogen content of grasses and to decrease digestibility (Brown *et al.*, 1963).

Although leguminous plants such as alfalfa, lucerne and white clover constitute valuable feed for ruminants, it has long been appreciated that ingestion can seriously affect the life of the animal as they can cause excessive fermentation and the retention of large volumes of gas (bloat). Bloat is due to the formation of stable, proteinaceous foams in the rumen (Mangan, 1959). Plants which produce tannins (proanthocyanidins) are known to be capable of destabilizing or suppressing the formation of such foams, probably by the tannins forming complexes with the proteins (Jones and Mangan, 1977). The minimum amount of proanthocyanidin necessary for the prevention of bloat has been shown to be very low (Li *et al.*, 1996). If the genes responsible for the formation of proanthocyanidins in the plant could be transferred to forage species such as white clover, the dangers of bloat development could be substantially diminished. Already several genes involved in the biosynthesis of proanthocyanidins have been identified and cloned (G. Tanner, personal communication).

2.4.4   Trace materials in soils and pastures

It is being increasingly recognized that many animal ailments can be explained on the basis of dietary deficiencies or excesses of biologically potent materials which are present only in minute quantities in soils and pastures. The great importance of traces of certain elements in the nutrition of ruminants was first realized about 60 years ago when workers in western and southern Australia discovered that a deficiency of cobalt in the soils of certain areas was the cause of various wasting and nervous diseases in cattle and sheep which had been known for many years by set-tlers. Animals, after grazing for several months, would lose their appetite and finally die in the midst of rich pasture (Russell and Duncan, 1956). Spraying land with cobalt sulphate (only 5 oz per acre, 11 g/ha, annually), or the giving of cobalt orally, eliminated the condition. Since then, soil areas in many other countries have been found to be deficient in cobalt. The ingestion of liver was long known to be effec-tive in cases of pernicious anaemia. With the isolation from liver of an active prin-ciple, vitamin B_{12}, and the demonstration that its molecule contained 4 per cent of cobalt (Smith et al., 1948), it appeared that anaemia might be a significant aspect of cobalt deficiency syndromes. This is still in doubt, however. Several of these cobalt-deficiency diseases involve a concomitant deficiency of copper; but there are other ailments affecting stock where the primary deficiency is one of copper, e.g. sway-back in lambs, a nervous affliction arising in the United Kingdom. Copper is impor-tant for blood formation; but it is also an essential part of enzymes which oxidize phenols in plants. Polymerized phenols (e.g. leuco-anthocyanins) form lignin (Bate-Smith, 1957); and plants having a high lignin content resist attack during digestion by the ruminant micro-organisms. Seasonal changes in the quantities of polyphe-nols in plants, due to the availability of copper, may thus affect ruminants indirectly (Greene, 1956). Copper deficiency inhibits the action of various amine oxidases, and, thereby, the structural integrity of elastin and collagen in connective tissues (Mills, 1979: and cf. § 3.2.1).

An excess of molybdenum in the soil is said to enhance copper deficiency in animals; conversely, a deficiency of molybdenum may cause toxicity through making excess copper available (Russell and Duncan, 1956). Numerous ailments due to an excess or deficiency of elements in the soil and pasture are known. An excess of selenium interferes with metabolism by displacing sulphur from the essential –SH groups of dehydrogenase enzymes; an excess of potassium, by interfering with the accumulation of sodium, alters the ionic balance of body fluids and may cause hypersensitivity to histamine. Conditions such as grass tetany and milk fever, wherein the magnesium or calcium content of blood serum is especially low, are said to arise in this way. Muscular tissue tends to accumulate caesium-137 from radioac-tive fallout (Hanson, 1961). In Cu, Co and Se deficiencies, and in Cu and Mo toxi-cities of grazing ruminants, the relation of the disease conditions to the geochemical environment has been established by chemical analyses of feeds and animal tissues (Underwood, 1977).

Toxicity in stock may also arise by the excess ingestion of trace organic sub-stances. Thus, lambs in New Zealand suffer from facial eczema if they graze pasture including red clover. Under certain environmental conditions the latter harbours the fungus *Penicillium chartarum* which elaborates a toxin deranging liver metabo-lism. As a result, haematin compounds are broken down abnormally, producing the photosensitive pigment phyloerythrin which circulates in the blood, causing the skin to react severely to sunlight (Thornton and Ross, 1959). The ingestion of sweet clover may cause haemorrhagic disease because of its content of dicumarol. Some pastures, particularly subterranean clover in the green stage, may contain sufficient isoflavones or flavones of oestrogenic potency to affect the reproductive activity of grazing ewes (Flux et al., 1961). The oestrogenic potency of pasture can be deter-mined by the increase in length of wethers' teats (Braden et al., 1964).

Fluoracetate, found in the leaves of certain plants in Africa and Australia, can be toxic to grazing sheep and cattle, since it blocks the metabolic pathway for oxidation (Peters, 1957). Of particular interest is the finding that some amino acids may be toxic for stock (Ressler, 1962). These are found in plants but are not derived from proteins. Some sufficiently resemble the normal amino acids from proteins to compete with the latter, thus disrupting metabolism.

Pastures, of course, may also be deficient in trace organic substances. Thus muscular dystrophy occurs in sheep, cattle and pigs (especially young ones) when there is insufficient alpha tocopherol, particularly if the diet contains unsaturated fat (Blaxter and McGill, 1955). Many other examples could be given. Apart from naturally occurring excesses or deficiencies of minerals and organic matter in soils and feed, an artificial hazard has arisen, since the early 1960s, from the extensive use of pesticides. It is now appreciated more fully that the benefits these have conferred in conservation of fodder and in the diminution of worry and wastage of stock caused by parasites and insects, must be balanced against the effects of ingestion of pesticide residues by the human consumer of the meat and the undesirable changes in the ecological pattern which have been observed. Probably the most extensively used pesticide has been the organochloride compound, DDT; but other organochlorine substances such as dieldrin, organophosphorus compounds such as malathion, and organomercurials, have also been used extensively. The main danger arises from the persistence of such substances. They resist degradation to nontoxic derivatives, and can gradually build up on soils. Pesticide residues can accumulate in the flesh of animals feeding in such areas; but in the United Kingdom lard is so far the only meat product in which more than acceptable levels of pesticide residues have been regularly detected (Anon., 1967).

The dangers of ingesting methylmercury through environmental pollution have been recognized. Although fish appears to be the major foodstuff from which people derive this toxin, meat and more particularly pork can be implicated (Curley *et al.*, 1971).

2.4.5   Unconventional feed sources

The gravity of the world protein shortage has stimulated the search for more effective ways of producing protein of high biological value. One approach to this problem is to ensure the fullest utilization of the food which ruminants are given. The rumen microflora can degrade essential amino acids ingested by cattle and sheep and thus prevent their absorption. Treatment of the feed with formalin protects amino acids during their passage through the rumen and makes them available to the animal when they reach the small intestine (McDonald, 1968).

Another possibility is to recycle manure. Fresh feedlot manure can be mixed with concentrates and fed successfully to cattle with a considerable saving in feed per unit of beef produced (Anthony, 1969). The manure is removed daily, mixed with fresh hay, and fermented by rumen micro-organisms. Yeast can also be produced on fluidized manure, 70 per cent of the dry matter of the latter being incorporated in the yeast, which can then be eaten by stock.

It has been demonstrated that yeast can ferment mineral hydrocarbon fractions to produce protein of high biological value which is as effective for the growth of stock as soya meal or fish protein concentrates (Champagnat, 1966; Shacklady, 1970). Bacteria can also subsist on crude petroleum; and, like yeast, they can convert it into protein of high biological value for animal feeding and, simultaneously, upgrade the hydrocarbon to produce high-grade domestic fuel.

Although the process is currently prohibitively expensive, algae of various types can subsist on waste and, by direct use of light, produce high-quality protein feed.

2.5   Exogenous manipulation

As we have seen, the present forms of meat animals have arisen from the long-term direct and indirect effects of natural factors on their genetic potential; and, over a relatively short period, from the artificial selection by man of desired variants. Deliberate, scientific manipulation of domestic livestock to alter their reproduction, growth and development along largely predetermined lines is now possible.

2.5.1   Reproduction control
2.5.1.1   *Fertility*
Two main approaches have been made to augment the fertility of female animals, namely, increasing the number of young produced at any parturition (through raising the plane of nutrition or through the administration of gonadotrophic hormones) and inducing ovulation in those which fail to ovulate naturally due to seasonal or post-partum effects (through the administration of hormones and, in sheep, by manipulation of the diurnal light/dark ratio: Yeates, 1949; Laing, 1959).

The number of ova produced at one time may be increased above that characteristic of the species (superovulation) by hormone administration (Hammond *et al.*, 1942; Stewart *et al.*, 1976). The process may be combined with ova transplantation, whereby fertilized ova from genetically desirable donors are implanted into the uteri of recipient females which have no traits worth propagating and which act merely as incubators (Hafez, 1961). The oestrus cycles of donors and recipients must be synchronized (cf. § 2.5.1.2). Experiments involving the transfer of ova between ewes of the large Lincoln breed and of the small Welsh Mountain breed have shown that the latter can respond adequately to the demands made by embryos of greater size (Dickinson *et al.*, 1962).

For practical purposes, the application of superovulation is confined to cattle and sheep. It is achieved by the use of gonadotrophins, usually pregnant mare serum gonadotrophin (PMSG), administered in the follicular phase of the oestrus cycle. Provided that the dose of PMSG is not excessive, cattle may produce a high proportion of twins without complications (McCaughey and Dow, 1977) but variability of response, leading to undesirably large litter sizes, has been the more general experience (Gordon *et al.*, 1962). Superovulation, with ova transfer, can now be applied practically in various situations with cattle (Gordon and Boland, 1978) and commercial organizations exist for its exploitation.

In sheep, an increase in the rates of twinning by superovulation has also been achieved (Pállson, 1962) but superovulation with ova transplantation is unlikely to prove advantageous in this species for economic reasons.

There could also be benefits, in terms of increased production, if ovulation could be induced more frequently than normal. In the seasonally anoestrus ewe, fertile oestrus may be induced by treatment with progesterone (or one of its analogues) followed by PMSG (Cognie *et al.*, 1975). Artificial reduction of the hours of daylight exposure, although effective, is difficult to apply under prevailing conditions of sheep husbandry. In cattle, the occurrence of ovarian follicular cysts, and a number of other conditions, may prevent the desired early return to oestrus and ovulation after calving. Recent advances in diagnostic techniques offer the promise of practical application.

A serious problem in commercial pig production is infertility in the sow (Pomeroy, 1960). It may be caused by infection of the reproductive tract, or by abnormalities in the latter which lead to mechanical obstruction. It may arise from failure of mating in sows with normal reproductive tracts, or through death of the embryos or newly-born pigs. Sows which fail to mate may have atrophied or encysted ovaries. On the other hand, the ovaries may appear quite normal and, in such cases, the cause of infertility may be excessive fatness (the fat possibly absorbs oestrogens: Hammond, 1949). Failure of fertilization may be due to the sows being served at the wrong phase of oestrus (or where ovulation is unusually late) or through the boar being sterile. Work on boar taint (Patterson, 1968a, b) has suggested another approach which makes insemination more effective. A substance, 3α-OH–5α-androst-16-ene, which is closely related to that responsible for boar odour, is produced in the submaxillary gland of entire male pigs. It is released in considerable quantity when boars salivate excessively just prior to mating. Its musk-like odour stimulates the female. As an aerosol spray it offers a valuable adjunct to present means of encouraging fertilization.

The lactational anoestrus period of the sow represents a commercially important limitation on the efficiency of reproduction; but, in spite of many attempts, no technique of wide applicability has been developed to challenge early weaning as a means of overcoming it.

Although many factors are known to affect fertility in the male, it is difficult to enhance the fertility of normal individuals. No increase in sperm production has yet been demonstrated following the injection of hormones into male animals (Emmens, 1959); on the other hand, the injection of androgens or of pregnant mare serum gonadotrophin will stimulate sexual drive. The quantity and quality of semen appear to be enhanced by psychological conditioning of the animal (Crombach *et al.*, 1956).

2.5.1.2 *Artificial insemination and synchronized oestrus*
Artificial insemination was first developed on a large scale in the former Soviet Union: 16 million ewes are said to have been treated as early as 1932 (Emmens, 1959). There is little doubt that it is the most effective method to improve and multiply superior meat-producing animals. For example, under natural breeding conditions a proven bull can sire only 25–30 calves per year: artificial introduction of the semen may permit the siring of 5000 calves annually (Hill and Hughes, 1959). The efficiency of the semen depends on the mobility, morphology and numbers of the spermatozoa: it is protected by glycerol; and indeed semen may then be frozen and maintained at $-80\,°C$ for as long as 5 years (Emmens, 1959). It is thus possible to successfully inseminate females in distant areas and to produce offspring long after the death of the father. Work in Scandinavia has shown that the female- and male-determining spermatozoa can be separated by centrifugation, thus permitting preselection of the sex of the offspring.

The application of artificial insemination, however, is limited because, within groups, females are randomly cycling. It would clearly be advantageous if all were to ovulate at a predetermined time. The synchronization of ovulation (and oestrus) has been achieved by administering hormones of one or other of two classes, namely, progestins (progesterone – the natural hormone produced by the corpus luteum – and its analogues) and prostaglandins (prostaglandin $F_{2\alpha}$ – found in the uterine wall – and its analogues). The use of both is related to the fact that a fall in the level of blood progesterone is followed, at a consistent interval, by oestrus and ovulation.

Progestins are administered so as to maintain high levels in the blood after endogenous progesterone has fallen. If such administration is continued for a sufficient time in a group of randomly cycling females, sudden withdrawal of the drug will precipitate oestrus in all its members. The prostaglandins, however, are employed to destroy the corpus luteum during the luteal phase of the reproductive cycle and promote a synchronized fall in progesterone, oestrus following thereafter. Since they are only effective when the corpus luteum is secreting progesterone, a single administration of the hormone may fail to achieve complete synchronization of oestrus in any group (Crichton, 1980). Thus, in cattle, treatment with prostaglandin is ineffective when administered during the first 4–5 days and the last 4–5 days of the oestrus cycle (Rowson *et al.*, 1972). This problem has been overcome by giving two prostaglandin doses 10–12 days apart (Cooper, 1974).

The same principle has been applied to sheep (Haresign, 1976), but pigs are resistant to the administration of most progestins and prostaglandins. Progestin administration has led in many cases to the development of cystic ovarian follicles. Prostaglandin is effective during only a short part of the oestrus cycle in the pig (Guthrie and Polge, 1976). The use of methallibure, then the only effective synchronizing agent in the pig, was terminated because of suspected teratogenic effects.

2.5.2   Growth control

2.5.2.1   *Hormones and tranquillizers*

As is well known, the muscles of male animals tend to be larger than corresponding muscles in females; and castration in the male reduces the efficiency of weight gain in comparison with entire animals. This is not merely a reflection of differences in overall body size. Sex is an important determinant of muscle growth. In this context, the involvement of the steroid hormones from the testis and ovary, and their interaction with hormones from the pituitary gland, the pancreas and the hypothalamus, has been elucidated over the past 50 years. Thus, in 1954 Burris *et al.* demonstrated that male hormones (androgens) stimulated protein synthesis in cattle. Females tended to respond more to androgens than males, both in growth rate and feed conversion (Burris *et al.*, 1954; Andrews *et al.*, 1954). In 1952, Baird *et al.* showed that the content of growth hormone (somatotrophin) in the pituitary of rapidly growing pigs was markedly higher than that of slowly growing pigs. The daily injection of growth hormone into pigs significantly increased the total protein content over that of non-injected controls (Turman and Andrews, 1955). Since somatotrophins can now be produced in quantity by bacteria into which the genes concerned can be incorporated by recombinant DNA, objection to their use may diminish.

Gonadectomy has long been used to facilitate control of male animals, and this practice is likely to continue. The removal of the testes, however, also eliminates the protein-accreting (anabolic) benefits of the gonadal hormones; but this aspect can be restored by the administration of either male hormones (androgens) or female hormones (oestrogens). These anabolic agents are now applied to entire male and female animals as well as to castrates. The use of both androgens and oestrogens is believed to be necessary for maximum growth response (Heitzman, 1976).

Anabolic agents include steroid hormones which are naturally present in the body, such as testosterone and oestradiol-17β, synthetic steroids such as trenbolone, and certain non-steroids such as diethylstilboestrol (Crichton, 1980). Administration as removable implants is the preferred mode of usage. According to Hammond (1957), the best time to implant oestrogen pellets is *ca.* 100 days before slaughter. Before puberty, growth hormone is probably the major endocrine regulator of growth (Kay and Houseman, 1975). At least some of the anabolic sex hormones appear to operate by increasing the secretion of growth hormone from the pituitary; and the latter probably accounts for the raised levels of plasma insulin in treated animals (Crichton, 1980). It is now believed that most of the influence of pituitary growth hormone in stimulating growth is effected via a series of serum peptides, somatomedins (Daughaday *et al.*, 1972). Indeed most other hormones which affect growth (e.g. androgens, oestrogens, thyroid hormone and glucocorticoids) probably operate by altering the production or action of somatomedins (Spencer, 1981). Insulin, however, appears to operate in a different manner, having a marked influence on fat deposition (Gregory *et al.*, 1980).

The somatomedins are also referred to as 'insulin-like growth factors (IGF) I and II'. They are peptides composed of 70 amino acids and are structurally related to insulin. IGF-I is produced in large amounts by the liver, but all tissues produce some (Bass and Clark, 1989). It is believed that pituitary growth hormone (somatotrophin) has two distinct actions: firstly it stimulates the production of IGF-I; and, secondly, it affects precursor cells which are stimulated to differentiate whereas IGF-I only stimulates an increase in cell numbers.

The two insulin-like factors, produced locally in the muscular tissue, appear to control several fundamental processes, such as the proliferation of myofibrils and their differentiation (Magri *et al.*, 1991). Although such control is complex, it can be anticipated that manipulation of the components of the system will become possible and will provide a powerful means of increasing the efficiency of muscle growth in animals.

Hormones apparently initiate their action by combining with receptors located either on the surface membrane, or within the cells, of target organs, whereby the stimulus is amplified. Thus, receptors for oestrogens, androgens and thyroid hormones are intracellular, whereas those for ACTH, insulin, adrenalin and noradrenalin are on the cell surface. Some surface receptors are phospholipid in nature. On stimulation, they activate phospholipase A_2, which releases arachidonic acid. The latter is converted into prostaglandin $F_{2\alpha}$, which fosters the synthesis of protein, and prostaglandin E_2, which fosters the breakdown of protein (Palmer *et al.*, 1982). Both IGF-I and IGF-II have specific cell surface receptors. Other surface receptors are proteins that have seven helical polypeptide strands and link, across the width of the cell membrane, with 'guanylyl nucleotide binding proteins' (G proteins) on the inner surface. Following intramolecular transformation and migration along the inner surface, they cause adenyl cyclase to produce cyclic AMP and, thereby, activate a cascade of protein kinases which phosphorylate cytoplasmic enzymes; and also phosphatases which remove the phosphate groups (Mann, 1999). Ca^{++} ions, released from organelles within the sarcoplasmic reticulum (and mitochondria) promote the phosphorylation. A single molecule binding to an appropriate receptor on the cell surface has its effect very greatly magnified through the cytoplasmic cascade. Because Ca^{++} ions diffuse slowly through the fluid of the cytoplasmism, highly localized Ca^{++} signals, or intracellularly distant ones, can be generated, and varied with the duration and concentration of the hormone stimulating the cell surface receptor (Petersen et al., 1994; Petersen, 1999). (The importance of Ca^{++} ions in stimulating muscular contraction, in controlling the action of the protein components involved therein and in activating the calpain/calpastatin system whereby proteins undergo catabolism and anabolism, will be referred to in §§ 3.2.2, 4.2.1, 4.3.5.1 and 5.4.2).

The catecholamine receptors are classified as α, β_1 and β_2 adrenergic, on the basis of their binding pattern and on the response they elicit. Thus, α receptors are associated with vasoconstriction of the uterus, β_1 receptors with increased cardiac force and rate, adipocyte lipolysis and calorigenesis and β_2 receptors with relaxation of the bronchi, trachea, vascular smooth muscle and the uterus; and with contraction of striated muscle.

β_2 agonists, that is substances which bind to β_2 receptors and thereby promote metabolism, are of particular interest in relation to meat animals. When β_2 agonists (such as cimaterol, clenbuterol and ractopamine), are included in the diet they cause a marked repartitioning between fat and protein whereby animals become leaner. The effect is achieved by increasing lipolytic action in adipose tissue and by reducing protein breakdown in muscular tissue (Buttery, 1983), whereas most other anabolic agents, including the sex hormones, act by a less selective increase of tissue components. Some effects of including cimaterol at different levels in the diets of pigs are shown in Table 2.6.

Since anabolic agents can affect the ratio of muscle to fat in the body, and thus the chemical composition of meat, it might be anticipated that by accelerating the growth rate, these agents would foster an enhanced proportion of mature collagen and thus decrease tenderness. Few consistent reports on the eating quality of the meat from animals given anabolic agents have been reported (Patterson and Salter, 1985; Renerre et al., 1989). β agonists have been said to cause a significant increase in toughness (Bailey, 1988). Thus, the administration of cimaterol to cattle had this effect (Fiems et al., 1989): there being a lowering of the concentration of collagen in the muscle, but a concomitant decrease in the percentage of heat-soluble collagen, suggesting that the treatment had accelerated the maturation of the tissue (Dawson et al., 1990, cf. §§ 3.2.1 and 10.3.2). In respect of experimental studies with lambs, however, although the increase in toughness was proportional to dose, the

Table 2.6   Some effects of dietary cimaterol on pig carcasses (after Jones et al., 1985)

Parameter	Dietary cimaterol (ppm)			
	0	0.25	0.50	1.0
Semitendinosus (wt. g)	382	418	439	426
Biceps femoris (wt. g)	1340	1440	1460	1490
L. dorsi (cross-sect. cm²)	29.85	31.96	33.96	33.42
Fat depth				
above first rib (cm)	3.85	3.54	3.72	3.48
above last lumbar vertebra (cm)	3.00	2.73	2.65	2.47

Means of 40 pig carcasses at each dietary level.

shear force of the cooked meat was still within an acceptable range (Beerman *et al.*, 1990). These workers also demonstrated that electrical stimulation could improve the tenderness of lambs fed β agonists, although this action was not due to the avoidance of 'cold-shortening': it appeared to be related to overcoming an inhibition of calpain activity (cf. § 5.4.2) with which the administration of β agonists is associated.

The β agonists appear to stimulate the action of the calpain inhibitor, calpastatin (Parr *et al.*, 1992) and naturally occurring levels of calpastatin appears to correlate inversely with the rate of post-mortem tenderizing in meat (Koohmaraie *et al.*, 1991). In later chapters (cf. §§ 5.4.2 and 10.3.3.2) the increase in tenderness which develops when muscles are held for some time post-mortem (in the absence of microbial spoilage) – the phenomenon of 'conditioning' or 'ageing' – will be considered further. More or less subtle proteolytic changes are responsible. These are caused by the action of calcium-dependent proteinases (including calpains) and other proteinases, cathepsins, which are normally held within organelles (lysosomes). Protein turnover *in vivo* is a dynamic process, however, and it has always seemed unlikely that such enzymes function merely as catabolic agents in post-mortem muscle. It is now evident that at least the calpain system is implicated in anabolic activity in muscle (Croall and Demartino, 1991). The mechanism involves inhibition of the turnover of myofibrillar proteins by calpastatin, which opposes proteolytic action by the calpains (cf. §§ 4.3.5 and 5.4.2). The calpastatin/calpain system is thus important in the *in vivo* synthesis of muscle proteins (Goll *et al.*, 1992).

The growth promoter, ractopamine, causes a marked increase in muscle mass in cattle and pigs, as manifested, for example, by the greater cross-sectional area of *l. dorsi*. The effect can be correlated with an increase in that mRNA which is responsible for the synthesis of myosin light chain 1/3 (Smith *et al.*, 1989) and that responsible for the synthesis of G-actin (Merkel *et al.*, 1990).

Oestrogen administration may enhance growth indirectly by the stimulation it gives to the animal's antibacterial defences. It has been shown to increase phagocytosis (i.e. ingestion of bacteria by white blood corpuscles) and the level of serum globulin (Nicol and Ware, 1960). It has also been used to eliminate boar taint from the flesh of mature pigs (Deatherage, 1965).

Reports, at one time prevalent in the UK, that implantation of oestrogens caused dark-cutting beef could not be substantiated (§ 10.1.2: Lawrie, 1960a, b). Since one effect of hormone implantation is to increase the amount of lean meat and to decrease the amount of fat in carcasses (Lamming, 1956), the relative absence of subcutaneous fat would facilitate the observation of the underlying deoxygenated muscle – in which the purplish-red colour of reduced myoglobin would predominate – and this may explain earlier impressions.

It would seem that the growth of muscles is *differentially* stimulated by hormones (Kochakian and Tillotson, 1957). Various steroids, with more or less androgenic activity, produce growth stimulation in the muscles of the neck, head, chest, shoulder, back and abdominal wall which is proportionally much greater than the concomitant increase in body weight. Indeed the hypertrophy of the neck muscles in the bull is reflected by a relative scarcity of lean musculature in the proximate pelvic limb, which is a desirable portion of the carcass. This difference is accentuated by the implantation of steers with a combination of trenbolone acetate and oestradiol, and of bulls with zeranol; but it is more than compensated for by the overall increase in saleable meat (Wood, 1984a). Muscles appear to differ in their responsiveness to β agonists. Thus, *l. dorsi* and *vastus lateralis* muscles of steers treated with cimaterol showed significant increases in weight and protein content in comparison with corresponding control muscles, whereas the *semitendinosus* did not respond (Dawson *et al.*, 1990). There is some evidence that cimaterol increases the proportion of anaerobic 'white' muscle fibres.

Somatotrophin is said to increase the growth of *masseter, quadriceps, supraspinatus* and diaphragm preferentially (Greenbaum and Young, 1953). These are 'red' muscles in respect of their enzyme complement (Talmant *et al.*, 1986). Yet the protein turnover in 'red' muscles is less affected by hormones than that in 'white' muscles (Odebra and Millward, 1982; Ouali *et al.*, 1988). There is certainly a wide range in the fractional synthetic rate for protein turnover between muscles (Harper and Buttery, 1988), that in the heart being particularly high (Davis *et al.*, 1981).

Since surgical gonadectomy is irreversible, it may be associated with ancillary problems and does not permit a graded effect, the suppression of the gonads by some other means has been sought. Such could permit control of aggression and fertility (and of boar taint in pigs) whilst retaining sufficient circulating androgens and oestrogens to produce desired anabolic effects. An immunological approach shows promise. Antibodies can be provoked by administered hormones even when these are identical with those naturally present in the body (Van Loon and Brown, 1975). Thus, 'immunogonadectomy' is feasible.

Antibodies against the steroid hormones tend to cross-react with steroids other than those against which control is sought. Antibodies against the gonadotrophins of the pituitary, however, can be more specifically produced. Thus, immunization with gonadotrophins produces castration-like effects (Lunnen *et al.*, 1974); but, since synthetic gonadotrophins are not yet available, and extraction from the pituitary gland of natural gonadotrophins is exceedingly costly, their use is impracticable (Crichton, 1980). On the other hand, Gn-RH, the gonadotrophin-releasing hormone from the hypothalamus (see below), is a decapeptide and has been synthesized (Matsuo *et al.*, 1971). Its injection produces antibodies which lead to testicular atrophy in the male and the cessation of the reproductive cycle in the female. As yet it is not known to what extent the anabolic effects of the sex hormones are retained (Crichton, 1980). Somatostatin, a hormone now known to be present in many tissues, suppresses the release of growth hormone from the pituitary (Brazeau *et al.*, 1973), and also affects the release of insulin, glucagon and thyroid-stimulating hormone (Spencer, 1981). Since somatomedins are produced in many tissues, it may be that somatostatin operates by inhibiting the release of those peptides locally. Immunization against somatostatin, therefore – if it could be developed – should release this inhibition, increase the levels of somatomedins and stimulate growth. This effect has now been demonstrated in practice with sheep (Spencer and Williamson, 1981), but effects are still inconsistent.

It has been reported that the injection of antibodies against adipose cell membranes has been successful in reducing subcutaneous fat in lambs and pigs (Nasser and Hu, 1991; Kestin *et al.*, 1993).

Another immunological approach to growth stimulation depends on the production in one species of antibodies to the growth hormone of another. The antibodies to the growth hormone of the second species (idiotypes), when injected into the latter, stimulate the production of anti-idiotypes which are, in effect, the original growth hormone. Anti-idiotypes can be produced which resemble only part of the growth hormone and thus mimic only that function desired. By immunizing an animal with antigrowth hormone antibody, a longer-term effect will be achieved than that by the growth hormone itself and thus avoid the need for daily injections (Morrison, 1986; Flint, 1987).

An interesting hormonal application of immunology is the suppression of boar taint by the production of an antibody to the steroid 5α-androst-16-en-3-one, which is the substance responsible (Claus, 1975) (cf. § 10.4.3). Immunization of boars with conjugates between androstenone and bovine serum albumen produces a significant lowering of the steroid in the adipose tissue in comparison with untreated controls (Williamson et al., 1985). Stilboestrol administration is also effective (Deatherage, 1965). Pigs selected for a high lean/feed conversion ratio, and for lean growth rate, on ad libitum feeding, had a higher score for skatole odour than those for lean growth rate on a restricted diet (Cameron et al., 2000a,b).

Reference has already been made to a condition, apparently controlled by a single gene and presumably mediated by hormones, in which there is marked increase in the proportion of muscular tissue in the animal (doppelender hypertrophy – McKellar, 1960). The overall hypertrophy of the musculature is about 2–3 fold, but it is selective for individual muscles, being most pronounced in those which are the latest to develop (Pomeroy and Williams, 1962). The musculature which develops is quite normal (Lawrie et al., 1963b) and if the nature of the growth stimulus could be elucidated and regulated, it might be possible to produce animals of enhanced meat content.

In this context, it is of interest to note that, if a DNA fragment containing the promoter of mouse metallthionein-I gene fused to the structural gene for rat growth hormone is injected into the pronuclei of fertilized mouse eggs, the mice which develop from the latter will have very high levels of the fusion m RNA in their liver and of growth hormone in their blood serum (Palmiter et al., 1982). Gigantism is thus produced. Such results strongly suggest that it will eventually be possible to control specifically the growth of muscular tissue in directions conducive for desirable meat quality by deliberate manipulation of the coding genes. Nevertheless the transfer of gene sequences for the expression of growth hormone, even when enhanced growth is achieved, has tended to be associated with concomitant detrimental effects, such as lameness and ulceration of the digestive tract (Pursel et al., 1990).

Immunological approaches to growth stimulation and control, and to specific relative increments in muscular tissue, in relation to bone and fat, also appear likely to progress in the near future.

The psychological as well as the physiological status of animals is now known to be mediated by means of hormones (Selye, 1950; Himwich, 1955), the hypothalamus exerting hormonal control of the pituitary and thereby of other endocrine glands. One feature of its control is the elaboration by the hypothalamus of several 'releasing' hormones which appear to be peptides of relatively low molecular weight. The production of luteinizing and follicle-stimulating hormones by the pituitary is elicited by a specific decapeptide (Crichton, 1972); and another is responsible for releasing thyrotrophin from this gland. An imbalance at various points in this system could cause stress and, if chronic, the so-called diseases of adaptation. The hypothalamus also exerts its influence through the production of the hormone serotonin; and the action of the latter can be inhibited by the drugs reserpine and chlorpromazine which tranquillize the organism (Udenfriend et al., 1957). Stress susceptibility would be expected to interfere with growth and development; and, in fact, it has been shown that its control by low doses of tranquillizers increases the rate of weight gain and the feed conversion efficiency of fattening cattle and sheep (Sherman *et al.*, 1957, 1959; Ralston and Dyer, 1959), whereas artificially induced stress depresses weight gains (Judge and Stob, 1963).

Removal of the thymus gland from male hamsters, but not from females, causes a regression in growth which appears to be mediated in some way through wasting of the spleen (Sherman and Dameshek, 1963), but the precise mechanism has not been elucidated so far. A recently identified protein, leptin – which is present primarily in fat cells influences hypothalamic control of thermoregulation (Hamman and Matthaei, 1996), maintains energy expenditure after a reduction in feed intake (Stehling *et al.*, 1996) and stimulates adipocyte lipolysis (Fruhbeck *et al.*, 1997) – is likely to be increasingly involved in controlling the growth of meat animals.

2.5.2.2 Antibiotics

For some years, in the USA particularly, there were reports that the feeding of antibiotics produced better growth rates in pigs (Cuff *et al.*, 1951) and calves (Volekner and Casson, 1951). In the United Kingdom a scientific committee, appointed jointly by the Agriculture and Medical Research Councils (Anon., 1962), accepted the view that the practice of feeding antibiotics has been associated with improved growth rates and feed conversion efficiency in young pigs and poultry, and believed it could be extended to calves, but not to adult livestock. Penicillin, chlortetracycline and oxytetracycline are permitted in animal feeding-stuffs without veterinary prescription. It would appear that the beneficial effects of feeding antibiotics may be due to their control of subclinical infections. In young ruminants, however, it is also possible that they may be effective by increasing food utilization more directly since they enhance the digestion of starch through depressing the microbial activity responsible for gas production (Preston, 1962). In this context, it is known that bloat in cattle can be prevented by antibiotics in the feed (Johns *et al.*, 1957). The danger of depressing the activity of the rumen micro-organisms, normally responsible for the digestion of cellulose, seems small (Hardy *et al.*, 1953–4). More recently, a special class of antibiotics, such as monensin and lasalocid, which are referred to as rumen modifiers, have been shown to improve growth by selecting against Gram-positive bacteria, whereby the pattern of volatile fatty acids produced from the feed is altered and loss of energy as hydrogen and methane is reduced (McCutcheon, 1989).

There is no evidence to suggest that the animals suffer any ill effects or that other than traces of antibiotic remain in the carcass after slaughter. On the other hand, there are indications that antibiotic-resistant strains of certain pathogenic micro-organisms (e.g. *Salmonella* spp.) may become established in farm animals given antibiotics either therapeutically or as feed additives (Anon., 1962). 'R factors' – entities which confer resistance to various antibiotics – can be transferred between bacterial cells. It now appears that transfer can occur in the rumen of sheep even in the complete absence of antibiotics. This was shown when sheep were starved for 2–3 days after being inoculated with large numbers (10^9) of donor and recipient micro-organisms (Smith, 1973).

2.5.2.3 Sterile hysterectomy

Much retardation of growth and development in animals is indirectly caused by disease in young pigs; for example, virus pneumonia and atrophic rhinitis are particularly troublesome. One way to overcome these losses is to remove unborn pigs from the uterus just before term under aseptic conditions and to rear them away from possible infection (Betts, 1961). Such animals are free from the natural pathogens of the species. They are referred to as minimal disease pigs in the United Kingdom and as specific pathogen-free pigs in the USA. Swine repopulation with such pigs appears to be effective in the USA in eradicating respiratory diseases.

Chapter 3

The structure and growth of muscle

3.1   The proportion of muscular tissue in sheep, cattle and pigs

As normally prepared for the meat trade, the carcasses of sheep, cattle and pigs represent those portions of the body remaining after the removal of the blood, the head, feet, hides, digestive tract, intestines, bladder, heart, trachea, lungs, kidney, spleen, liver and adhering fatty tissue. The joints into which beef carcasses are commonly split are shown in Fig. 3.1. On the average, about 50, 55 and 75 per cent of the live weight of sheep, cattle and pigs, respectively, remains on the carcass (Gerrard, 1951). Cattle exhibiting the doppelender condition dress out at about 5 per cent higher than typical animals of the same breed, sex and weight (Boccard, 1981). The carcass itself consists substantially of muscular and fatty tissues, of bone and of a residue which includes tendon and other connective tissue, large blood vessels, etc. According to Callow (1948), who dissected a series of cattle and analysed the data of Pállson (1940) and of McMeekan (1940, 1941) on sheep and pigs respectively, the weight of muscular tissue ranged from 46 to 65 per cent of the carcass weight in sheep, from 49 to 68 per cent in cattle, and from 36 to 64 per cent in pigs. Its proportion varied in a roughly inverse manner with that of fatty tissue, the latter being determined, in turn, by such factors as age, breed and plane of nutrition: there are no clear-cut differences between species. The effect of age may be seen from Table 3.1 (Cuthbertson and Pomeroy, 1962), which indicates that the proportion of muscular tissue is high, and that of fatty tissue low, in pigs aged 5 months, in comparison with those aged 6 or $7^{1}/_{2}$ months.

The proportion of bone also decreases as the animal grows older. Sheep and cattle show a similar trend (Callow, 1948).

The effects of breed and level of nutrition are shown in Table 3.2.

It is clear that the percentage of muscular tissue is lower, and that of fat higher, in animals on a high plane of nutrition than in those on a low plane. Moreover, the proportion of muscular tissue is relatively low in Shorthorns and relatively high in Fresians.

With current trends for smaller families, large, fat joints are no longer in demand and overfat carcasses represent wasteful growth. In the United Kingdom, the Meat

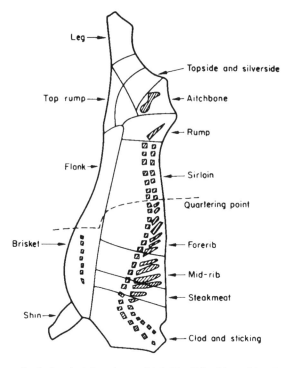

Fig. 3.1 Location of wholesale joints into which English sides of beef are commonly cut.

Table 3.1 The proportion of muscular and fatty tissue and of bone in pig carcasses (after Cuthbertson and Pomeroy, 1962)

Age (months)	Per cent muscular tissue	Per cent fatty tissue	Per cent bone
5	50.3	31.0	10.4
6	47.8	35.0	9.5
7.5	43.5	41.4	8.3

and Livestock Commission has classified large numbers of cattle and sheep carcasses on the basis of level of fatness. The data, together with information from dissection, indicated that, in 1976, approximately 48,000 tonnes of waste fat was produced by cattle (Kempster and Harrington, 1979) and, in 1977, approximately 26,000 tonnes of waste fat by sheep (Kempster, 1979).

In an assessment of the trends in fatness in British cattle, sheep and pigs in the decade 1974–1984, Kempster *et al.* (1986) found that there had been no change in the composition of cattle and sheep carcasses, although the mean weight of the former had increased by 8 per cent and that of the latter had decreased by 3½ per cent. On the other hand, there had been a fall in the lipid content of the average

Table 3.2   The proportion of muscular and fatty tissue and
of bone in cattle carcasses (after Callow, 1961)

Breed	Plane of nutrition	Per cent muscular tissue	Per cent fatty tissue	Per cent bone
Shorthorns	High–high	52.3	33.9	11.1
	Medium–medium	55.8	29.3	12.5
Herefords	High–high	54.5	31.5	11.7
	Medium–medium	58.0	27.7	12.2
Friesians	High–high	59.0	26.1	12.5
	Medium–medium	62.3	21.6	15.2

pig carcass from 27 to 22 per cent, this being associated with a slight increase in
average carcass weight.

On average, 30–40 per cent of the live weight of the three domestic species consists of muscular tissue. Under the microscope such muscle is seen to be crossed by
parallel striations: because it is associated directly or indirectly with the movement
of the skeleton and is under control by the higher centres, it is also referred to as
skeletal or voluntary muscle. There is also a minor amount of unstriated, involun-
tary muscle associated with the intestines, glands, blood vessels and other members,
but detailed consideration will not be given to these muscles in the present volume.

The musculature of sheep, cattle and pigs consists of about 300 anatomically distinct units (Sisson and Grossman, 1953). The approximate location of those mentioned in the text is indicated in Fig. 3.2. Muscles vary both superficially and
intrinsically. They differ in overall size, in shape (which may be triangular, fan-like,
fusiform, long or short, broad or narrow), in attachments (to bone, cartilage or lig-
aments), in blood and nerve supply, in their association with other tissues and in
their action (which may be fast or slow, prolonged or intermittent, simple or in
complex association with other muscles). In short, muscles are highly differentiated
from one another for the performance of numerous types of movement. Some idea
of the size which individual mammalian muscles can attain is given in Fig. 3.7(a) in
which the two *l. dorsi* muscles of a Fin whale have been exposed by removal of the
overlying layer of blubber. The combined weight of the two *l. dorsi* and two *psoas*
muscles of a 25 m (82 ft) long Blue whale may be more than 22 tonnes (Sharp and
Marsh, 1953). Notwithstanding their differentiation, a basic structural pattern is
common to all muscles.

3.2   Structure

3.2.1   Associated connective tissue

Surrounding the muscle as a whole is a sheath of connective tissue known as the
epimysium: from the inner surface of the latter, septa of connective tissue penetrate

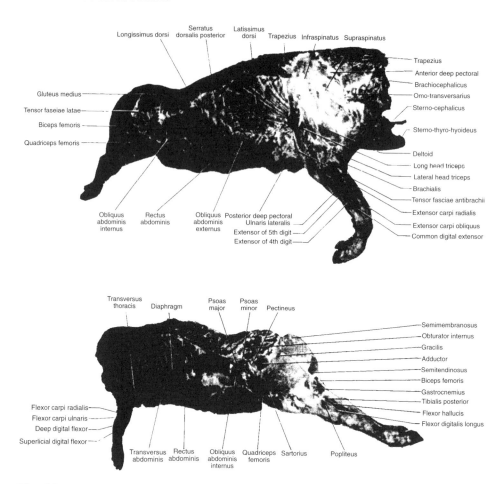

Fig. 3.2 Approximate location of various muscles in carcass. (After Cuthbertson and Pomeroy, 1962.) (Courtesy of the late Dr R. W. Pomeroy and Mr A. Cuthbertson.)

into the muscle, separating the muscle fibres – its essential structural elements, see below – into bundles: these separating septa constitute the perimysium, which contains the larger blood vessels and nerves. From the perimysium a fine connective tissue framework passes further inwards to surround each individual muscle fibre. The connective tissue round each fibre is called the endomysium. The distinction between the perimysial and endomysial connective tissue is clearly illustrated in Fig. 3.3 (Nishimura *et al.*, 1995). The latter is an amorphous, non-fibrous sheath which is associated with fine connective tissue fibres. These were once referred to as reticulin and regarded as a protein distinct from collagen (Windrum *et al.*, 1955) but are now recognized to be collagenous. A structure, referred to as the basement membrane, links the collagenous fibres of the endomysium to the muscle cell membrane. Only about 40 per cent of the dry weight of the basement membrane is collagen, the remainder consisting of complex polysaccharides (proteoglycans and glycoproteins) (Bailey and Light, 1989). Some workers include the basement membrane in the term 'endomysium', others use the term 'sarcolemma' to include the muscle cell

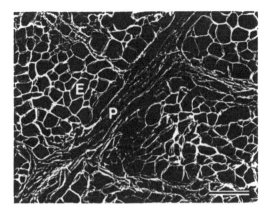

Fig. 3.3 Scanning electron micrograph of bovine *semitendinosus* muscle immediately post-mortem from which muscle fibres have been removed to show network of endomysial connective tissue (E) and layers of perimysial connective tissue (P). The bar represents 200 μm. (Reprinted from Nishimura, Hattori and Takahashi, 1995, with kind permission from Elsevier Science Ltd and courtesy of Prof. K. Takahashi.)

membrane and the basement membrane. Scanning electron microscopy clearly illustrates the distinction, however, between the endomysium/basement membrane/sarcolemma complex on the one hand, and the myofibrils on the other (cf. Fig. 3.4: Rowe, 1989).

The size of the muscle fibre bundles determines the texture of the muscle (Hammond, 1932a; Walls, 1960). In muscles capable of finely adjusted movement, as in those which operate the eye, the texture is fine, whereas in those performing grosser movements it is coarse. Nevertheless, it is of interest to note that the *proportion* of connective tissue is higher in muscles of the former type (Fernand, 1949). The relative proportions of connective tissue and muscle fibres vary between muscles and, in part, account for the relative toughness of meat.

Muscle fibres do not themselves directly attach to the bones which they move or in relation to which their force is exerted; the endomysium, perimysium and epimysium blend with massive aggregates of connective tissue (or *tendons*) and these attach to the skeleton, but the precise mode of connection between the contractile proteins and tendon is not yet clear. Microscopic examination of tendons shows that the constituent bundles of collagen fibrils have a regular 'crimp' at intervals of *ca.* 100 μm. It seems possible that this feature permits the collagen to take up the initial shock of movement by straightening the crimp before the full load is sustained by the tendon (Bailey and Light, 1989). The ultimate tensile strength of tendon, at an extension of 10–15 per cent, is *ca.* 100 N nm^{-2} compared with *ca.* 220 N nm^{-2} for aluminium (Bailey, 1989).

Connective tissue includes formed elements and an amorphous ground substance in which the formed elements are frequently embedded. The latter consist of the fibres of collagen, which are straight, inextensible and non-branching; and of elastin, which are elastic, branching and yellow in colour.

The general structure of collagen, at various levels of organization, is shown in Fig. 3.5 (Gross, 1961). Collagen is one of the few proteins to contain large quantities of hydroxyproline – about 12.8 per cent in warm-blooded mammals (Bowes,

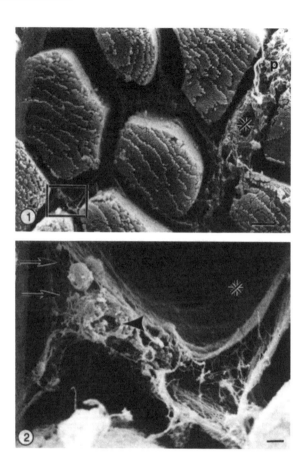

Fig. 3.4   (1) Scanning electron micrograph of bovine *semitendinosus* muscle in TS, showing separation of myofibrillar proteins from the surrounding sarcolemma/basement membrane/endomysium complex. P = perimysium, * = blood vessel. Bar represents 10 μm. (2) Area in (1) enlarged. * = sarcolemma, → = collagen fibrils, ▲ = blood capillary. Bar represents 1 μm. (Reproduced by kind permission of Dr R. W. D. Rowe and Elsevier Applied Science Publishers Ltd.)

Elliott and Moss, 1957). The polypeptide chains of its primary structure each have the repeating sequence glycine-proline-hydroxyproline-glycine-one of the other amino acids. It was shown (Piez, 1965, 1968) that one chain in three had a somewhat different amino acid composition from the other two; and the two chain types were referred to as a_1 and a_2. Subsequently at least 12 different forms of collagen have been isolated and identified, each having a unique sequence of amino acids in the primary polypeptide chains and associations with different carbohydrate molecules. The chain sequences are determined by different genes, these often being located on different chromosomes (Martin *et al.*, 1975; Bailey and Light, 1989; and cf. § 3.3.1). The three α chains are able to pack together very closely in forming the triple α helix of the tropocollagen molecule because each third residue is glycine in which the side chain consists only of hydrogen. The amounts of proline and of hydroxyproline are directly related to the thermal stability of the triple helix; and this is important for the eating quality of meat. The N- and C-terminals, non-

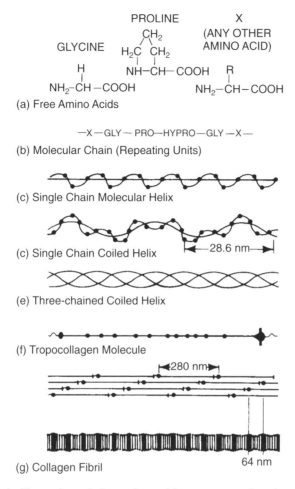

helical regions of the α chains are only 20–25 amino acid residues in length. The N-terminal region always has a lysine residue at position 9. The α chains of type III collagen only contain cysteine at the C-terminal end; the contents of hydroxylysine vary markedly, being greater in the α chains of types II and IV (cf. Table 3.3).

Hydroxylation by specific enzymes takes place on the nascent chains prior to their release from the membrane-bound ribosomes (Miller and Udenfriend, 1970). The enzymes require ascorbic acid and oxygen for their activity. Hydroxylysine is necessary for the addition to collagen of small amounts of carbohydrate (Bailey and Robins, 1973a). Ascorbic acid is believed to control the conversion of proline, after it has already been incorporated into the polypeptide chain, into hydroxyproline (Stone and Meister, 1962). In ascorbic acid deficiency, an elastin low in hydroxyproline is formed (Barnes *et al.*, 1969).

Table 3.3   Types of collagen (after Sims and Bailey, 1982; Bailey and Light, 1989)

Type	Molecular composition	Tissue distribution	3-Hydroxyproline content (residues/1000)	Hydroxylysine content (residues/1000)	Glycosylation of hydroxylysine (%)
I	$[\alpha_1(I)]_2\alpha_2$	Tendon, bone dentine, dermis, muscle	1	6–8	20
II	$[\alpha_1(II)]_3$	Cartilage, vitreous humour, intervertebral discs	2	20–25	50
III	$[\alpha_1(III)]_3$	Foetal dermis, cardiovascular system, synovial membrane, viscera, muscle	1	6–8	15–20
IV	$\alpha_1(IV)[\alpha_2(IV)]_2$	Basement membrane, lens capsule and kidney glomeruli	10	60–70	80
V	$\alpha_1(V)[\alpha_2(V)]_2$ and $[\alpha_3(V)]_3$	Placental membrane, cardiovascular system, lung; minor component of most tissues	2–3	6–8	–

In its secondary structure, the greater part of each chain is arranged as a left-handed helix (Fig. 3.5(c)), and three of these intertwine to form a right-handed super helix (Fig. 3.5e, f), which is the tropocollagen molecule or collagen protofibril.

Tropocollagen molecules self-assemble to form fibrils and these aggregrate to form fibres (Fig. 3.5g). Under the electron microscope a repeating band pattern, with a repeat unit of *ca.* 67 nm(D), is seen. This pattern is caused by the tropocollagen molecules, which are 4.4D in length, being quarter-staggered by D and overlapping their nearest neighbours by 0.4D (Bailey and Light, 1989). The fibres are laid down in a well-defined criss-cross lattice which is orientated at an angle to the long axis of the muscle (Rowe, 1974). Determination of the relative concentrations of the individual types has been greatly aided by reverse-phase, high-pressure liquid chromatography (Smolenski *et al.*, 1983). Types I, II and III are aggregated in fibres; but type IV is amorphous, being found in the basement membrane in which the fibres of the other collagen types are disposed (Bailey and Robins, 1976). Type V is found in the placental membrane but is a minor component of most tissues and is probably fibrous.

A number of filamentous collagens (e.g. types VI and VII) have been identified, but they are thought to be of only minor significance in meat (Bailey, 1989).

Only four of the types of collagen molecule are believed to be significant in muscular tissue (cf. Table 3.3: after Sims and Bailey, 1982; Bailey and Light, 1989). Types I and III are found in epimysial, perimysial and endomysial connective tissue, but the epimysium is mainly type I. There are significant amounts of type III in the perimysium. Although there are equivalent amounts of types I and III in the endomysium, the latter, substantially basement membrane, consists mainly of type IV collagen. Small amounts of type V are present in the endomysium and the perimysium (Bailey and Sims, 1977; Bailey *et al.*, 1979; Bailey and Light, 1989).

A major post-translational modification of collagen is the formation of both intramolecular (between α chains in the same molecule) and intermolecular cross-links. These are essential for the high mechanical strength of collagen, including that of non-fibrous type IV. These cross-links are covalent and of three kinds: disulphide bonds, which are confined to types III and IV because they alone contain cysteine (Bailey, 1974); divalent bonds formed between the α chains from lysine or hydroxylysine aldehydes (Bailey, 1968; Bailey et al., 1970), which are reducible in vitro; and more complex bonds, joining more than two α chains, which arise during the ageing of collagen (mature cross-links).

In a given tissue, the lysine-derived cross-links depend on the oxidation to aldehydes of specific lysine and hydroxylysine residues (in the nonhelical regions at both N- and C- terminal ends of the polypeptide chains) by lysyl oxidase, an enzyme which binds to the nascent collagen fibres (but not to soluble collagen monomers). The enzyme has copper as a prosthetic group. Both copper deficiency and the administration of β-aminopropionitrile inhibit the oxidizing reactions. Various diseases of connective tissue can be related to failure of one or other of the biochemical stages involved in cross-link formation (Bailey and Robins, 1973b; Bailey and Light, 1989).

The cross-link produced from two lysine aldehydes is an aldol and can only be intramolecular since it depends on their exact juxtaposition in two α chains of the same collagen molecule, but only at the N-terminal ends (Bailey and Light, 1989). Such cross-links thus cannot affect the stability of collagen fibres, which depends on intermolecular bonds. In one type, the lysine aldehyde in one α chain can react with the ε-amino group of the hydroxylysine in the overlapped adjacent molecule of the collagen fibril to form a stable aldimine link (dehydrohydroxylysinonorleucine). It is reducible to, and may be isolated as, hydroxylysinonorleucine, and is heat-labile. A second type of stable cross-link can be formed between the hydroxylysine aldehyde in the non-helical region of one molecule and the hydroxylysine of an adjacent molecule, to form a group which, after Amadori rearrangement, becomes an oxo-imino link (hydroxylysino-5 oxonorleucine). This is reducible to acid-stable dihydroxylysinonorleucine. Tissues in which collagen is relatively little hydroxylated will tend to contain high ratios of aldimine to oxoimino intermolecular cross-links; and conversely (cf. Table 4.24).

During maturation of the animal's tissues, collagen becomes much more resistant to breakdown. This is not due to an increase in the number of intermolecular cross-links – indeed they may decrease – but to the formation of nonreducible links involving three or more chains whereby a three-dimensional network is generated and high tensile strength is developed. The nature of these mature cross-links is not yet clear, but there is evidence that hydroxyaldohistidine and pyridinoline are the structures responsible. Increasing glycosylation of lysine residues may also be significant in the maturation of collagen with age (Anon., 1981–83a). The increase of fibre diameter during maturation is an additional contributory factor (Perry et al., 1978).

Notwithstanding the correlation between the increase in thermally stable cross-links in muscle collagen and increased toughness in the meat as animals mature, the differences in toughness which are observed between a given muscle of animals of the same age cannot be readily explained. It has been shown that there is no correlation between toughness and the content of either immature (hydroxylysinonorleucine, dihydroxylysinonorleucine) or mature (hydroxylysylpyridinoline, histidinohydroxylysinonorleucine) cross-links in the perimysial connective tissue isolated from the longissimus lumborum muscles of different pigs of similar maturity (Avery et al., 1996).

On heating, when the collagen shrinks at *ca.* 65 °C to form gelatin, the nature of the cross-links it contains will determine the solubility, the extent of shrinkage and the tension developed during shrinkage. With increasing animal age, the formation of the mature, transverse cross-links causes a very marked increase in tension on heating which, as the temperature rises above *ca.* 65 °C, tends to remain high, whereas the collagen tension of younger animals greatly diminishes with increasing temperature. Horgan (1991) demonstrated that there was a linear relationship between the isometric force generated during the heating of tendon collagen from bovine *l. dorsi* muscles and the age of the cattle from which it was derived. On the other hand, the thermal transition temperature of tendon collagen was too variable to permit its use in estimating animal age. In a detailed study of the collagen in various goat muscles, however, Horgan *et al.* (1991) were able to demonstrate that the shrink temperature of intramuscular collagen increased with increasing animal age, and that this was reflected by a concomitant increase in the content of pyridinoline, which is a major end product of the cross-linking pathway involving hydroxyallysine (Eyre *et al.*, 1984). Differences in the thermal stability of intramuscular collagen between muscles were much less than that between the former and tendon collagen. The thermal lability of the latter was associated with a relatively low content of pyridinoline (Horgan *et al.*, 1991).

Elastic fibres, although associated with the walls of blood vessels, are generally minor constituents of the connective tissue of muscle (but cf. § 4.3.5); but since they shrink and toughen on heating (whereas collagen fibres are converted to soluble gelatin above *ca.* 80 °C in the presence of water), their contribution to meat texture cannot be ignored. Most of the elastic fibre consists of elastin (an apparently amorphous protein): the remainder is a microfibrillar protein containing many polar amino acids. Elastin contains about 40 per cent glycine, 40 per cent hydrophobic amino acids (including 18 per cent valine) and small amounts of proline and hydroxyproline (1.6 per cent: Partridge and Davis, 1955). Like collagen, elastin undergoes cross-linking under the action of lysyl oxidase; but a very large number of lysine residues are involved. Initially the mode of cross-linking is similar to that of collagen, aldols and aldimines being formed (see above). These then condense to form 1:2-dihydrodesmosine and isodesmosine; and these are oxidized to desmosines (Thomas, Elsden and Partridge, 1963). Desmosines are salts of tetramethyl-substituted pyridine and are responsible for the yellow colour and fluorescence of elastin (Partridge, 1962).

Elastin, again like collagen, becomes more insoluble with increasing animal maturity, although there is no direct evidence for further cross-linking. Tissues contain an elastase which can proteolyse elastin: it is secreted by the macrophages.

Rowe (1986) demonstrated that coarse elastin fibres (dia. 5–10 µm) are found in epimysium and perimysium parallel to the long axis of the muscle fibres. Finer elastin fibres (dia. 1–2 µm) are also present, but these tend to be parallel to the network of collagen fibres and thus to be orientated at an angle to the muscle fibre axis. The quantity of coarse elastin fibres is greater in both perimysium and epimysium of *semitendinosus* than of *l. dorsi*, although the amount of fine elastin fibres is similar in both muscles. The general arrangement of elastin and collagen fibres, as postulated by Rowe (1986), is schematically shown in Fig. 3.6.

The structure and arrangement of elastin fibres differ with their origin (Partridge, 1962). In the *ligamentum nuchae* of the ox, thick elastic fibres make up the greater part of the tissue, these being separated by a proteoglycan ground substance: the

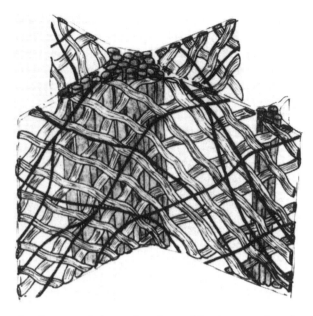

Fig. 3.6 Schematic diagram of the perimysium of bovine muscles, showing the relative organization of the crimped collagen fibres (in characteristic criss-cross pattern), thin elastin fibres (solid black), approximating the collagen network in orientation and longitudinally aligned, thick elastin fibres at junction regions and in the thick perimysial sheets. (Rowe, 1986; reproduced by kind permission of Elsevier Applied Science Publishers Ltd.)

elastin fibres of arteries and veins are additionally associated with fibres of collagen and smooth muscle; and the elastin fibres of elastic cartilage form a lace-like network of fibrils with large numbers of chondrocyte cells (i.e. those found in connective tissue).

A minor proportion of intramuscular connective tissue consists of proteoglycans and glycoproteins. Proteoglycans are large molecules in which a protein core is covalently linked to glycosaminoglycan chains, the latter being repeating disaccharide units of hexosamine and a hexuronic acid, associated with sulphate esters. The constitution of different types of proteoglycans is shown in Table 3.4. Their strong negative charge causes mutual repulsion of the chains and thus an extended structure, whereby they are able to bind considerable amounts of water. It seems likely that the proteoglycans may control the size and alignment of connective tissue fibres and thus contribute to meat texture (Bailey and Light, 1989).

Glycoproteins (e.g. laminin and fibronectin) are also large protein-polysaccharide molecules found in the extracellular matrix. They are believed to anchor cells to basement membranes and other types of collagen.

3.2.2   The muscle fibre

The essential structure unit of all muscles is the fibre. Fibres are long, narrow, multinucleated cells which may stretch from one end of the muscle to the other and may attain a length of 34 cm, although they are only 10–100 μm in diameter (Walls, 1960). In healthy animals the diameters of muscle fibres differ from one muscle to another

Table 3.4   Constitution and location of proteoglycans (after Bailey and Light, 1989)

Proteoglycan	Disaccharide repeating unit	Location
Chondroitin sulphate	glucuronic acid and *N*-acetyl galactosamine	cartilage: intravertebral disc
Dermatan sulphate	iduronic acid and *N*-acetylgalactosamine	skin: tendon
Heparin sulphate and heparin	glucuronic acid and *N*-acetylglucosamine + iduronic acid and *N*-acetylglucosamine	basement membrane: fibrous tissues
Keratan sulphate	galactose and *N*-acetylglucosamine	cartilage: cornea: intravertebral disc
Hyaluronic acid	glucuronic acid and *N*-acetylglucosamine	all connective tissues: vitreous body

Table 3.5   Mean fibre diameter of three muscles from lambs of different age and on two planes of nutrition (Joubert, 1956)

Age (days)	Muscle (diameter µm)		
	L. dorsi	*Rectus femoris*	*Gastrocnemius*
High plane			
0	9.0	10.4	10.9
60	31.7	33.8	35.8
290	48.2	49.5	45.5
Low plane			
0	7.3	8.3	8.7
60	17.3	19.8	21.3
290	35.0	36.3	39.5

and between species, breeds and sexes (Hammond, 1932a; Joubert, 1956). They are increased by age, plane of nutrition and training (Joubert, 1956; Goldspink, 1962a), by the degree of *postnatal* development in body weight rather than by the body weight itself (Joubert, 1956), and by oestradiol administration (McDonald and Slen, 1959). Some of these effects are shown in Tables 3.5 and 3.6.

Garven (1925) had noticed that a given muscle contained fibres of varying diameter, the smaller ones being more peripheral and the large ones more central in their distribution; and Hammond (1932a) observed that muscles which were more pigmented had relatively more small diameter fibres containing pigment than large non-pigmented ones. The significance of the associated chemical differences will be considered later. Goldspink (1962a, b), in studying the *biceps femoris* of the mouse, showed that the distribution of fibre diameters observed is not normal and that such fibres can exist in two phases – one having a small diameter (20 µm) and the other a large one (40 µm): there are few fibres of intermediate cross-section. The small, parallel units (myofibrils), of which the fibres themselves are principally composed,

Table 3.6 Effect of species on muscle fibre diameter at birth and maturity, showing the importance of rate of postnatal development in body weight (after Joubert, 1956)

Species	Birth			Maturity		
	Body weight (kg)	Fibre diameter (μm)	Body weight (kg)	(rel. incr.)	Fibre diameter (μm)	(rel. incr.)
Sheep	4.2	11.3	113.5	2687	50.4	446
Cattle	30.2	14.3	817.2	2707	73.3	511
Pigs	1.3	5.3	236.1	17660	90.9	1705

(Relative increases: Birth = 100)

have the same diameter irrespective of the size or development of the fibre (Davies, 1989).

Surrounding each fibre, and underneath the connective tissue of the endomysium, is a sheath, the sarcolemma, which was once thought to be structureless but was later shown by the electron microscope to represent a double membrane of which the components are about 50–60 Å apart (Robertson, 1957). Within the sarcomere are the myofibrils. These are surrounded by a fluid phase, the sarcoplasm. In the latter are found certain formed structures, the organelles, which include mitochondria (the micro-environments which couple respiration with the synthesis of ATP), lysosomes (which contain various catabolic enzymes) (De Duve, 1959a), peroxisomes (which contain fatty acyl oxidases and catalase) (De Duve and Baudhuin, 1966) and the sarcoplasmic lipid bodies: the sarcoplasm is also penetrated by a series of narrow tubes (the sarcotubular system) (Bennett, 1960), and contains numerous dissolved or suspended substances. The muscle cell nuclei are generally found just below the sarcolemma.

Scopes (1970) has pointed out that there can be no sarcoplasmic proteins *within* the myofibrils since their molecular size would interfere with the contraction mechanism. Their concentration in the interfibrillar fluid, therefore, must be 25–30 per cent. In a muscle such as the *psoas* of the rabbit, the actual size of which is shown in Fig. 3.7(b), there would be about 20,000 fibres: the grouping of such fibres is represented in Fig. 3.7(c), which shows pig *l. dorsi* muscle taken at 20 × magnification in longitudinal (LS) and transverse (TS) section. When the magnification is increased to 200× it is possible to see (LS) that the fibres are crossed by parallel striations: in TS, at this magnification, the varying size and shape of the individual fibres is evident (Fig. 3.7d). At 2000× magnification (Fig. 3.7e) it can be seen in TS that each individual fibre is composed of a number of smaller units, the myofibrils; in LS the individual myofibrils are not so apparent, but details of the cross-striations are visible. Thus, the dark or *A*-band has a central clear area (the *H*-zone) and the light or *I*-band has a central dark division (the *Z*-line). The distance between two adjacent *Z*-lines is the functional unit of the myofibril: it is known as the sarcomere. The sarcomeres of the myofibrils are shown in Fig. 3.7(f) at 20,000× magnification, in LS and TS. It can now be observed that the myofibril is itself composed of numerous parallel filaments. Some of these extend from the *Z*-lines to the edge of the *H*-zone: others traverse the entire width of the *A*-band. When, finally, the

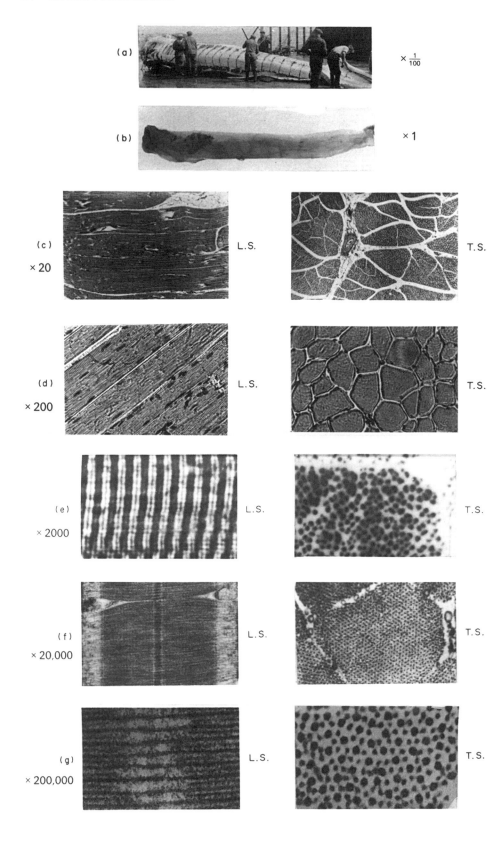

(a) $\times \frac{1}{100}$

(b) $\times 1$

(c)   L.S.   T.S.
$\times 20$

(d)   L.S.   T.S.
$\times 200$

(e)   L.S.   T.S.
$\times 2000$

(f)   L.S.   T.S.
$\times 20,000$

(g)   L.S.   T.S.
$\times 200,000$

magnification is increased to 200,000 × (Fig. 3.7g), those filaments which traverse the
A-band are seen to be relatively thick; those which stop at the edge of the H-zone
are relatively thin. The thick and thin filaments consist of molecules of the contractile proteins myosin and actin respectively (Hanson and Huxley, 1953, 1955; Huxley
and Hanson, 1957; Huxley, 1960). It may be noted in TS that each myosin filament
is surrounded by six actin filaments in hexagonal array. The LS of Fig. 3.7(g) also
indicates that there are small projections between the myosin and actin filaments.
The projections are so arranged that each sixth one is on the same radial plane of
the cylinder of the myosin filament and aligned opposite to one of the six surrounding actin filaments. The actin filaments consist of two helically wound strands
composed of sub-units which appear to be alike and are approximately spherical
(Hanson and Lowy, 1963). This general arrangement is shown diagrammatically in
Figs 3.8(a), (b). A strand of the minor myofibrillar protein, tropomyosin, runs along
on each side of the actin polymers, and, at 38.5 nm intervals along the thin filaments,
there is located another protein troponin, which itself consists of three units, T, C
and I (Schaub and Perry, 1969). Troponin T binds to tropomyosin. In the ultimate
analysis, the myosin filaments represent the lateral aggregation of the individual
tadpole-like molecules of myosin. The latter aggregate – with the 'tails' towards one
another and the 'heads' directed towards the Z-lines (Fig. 3.8(c) – until a cylinder
with tapering ends, and about 1.5 μm in length, is formed. This is the myosin filament (Huxley, 1963) and comprises about 200 myosin molecules. The myosin molecule itself is about 1500 Å in length. Each molecule consists of two apparently
identical units. Each unit has a long 'tail' (light meromyosin), a 'collar' (heavy
meromyosin S–2) and a 'head' region (heavy meromyosin S-1) (Fig. 3.8d) There is
additional protein associated with the latter (Lowey et al., 1969; Gergely, 1970).
Three light sub-units have been found in the 'head' region (Gershman et al., 1969).
Digestion of myosin with trypsin yields the heavy and light meromyosin fragments.
Acetylation yields three small fragments of MW 20,000 (Locker and Hagyard,
1968); and a further multiplicity of subunits is formed on treatment with 8 M urea.
One of these sub-units contains N^ε-methyl lysine, and, indeed, 3-methylhistidine has
also been found as part of the myosin molecule (Hardy et al., 1970). Methylation of
these two amino acids in myosin occurs after peptide bond synthesis. The points of
attachment of the light meromyosin shaft and of the head of the myosin molecule
(S-1) to the intermediate region (S-2) are susceptible to attack by proteolytic

---

Fig. 3.7   Comparative aspects of muscle structure. (a) Fin Whale *l. dorsi* muscles exposed
after flensing (×1/100). (Courtesy of the late Dr J. G. Sharp.) (b) Rabbit *psoas* muscle. Actual
size (about 20,000 fibres) (×1). (c) Pig *psoas* muscle. LS and TS showing arrangement of
muscle fibres in bundles (×20). (Courtesy C. A. Voyle.) (d) Pig *l. dorsi* muscle. LS and TS
showing several fibres. Note cross-striations and nuclei in LS (×200). (Courtesy C. A. Voyle
and D. J. Restall.) (e) Ox *sterno mandibularis* muscle. LS – portion of single fibre, showing
dark A-band, with central H-zone, and light I-band, with dark Z-lines between sarcomeres.
TS – portion of single fibre showing individual myofibrils (×2000). (Courtesy N. King and D.
J. Restall.) (f) Rabbit *psoas* muscle. LS of myofibrils, showing thin rods of actin extending
from Z-lines to border of H-zone and thick rods of myosin, extending across the A-band. TS
of myofibrils showing array of myosin rods (×20,000). (Courtesy Dr H. E. Huxley, FRS.)
(g) Rabbit *psoas* muscle. LS edge of H-zone, showing apparent projections between myosin
rods (thick) and actin rods (thin). TS showing thick myosin rods surrounded by six thin actin
rods (×200,000). (Courtesy Dr H. E. Huxley, FRS.)

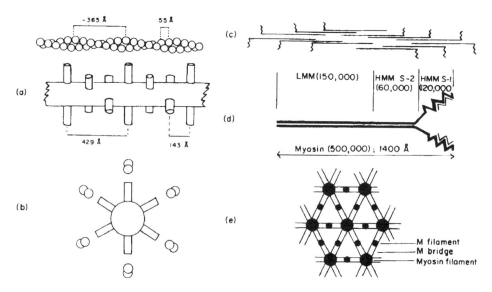

Fig. 3.8 Ultrastructure of muscle. (a) Diagram, based on X-ray analysis, showing part of a myosin filament, with one double helix of G-actin monomers above. Note that pitch of the helix, and monomer repeat distance, differ from repeat distances of the H-meromyosin heads on the myosin filament. (After Huxley, 1969; courtesy Dr H. E. Huxley, FRS, and American Association for the Advancement of Science.) (b) Diagrammatic cross-section of myosin filament showing position of six surrounding actin filaments. (c) Sketch showing mode of aggregation of myosin molecules in forming myosin filament. Note opposite polarity of molecules on each side of *M*-zone. (After Huxley, 1969; courtesy Dr H. E. Huxley, FRS, and American Association for the Advancement of Science.) (d) Sketch showing one myosin molecule: note double-stranded form and relative proportions of light meromyosin (LMM) and heavy meromyosins (HMMS-1, HMMS-2). (After Lowey *et al.*, 1969; courtesy Dr Susan Lowey.) (e) Diagrammatic cross-section of myofibril in region of *M*-zone, showing myosin filaments within a network of *M*-filaments and *M*-bridges. (After Knappeis and Carlsen, 1968; courtesy Prof. G. G. Knappeis.)

enzymes. This suggests that these junction points could act as hinges, permitting S-1 and S-2 to swing out from the shaft towards the actin filaments (Lowey, 1968), and could mean that the S-2 portion can always attach to actin in exactly the same orientation (Huxley, 1971). The S-2 'collar' region is up to 50 nm in length: the S-1 'heads' of the myosin molecules appear to be curved permanently (Anon., 1981–83b). Davey and Graafhuis (1976a) have shown by electron micrography that the light meromyosin backbone of the myosin filaments is arranged as a right-hand tertiary coil, comprising three secondary coils – each of which is composed of three primary strands in a left-hand helix. This evidence supports X-ray data indicating that the heavy meromyosin heads of the myosin molecules are regularly disposed along the myosin filament to give nine in the helical repeat distance. There appears to be no need to postulate the existence of a protein to act as a central core for the myosin filaments.

Electron micrographs have revealed the fine structure of the *M* zone in the centre of the myosin filaments. There are three to five parallel striations running perpendicular to the long axis. These *M*-bridges appear to link the myosin filaments to their six nearest neighbours. The *M*-bridges themselves are linked by thin filaments running between those of the myosin and parallel to the latter (Fig. 3.8e: Knappeis and Carlsen, 1968). It would thus appear that the bridge-filament lattice of M substance keeps the myosin filaments centrally aligned in the sarcomere.

Another feature which appears to reflect the basic structural skeleton of the sarcomere was discerned by the electron microscope. So-called 'gap filaments' may be seen in the spaces which develop in muscles when these have been stretched beyond the point of overlap of A and I filaments (Carben et al., 1965). It was suggested (Locker and Leet, 1975) that each gap filament starts as the core of an A filament in one sarcomere, extends through (and may be attached to) the Z-line and terminates as the core of the aligned filament in the adjacent sarcomere; but King (1984) suggested they are more peripheral in location. However elusive, gap filaments may have some significance for meat quality (Davey and Graafhuis, 1976b; Locker, 1976). They appear to correspond to the series of elastic elements long postulated as components of muscle by physiologists, and an elastic protein, referred to as 'connectin', has been characterized by Muruyama et al. (1977a, b). It contains 5 per cent lipid and 1 per cent carbohydrate. Connectin is present throughout the sarcomere of skeletal muscle (Maruyama et al., 1979).

Wang and Williamson (1980) resolved connectin into two proteins, titin and nebulin, of very large molecular weight, which are believed to constitute, respectively, 10 per cent and 5 per cent of the myofibrillar proteins. Titin is probably the major protein of the 'gap filaments' which are thus more appropriately referred to as 'T-filaments' (Locker, 1987), whereas nebulin was originally identified histochemically as thin, continuous transverse arrays ('N_2' lines) near the boundary of the A-I zones and on each side of the Z-lines (Yarom and Meiri, 1971). More recently nebulin has been shown to bind along the thin (actin) filaments (Wang and Wright, 1988). The ends of the latter are 'capped' by a protein (cap-Z: Cassella et al., 1987) and associated with other proteins, zeugmatin (Maher et al., 1985) α-actinin and tropomodulin (cf. Fig. 3.9). The use of monoclonal antibodies against titin has suggested that T-filaments consist of single giant molecules of titin. There are six of these in each half-sarcomere, commencing at the M-line, winding helically round the A-filament and passing to the adjacent Z-line, (Wang, 1985; Maruyama, 1986).

The molecular weight of titin has been reported as 2.8 megadalton (Maruyama, 1986) and as 1.4 megadalton (Harding and Bardsley, 1986). If the former, according to Locker (1987), its length would be about 7.8 µm, which is equivalent to half a sarcomere length and approximates to the maximum extension possible with beef muscle fibres. There is considerable speculation respecting the manner of folding of titin in muscle at rest and in contraction. Titin has a relatively high denaturation temperature as assessed by differential scanning calorimetry, viz. 75.6 °C and 78.4 °C for porcine and bovine titin respectively (Pospiech et al., 2002).

Another salt-insoluble protein, desmin (Granger and Lazarides, 1978), appears to form a network of collars within the plane of the Z-line, and may be responsible for maintaining the alignment of adjacent sarcomeres. Desmin and connectin ('T-filament' or 'gap-filament' protein) thus constitute a filamentous cytoskeleton in muscle which is additional to the well-established system of actin and myosin filaments. The proteins vinculin and talin are also components of the cytoskeleton and attach actin to the cell membrane (Geiger, 1979; Tidball et al., 1986). As with desmin, vinculin is arranged as rib-like bands (or costameres) around the muscle fibre at the Z-line. (In 2001 Takada et al. identified a further protein, myozenin

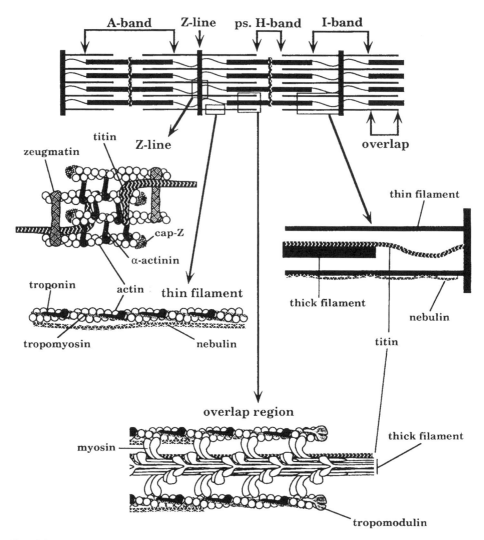

Fig. 3.9 Schematic diagram illustrating the relationship between nebulin, titin and the thin (actin) and thick (myosin) filaments, and showing how the actin filaments interact with titin and other proteins at the Z-line. (Reprinted with modification by Dr Darl Swartz from Proc. 47th Annual Recip. Meat Conference, June 12–15, 1994, after Swartz and Greaser, 1995; by kind permission of Dr Darl Swartz and Dr Marion Greaser and the American Meat Science Association.)

(calsarcin 2) which binds to alpha actin and may regulate the spacing of the actin filament.)

In muscle there is a continuous attachment of the cytoskeleton to the extracellular matrix, involving two groups of proteins, the dystrophins and the integrins. Integrins are dimeral proteins consisting of an α and β chain. There are at least 24 combinations of the two chains (Hynes, 1992). The β chain is responsible for the attachment of the cell membrane to the cell body (Van der Flier and Sonnenberg, 2001). The integrin which is believed to be most significant in linking the cell body to the membrane matrix is $\alpha7\beta1$ (Belkin and Stepp, 2000), which binds to laminin in the membrane. Integrin adhesion complexes on the cell surface also contain many signalling proteins such as talin, α-actinin, vinculin and several kinases (Lawson, 2004).

The mitochondria of skeletal muscle are particles having a fine internal membranous structure: they are located between the myofibrils in longitudinal rows or situated at the Z-line and are especially prevalent in active muscles (Paul and Sperling, 1951).

Another type of organelle, the peroxisome, was first described by DeDuve and Baudhuin in 1966. These are distinct from lysosomes and mitochondria and are characterized by their content of fatty acyl-CoA oxidase, which form H_2O_2 during the oxidation of fatty acids, and of catalase, which destroys it. They are particularly plentiful in heart muscle and in other tissues which vigorously oxidize fat. Peroxisomes appear to oxidize long chain fatty acids (having 20–26 carbon atoms in the chain) more effectively than mitochondria (Tolbert, 1981).

A separate network – the sarcotubular system – appears to surround each myofibril and the level of the myofibril at which it occurs is characteristic for different muscles (Porter, 1961). It was first extensively studied by Veratti (1902), forgotten for 50 years and rediscovered by electron microscopists, when it was shown to consist of two series of tubules along which it has been presumed chemical control may be swiftly and intimately exerted over muscle function (Bennett, 1960). Longitudinal tubules – to which the term 'sarcoplasmic reticulum' is now taken to refer – run parallel to the myofibrils, being linked at intervals along the sarcomeres, and unite to form a terminal sac usually* before each Z-line. Between the pairs of terminal sacs (from adjoining sarcomeres) of the longitudinal elements a second series of tubules runs transversely across the fibrils, apparently as invaginations of the sarcolemma (the 'T'-system).

The junction between the transverse tubules and the longitudinal tubules on each side, referred to as a 'triad', involves a coupling mechanism. Associated with the 'T'-system tubules (and the calcium channels of the sarcolemma) is a large five-unit protein complex, the L-type calcium channels (LTC). The LTC are fundamentally involved in various cellular processes which are regulated by Ca^{++} ions, including the initiation of contraction in skeletal muscle, and they are most prevalent in the latter (Campbell et al., 1988). The α_1 and β sub-units of LTC are phosphorylated by cyclic AMP-dependent protein kinase (Curtis and Catterall, 1988): phosphorylation is a prerequisite for ion flux through the channels. The α_1 sub-unit of LTC is itself sufficient to form a functional ion channel (Perez-Reyes et al., 1989). (LTC reacts

---

* In typical mammalian muscles. In slow-acting muscles the longitudinal tubules of the sarcoplasmic reticulum continue without interruption between successive sarcomeres (Smith, 1966; Page, 1968).

with dihydropyridine and is referred to as the dihydropyridine receptor.) It is believed that degradation by calcium activated protease (calpains) and the inhibition by calpastatin is physiologically important in controlling the initiation of muscular contraction (Remanin *et al.*, 1991).

Projecting from the membrane of the sarcoplasmic reticulum is another large protein complex, the junctional foot protein (JFP), so-called because of its morphology and its location at the triad junction. During muscular contraction the 'T' tubules transmit the excitatory impulse from the depolarized sarcolemma to the sarcoplasmic reticulum, from which Ca^{++} ions are released. They are sequestered therein by acidic proteins of which the most important is calsequestrin, each molecule of which can bind 40 Ca^{++} ions (MacLennan, 1974). Among the agents which induce the release of Ca^{++} ions is the plant alkaloid, ryanodine (cf. § 2.2), the protein with which it reacts being accordingly referred to as the ryanodine receptor, but which appears to be identical to the junctional foot protein.

The calcium-dependent proteinase system (Croall and Demartino, 1991: and cf. §§ 4.3.5 and 5.4.2) is apparently capable of interacting by the excitatory coupling mechanism of muscle via both LTC and JFP (Brandt *et al.*, 1992; Seydl *et al.*, 1995).

In Fig. 3.10 the sarcoplasmic reticulum in pig *l. dorsi* muscle is shown at a magnification of 2000×. The structure stains with osmic acid – silver nitrate. Although in LS the appearance presented is similar to that in the LS of Fig. 3.7e – which shows

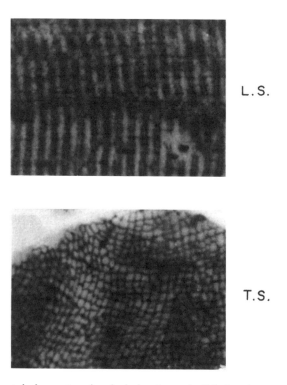

L.S.

T.S.

Fig. 3.10   The sarcotubular system in pig *l. dorsi* muscle. LS showing regular arrangement of transverse strands of reticulum at A–I band junction. TS showing reticulum surrounding each myofibril within a fibre (×2000). (Courtesy C. A. Voyle.)

muscle at the same magnification when stained by Heidenhain's reagent – the two TS are quite different. It is clear that the myofibrillar protein stains with Heidenhain, the myofibrils being separated by unstained areas: in Fig. 3.10 the myofibrils do *not* stain, but the sarcoplasmic reticulum surrounding each myofibril does so.

Like other biological membranes, the sarcoplasmic reticulum is largely phospholipid in nature, but, as an integral component of the membrane, there is intrinsic protein (about 1–10 per cent of the phospholid content). This is predominantly the enzyme ATP-ase (Martonosi and Beeler, 1983). On relaxation it is responsible for pumping Ca^{++} ions from the cell interior into the sarcoplasmic reticulum tubes. In porcine muscle the ability of the sarcoplasmic reticulum to regulate Ca^{++} ion concentration is less in summer than in winter; and this contributes to the lower eating quality of pork at that time of year (Küchenmeister *et al.*, 2000).

3.3   The growth of normal muscle

The initiation and growth of muscle automatically implies (1) the synthesis of those complex protein molecules which are specific for the tissue and the secretion of their necessary components (amino acids), (2) the precise alignment of the specific proteins into the structural element peculiar to muscle (fibres) and (3) the subsequent differentiation and development of the fibres according to muscle type and function, all these processes being subject to the overall requirement of perpetuating the pattern of the parent body. Various hormones are known to expedite the growth of biological tissues and to exert control over their function either directly or indirectly: their general mechanism of action is upon the enzyme proteins which control the rate of chemical reactions whether these be synthetic or otherwise (Villee, 1960). In some cases hormones are known to act by making substrate molecules more accessible, e.g. insulin (Levine and Goldstein, 1955). One might thus explain the accretion of the amino acids required for protein building. An explanation of how these amino acids are built into the exact and reproducible position which they occupy in the polypeptide chains of which proteins are constructed has only been possible since the 1960s (Perutz, 1962).

3.3.1   Fundamental basis of protein organization and replication in biological tissues

While hormones may expedite the building of proteins through their action on the synthesizing enzymes, the question of how the enzymes themselves are synthesized naturally arises. The genes on the chromosomes of the cell nucleus are responsible. One gene controls the synthesis of the structure specific for one or part of one enzyme protein: it also effects the synthesis required for self-replication, thus ensuring the perpetuity of its own structure. The genes are not themselves proteins: they consist of nucleic acid, generally, deoxyribonucleic acid (DNA). DNA consists of two chains of nucleotides coiled around each other to form a double helix (Watson and Crick, 1953). Although there are only four different nucleotides in the chains (adenosine, guanosine, cytidine and thymidine), they occur many times and are arranged in a complicated sequence. This sequence in the two chains is identical but complementary, being followed in opposite order in each.

In replication one chain of the parent double helix is transferred in forming each of the daughter double helices. The biosynthesis of the second daughter chain is catalysed by an enzyme which uses as substrate the four component deoxyribonucleoside triphosphates: it uses the daughter DNA chain as a basis for building the second complementary chain (Perutz, 1962).

It has long been clear that enzyme molecules are not synthesized by the DNA of the gene directly, the DNA being in the nucleus and protein synthesis occurring in the cytoplasm, where another type of nucleic acid involving ribose instead of deoxyribose occurs (ribonucleic acid-RNA). A multiprotein enzyme complex, mRNA polymerase II, is responsible for selecting the DNA sequences in the nucleus and transcribing them into messenger RNA (mRNA); and it appears that about 50 proteins must associate with a portion of DNA before it can be transcribed to produce the mRNA appropriate for the synthesis of the protein required (Mann, 1999). Activation of the associating proteins depends on the receipt of a signal from the cell exterior, transduced through a cascade of cytoplasmic reactions, in which Ca^{++} ions play an essential role by promoting the action of cyclic AMP-dependent enzymes (Petersen, 1999). The mRNA carries the genetic message from the DNA on the gene in the nucleus to particles in the cytoplasm known as ribosomes. Activated amino acids are brought to the ribosome by other forms of RNA (transfer RNA), there being at least one molecule of transfer RNA for each amino acid. It seems likely that growth hormone regulates the rate of protein biosynthesis by controlling the synthesis of messenger RNA (Korner, 1963). One function of insulin appears to be to accelerate the translation of the messenger RNA for ribosomal and sarcoplasmic proteins in particular (Kurichara and Wool, 1968).

Each molecule of transfer RNA contains a code in the form of a sequence of the nucleotides already referred to, which permits it to pair with a sequence of complementary nucleotides on the messenger RNA, held in the ribosomes, and thus to add the correct amino acid to the growing polypeptide chain. It was established that a specific sequence of three nucleotides determines the selection of a given amino acid (Crick et al., 1961). The four different nucleotides on the DNA chains, and hence on the protein-building RNA chains, if taken three at a time permit 64 different sequences. Since only about 20 amino acids are found in proteins, it would appear that each one may be determined by more than one sequence. The sequence of nucleotides in nucleic acid is colinear with the sequence of amino acids on the polypeptide chain which it determines (Whitmann, 1961). The fundamental mechanism for the control of enzyme synthesis (and, thereby, of growth and development) by DNA and RNA had just been elucidated when the first edition of this book was being written. As revealed then, the genetic code was elegant and logical. Since then, however, much further information has become available which indicates that the code embodies major complexities. Only a relatively small proportion of the nucleotide sequences of the genome code for proteins (exons). They are flanked by other segments of DNA (introns) which, although they are transcribed by mRNA, are excised and destroyed prior to translation of the exons. By far the greatest proportion of the nucleotide sequences are non-coding and their function is still not clearly understood. These non-coding sequences include segments which are important for identifying species (and even the individual) (cf. § 4.3.1). The sequence of nucleotides on the genes may be used to code for more than one protein by altering the base at which reading is initiated (Crick, 1979; Chambon, 1981), and it can be interrupted by the insertion of nucleotides from other genes or viruses.

Again, most cells contain large numbers of copies of certain of their genes. These may represent material from which new genes can be developed, or the capacity to swiftly increase the production of a particular protein, in response to extraneous circumstances. There appears to be some higher order of interaction between genes in controlling cells and tissues (Cavalier-Smith, 1980; Taylor, 1992). Although much of the genetic material is remarkably stable in evolution, albeit subject to chance mutation, other portions are dynamic and susceptible to purposeful change.

In respect of muscular tissue, polysomes which synthesize myofibrillar proteins specifically have been identified (Heywood and Rich, 1968). These comprise 50–60 ribosomal particles. Such are large enough to code for a protein of MW 170,000 to 200,000, i.e. the mass of the main myosin subunit. These polysomes appear to correspond to that species of RNA with 26 S sedimentation characteristics. Actin and tropomyosin are synthesized by smaller polysomes: and there seems to be a correspondence between the size of mRNA and the muscle protein for which it codes (Heywood, 1970). In embryogenesis the synthesis of actin precedes that of myosin and the latter proceeds before tropomyosin is synthesized. Sarcoplasmic proteins turn over at a faster rate than those of the myofibrils (Burleigh, 1974). In muscle development, the quantity of mRNA coding for the heavy chain segment of myosin increases from about 1000 to about 3500 copies per nucleus as the muscle becomes fully differentiated (Young and Achtymuchuk, 1982). Reflecting this increase, the rate of synthesis rises from about 1000 to about 30,000 molecules per minute per nucleus (Young and Denome, 1984).

The complete DNA sequence of bases of the exons and introns for the myosin heavy chain of rat embryonic skeletal muscle has been defined (Strehler *et al.*, 1986). It contains 24,000 bases and codes for an mRNA of 6035 nucleotides, of which 5820 are responsible for the sequence of 1939 amino acid residues in the molecule of myosin heavy chain. At least 7 different genes coding for myosin heavy chains have been identified (e.g. for embryonic, neonatal and adult skeletal muscle, cardiac and ocular muscles) (Mahdavi *et al.*, 1986). Although closely related, the heavy myosin chains for which these genes code must differ structurally in the region which binds, hydrolyses and expels ATP. The mode of contraction of a muscle in terms of its speed, force generated and other parameters, is influenced by the type of myosin heavy chain it contains and presumably also by the types of myosin light chain, actin, tropomyosin and troponins present (Young and Davey, 1981: cf. Table 4.32), which are also encoded by multigene groups (Nadal-Ginard *et al.*, 1982). Further aspects of the differences between myosin heavy chain isoforms are considered below (cf. § 4.3.5.2). Isoform identification by DNA analysis is likely to elucidate the nature of various pathological conditions in muscle (cf. § 3.4.1) and to suggest how gene manipulation could correct them. Thus, in the *l. dorsi* muscles of pigs carrying the halothane positive gene, and exhibiting the PSE condition (cf. § 2.2), there is a greater proportion of Type II (fast-twitch, glycolytic) fibres than of Type I (slow twitch, oxidative) (cf. § 4.3.5.2), this being correlated with the proportion of fast-glycolysing myosin heavy chain present (Depreux *et al.*, 2000).

In respect of the connective tissue proteins, synthesis is similar. Since there are at least 15 different chains in collagen, however, there must be at least that number of distinct genes coding for the protein. Essential for the formation of the triple helix is the hydroxylation of proline (ascorbic acid is required for the activity of the enzyme responsible, 4-prolyl hydroxlase). A second important post-translational modification is the hydroxylation of certain lysine residues which determine the nature of the cross-links in the non-helical, terminal regions of the chains and the stability of the connective tissue. The completion of the synthesis of the collagen molecule also involves glycosylation of a proportion of the hydroxylysine residues, and the formation of the triple helix. An understanding of complex biological systems has hitherto been derived from studying their components in isolation; and from such observations the nature of their dynamic functions *in vivo* have been deduced, albeit with considerable accuracy. The revelation of the genetic code some 50 years ago has led to the emergence of a much more fundamental, intimate and precise appreciation of biology. Thus, the ability to identify the complete nucleotide sequence in the genetic complement (genome) of cells has been coupled with mathematical means of organizing and interpreting the immense volume of data generated by sophisticated computers (bioinformatics). Techniques are becoming available to determine *simultaneously*, in a cell or organism, all the possible modes of transcription of nuclear DNA by mRNA (transcriptomics), the entirety of the protein species and isoforms expressed thereby (proteomics) and all the metabolites arising during functioning (metabolomics). These developments, collectively referred to as 'genomics', have mainly occurred since the last edition of this book was published in 1998. It is now becoming possible to relate the phenotypic characteristics, which distinguish species, breeds and even individual muscles, to the operation of specific genes. Genomics provides a new approach to understanding and positively controlling the attributes of eating quality and nutritive value in meat (Eggen and Hocquette, 2003) (cf. Chapter 10).

3.3.2   General origins of tissues

Although the way in which cells perpetuate their proteins is becoming clearer, and how histones suppress those genes which are not permitted to code for proteins, it is not as yet known how such proteins are organized spatially to produce the cell's formed elements. The sciences of cytology and embryology, however, provide some clues.

It is no longer possible to regard the initial cell from which embryonic growth occurs as having a completely undifferentiated protoplasm: the spherical symmetry of the egg is transitory and disguises the heterogeneous nature of its contents. Either from its position in the ovary or because it possesses within itself incipient polar organization, the egg is already highly polarized (Picken, 1960). The embryo becomes differentiated into 'head' and 'tail' regions through morphogenetic stimuli from a primary inductor (Needham, 1942). Holtfreter (1934) showed that the adult tissues from members of all phyla, if implanted into the embryonic body cavity formed at an early stage in the mass of dividing cells, induced the formation of a secondary embryo. It has since been shown that the primary inductor is extractable from such tissues. Several substances act as hormones controlling the form of growing tissue (e.g. ribonucleoproteins), but it seems likely that the natural primary organizer of the embryo is steroid in nature: it is thus of interest that the unorganized differentiation of most carcinoma should be caused by compounds belonging to the same chemical family as the steroids (Cook, 1933, 1934). There is strong evidence that retinoic acid is an important morphogen for anterior/posterior location and that it acts via 'homoeotic selector gene complexes' (HOM) of which at least four have been identified in mammals (Lewis and Martin, 1989). The further a gene lies from the beginning of the HOM, the more posterior is its domain of expression along the body axis. It is likely that there is a hierarchy of tissue organizers, and that, after the primary induction of 'head' and 'tail' regions, secondary or tertiary inductors evoke the production of different types of tissue in the embryo and of different organs and structures within these types. Significant advances in the techniques available for micromanipulation have made it possible to isolate individual cells from early embryonic tissue which have the capacity to replicate as identical copies *in vitro*, being potentially immortal. These 'stem cells' also have the capacity to develop into every tissue (pluripotent) of the body. Stem cells can also be separated from certain tissues in the adult body; but these are usually only capable of forming the particular tissue from which they are isolated (Alison *et al.*, 2002; O'Donoghue and Fisk, 2004). It is envisaged that stem cells will permit regeneration of damaged or diseased tissue in human medicine, and effect cures for hitherto untreatable conditions. In the present context, it is feasible that stem control of the muscles of meat animals will be exploited as a factor in preferentially developing those locations with organoleptic characteristics desired by the consumer.

3.3.3   Development of muscular tissue
Skeletal muscles arise in the embryo from the mesodermic somites, i.e. from the third (and central) germinal layer of the embryo. The somite cells begin to form along each side of the embryonic axis 2–3 weeks after conception. From the somites muscle cells arise in about 40 groups (myotomes). Initially these consist of a mass of closely spaced and undifferentiated cells of fusiform shape. As development proceeds, two types of cell can be distinguished, one acquiring the morphology of primitive branching connective tissue cells, and the other that of primitive muscle cells (myoblasts). These are bipolar spindle-shaped cells and fuse to form myofibrils, a process which has been observed by time-lapse cinematography (Bachmann, 1980). The latter at first multiply by mitotic division whereby the nuclear material is divided equally between mother and daughter cells. Later they elongate, become multinucleated and divide amitotically. Within a muscle fibre, the number of myofibrils increases, during embryonic development, from a single original fibril (Maurer, 1894) by longitudinal fission (Heidenhain, 1913). The fibrils produce the fibre by forming first a hollow tube and then filling its interior (Maurer, 1894). The first fibrils formed are unstriated even when several are in parallel and dots delineating each sarcomere (the subsequent Z-lines) appear before the fibrils form a tube (Duesberg, 1909). In mammals, the myofibrils just beneath the sarcolemma, i.e. those forming the periphery of the muscle fibre tube, are the first to become striated. Muscle nuclei are originally located in a central position but eventually migrate to the periphery of the fibre, later becoming flattened against the sarcolemma as the number of myofibrils increases.

More recent views on the histogenesis of skeletal muscle, based on modern technology have confirmed earlier understanding, and have provided much more information on how the specific myofibrillar proteins which are now known (§§ 3.2.2 and 4.1.1) are involved (Stockdale, 1992). The first myofibrillar protein to develop is titin. This is followed, in turn, by myosin, actin, actinin, nebulin and the various other proteins which regulate the contractile machinery (Colley *et al.*, 1990). The assembly of the myofibrillar proteins into mature myofibrils must involve a higher order structure and may include proteins which promote extracellular (e.g. integrins: Hynes, 1992) and intracellular (e.g. cadhesins: Knudsen *et al.*, 1970) cell adhesion.

The establishment of skeletal muscles during embryonic development involves the commitment of mesodermal cells to a specific myogenic pathway. Further proliferation and the subsequent differentiation into muscle cells depends on the interplay of many inter- and intracellular proteins which are induced by the local cell environment (Olson, 1992). It is feasible that a single molecular event could initiate this chain of changes, activating a 'master gene' which, via a cascade of gene expressions, would activate other groups of genes. A number of genes in the group of factors regulating myogenesis, have been identified (Aurade *et al.*, 1994). They produce so-called basic helix–loop–helix (bHLH) proteins. The latter bind DNA and when they (or the parent gene) are introduced into fibroblasts, they convert them into muscle cells (Olson and Klein, 1992). The location at which they bind to DNA appears to be the promotional region and since most of the myofibrillar contractile genes have this sequence, they can all be activated.

Generally, after the second half of intrauterine life, muscles increase in size not by augmenting the number of their constituent fibres but by increasing the size of the latter (Adams *et al.*, 1962). Nevertheless, Goldspink (1962b) found that the fibres of the *biceps brachii* of the mouse increased in number for some time after birth: subsequent muscle development with age or with exercise is due not to a further increase in fibre numbers, but occurs because there is a greater number of fibres having a large diameter (Goldspink, 1962a), and those fibres which acquire a large diameter do so by an increment in the number of their constituent myofibrils (Goldspink, 1962c). Inanition is associated with a redistribution in the population of muscle fibre diameters, those of small diameter increasing in number at the expense of those of large diameter (Goldspink, 1962b).

During early development, fibrils grow in length from each end in complete sarcomere units (Holtzer *et al.*, 1957), i.e. the number of sarcomeres per fibril increases: after birth, the number of sarcomeres per fibril tends to remain constant and increase in fibre length is achieved by increasing the width of existing sarcomeres (Goldspink, 1962c). There is a concomitant increase in the degree of overlap of actin and myosin filaments with increasing age. This explains why the muscles of the young animal cannot develop much power (Goldspink, 1970).

For the musculature generally, the greatest rate of increase and weight occurs in the immediate postnatal period: the rate tends to diminish as growth continues.

In sheep (Hammond, 1932a; Pállson, 1940) and pigs (McMeekan, 1940) it has been shown that there is a greater rate and amount of postnatal growth in the musculature of the head and trunk, as one proceeds from the fore to the hind end of the body, and in the musculature of the limbs, as one proceeds from the feet towards the body: the latest maturing region is that where these waves of growth meet at the junction of the loin and the last rib, as has been mentioned in connection with body growth generally. Hammond (1932a) studied how the main groups of muscles in the hind limb of the sheep developed with increasing age and showed that those in the 'leg' portion were relatively better developed at birth than those in the 'thigh' portion; muscles of the latter group matured later. Sex also influenced the relative development of the groups, the 'thigh' muscles (i.e. those of the upper portion of hind leg) being relatively more developed than those of the 'leg' (i.e. those of the lower portion of hind leg) in males than in females. Again, the level of fatness influences the issue. The 'thigh' muscles are capable of depositing more fatty tissue between their fibres than the 'leg' muscles.

The postnatal growth of individual muscles is determined by the relative maturity at birth of the area in which they are found. Thus, those in the 'thigh' show a greater development than those in the 'leg' (Hammond, 1932a). Joubert (1956), on the basis of fibre diameter measurements, compared the *l. dorsi* from the loin – a late developing area – with the *rectus femoris* from the 'thigh' and the *gastrocnemius* from the 'leg' of the hind limb. At birth the *gastrocnemius* possessed the largest fibres and the *l. dorsi* the smallest. Those of *l. dorsi*, on the other hand, showed greatest relative increase during postnatal life while the fibres of *gastrocnemius* increased least at this time. In mature animals the early developing *gastrocnemius* increased most under a high plane of nutrition: on a submaintenance diet the fibres of *l. dorsi* were most severely retarded.

In the pig at birth the weights of lumbar and thoracic *l. dorsi* and of the neck muscles represent, respectively, 2.6, 3.6 and 4.6 per cent of the total muscle weight: 100 days later the respective values are 3.9, 5.4 and 4.8 per cent (Cuthbertson and Pomeroy, personal communication). These data again demonstrate that muscles located nearer to the rear of the animal – here the two portions of *l. dorsi* – have a greater postnatal development than those in the fore end.

Within a given area, however, individual muscles vary considerably in their rates of growth (Hammond, 1932a). So many factors interact in producing growth in individual muscles that generalizations are difficult, but the largest muscles have the greatest rate of postnatal development, this being possibly related to muscle function. In classifying the relative growth patterns of muscles in cattle, Butterfield and Berg (1966) have pointed out that most muscles show more than one growth phase.

As in the case of the animal as a whole, it is normal for muscles to lay down both intracellular and extracellular fat. This occurs as a result of age or because of a high plane of nutrition (Helander, 1959).

3.4   Abnormal growth and development in muscle

A variety of factors can cause abnormal growth and development in muscle. Such may be superficially manifested by an unusual increase or decrease of normal muscular tissue or by the production of atypical tissue, which may be accompanied by an overall increase or decrease in size. At the present moment, most of these abnormalities would automatically preclude the affected musculature from consumption as meat on aesthetic grounds or on those of public health; but they should be considered, since they help to indicate the nature of muscle, and it is conceivable that one day some may be utilized deliberately to produce desired qualities in meat.

3.4.1   Genetic aspects
The muscular hypertrophy of 'doppelender' cattle and the stunted growth of 'snorter' dwarf cattle have already been mentioned as being due to recessive genes. In neither case is the musculature other than entirely wholesome as meat.

Imperfections in embryogenesis account for a number of conditions where there is anomalous development in muscle or where muscles fail to develop at all. Club-foot exemplifies one of the conditions where development is faulty: histological examination shows that the muscle fibres are of uniformly small diameter. When congenital absence of muscles occurs, those which most frequently fail to develop are the pectorals (Bing, 1902). The ingestion by pregnant women of tranquillizers such as thalidomide has emphasized the susceptibility to mutation of the genes controlling muscular development.

Many other abnormalities of development have been shown to be heritable. These include various diseases in muscle where inflammation by infecting organisms is not involved (muscular dystrophy). The heritable types are characterized by the relative absence of regenerative activity: they are largely degenerative, and the chain of causes is unknown (Adams et al., 1962). The muscle fibres have a greatly lessened capacity to retain creatine (Ronzoni et al., 1958) and potassium (Williams et al., 1957), more collagen (Vignos and Lefkowitz, 1959) and a decreased ability to produce lactic acid by glycolysis due to lowered contents of aldolase, phosphorylase and creatine kinase (Dreyfus et al., 1954; Ronzoni et al., 1958). In Duchenne muscular dystrophy the electrophoretic pattern of the lactic dehydrogenase enzymes reverts to that of the embryonic muscle (Emery, 1964). The contractile proteins are relatively unaffected. Two as yet unidentified proteins have been detected in the urine of those suffering from the disease (Frearson et al., 1981) and the gene responsible for the condition has been shown to produce a protein, dystrophin, associated with the sarcolemma (Zubrzycha-Gaarn et al., 1988), which it strengthens by fixing portions of the cytoskeleton to the surface membrane.

The most important histological feature of the dystrophies is the disappearance of muscle fibres, which proceeds, in phases of hypertrophy, atrophy, splitting and fragmentation, to degenerating myoblasts.

A series of heritable glycogen storage diseases are known (Cori, 1957). Certain of these are characterized by the deposition of large quantities of glycogen in muscle. They are distinguished from one another by various genetically determined deficiencies of glycolytic enzymes, on account of which the accumulation occurs. In one condition 'debranching enzyme' is deficient and the structure of the glycogen deposited is abnormal: in another, phosphorylase is deficient, but the glycogen is normal. It should be mentioned, however, that the glycogen concentration in the muscles of the new-born pig is normally very high (about 7 per cent compared with 1–2 per cent in the muscles of older animals: McCance and Widdowson, 1959).

Familial periodic paralysis, in which potassium accumulates (McArdle, 1956), and a spontaneous discharge of muscle pigment with myoglobinuria (Biörck, 1949), exemplify two other types of heritable abnormalities of muscle growth.

3.4.2 Nutritional aspects

Reference has already been made to the differential effects which animal age and the general plane of nutrition have on the development of various groups of muscles. Provided the diet is qualitatively adequate such growth is normal, but the absence or excess of specific substances can cause atypical development.

Dystrophic muscle which is superficially white, and may be exudative, can arise in cattle, sheep and pigs through a deficiency of vitamin E: the latter appears to be essential for the integrity of muscle (Blaxter and McGill, 1955). Histologically, there are distinct pathological features such as hyaline degeneration (cell transparency) and phagocytosis. Characteristically, long segments of fibres show 'coagulation necrosis': areas of regeneration are found concomitantly (West and Mason, 1958). Biochemically, there is an increase of γ-myosin (Kay and Pabst, 1962). Vitamin E-dystrophic muscle has a greater capacity for proteolytic breakdown, which may be attributed, in part, to increased dipeptidase activity (Weinstock *et al.*, 1956); a lowered capacity for respiration (Schwartz, 1962); a greater content of connective tissue protein, fat and water; and a lower content of total nitrogen (Blaxter and Wood, 1952). Structural changes in myosin are also induced by vitamin E-dystrophy (Lobley *et al.*, 1971). The content of 3-methyl histidine is lowered, one of the soluble subunits of myosin (Perrie and Perry, 1970) apparently fails to be synthesized and its Ca^{++}-activated ATP-ase is markedly depressed. A diet containing appreciable quantities of unsaturated fatty acids, especially linoleic acid, predisposes to vitamin E deficiency (Lindberg and Orstadius, 1961), emphasizing the antioxidant aspects of the role of vitamin E. A dietary absence of selenium also produces muscular dystrophy, which vitamin E counteracts; here its role is unknown (Blaxter, 1962).

As in the case of vitamin E deficiency, it seems possible that an excess of dietary vitamin A increases the proteolytic activity of muscle, perhaps by increasing the permeability of the membrane within which catheptic enzymes are contained (Fell and Dingle, 1963).

The ingestion of specific toxins, such as the diterpenes pimaric and abietic acids, is also thought to be responsible for the occurrence of white, exudative muscle and myoglobinuria ('Haff disease'; Assman *et al.*, 1933).

3.4.3 Physiological aspects

A white, exudative appearance is a superficial symptom of many abnormalities in muscle growth which are directly attributable to genetic or nutritional factors; as we have seen, microscopic examination reveals pathological features. Much interest has been shown in a condition in the muscles of pigs which resembles the nutritional or genetic dystrophies superficially, but in which virtually no pathological changes can be observed. According to Bendall and Lawrie (1964) its most immediate cause is physiological. It has been suggested that overintensive selection for high feed conversion efficiency and for leanness in pigs (e.g. Danish Landrace, Piétrain) has inadvertently also selected for pigs having an excess of growth hormone (GSH) and a deficiency of adrenocorticotrophic hormone (ACTH) in the pituitary and hyperthyroidism (Ludvigsen, 1954). According to Wood and Lister (1973), however, such pigs possess instead an impaired capacity to deposit fat. The leanness of stress-sensitive, PSE-susceptible pigs may be due to an enhanced capacity to mobilize fat associated with impaired insulin metabolism and a greater sensitivity to the action of catecholamines (especially norepinephrine) of the body stores of fat (Wood *et al.*, 1977).

Despite the absence of pathological features the condition has been referred to as 'Muskel-degeneration' (Ludvigsen, 1954) and 'la myopathic exudative dépigmentaire du pore' (Henry, Romani and Joubert, 1958); the original name 'wässeriges Fleisch' (Herter and Wildsdorf, 1914) or the description 'pale, soft, exudative' (PSE) musculature (Briskey, 1964) are more appropriate. Nevertheless, the histological features are uncommon and resemble those which can be artificially produced by a fast rate of post-mortem glycolysis. In longitudinal section there is frequently an alternate array of strongly contracted and adjacent passively kinked fibres, or irregularly spaced bands of dark-staining protein deposits running across the fibres (Lawrie *et al.*, 1958; Bendall and Wismer-Pedersen, 1962). These bands penetrate into the depth of the fibre (Fig. 3.11) and appear to consist of denatured sarcoplasmic protein,

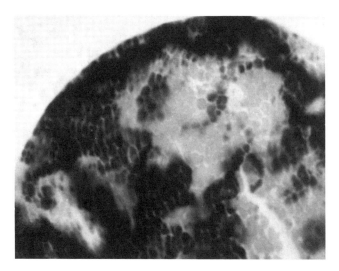

Fig. 3.11   Cross-section of muscle fibre from pig affected by so-called White Muscle disease (PSE muscle) showing irregular disposition of dark-staining myofibrils, presumably coated with a precipitate of denatured sarcoplasmic proteins (×2000). (Courtesy C. A. Voyle.)

which has precipitated on the myofibrils, lowering the extractability of the latter (Bendall and Wismer-Pedersen, 1962). The sarcoplasmic precipitate includes the enzyme creatine kinase (Scopes and Lawrie, 1963).

In pork which exhibits the PSE condition, there may be an unusually fast rate of pH fall during post-mortem glycolysis, with a normal ultimate pH (*ca.* 5.5) or an unusually low ultimate pH (*ca.* 4.8) (Lawrie *et al.*, 1958). Halothane anaesthesia induces a fast rate of pH fall in the muscles of stress-susceptible pigs (malignant hyperthermia; Gronert, 1980). The excess anaerobic release of Ca^+ ions by the mito-chondria of the muscles of stress-susceptible pigs has thus been postulated as a 'trigger' for the PSE condition, and also for malignant hyperthermia (Cheah and Cheah, 1976). Subsequently Cheah and Cheah (1981a, b), in endeavouring to eluci-date the mechanism more fully, suggested that unsaturated fatty acids from mito-chondrial membranes, released by endogenous phospholipase A_2, would cause the sarcoplasmic reticulum to release Ca^{++} ions, under the influence of an unusually high level of calmodulin (cf. § 4.1.1) within the mitochondria (Cheah *et al.*, 1986). Sig-nificant increases in the levels of calmodulin, fatty acids and phospholipase a_2 within mitochondria, and of Ca^{++} ions both in the sarcoplasm and within mitochondria, are observed in adult halothane-sensitive pigs when compared with young halothane-sensitive pigs and both young and adult halothane-insensitive pigs (Cheah *et al.*, 1986). Eikelenboom and Nanni Costa (1988) reported that the incidence of PSE was more frequent, and its manifestation more serious, in halothane-negative pigs than in a comparable group of pigs which responded to the drug. The occurrence of PSE in the former group was associated with a low ultimate pH. It is thus believed that at least two distinct genes are responsible for the PSE condition (Sellier, 1988) (cf. § 2.2). One is associated with a fast rate of post-mortem pH fall (low pH_1) and with halothane sensitivity; the other is associated with an abnormally low ultimate pH, although the rate of pH fall may not be unusually fast. The high glycolytic potential which this latter condition signifies is found especially in the Hampshire breed of pigs, from which the halothane sensitivity gene is virtually absent (Monin and Sellier, 1985), whereas the Piétrain breed tends to be halothane sensitive and to exhibit a low pH_1 (cf. § 2.2). The muscles of the latter have a higher proportion of fast-twitch, glycolytic 'white' fibres (type II) than those of other pig breeds. The muscles of Hampshires tend to have more slow-twitch, 'red' fibres (type I) than those of other breeds.

In a recent investigation of the *l. dorsi* of pigs Danish investigators demonstrated that the *in vivo* levels of glycogen were higher, and the 'resting' pH distinctly lower, in animals which were either homozygous or heterozygous for halothane sensitivity. They proposed that *in vivo* assessment of the level of muscle glycogen was a more effective predictor of PSE pork than sensitivity to halothane.

Essen-Gustavsen *et al.* (1992, 1994) found that, in the *l. dorsi* muscles of halothane-sensitive (*nn* genotype) pigs, the fibres were of greater cross-sectional area and the density of the capillaries lower, than in the corresponding muscles of halothane-resistant (NN genotype) individuals. Post-mortem, the muscles of the halothane-sensitive pigs had twice as many glycogen-depleted type II fibres as had those of the NN genotype, as well as the highest reflectance values (paleness) and exudation (Essen-Gustavsen *et al.*, 1992, 1994; Fiedler *et al.*, 1999). The muscles of heterozygotic pigs which, although non-responsive to halothane, carry susceptibility to malignant hyperthermia (N*n*), are also characterized by a rapid rate of post-mortem glycolysis (and high ATP turnover). They can be identified by ^{31}P-NMR (Moesgaard *et al.*, 1995). Deterioration in the power of the sarcoplasmic reticulum to accumulate Ca^{++} ions, measured in muscle homogenates by a Ca^{++}-sensitive electrode, can be used to identify NN, N*n* and *nn* genotypes (Cheah *et al.*, 1994).

In vivo susceptibility to malignant hyperthermia and halothane anaesthesia can be detected by DNA polymerase in blood samples (Houde and Pommier, 1993: cf. § 2.2).

Either a fast rate of pH fall or a very low ultimate pH would tend to denature muscle proteins and lower their capacity to hold water – in the former case because a relatively low pH would be attained whilst the temperature of the carcass was still high (Bendall and Wismer-Pedersen, 1962). Arakawa *et al.* (1970) have shown that the regulatory functions of α-actinin and tropomyosin-troponin are markedly lowered in such conditions. Penny (1969) demonstrated (in porcine *l. dorsi*) that the lower the pH at 90 min post-mortem, the lower the water-holding capacity, the ATP-ase activity and the extractability of the myofibrillar proteins. It is evident that a fast rate of pH fall post-mortem is also effective in denaturing the *contractile* proteins.

Using a predictive model, Offer (1991) showed that denaturation of myosin is likely to be the predominant cause of exudation in PSE muscle and that the severity of the denaturation increases with *both* rate and extent of post-mortem pH fall.

A high environmental temperature, struggling immediately before slaughter and delayed cooling of the carcass cause the condition to be manifest (Bendall and Lawrie, 1964). Briskey (1969) and Kastenschmidt (1970) presented biochemical evidence, however, which suggested that the degree of struggling at death is *not* the main reason for a high rate of post-mortem glycolysis in stress-susceptible pigs. They believed that the inherent constitution of the muscles in the latter is such that they are more readily made anoxic post-mortem and hence encourage a fast rate of lactic acid production. The efficacy of relaxant doses of magnesium sulphate administered pre-slaughter in slowing post-mortem ATP breakdown (Howard and Lawrie, 1956) and in preventing PSE (Briskey, 1969; Sair *et al.*, 1970) may be due to a vasodilatory action. Although the pre-slaughter injection of magnesium sulphate slows the rate of ATP breakdown in pigs of Piétrain, Landrace and Large White breeds, and there is a marked reduction in exudation post-mortem in the former two breeds, this is less so with the meat from Large White animals, suggesting that there may be influences other than the rate of post-mortem glycolysis affecting waterholding capacity. The Ca^{++}-accreting ability of the sarcoplasmic reticulum from muscles showing exudative character is said to be less than that of normal porcine muscle (Greaser *et al.*, 1969b). This feature would exacerbate any tendency for a fast rate of pH fall post-mortem and this, in turn, increase the damage to the sarcoplasmic reticulum (Greaser *et al.*, 1969a; Greaser, 1974). Mitochondria from *l. dorsi* muscles of stress-susceptible Piétrain and Poland China pigs release Ca^{++} ions anaerobically at twice the rate of those in stress-resistant pigs.

Prophylaxis is said to be made possible by giving cortisone to stress-susceptible pigs (Ludvigsen, 1957). At rest the level of 17-hydroxycorticosterone in the blood serum is 50 per cent greater in stress-susceptible than in normal pigs (Topel, 1969); and, on exposure to stress, the level falls in the former, whereas it rises in normal pigs. The intravenous injection of aldosterone induces a pale, soft exudative condition in pig musculature. This effect is prevented if an oral drench of aldactazide (a competitive inhibitor of aldosterone) is administered 30 min beforehand (Passbach *et al.*, 1969).

Another index of the potentially stress-susceptible pig is the presence of lactic dehydrogenase isozyme V in the blood serum: usually isozyme I is found. A high isozyme V/I ratio in blood samples would indicate a tendency to develop PSE post-mortem (Addis, 1969). There are said to be enhanced levels of creatine phosphokinase in the blood serum of stress-susceptible pigs (Allen and Patterson, 1971), and of glucose-6-phosphate in muscles sampled by biopsy (Schmidt *et al.*, 1971). These observations are compatible with a physiological explanation of the condition.

Wismer-Pedersen (1969b) has shown that the proteins from watery pork have a much lower capacity to form stable emulsions than those from normal porcine muscles. He suggests that the greater insolubility of sarcoplasmic proteins in such meat (see above) prevents myofibrillar proteins forming the strong membrane round fat globules to which, he believes, emulsion stability is due when sausage meat is heated. It seems likely, however, that the connective tissue proteins also affect the stability of sausage meat emulsions. It is of interest, in this regard, that the epimyseal connective tissue appears to have significantly more salt-soluble collagen and a greater amount of heat-labile collagen, when derived from watery pork than from the flesh of normal pigs (McLain *et al.*, 1969).

Henry *et al.* (1958), Lawrie (1960) and Scopes and Lawrie (1963) attributed the paleness largely to the absence of myoglobin, whereas Wismer-Pedersen (1959a) and Goldspink and McLoughlin (1964) attributed it to denaturation of myoglobin, while Hector *et al.* (1992) suggested in electically stimulated beef paleness in the *semimembranous* muscle was due to partial denaturation of myosin.

Bendall and Swatland (1988) published a comprehensive review of the relationship between pH and the physical aspects of pork quality. They concluded that, whilst the 'pH$_1$ index' (i.e. the percentage of carcasses with pH$_1$ less than 6.0 at 45 minutes post-mortem) gives some indication of the likelihood of PSE developing at 24 hours post-mortem, many of the studies of the condition have failed to separate the genetic causes of low pH_1 from environmental ones such as pre-slaughter handling.

Complete disuse of muscles causes a physiological atrophy. Histologically, there is a reduction in the mean diameter of muscle fibres (Tower, 1937, 1939). Conversely, continuous training increases the size of muscles (Morpurgo, 1897). This reflects an increase in the number of fibres which have a large diameter rather than an increase in the width of all the component fibres (Goldspink, 1962a, b). The muscle fibres increase in diameter both by the elaboration of new myofibrils and by an increase in sarcoplasm (Morpurgo, 1897). Such coarsening of texture would tend to make the muscles tougher as meat (Hiner *et al.*, 1953).

Physiological hypertrophy is also a reflection of hormonal activity. Much of this is 'normal', such as the effect of androgens (male hormones) in increasing the muscle size of males. On the other hand, disorders of the pituitary, thymus, thyroid and adrenal cortex glands are frequently associated with excessive or stunted muscular development. Little analytical work has been done on such material. Over-hydration of muscle has been noted in impaired adrenal function (Gaunt, Birnie and Eversole, 1949) and fatty infiltration in hyperthyroidism (Adams *et al.*, 1962).

3.4.4 Various extrinsic aspects

Atrophy is a common reaction of living muscle to injury. This response may follow directly from crushing or cutting of the muscle substance, ionizing radiation, excessive heat or cold, or high voltage electricity; or, indirectly, by section of or damage to the muscles' blood supply, nerves or tendons. These circumstances are fully discussed by Adams *et al.* (1962). There is generally some degree of reversion to the primitive foetal muscle structure (Denny-Brown, 1961). Histologically, the reaction ranges from a cloudy swelling of the sarcoplasm to total dissolution of the muscle fibre. Regardless of the precipitating cause, an orderly series of changes can be observed. There is firstly an enlargement of the nuclei and a tendency for these to migrate centrally to form rows. Next there is an accumulation of granular sarcoplasm around the nuclei and then, depending on the extent of the injury, regeneration or degeneration occurs, i.e. budding of new muscle tissue or fragmentation and splitting of the tissue into spindle cells, respectively (Denny-Brown, 1961). The excessive deposition of lipid in muscular tissue has been attributed to various cases, including injury and vascular abnormalities (Adams *et al.*, 1962).

Intermediate doses of ionizing radiation (*ca.* 1000 rad*) before slaughter increase the waterholding capacity of the subsequent meat and decrease its catheptic activity. Large doses (*ca.* 5000 rad) may cause disappearance of cross-striations and vacuolation of the sarcoplasm (Warren, 1943), oedema (Wilde and Sheppard, 1955), a rise in sodium and a fall in potassium and aldolase (Dowben and Zuckerman, 1963). The effects of massive, megarad, doses on muscle *in vitro* will be considered in a later chapter.

Crushing (Bywaters, 1944) and high voltage electricity (Biørck, 1949) cause substantial changes in muscles. They may lose most of their myoglobin, potassium and other soluble sarcoplasmic material.

---

* A rad is a measure of the dose of irradiation sustained and may be defined as an energy absorption of 100 ergs/g of material. 100 rads = 1 gray.

A series of inflammatory conditions in muscle (myositis) is known and in each there is destruction of the muscle fibres and proliferation of connective tissue (Adams *et al.*, 1962). The inflammatory agent may be parasitic (e.g. trichinosis in pork), bacterial (e.g. spontaneous acute streptococcal myositis), viral (e.g. Bornholm disease, caused by Coxsackie Group B virus) or metabolic (e.g. various rheumatic conditions).

Muscle tissue itself rarely elaborates into carcinomata: these are found generally as invasions of muscle by direct extension of a primary growth in another tissue. In such cases compression atrophy may result. Techniques employed in functional genomics (cf. § 3.3.1) have identified gencs which arrest the development of rhabdomyosarcoma, a highly malignant tumour of muscle (Astolfi *et al.*, 2001).

Chapter 4

Chemical and biochemical constitution of muscle

4.1 General chemical aspects

In a broad sense the composition of meat can be approximated to 75 per cent of water, 19 per cent of protein, 3.5 per cent of soluble, non-protein, substances and 2.5 per cent of fat, but an understanding of the nature and behaviour of meat, and of its variability, cannot be based on such a simplification. On the contrary, it must be recognized that meat is the post-mortem aspect of a complicated biological tissue, viz. muscle, and that the latter reflects the special features which the function of contraction requires, both in the general sense and in relation to the type of action which each muscle has been elaborated to perform in the body.

As outlined in Chapter 3, the essential unit of muscular tissue is the fibre which consists of formed protein elements, the myofibrils, between which is a solution, the sarcoplasm, and a fine network of tubules, the sarcoplasmic reticulum, the fibre being bounded by a very thin membrane (the sarcolemma) to which connective tissue is attached on the outside. The spatial distribution, between these structural elements, of the 19 per cent of protein in the muscle is shown in Table 4.1 (compiled from various sources), together with other data on the chemical composition of a typical adult mammalian muscle, after rigor mortis but before marked degradative changes. The principal free amino acids in fresh muscle are α-alanine, glycine, glutamic acid and histidine (Tallon *et al.*, 1954).

4.1.1 Muscle proteins

The proteins in muscle (Table 4.1) can be broadly divided into those which are soluble in water or dilute salt solutions (the sarcoplasmic proteins), those which are soluble in concentrated salt solutions (the myofibrillar proteins) and those which

Table 4.1 Chemical composition of typical adult mammalian muscle after rigor mortis but before degradative changes post-mortem (after Lawrie, 1975; Greaser, Wang and Lemanski, 1981)

Components		Wet % weight
1. Water		75.0
2. Protein		19.0
(a) Myofibrillar		11.5
myosin[1] (H- and L-meromyosins and several light chain components associated with them)	5.5	
actin[1]	2.5	
connectin (titin)	0.9	
N2 line protein (nebulin)	0.3	
tropomyosins	0.6	
troponins, C, I and T	0.6	
α, β and γ actinins	0.5	
myomesin, (M-line protein) and C-proteins	0.2	
desmin, filamin, F- and I-proteins, vinculin, talin, etc.	0.4	
(b) Sarcoplasmic		5 5
glyceraldehyde phosphate dehydrogenase	1.2	
aldolase	0.6	
creatine kinase	0.5	
other glycolytic enzymes especially phosphorylase	2.2	
myoglobin	0.2	
haemoglobin and other unspecified extracellular proteins	0.6	
(c) Connective tissue and organelle		2.0
collagen	1.0	
elastin	0.05	
mitoehondrial, etc. (including cytochrome c and insoluble enzymes)	0.95	
3. Lipid		2.5
neutral lipid, phospholipids, fatty acids, fat-soluble substances	2.5	
4. Carbohydrate		1.2
lactic acid	0.90	
glucose-6-phosphate	0.15	
glycogen	0.10	
glucose, traces of other glycolytic intermediates	0.05	
5. Miscellaneous Soluble Non-Protein Substances		2.3
(a) Nitrogenous		1.65
creatinine	0 55	
inosine monophosphate	0.30	
di- and tri-phosphopyridine nucleotides	0.10	
amino acids	0.35	
carnosine, anserine	0.35	
(b) Inorganic		0.65
total soluble phosphorus	0.20	
potassium	0.35	
sodium	0.05	
magnesium	0.02	
calcium, zinc, trace metals	0.03	
6. Vitamins		
Various fat- and water-soluble vitamins, quantitatively minute.		

[1] Actin and myosin are combined as actomysosin in post-rigor muscle

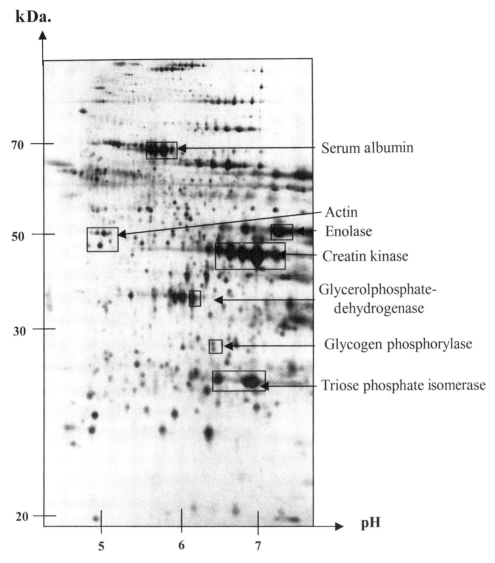

kDa.

70 — Serum albumin

Actin
50 — Enolase
Creatin kinase

Glycerolphosphate-
dehydrogenase

30 — Glycogen phosphorylase

Triose phosphate isomerase

20 — pH

5   6   7

Fig. 4.1 Proteins separated by two-dimensional electrophoresis from exudates of porcine muscle. (Reproduced from *Meat Science*, Vol. 71, E. Bendixen, 'The use of proteomics in meat science', pp. 138–149. Copyright 2005, with permission from Elsevier Science and Dr E. Bendixen.)

are insoluble in the latter, at least at low temperature (the proteins of connective tissue and other formed structures).

The sarcoplasmic proteins are a mixture of several hundred molecular species the complexity of which has been shown by modern proteomic techniques, such as two-dimensional electrophoresis (Bendixen, 2005; cf. Fig 4.1). Several of the sarcoplasmic proteins are enzymes of the glycolytic pathway and may be present as in more than one form (isozymes). Thus there are five tetrametric isozymes of lactic dehydrogenase (Dawson *et al.*, 1964). Many of the sar-coplasmic proteins were characterized as glycolytic enzymes many years ago (Scopes, 1964, 1966) and crystallized (Scopes, 1970). Apart from the differences in charge implicit in Fig. 4.1, the sarcoplasmic proteins differ in various other param-eters including their relative susceptibility to denaturation (Bate-Smith, 1937b; Scopes, 1964), but their individual characteristics as proteins will not be considered here. The fact that the sarcoplasmic proteins are soluble at low ionic strength *in vitro* tended to obscure the possibility that, *in vivo*, they might be present in a less mobile phase. It is now known that the glycolytic enzymes (which constitute the major pro-portion of the sarcoplasmic proteins) are bound to the myofibrillar protein, actin, *in vivo* – a feature which may well assist in the orientation and control of enzymic reactions in the muscle (Clarke *et al.*, 1980; Trinick and Cooper, 1982). The propor-tion of each glycolytic enzyme which is bound increases on stimulation of glycoly-sis (e.g. electrical) and decreases when such stimulation ceases (Starlinger, 1967; Morton *et al.*, 1988). The enhanced binding on stimulation to activity is especially marked in respect of phosphofructokinase, triose phosphate dehydrogenase, aldolase, phosphopyruvate hydratase and phosphopyruvate kinase (cf. Fig. 4.3). The binding of aldolase and of triose phosphate dehydrogenase to F-actin produces a marked increase in the viscosity of the latter. In model systems, the ratio of aldolase bound to actin is 1:14, increasing to 2:14 in the presence of tropomyosin, and to 3:14 when troponins are also present. These ratios are significant insofar as there are 14 units of G-actin in each helical repeat of the F-actin filament (Stewart *et al.*, 1980). Morton *et al.* (1988) concluded that, in each repeat of the F-actin helix, there are two binding sites. Both of these can be occupied by aldolase, but only one of them is preferentially occupied by triose phosphate dehydrogenase. It is evident, however, that some of the glycolytic enzymes bind not to F-actin, but to enzymes which are already bound to the latter (Stephen *et al.*, 1980). Thus triose phosphate isomerase binds to aldolase and triose phosphate dehydrogenase (both of which are bound to F-actin). It is interesting to note that these enzymes are arranged spatially in their correct sequence in the glycolytic pathway, the clusters of enzymes forming separate metabolic compartments wherein metabolites are transferred from the enzymes which generate them to those enzymes which utilize them ('metabolic channelling'; Welch, 1977). Phosphorylase molecules are bound to glycogen at two points (Goldsmith *et al.*, 1982).

Glycolytic enzymes bind to other locations in the muscle cell apart from F-actin, including the sarcolemma, the sarcoplasmic reticulum and the membranes of the nuclei and mitochondria (cf. § 3.2.2). Much of the AMP-deaminase activity of muscle is located at the ends of the myosin filaments on the A/I junction (Trinick and Cooper, 1982). Phosphorylase *b* appears to be localized both at the Z-disk and the M-line (Maruyama *et al.*, 1985). The M-line is also the location of creatine kinase (see below).

Myosin is the most abundant of the myofibrillar proteins. Its identity was a some-what confused issue for nearly 100 years from 1859, when the name was first given to a substance in muscle press juice which formed a gel on standing (Bailey, 1954). The molecule of myosin, which has a molecular weight of about 500,000, is highly asymmetric, the ratio of length to diameter being about 100:1. Because of its high content of glutamic and aspartic acids, and of dibasic amino acids, it is highly charged and has some affinity for calcium and magnesium ions. Myosin, as indicated above (§ 3.2.2), is composed of two heavy polypeptide chains and four light polypeptide chains. Heavy meromyosin (HMM) and light meromyosin (LMM) are proteolytic fragments of myosin. LMM contains parts of both heavy chains and HMM comprises parts of the two heavy chains and the four light chains (M. L. Greaser, personal communication). H-meromyosin, which contains all the ATP-ase and actin-combining properties of myosin, is sited on the periphery of the myosin filaments. The properties depend upon free-SH groups in the molecule (Bailey, 1954). Tropomyosin was discovered in 1946 by Bailey. Once extracted from muscle it is soluble at low ionic strength but *in situ* is extracted only at high ionic strength. Its amino acid composition is similar to that of myosin (Bailey, 1954) and like the latter, there are few free amino groups: it appears to be a cyclopeptide (a chain of amino acids forming a closed figure). It was suggested that actin filaments are attached to the Z-line by a meshwork of tropomyosin (Huxley, 1963), and the tropomyosin extends along the helical groove in the actin filament (cf. § 3.2.2). The actin filaments are attached to the inner surface of the plasma membrane and the Z-line by vinculin, a lipid-binding protein (Geiger, 1979). At less specific sites the cell proteins appear to be attached to the membrane by a complex of the protein integrin (Lawson, 2004) (cf. § 3.2.2).

The other major protein of the myofibril is actin (Straub, 1942). It can exist in two forms, G-actin, which consists of relatively small globular units having a molecular weight of about 42,000, and F-actin, in which these globular units are aggregated end to end to form a double chain (cf. Fig. 3.8). G-actin polymerizes into F-actin in the presence of salts and small amounts of ATP. It is F-actin which combines with myosin to form the contractile actomyosin of active or pre-rigor muscle and the inextensible actomyosin of muscle in rigor mortis (cf. § 4.2.1). The interrelation of actin, myosin and ATP is complex (Bailey, 1954) and will not be discussed in detail.

Relatively smaller quantities of other proteins, which are associated with the myofibrils, have been isolated, and functions have been assigned to some of them (Ebashi and Endo, 1968; Schaub and Perry, 1969; Maruyama, 1970). Thus the troponin complex promotes the aggregation of tropomyosin, binds calcium and prevents actomyosin formation; α-actinin promotes the lateral association of F-actin; β- and γ-actinins inhibit polymerization of G-actin. Tropomyosin B is the term now given to the protein remaining after troponin has been removed from tropomyosin as it occurs naturally. Because of its high content of α-helix, tropomyosin B is capable of contributing mechanical stability to the muscle filaments. The proteins of the M-line substance represent at least two molecular species (Porzio, Pearson and Cornforth, 1979). One of these (myomesin) promotes the lateral polymerization of L-meromyosin but not that of H-meromyosin. It has a sub-unit weight of 165,000 dalton and may bind creatine kinase to the M-line (Mani and Kay, 1978). Indeed creatine kinase is believed to be the principal component of the M-bridge (cf. Fig. 3.8) (Wallimann and Eppenberger, 1985). The proteins of the M-line appear to be essential for controlling the polarity of the myosin molecules in each half of the sarcomere.

There can be three types of light chain protein associated with myosin (Perry *et al.*, 1975). As indicated in § 3.2.2, the myosin molecule has a double shaft (light meromyosin) and a double head (heavy meromyosin); and there are two light chain molecules associated with each head of the myosin molecule. One of these has a MW of 18,000 daltons and is referred to as the DTNB light chain (since it is released by treating myosin with 5,5-dithiobis (2-nitrobenzoic acid)), the other being referred to as the alkaline light chain (since it is released from myosin by alkali). The latter occurs in two forms, one having a MW of 25,000 daltons (alkali 1), the other a MW of 16,000 daltons (alkali 2); but only one form is present in a given myosin molecule. It appears that, in any population of myosin, 50 per cent of the molecules are associated with two molecules of alkali 1 light chain and the other 50 per cent with two molecules of alkali 2 light chain; but all the myosin molecules are associated with two molecules of DTNB light chain. This explains why, on electrophoresis, three light chains separate, although there are only two types in any myosin molecule (Lowey and Holt, 1972).

The 18,000 MW light chain component is also referred to as the 'P' light chain since it is the specific substrate for the enzyme myosin light chain kinase (Frearson and Perry, 1975). Another enzyme, myosin light chain phosphatase (Morgan *et al.*, 1976), specifically removes phosphate from the 'P' light chain. Changes in the phosphorylation status of the 'P' light chain of myosin have been correlated with the physiological state of muscle (Frearson *et al.*, 1976), i.e. the interaction of myosin with actin. Myosin light chain kinase has a MW of 80,000 dalton. For its activation, an equimolar concentration of an acidic protein, calmodulin, is required (Nairn and Perry, 1979). Calmodulin resembles troponin C in binding Ca^{++} ions, but the latter is only marginally effective in activation.

The pattern of myosin light chains varies with the source of the myosin. Thus 'red' fibres usually contain only two, and 'white' fibres three, different light chains (Gauthier *et al.*, 1982).

Troponin is composed of three major members, referred to as C, I and T, which are concerned with the contractile process (Greaser and Gergely, 1971; Schaub *et al.*, 1972). The amino acid sequence in each has been determined. Troponin C (MW 18,000, 159 amino acid residues) has four binding sites for Ca^{++} ions and forms an equimolar complex with troponon I. It is phosphorylated neither by 3,5-cyclic-AMP-dependent protein kinase nor by phosphorylase *b* kinase (Perry *et al.*, 1975). Troponin I (MW 21,000, 179 amino acid residues) inhibits actomyosin ATPase. It can be phosphorylated by both enzymes (at serine and threonine residues, respectively (Cole and Perry, 1975). Troponin T (MW 30,000, 259 amino acid residues) binds to tropomyosin and troponin C: it is phosphorylated by phosphorylase *b* only.

Antigen–antibody techniques have confirmed that tropomyosins α and β, troponins C, T and I, and the alkali light chain proteins differ in 'red', 'white' and cardiac muscles (Dhoot and Perry, 1979). Immunofluorescence has revealed 11 transverse bands (about 70 Å wide and 430 Å apart) in each half of the myosin filaments (Trinick *et al.*, 1983–85). These bands represent the locations of C-, X- and H-proteins on the myosin. The pattern of their distribution differs between 'red' and 'white' fibres, X-protein predominating in the former and C-protein in the latter.

Apart from desmin (cf. § 3.2.2), small quantities of additional proteins have been isolated from skeletal muscles which also appear to be involved in forming the Z-line lattice (Ohashi and Maruyama, 1979), such as eu-actinin, filamin, synemin, vimentin and zeugmatin. Other minor proteins of the myofibrillar group include F-protein, which binds to myosin and from which it can be detached by C-protein (Miyahara *et al.*, 1980); I-protein, which appears to inhibit the Mg-activated ATPase of actomyosin in the absence of Ca^{++} ions (Maruyama *et al.*, 1977a, b); and para-tropomyosin, which is located at the A/I junction *in vivo*, but migrates post-mortem to the actin filaments as the concentration of free Ca^{++} in the cell increases (Hattori and Takahashi, 1988).

Connectin (variously referred to as 'gap filament', T-filament or as a mixture of titin and nebulin), nebulin and desmin have already been described as forming a filamentous cytoskeleton in muscle (§ 3.2.2); their arrangement in the sarcomere, and their relationship with myosin and C-protein, are also considered in some detail. The very large MW of titin and nebulin makes them susceptible to destruction by ionizing radiation; and electron micrographs show that the integrity of the sarcomere is substantially disorganized by doses of the order of 5–10 kgray (Horowits et al., 1986). Such findings emphasize the importance of these large proteins in cytoskeletal structure.

Smooth muscle, which is found in varying amounts at different locations in the gastrointestinal tract, in the lungs and in the uterus, does not normally form part of the edible portion of the carcass. Although its operating mechanism is basically similar to that of striated muscle, and involves parallel filaments of myosin and actin, there are a number of differences in the proteins of the contractile machinery. Thus, the MW of smooth muscle myosin (600,000) is rather greater than that of striated muscle (Kotera et al., 1969), the ratio of actin to myosin is almost twice that of skeletal muscle (Somlyo et al., 1981) and there are some differences in the amino acid composition of both major and minor proteins (Carsten, 1968). As a reflection of these differences, the contractile proteins of smooth muscle are extracted at markedly lower ionic strength (Hamoir and Laszt, 1962).

Whereas skeletal muscle has a relatively constant ratio of thick to thin filaments (1:2), the ratio in smooth muscle varies from 1:5 to 1:25 depending on the specific tissue in which it is found (Cohen and Murphy, 1979). Because a considerably greater number of cross-bridges can form between actin and myosin in smooth muscle it can develop a maximal force greater than can skeletal muscle, despite having a five-fold lower content of myosin (Somlyo et al., 1981). The sarcoplasmic reticulum in smooth muscle is much less fully developed than in skeletal muscle, and this may be related to the much slower speed of contraction of the former. Titin has not been found in smooth muscle (Locker and Wild, 1986).

Calmodulins, with properties similar to those of troponin C and of bovine brain modulator protein, have been isolated from smooth muscle and other tissues (Grand et al., 1979). They contain small quantities of the unusual amino acid, trimethyllysine. Virtually all of the Ca^{++}-binding protein of smooth muscle is of this type. Although calmodulins are also present in striated muscle, the Ca^{++}-binding protein in the latter is predominantly troponin C.

The distribution of amino acids in the various myofibrillar proteins of the rabbit is shown in Table 4.2. It will be noted that there is a higher content of aromatic residues in actin, the actinins, troponin C and H-meromyosin than in the other proteins. Actin and the actinins also have relatively high contents of proline. The latter is particularly low in L-meromyosin and tropomyosin B.

Negishi et al. (1996) characterized the troponins of bovine muscle. The molecular weights of troponins C, I and T were, respectively, 19,500, 23,300 and 40,400 daltons. In respect of their amino acid composition, the bovine troponins showed differences similar to those in the rabbit.

In that portion of muscle which is insoluble in concentrated salt are the mitochondria, containing the enzymes responsible for respiration and oxidative phosphorylation, the formed elements of the muscle membrane (sarcolemma) and the collagen, reticulin and elastin fibres of connective tissues.

Table 4.2 Distribution of amino acids in myofibrillar proteins (as percentages of total residues) (after Bodwell and McClain, 1971; Wilkinson *et al.*, 1972; Cummins and Perry, 1973; Head and Perry, 1974)

Protein	Aromatic amino acids	Proline	Basic amino acids	Acidic residues
Actin	7.8	5.6	13	14
α-actinin	7.7	5.9	12	14
β-actinin	7.5	5.7	13	12
Tropomyosins B, α and β	3.0	0.1	18	36
Troponin C	7.8	0.6	10	33
Troponin I	3.7	2.8	20	28
Troponin T	5.3	3.5	22	32
Myosin	5.3	2.6	17	18
H-meromyosin	7.2	3.4	15	15
L-meromyosin	2.8	1.0	19	18

4.1.2 Intramuscular fat

Although the fat of adipose tissue generally consists of true fat (i.e. esters of glycerol with fatty acids) to an extent of more than 99 per cent, the fat of muscle, like that of other metabolically active tissues, has a considerable content of phospholipids and of unsaponifiable constituents, such as cholesterol (Lea, 1962). Only three or four fatty acids are present in substantial amounts in the fat of meat animals– oleic, palmitic and stearic (Table 4.6, p. 98); and of the four types of glyceride GS_3, GS_2U, GSU_2 and GU_3 (S and U represent saturated and unsaturated fatty acids respectively), certain isomers greatly predominate. In the majority of fats which have been examined by partial degradation using pancreatic lipase, saturated acids are found preferentially in the alpha or exterior positions of the glycerol molecule (Savary and Desnuelle, 1959). Pig fat is exceptional, however, in that un-saturated acids are usually found in the alpha position. Although the fatty acids are not randomly orientated in animal fats, they can be so arranged artificially by heating with an esterification catalyst. This causes a marked improvement in the texture and plasticity of the fat and other characteristics which are useful in baking (Lea, 1962).

The phospholipids – phosphoglycerides, plasmalogens and sphingomyelin – are more complex than the triglycerides. In the phosphoglycerides one of the three hydroxyl groups of glycerol is combined with choline, ethanolamine, serine, inositol or glucose. In the plasmalogens the second hydroxyl group of glycerol is esterified with a long-chain fatty aldehyde instead of with fatty acid; and in sphingomyelin the amino alcohol sphingosine is bound by an amide link to a fatty acid and by an ester link to phosphorylcholine. There are also present in muscular tissue complex sugar-containing lipids – glycolipids. The effect of such factors as species, age and type of muscle on the composition of the phospholipids is still little known (Lea, 1962). Of the total phospholipids in beef muscle, lecithin accounts for about 62 per cent, cephalins for 30 per cent and sphingomyelin for less than 10 per cent (Turkki and Campbell, 1967). The metabolism of the phospholipids, especially that of the phosphoinositides, appears to be important in morphogenesis, protein turnover and contraction in muscular tissue (Campion, 1987). Accompanying the triglycerides are small quantities of substances which are soluble in fat solvents, e.g. vitamins A, D, E and K and cholesterol derivatives.

4.2   Biochemical aspects

4.2.1   Muscle function *in vivo*

The thick filaments which were apparent in Fig. 3.7g consist essentially of myosin and the thin filaments of actin. The latter are continuous through the *Z*-line, but do not traverse the *H*-zone which bounds each sarcomere; the myosin filaments traverse the *A*-band only. The three-dimensional aspect of this arrangement was briefly outlined in Chapter 3 and shown diagrammatically in Fig. 3.8 when it was indicated that there are six straight rows of projections running longitudinally along the side of each myosin filament, the sets of projections being symmetrically distributed around the periphery of the latter, so that one set of projections is opposite one of the six filaments of actin which surround each myosin filament (Fig. 3.7g). The two-dimensional aspect of this arrangement is shown in Fig. 4.2 (Bendall, 1969). This shows, in diagrammatic form, a longitudinal section of the sarcomeres (a) at rest length, (b) when extended and (c) during contraction. The thick myosin filaments are depicted with (above and below) two of the six actin filaments with which they are associated. The degrees of interdigitation, and of linkage between the myosin heads and the actin, in each condition will be apparent.

Figure 4.2 also depicts the patterns seen in cross-sections corresponding to these sarcomere lengths. It will be apparent that, as the muscle is extended beyond rest length, it becomes narrower and the hexagonal array of myosin and actin filaments becomes tighter. Conversely, when the muscle contracts, the cross-sectional area, and the distance apart of the myosin and actin rods, both increase. In muscle at rest in the living animal (or in the pre-rigor state in the dying muscle), the beads comprising the actin filaments are prevented from combining with the corresponding projections on the myosin by the magnesium complex of adenosine triphosphate ($MgATP^{2-}$): the latter acts as a plasticizer.

In the myofibril the contractile proteins are associated with the regulatory complex of troponins (troponins C, T and I) and tropomyosin (§ 4.1.1). These confer sensitivity to Ca^{++} ions upon the hydrolysis of $MgATP^{2-}$ by the actomyosin ATP-ase (Perry, 1974).

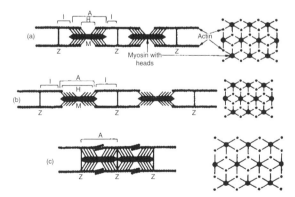

Fig. 4.2   Schematic representation of fine structure of ox muscle in longitudinal and cross-sections. The length of the sarcomeres has been reduced tenfold in proportion to the thickness of the filament. The beaded nature of the actin rods and the 'heads' in those of myosin are indicated. (a) Sarcomeres at rest length *ca.* 2.4 μm; (b) sarcomeres stretched to length *ca.* 3.1 μm; (c) sarcomeres contracted to length *ca.* 1.5 μm. (Courtesy the late Dr J. R. Bendall.)

A possible sequence of events in contraction may be outlined. Activation of the muscle usually is the result of a nerve stimulus arriving at the motor end plate, whereby the polarization of the sarcolemmal surfaces is reversed. The sarcolemma temporarily loses its impermeability to potassium and sodium ions, and Ca^{++} ions dissociate from the calsequestrin by which they are normally bound in the sarco-tubular system, and equilibrate with those in the sarcoplasm. As already indicated (§ 3.2.2), two large protein complexes, the junctional foot protein and the L-type calcium channel subunit are involved in the calcium release mechanism. As a result, the Ca^{++} ion concentration rises from about $0.10\,\mu M$ to $10\,\mu M$. This saturates tro-ponin C, the calcium-binding member of the troponin complex, causing a configu-rational change whereby the inhibitory protein, troponin I, no longer prevents actin from interacting with the $MgATP^{2-}$ on the H-meromyosin heads of the myosin mol-ecule. The contractile ATP-ase in the vicinity of the linkage is thus strongly acti-vated, splitting $MgATP^{2-}$ to $MgADP^{-}$ at a high rate and providing the energy for the actin filament to be pulled inwards towards the centre of the sarcomere, i.e. the portion of the myofibril involved contracts. The link between actin and myosin is simultaneously broken, although tension will remain as there are 5.4×10^{16} cross-links per millilitre of muscle; and some will be bearing tension at any given moment in contracting muscle. The $MgADP^{-}$ on myosin is recharged to $MgATP^{2-}$, either by direct exchange with cytoplasmic ATP, by the action of ATP:creatine phospho-transferase or by the action of ATP:AMP phosphotransferase.

The process is repeated so long as an excess of Ca^{++} ions saturates troponin C and myosin cross-bridges link with the myosin-binding sites on actin at successively peripheral locations as the interdigitation continues.

Huxley (1971) has pointed out that interdigitation of the actin and myosin fila-ments may involve tilting movements of the cross-bridges between the S-1 subunit of myosin and the actin – if one can assume that some structural changes alter the angle of the bridging material and that some system of forces keeps the filament separation approximately constant over short axial distances. Elliot (1968) has pos-tulated that it is the disturbance of dipole–dipole and van der Waals forces which determines the distances between actin and myosin filaments during contraction.

In order to explain how the myosin cross-bridges are able to link with the myosin-binding sites on actin, Davies (1963) postulated that $MgATP^{2-}$ was bound to the H-meromyosin cross-bridge by a polypeptide chain. Its helices were extended at rest by mutual repulsion generated between the negative charge on $MgATP^{2-}$ at one end and a net negative charge at the other, where the polypeptide joined the H-meromyosin. On stimulation, Ca^{++} ions annulled the negative charge on $MgATP^{2-}$ and this eliminated the repulsive effect on the coils of the polypeptide, causing it to assume the α-helical configuration – by the energy of formation of about 46 hydro-gen bonds – and, through the link with actin, pulling the latter inwards by an amount equivalent to the distance between successive myosin-binding sites on actin. On the $MgADP^{2-}$ being recharged to $MgATP^{2-}$, the polypeptide was re-extended, but now to a position opposite to the next distal myosin binding site on actin.

When the stimulus to contract ceases, the concentration of Ca^{++} ions in the sar-coplasm is restored to rest level ($\sim 0.10\,\mu M$), being reabsorbed into the sarcotubu-lar system by the sarcoplasmic reticulum pump which depends upon ATP for the necessary energy. The sarcoplasmic reticulum has itself some ATP-ase activity (Engel, 1963). ATP is also needed to restore the differential distribution of sodium between the two surfaces of the sarcolemma which provides the action potential on nerve stimulation. This probably requires only about one-thousandth, and the calcium pump one-tenth, of the energy required in contraction *per se* (Bendall, 1969).

Being no longer saturated with Ca^{++} ions, troponin C and troponin I return to their resting configurations whereby the latter prevents interaction of myosin and actin (Schaub *et al.*, 1972). It is feasible that the gap filaments (Carben *et al.*, 1965; Locker and Leet, 1975), which have some elasticity, assist in pulling the actin filaments outwards from their interdigitation with those of myosin so that the sarcomeres' resting length is re-established. It was also suggested by Rowe (1986) that the fine elastin fibres which he found to parallel the criss-cross weave of epimysial and perimysial collagen fibres in muscle could store energy during either elongation or contraction and thus assist in restoring the muscle to rest length during relaxation.

Apart from the major involvement of an elevated sarcoplasmic concentration of Ca^{++} ions in the interaction of actin and myosin during muscular contraction such also activates myosin light chain kinase (which phosphorylates one of the light chain components of myosin) and phosphorylase *b* kinase (which phosphorylates troponins T and I) (cf. § 4.1.1). Moreover, Schaub and Perry (1969) have shown that purified troponin requires the addition of tropomyosin for its inhibitory action on actomyosin ATP-ase. It may be presumed that these subsidiary proteins are important in the response of the contractile system (Perry *et al.*, 1975). It has been demonstrated that muscular contraction changes the angles of the lattice of the perimysial connective tissue and the crimp length of the collagen fibres (Rowe, 1974). Collagen, therefore, may have a more positive role in contraction than has been supposed hitherto. Most other aspects of muscle contraction concern the mechanism of ensuring an adequate supply of ATP. In this both the soluble and insoluble proteins of the sarcoplasm play an essential role (Needham, 1960). The most immediate source of new ATP is resynthesis from ADP and creatine phosphate (CP), by the enzyme creatine kinase, which is one of the soluble proteins of the sarcoplasm:

$$ADP + CP \rightleftarrows creatine + ATP.$$

The technique of nuclear magnetic resonance, being non-destructive, permits biochemical changes to be followed in intact tissues (Gadian, 1980). The resonance pattern of ^{31}P has proved particularly useful. It has confirmed, for surviving muscle, that the level of CP quickly falls in tetanic contraction and under anaerobic conditions postmortem, that the initial ATP level is sustained until CP is depleted and that (in tetanus) the force developed is proportional to the rate of ATP hydrolysis (Dawson *et al.*, 1978). The technique also confirms that the initial intramuscular pH is *ca.* 7.1 (Gadian, 1980). But *in vivo* the major source of ATP is its resynthesis from ADP by respiration, whereby muscle glycogen (or in some cases fatty acids) is oxidized to carbon dioxide and water. When energy is needed in excess of the power of the respiratory system to generate ATP, the process of anaerobic glycolysis, whereby glycogen is converted to lactic acid, can do so – although much less efficiently. The mechanism of ATP resynthesis by respiration or by anaerobic glycolysis is complicated: a highly simplified outline is given in Fig. 4.3 (after Baldwin, 1967). It will be seen that much more ATP can be resynthesized by respiration than by anaerobic glycolysis. Most of the enzymes needed to convert glycogen to lactic acid (and many other substances) equilibrate between free solution in the sarcoplasm and attachment to structural elements of the cell such as the F-actin filaments (cf. § 4.1.1). The sarcoplasm, includes the muscle pigment, myoglobin, which is quite

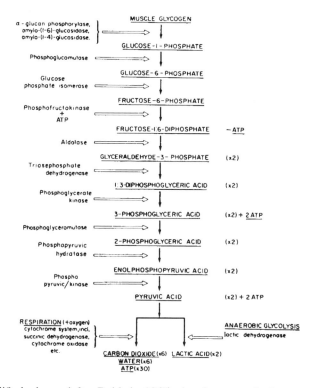

Fig. 4.3 Simplified scheme (after Baldwin, 1967), showing stages in the conversion of muscle glycogen to carbon dioxide and water, by respiration in the presence of oxygen, and to lactic acid under anaerobic conditions. The yields of ATP are indicated. Substrates in capitals, enzymes in lower-case.

distinct from the haemoglobin of the blood (Theorell, 1932). It exists in two molecular species in horse (Boardman and Adair, 1956) and five species in the seal (Rumen, 1959); and it appears to act as a short-term oxygen store in muscle (Millikan, 1939). The oxygen-utilizing enzymes of respiration, in particular the cytochrome system, and those required to convert pyruvic acid to carbon dioxide and water and to form ATP, are located in insoluble particles, the mitochondria. These are distributed in the sarcoplasm (Cleland and Slater, 1953; Chappel and Perry, 1953). Most of the glycolytic and respiratory enzymes require co-factors which are either vitamins or trace metals; and the composition of muscle reflects this requirement. Non-contractile myosin ATP-ase is responsible for the small degree of contractility necessary to maintain tone of resting muscle and body temperature: it also causes the depletion of ATP, and hence rigor mortis, after death (Bendall, 1973).

4.2.2 Post-mortem glycolysis
Since it reflects the basic function of muscle, it is appropriate to consider at this point the irreversible anaerobic glycolysis which occurs when oxygen is permanently removed from the muscle at death, although the more general consequences of circulatory failure will be outlined in Chapter 5. The sequence of chemical steps by which glycogen is converted to lactic acid is essentially the same post-mortem as *in*

vivo when the oxygen supply may become temporarily inadequate for the provision of energy in the muscle; but it proceeds further. Except when inanition or exercise immediately pre-slaughter has appreciably diminished the reserves of glycogen in muscle, the conversion of glycogen to lactic acid will continue until a pH is reached when the enzymes effecting the breakdown become inactivated. In typical mammalian muscles this pH is about 5.4–5.5 (Bate-Smith, 1948; Ramsbottom and Strandine, 1948). An initial level of *ca.* 600 mg glycogen/100 g muscle is required to attain this pH; but muscles which have an ultimate pH of 5.4–5.5 after post mortem glycolysis may still contain as much as 1800 mg residual glycogen/100 g muscle, as in bovine *l. dorsi* (Howard and Lawrie, 1956). There is some evidence that residual glycogen of this order increases the water-holding capacity and tenderness of the muscles when cooked (Immonen *et al.*, 2000).

It is generally considered that there will be no residual glycogen if the pH fails to fall to 5.4–5.5 during post-mortem glycolysis; but certain atypical muscles (as in the horse) may have as much as 1000 mg/100 g when the ultimate pH is above 6.0 (Lawrie, 1955).

Findings by Lawrie (1955) and Lawrie *et al.* (1959) indicated that glycogen exists in more than one form in muscles, distinguished by their structure and/or by their relative accessibility or susceptibility to metabolic change.

It is now believed that glycogenesis in muscle is initiated by a protein, glycogenin, the molecule of which combines autocatalytically with eight glucose molecules to form a glycosyl-protein. Using the latter as primer, the enzyme proglycogen synthase forms proglycogen; and macroglycogen synthase and branching enzyme convert the proglycogen into larger aggregates of macroglycogen. The major factor limiting glycogen storage in muscle is thus probably the availability of glycogenin (Alonso *et al.*, 1995). There is evidence that macroglycogen is preferentially metabolized during aerobic stress, whereas proglycogen is utilized during anaerobic stress and during postmortem glycolysis (Rosenvold and Andersen, 2003).

The higher stores of glycogen found in the muscles of pigs carrying the RN$^-$ gene are due, primarily, to their content of macroglycogen (Rosenvold and Andersen, 2003).

^{13}C and proton-NMR studies have confirmed that lactic acid production is virtually the only event causing the pH fall during post-mortem glycolysis (Lundberg *et al.*, 1986).

The final pH attained, whether through lack of glycogen, inactivation of the glycolytic enzymes or because the glycogen is insensitive (or inaccessible) to attack, is referred to as the ultimate pH (Callow, 1937). Because it is generally about 5.5, which is the iso-electric point of many muscle proteins, including those of the myofibrils, the water-holding capacity is lower than *in vivo*, even if there is no denaturation. This is reflected in the low electrical impedance of the muscle at normal ultimate pH, and its high electrical resistance at high ultimate pH (Callow, 1936), and this has been related to the mechanical resistance of the muscle fibres (Lepetit *et al.*, 2002). Both the *rate* and the *extent* of the post-mortem pH fall are influenced by intrinsic factors such as species, the type of muscle and variability between animals; and by extrinsic factors such as the administration of drugs preslaughter and the environmental temperature. The effect of species in a given muscle and at a given temperature is illustrated in Fig. 4.4a and the type of muscle in Fig. 4.4b. Variation in time taken for the pH of the *l. dorsi* of different pigs from 6.5 to an ultimate of 5.5 at 37 °C is considerable (Fig. 4.5); and such differences have been

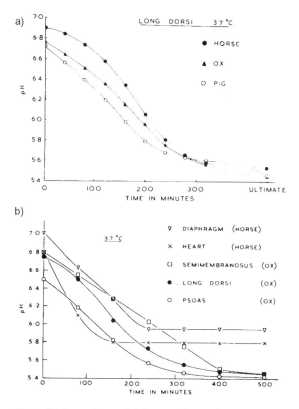

Fig. 4.4   The effect of (a) species and (b) type of muscle on the rate of post-mortem pH fall at 37 °C. Zero time is 1h post-mortem.

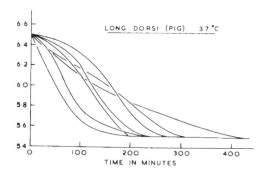

Fig. 4.5   Variability in the rate of post-mortem pH fall in *l. dorsi* muscles at 37 °C between individual pigs.

observed between specific muscles of different beef animals at constant temperature (Howard and Lawrie, 1957a; Bendall, 1978 a, b). The intravenous administration of relaxing doses of magnesium sulphate before slaughter will slow the subsequent rate of post-mortem glycolysis; injection of calcium salts (Howard and Lawrie, 1956) and of adrenaline and noradrenaline (Bendall and Lawrie, 1962) will accelerate the rate. Insulin shock (Hoet and Marks, 1926) and the injection of

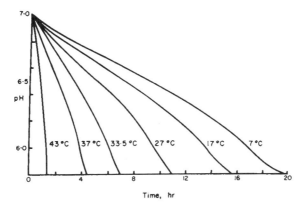

Fig. 4.6   The effect of environmental temperature on the rate of post-mortem pH fall in beef
l. dorsi (Marsh, 1954). (Courtesy Prof. B. B. Marsh.)

adrenaline (subcutaneously: Cori and Cori, 1928), tuberculin (Howard and Lawrie, 1957a) and tremorine (Bendall and Lawrie, 1962) will produce a high ultimate pH through depleting glycogen reserves, but by a different mechanism in each case.

The rate of post-mortem glycolysis increases with increasing external temperature above ambient (Bate-Smith and Bendall, 1949; Marsh, 1954 – cf. Fig. 4.6). Contrary to expectation, however, the rate of post-mortem glycolysis also increases as the temperature at which it occurs falls from about 5° to 0°C (Newbold and Scopes, 1967). Indeed Smith (1929) found that the rate was even greater at –3°C than at 0°C, i.e. as the system was freezing.

It will be obvious that, in the carcasses of meat animals, various muscles will have different rates of fall of temperature postmortem, according to their proximity to the exterior and their insulation. As a result, the rates of post-mortem glycolysis will tend to be higher in muscles which are slow to cool; and vice versa. Some of the observed differences in the rates of post-mortem glycolysis in beef muscles *in situ* are certainly due to this factor (Bendall, 1978 a, b). It has been suggested, however, that it is the temperature of the muscle after excision, rather than that in the carcass beforehand, which mainly determines the rate of post-mortem glycolysis (Monin *et al.*, 1995).

It has been suggested that the exceedingly fast rate of pH fall in the musculature of pigs affected by PSE (§ 3.4.3), in which the pH may have fallen to about 5.4 in 40 min (Ludvigsen, 1954; Briskey and Wismer-Pedersen, 1961; Bendall *et al.*, 1963), may reflect the development of an abnormally high temperature post-mortem (Bendall and Wismer-Pedersen, 1962) or immediately pre-slaughter (Sayre *et al.*, 1963b). Elevated levels of Ca^{++} ions are found in the muscles of such pigs post-mortem (Cheah *et al.*, 1984). These would activate the ATP-ase of actomyosin and hence accelerate the rate of post-mortem glycolysis. On the other hand, there is no doubt that the rate of pH fall is abnormally fast even at normal temperature in some pigs and may proceed to an exceptionally low ultimate pH, e.g. 4.7 (Lawrie *et al.*, 1958). It is known that the complement of the enzymes which effect glycolysis varies between individual animals. As indicated above (§ 3.4.1), various genetically determined glycogen storage diseases are known (Cori, 1957) in which certain enzymes in the muscle are deficient. Since it seems feasible that the so-called PSE condition is heritable, a genetically controlled excess or imbalance of glycolytic enzymes could well be implicated, as well as morphological changes in the protein complexes (LTC and JFP), which control Ca^{++} release (cf. §§ 2.2 and 3.4.2). In the musculature of affected pigs post-mortem, certain sarcoplasmic proteins denature because of the high temperature–low pH combination (Scopes, 1964), in particular the enzyme creatine phosphokinase (Scopes and Lawrie, 1963). In its relative absence, there would be an excess of ADP and inorganic phosphate in the muscle and this could accelerate post-mortem glycolysis. It was shown that electrical stimulation of beef carcasses immediately post-slaughter (Harsham and Deatherage, 1951) and excising pig muscle (Hallund and Bendall, 1965) enhanced ATP-ase activity, causing a faster than normal rate of pH fall during post-mortem glycolysis (cf. § 7.1.1.2). Pressures of the order of 150 MPa also greatly accelerate post-mortem glycolysis (Horgan and Kuypers, 1983), an effect which is probably due to their action on the Ca^{++}-activated ATP-ase of the sarcoplasmic reticulum, whereby Ca^{++} ions are released from the latter and stimulate myofibrillar ATP-ase.

4.2.3   Onset of rigor mortis

As post-mortem glycolysis proceeds the muscle becomes inextensible: this is the stiffening long referred to as rigor mortis. Its chemical significance has only recently been appreciated. Erdös (1943) showed that the onset of rigor mortis was correlated with the disappearance of ATP from the muscle: in the absence of ATP, actin and myosin combine to form rigid chains of actomyosin. The observations of Erdös were confirmed and greatly extended by Bate-Smith and Bendall (1947, 1949) and by Bendall (1951), who wrote a comprehensive review of the subject (1973). The loss of extensibility which reflects actomyosin formation proceeds slowly at first (the delay period), then with great rapidity (the fast phase): extensibility then remains constant at a low level. The time to the onset of the fast phase of rigor mortis (at a given temperature) depends most directly on the level of ATP which, in the immediate post-mortem period, is being slowly lowered by the surviving noncontractile ATP-ase activity of myosin (Bendall, 1973). Under local control the latter operates in an attempt to maintain body heat and the structural integrity of the muscle cell. The level of ATP can be maintained for some time by a resynthesis from ADP and creatine phosphate (CP). When the store of CP is used up, post-mortem glycolysis can resynthesize ATP; but only ineffectively (as has already been pointed out) and the overall level falls. It will be clear that this will happen sooner if there is little glycogen; but even with abundant glycogen the resynthesis of ATP by glycolysis cannot maintain it at a level sufficiently high to prevent actomyosin formation. Struggling at death will lower the initial pH and shorten the time until the fast phase, as will depletion of glycogen by other means (starvation, insulin tetany). On the other hand, an excess of oxygen, by stimulating respiration, will delay the onset of rigor mortis. Indeed thin portions of muscles (*ca.* 3 mm thick), if exposed to oxygen during post-mortem glycolysis, can produce ATP with such efficiency that not only is rigor mortis delayed but CP is resynthesized to above its *in vivo* level (Lawrie, 1950: unpublished). The pH of the muscle also tends to rise.

   With a knowledge of the temperature, the initial store of glycogen and the initial levels of ATP and CP, the time to onset of rigor mortis can be predicted accurately (Bendall, 1951). The initial pH alone gives a good general approximation (Marsh, 1954). The onset of rigor mortis is accompanied by lowering in water-holding

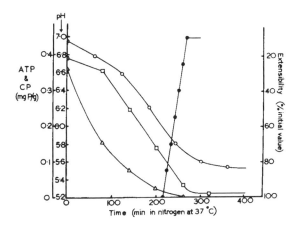

Fig. 4.7   The onset of rigor mortis in *l. dorsi* muscle of the horse. Changes in ATP, CP and extensibility as a function of time in nitrogen at 37 °C. Zero time: 1h post-mortem. Adenosine triphosphate (ATP) and creatine phosphate (CP) determined in trichloroacetic acid extracts made from muscles at various time intervals. ATP-P and CP-P expressed as mg P/g wet weight of muscle. Extensibility expressed as a percentage of initial value (Lawrie, 1953c). □–□, ATP; △–△, CP; ○–○, pH; ●–●, extensibility.

capacity. It is important to emphasize that this is not due solely to the drop in pH (and the consequent approach of the muscle proteins to their iso-electric point) or to denaturation of the sarcoplasmic proteins (Table 5.3, p. 144). Marsh (1952a) showed that, even when rigor mortis occurred at a high pH, there was a loss of water-holding capacity, due to the disappearance of ATP and to the consequent formation of actomyosin. Whilst the Marsh–Bendall factor (sarcoplasmic reticulum pump) is still operating in post-mortem muscle, readdition of ATP, at relatively high concentration, will cause swelling of muscle homogenates and restore the water-holding capacity towards its *in vivo* level. The plasticizing effect of ATP is thus associated with increased water-holding capacity. NMR studies have shown that there are at least two phases or environments for water in muscle, in each of which some of the water is 'bound' and the majority 'free'. During the onset of mortis the amount of 'bound' water in each environment remains constant, but there is progressive movement of 'free' water from one environment to the other – presumably reflecting the cross-linking of actomyosin (Pearson *et al.*, 1974).

An idea of the typical relative time relations of these changes is given in Fig. 4.7 for the *l. dorsi* muscle of the horse (at 37 °C). Once rigor mortis is complete, the muscle remains rigid and inextensible, provided it is kept free from microbial contamination: there is no 'resolution' of rigor (Marsh, 1954).

The patterns of rigor mortis onset can be classified (Bendall, 1960):

$$\begin{array}{cccc} \text{Pi} & \text{Pi} + \text{HN}_3 & \text{Pi} & \text{Ribose} \\ \uparrow & \uparrow & \uparrow & \uparrow \end{array}$$
$$\text{ATP} \longrightarrow \text{ADP} \longrightarrow \text{IMP} \longrightarrow \text{Inosine} \longrightarrow \text{Hypoxanthine}$$

(i)   Acid rigor:* characterized in immobilized animals by a long delay period and a short fast phase and in struggling animals by drastic curtailment of the delay period. At body temperature stiffening is accompanied by shortening.

(ii)   Alkaline rigor:* characterized by a rapid onset of stiffening and by marked shortening, even at room temperature.

(iii)   Intermediate type: characterized in starved animals by a curtailment of the delay period but not of the rapid phase: there is some shortening.

Shortening in rigor mortis involves only a fraction of the muscle fibres, is irreversible and is thus distinguished from physiological contraction (Bendall, 1960). Honikel *et al.* (1983) reported that some shortening occurs in all muscles (which are free to shorten) during post-mortem glycolysis at temperatures between -1 and $38\,^{\circ}C$. There is a minimum at $15–20\,^{\circ}C$: 'rigor shortening' increases at temperatures above $20\,^{\circ}C$ and 'cold-shortening' becomes more pronounced at temperatures below $10–15\,^{\circ}C$, if the pH is still above *ca.* 6.2 (cf. §§ 7.1.1.2 and 10.3.3.1). The former occurs just before marked loss of extensibility in rigor mortis, the latter relatively early. Insofar as ATP is required to enable the sarcoplasmic reticulum to pump Ca^{++} ions from the sarcoplasm, the fall in ATP during the onset of rigor mortis can be expected to foster leakage of Ca^{++} ions back into the sarcoplasm, whereby the contractile system is stimulated. This event probably explains the limited shortening which is observed during 'normal' rigor (Jeacocke, 1984), as well as the marked contraction in 'cold-shortening' when the ability of the sarcoplasmic reticulum pump to recapture Ca^{++} is greatly diminished by the low temperature (cf. §§ 7.1.1.2 and 10.3.3.1).

Using model systems and fluorescent-labelled proteins Swartz *et al.* (1993) endeavoured to elucidate the relationship between sarcomere shortening in rigor mortis and toughening in meat. Their studies led them to postulate that the toughness associated with very short sarcomeres is due to interactions between myosin filaments and not to actin–myosin bonds. They also suggested that, as the degree of saturation of the S-1 heads of myosin with ATP decreases post-mortem, the heads bind to the actin filaments and increase their affinity (via troponin C) for Ca^{++} ions, whereby sarcomere shortening can occur at a much lower Ca^{++} ion level than that *in vivo*.

The characteristics of rigor mortis, e.g. the levels of ATP and CP initially and at onset; the pH value initially, at onset and ultimately; the initial and residual stores of glycogen; the activities of ATP-ases and of the sarcoplasmic reticulum pump, will vary according to intrinsic factors, such as species and type of muscle, and extrinsic factors, such as the degree of struggling and temperature (cf. Tables 4.8 and 4.31).

Abnormal types of rigor mortis are known. Thus, if muscle is frozen whilst the ATP level is at the pre-rigor value, an exceedingly fast rate of ATP breakdown and of rigor onset ensues on thawing ('thaw rigor'). The muscle may contract to 50 per cent of its initial length and exude much fluid or drip (Chambers and Hale, 1932). It has been suggested that the contractile actomyosin ATPase, which is not normally responsible for ATP depletion post-mortem, is activated on freezing and thawing (Bendall, 1973). This reflects the inability of the sarcoplasmic reticulum pump to reabsorb calcium ions effectively in these circumstances (cf. §§ 4.2.1 and 10.3.3.1).

In contrast, a much delayed onset of rigor mortis has been observed in the muscles of the whale (Marsh, 1952b). The ATP level and the pH may remain at their high *in vivo* values for as much as 24h at $37\,^{\circ}C$. No adequate explanation of this

---

* Whereas exhausting exercise preslaughter will deplete glycogen reserves and lead to a high ultimate pH, struggling or *mild* exercise *immediately* before death will promote early and accelerated post mortem glycolysis and the production of lactic acid giving a low initial and ultimate pH.

phenomenon has yet been given; but the low basal metabolic rate of whale muscle (Benedict, 1958), in combination with its high content of oxymyoglobin *in vivo* (cf. § 4.3.1), may permit aerobic metabolism to continue slowly for some time after the death of the animal, whereby ATP levels can be maintained sufficiently to delay the union of actin and myosin in rigor mortis.

Besides the chemical changes which directly affect actomyosin formation, and thereby stiffening in rigor mortis, there are other concomitant reactions. Of these the most striking are in the pattern of nucleotides (Bendall and Davey, 1957). After ATP has been broken down to inorganic phosphate (Pi) and ADP, the latter is further dephosphorylated and deaminated to produce inosine monophosphate (IMP); and IMP is dephosphorylated to produce inosine. Ribose is then split from inosine, producing hypoxanthine (see above).

Some ADP is deaminated only, forming inosine diphosphate (Webster, 1953); and enzymes exist in muscle to resynthesize nucleotides, IMP being condensed with aspartic acid to form adenylsuccinic acid in the first instance (Davey, 1961). Ammonia liberation during rigor mortis can be related, therefore, to the onset of stiffening. Electrophoresis shows that the pre-rigor pattern consists of small amounts of ADP and relatively large amounts of ATP; post-rigor there is a large quantity of IMP, and traces of inosine, hypoxanthine, ADP, ATP, IDP and ITP (Bendall and Davey, 1957).

During and after post-mortem glycolysis some glycogen may be broken down by α-amylase to glucose and hexose-1,6-diphosphate instead of to lactic acid (Sharp, 1958; cf. Table 4.9).

4.3   Factors reflected in specialized muscle function and constitution

It is possible to classify muscles broadly as 'red' or 'white' (Needham, 1926). This superficial differentiation reflects both histological and biochemical differences. So-called 'red' muscles tend to have a greater proportion of narrow, myoglobin-rich fibres; so-called 'white' muscles have a greater proportion of broad, myoglobin-poor fibres (Denny-Brown, 1929; Hammond 1932a; George and Scaria, 1958). Differentiation of fibres into two types can be demonstrated histochemically by about half-way through the gestation period (Dubowitz, 1966). Gauthier (1969) gives a reminder that 'redness' and 'whiteness' do not invariably correspond to slow and fast contraction rates, respectively. The terms are useful, however. 'Red' fibres are rich in mitochondria (with abundant cristae) and have wide Z-lines. 'White' fibres are poor in mitochondria (and these have few cristae) and the width of the Z-lines is about half that in 'red' fibres. A (histologically) intermediate type of fibre is also found. The latter resemble 'red' fibres, but have larger diameter and a narrow Z-line (cf. § 4.3.5 below). It has been suggested that the narrow diameter of 'red' fibres reflects their preferential utilization of incoming protein precursors for energy-yielding purposes rather than for construction (Burleigh, 1974). Red muscles tend to operate over long periods without rest; in these it is found that the mitochondria, respiratory enzymes and myoglobin are present in greater quantity (Paul and Sperling, 1952; Lawrie, 1952a). 'White' muscles tend to operate in short fast bursts with frequent periods of rest and restitution: respiratory enzymes and myoglobin are present in relatively small amounts only; but lactic dehydrogenase activity is high (Dubowitz and Pearse, 1961). The concentration of the dipeptides, carnosine and anserine, which act as buffers in muscle (Bate-Smith, 1938), appears to bear an inverse proportionality to respiratory activity (Davey, 1960). Carnitine (β-hydroxy-γ-trimethylamino butyrate) is found in small quantities in all cells but especially in muscular tissue. Its concentration is associated with that of myoglobin both within and between species, where its function is to buffer excess production of coenzyme A in oxidative metabolism (Shimada et al., 2004).

It is now evident that the predominant factor which determines whether a muscle will develop a large proportion of glycolytic fibres and become fast-acting ('white') or become slow-acting ('red') is the nature of its nerve supply (Romanul and Van der Meulen, 1967). As early as 1948, Bach showed that, if the tendon of the 'red' *soleus* muscle of the rabbit was transplanted to the tendon of the pale posterior tibial muscle, the *soleus* lost myoglobin and became a fast-contracting 'white' muscle. Dhoot *et al.* (1981) subsequently showed that, 15–20 weeks after such surgery 90 per cent of the fibres of the previously 'red' *soleus* contained so-called fast forms of troponins C, T and I and 90 per cent of the previously 'white' *tibialis* muscle contained so-called slow forms of these regulatory proteins (cf. § 4.1.1). A detailed account of the trophic action of nerves on the nature of muscle fibres has been given by Swatland (1994).

Differences between muscles, however, are considerably more complex than their classification as 'red' or 'white' would signify. Both the dynamic (or biochemical) and static (or chemical) aspects of their constitution are elaborate. Their variability reflects the influence of a large number of *intrinsic* factors related to function. The most important of these are (i) species, (ii) breed, (iii) sex, (iv) age, (v) anatomical location of muscle, (vi) training or exercise, (vii) plane of nutrition and (viii) inter-animal variability, the nature of which is little understood. In addition, various *extrinsic* factors modify the behaviour of muscle in the immediate post-mortem period and during storage and processing; and its composition as meat. These are (i) food, (ii) fatigue, (iii) fear, (iv) pre-slaughter manipulation, and (v) environmental conditions at slaughter, in the immediate post-mortem period and during subsequent storage; but these are more appropriately considered in Chapters 5 and 7.

4.3.1 Species

Species is perhaps the most easily appreciated factor affecting the composition of muscle; but its effect is conditioned by the simultaneous operation of many of the intrinsic and extrinsic factors already mentioned. In comparisons between species it is thus desirable to choose definite values for these other variables.

The total nitrogen content (fat-free), which is referred to as the 'nitrogen factor', of the raw lean meat of the entire carcass is an important index for the analyst in assessing the meat content of foods. It differs to a numerically small but commercially significant extent between species. Moreover, because the meat industry reflects changes in consumer demand and social conditions, the types of pigs, cattle and sheep slaughtered vary in respect of such parameters as age, weight, sex and level of fatness (cf. Fig. 1.3). Thus, the nitrogen factor for pork has ranged from 3.6 (Anon., 1940) to 3.45 (Anon., 1961) and 3.50 (Anon., 1991). For beef the nitrogen factor has ranged from 3.4 (Anon., 1952) to 3.55 (Anon., 1963) and 3.65 (Anon., 1993a). The nitrogen factor for the lean meat of the entire carcass of the sheep is 3.50 (Anon., 1955, 1996).

Table 4.3 Chemical composition of *l. dorsi* muscle from mature meat animals

Characteristic	Rabbit[a]	Sheep[bc]	Species pig[d]	Ox[e]	Whale (Blue)[ef]
Water (% fat-free)	77.0	77.0	76.7	76.8	77.7
Intramuscular fat (%)	(2.0)	(7.9)	2.9	3.4	2.4
Intramuscular fat (iodine no.)	–	54	57	57	119
Total nitrogen (% fat-free)	3.4	3.6	3.7	3.6	3.6
Total soluble phosphorus (%)	0.20	0.18	0.20	0.18	0.20
Myoglobin (%)	0.02	0.25	0.06	0.50	0.91
Methylamines, trimethylamine oxide, etc.	–	–	–	–	0.01–0.02

[a] Bendall (1962). [b]Callow (1958). [c]Lawrie (1952a). [d]Lawrie *et al.* (1963a). [e]Lawrie (1961). [f]Sharp and Marsh (1953) and personal communication.

Some aspects of the chemical composition of the *l. dorsi* muscles from mature meat animals are compiled in Table 4.3. Although it is obvious that the contents of water, of total nitrogen and of total soluble phosphorus are similar in all five species, there are marked differences in the other characteristics. The fat of the *l. dorsi* of the Blue Whale has a very much higher iodine number than that of the other four. To some extent this is due to the krill on which whales subsist. Another peculiarity of whale meat is its high content of the buffers carnosine, anserine (Davey, 1960) and balenine*. This fact, together with the relatively high oxygen store held by the myoglobin, helps to explain the diving capabilities of this mammal. Its musculature is constrained to operate anaerobically for prolonged periods. Indeed, in the Sperm Whale and seal, in which the dives may be especially lengthy, the percentage of myoglobin may reach 5–8 per cent of the wet weight of the muscle (Sharp and Marsh, 1953; Robinson, 1939). Such muscle is almost black in appearance. The low myoglobin content of rabbit and pig muscle accords with the superficial paleness of the flesh of these animals. The myoglobin of pig muscle also differs qualitatively from that of ox muscle. The results obtained using freshly cut surfaces of *l. dorsi* suggest that the rates of oxygenation of myoglobin are fastest in pork, intermediate in lamb and slowest in beef (Haas and Bratzler, 1965). It had been supposed that balenine (as its name indicates) was exclusive to the whale (Cocks *et al.*, 1964); but Rangeley and Lawrie (1976) found β-alanyl–3-methyl-histidine in the muscles of mature pigs; and, subsequently, traces have been detected in the muscles of cattle, sheep and chicken (Carnegie *et al.* 1984).

The identification of the species of meat present in food products is important in ensuring the interests of consumers, and a variety of methods have been employed in analysing uncooked samples. Their relative efficacy has been compared by Winterø *et al.* (1990), who found that the sensitivity of detection of pork in mixtures with beef was 0.4 per cent by countercurrent immunoelectrophoresis, 0.5 per cent

* These are the dipeptides β-alanyl-histidine, β-alanylmethyl-histidine and β-alanyl-3-methyl-histidine, respectively.

by DNA hybridization, 1.0 per cent by immunodiffusion and 5.0 per cent by iso-electric focusing. Methods based on polymerase chain reactions, in which DNA fragments may be as small as 70 base pairs, are now increasingly used for species identification (Brodmann and Moor, 2003).

Another approach is by the use of 'direct probe' mass spectrometry. This is similar to pyrolysis mass spectrometry but the samples are volatilized at less than 300 °C. A plot of ions having mass : charge ratios of 36 against those of 74 can differentiate horse, beef, pork and lamb (Patterson, 1984). A major continuing problem with such methods is their inability to quantify the species present in processed foods. Nevertheless, King (1984) was able to differentiate species in meat heated up to 120 °C by isoelectric focusing of adenylate kinase. It is certain that further development of the DNA hybridization technique will prove effective in solving this important problem.

Species differences between the myoglobins of meat animals have elucidated the structure of these and of other proteins, and supported the view of Barlow *et al.* (1986) that all antigenic determinants involve some measure of three-dimensional topography. Thus, Levieux and Levieux (1996), using six monoclonal antibodies raised against bovine myoglobin, identified three topographical determinants involving amino acid residues which, although separated from one another sequentially on the protein chain, were closely associated through the folding of the myoglobin's eight alpha helices. In so far as four of these monoclonal antibodies react more strongly with oxymyoglobin than with metmyoglobin, Levieux and Levieux (1996) concluded they could elucidate the conformational changes in proteins during their modification by physical and chemical events.

Taylor *et al.* (1993a) have shown the potential of electro-spray mass spectroscopy for differentiating the haem pigments of sheep, cattle, pigs and horses, although the distinctive features were altered somewhat by autoclaving.

The ratios of the histidine peptides appear to be characteristic of species (Table 4.4) and have been used to identify them (Olsman and Slump, 1981; Concepcion-Aristoy and Toldra, 2004).

Species differences between the myosins of the *l. dorsi* muscles of ox, pig, sheep and horse were shown immunologically (Furminger, 1964) but the degree of cross-reaction was too high to permit clear identification of unknown meats. The tropomyosins of ox, sheep, pig and rabbit have different electrophoretic characteristics (Parsons *et al.*, 1969). This is true of the myofibrillar group generally (Champion *et al.*, 1970). Enzyme-linked immunosorbent assay (ELISA) procedures are most effective in identifying the species present in products containing unheated

Table 4.4　Typical ratios of histidine dipeptides in muscles from different species (after Carnegie *et al.*, 1984)

Species	Balenine:Anserine	Carnosine:Anserine
Horse	0	93
Pig	1.0	21
Ox	0.03	6
Sheep	0.02	1
Chicken	0.01	0.5
Kangaroo	0	0.1

meat, but quantitative assessment is limited, due to the variability in the levels of residual blood (Hitchcock and Crimes, 1985). Quantitative identification of species on the basis of the electrophoretic pattern of their protein is still possible after these have been heated to 120 °C for a 6 min (Mattey *et al.*, 1970).

Insofar as meat processing operations denature proteins, however, methods of identification, based on the much more robust DNA of the genes, have been developed in recent years. Restriction fragment length polymorphism (RFLP; now referred to as single nucleotide polymorphism, SNP), using a cDNA probe complementary to the sequence of bases in the α-actin multigene family, was employed by Fairbrother *et al.* (1998) to distinguish between beef, lamb, pork, horse, chicken and fish. The patterns obtained were the same whether the DNA was derived from fresh meat or from that heated to 120 °C. Although the sequence of bases in the exons of the α-actin nuclear genes is strongly conserved, between species, that in the introns of these genes is variable; and Lockley and Bardsley (2002), using primers based on the exons and introns of the α-actin nuclear genes, successfully distinguished closely related species. Not surprisingly, it is more difficult to distinguish between breeds within a species, but DNA extracted from meat and hybridized with probes recognizing species-specific satellite segments (Hunt *et al.*, 1997) or mitochondrial D-loop DNA (Partis *et al.*, 2000), have been effective. Verkaar *et al.* (2002) employed the RFLP technique, following amplification of the DNA by polymerase chain reactions, with probes complementary to both the mitochondrial cytochrome β gene and the centromeric satellite DNA, to distinguish such exotic breeds as benteng and zebu from one another and from taurine cattle.

Precise identification of DNA sources by DNA 'fingerprinting' has become possible from the work of Jeffreys *et al.* (1985). When an unknown sample of DNA is isolated, cleaved by a specific restriction endonuclease and hybridized with a probe (i.e. a single strand of DNA or RNA), a series of DNA fragments of varying length is produced (RFLP). These are separated (e.g. by gel electrophoresis) and visualized by colour development or autoradiography. The pattern revealed thereby appears to be unique to each individual. The probe is generated from a 'core' of repetitive in-line ('tandem') sequences of DNA. These tandem repeat sequences occur at numerous locations on the genome and are referred to as 'microsatellites' (Kirby, 1992). The oligonucleotide probes can be synthesized artificially or produced by the action of reverse trancriptase on RNA.

Wismer-Pedersen (1969a) has given data on the proximate composition of the canned meat from *l. dorsi* of a number of African ungulates. Of those studied the meat from wart-hog had the lowest water content (68.9 per cent) and the highest content of fat (4 per cent), whereas meat from the eland had the highest water content (75.2 per cent) and the lowest content of fat (1.9 per cent). The triglycerides of red deer have relatively high contents of myristic acid compared with those in cattle: those of moose contain more than 40 per cent of stearic acid (Garton *et al.*, 1971).

It might have been supposed from the data on *l. dorsi* in Table 4.3 that the intramuscular fat of pigs and cattle always had a comparable iodine number. The danger of such a generalization can be seen, however, from Table 4.5, in which data for the *psoas major* of boars and bulls are given.

In this case, despite an almost identical content of intramuscular fat in both species, that in the boar has a much higher iodine number. Generally the iodine number of the fat in the pig is higher than that of the ruminant. In Table 4.6, where

Table 4.5 Chemical composition of *psoas major* muscles from boars and bulls

Species	Breed	Sex	Age (months)	No.	Moisture (% fat-free)	Total intramuscular nitrogen (% fat-free)	fat (%)	Iodine number
Pig	Large White	Boar	5	5	77.1	3.5	1.6	71.4
Ox	Ayrshire	Bull	10	5	77.4	3.3	1.7	47.0

Table 4.6 Typical fatty acid composition of (a) fats and (b) muscular tissue from beef, lamb and pork sirloin steaks (after Enser *et al.*, 1996)

Fatty acid	Structure	Fatty acid (as % total fatty acids) Beef	Lamb	Pork
(a) Fats				
Palmitic	C16:0	26.1	21.9	23.9
Stearic	C18:0	12.2	22.6	12.8
Oleic	C18:1 (n-9)	35.3	28.7	35.8
Linoleic	C18:2 (n-6)	1.1	1.3	14.3
α-Linolenic	C18:3 (n-3)	0.48	0.97	1.43
Arachidonic	C20:4 (n-6)			
Eicosapentaenoic	C20:5 (n-3)	not detected		0.36
Dodecosahexanoic	C22:6 (n-3)			
(b) Muscular tissue				
Palmitic	C16:0	25.0	22.2	23.2
Stearic	C18:0	13.4	18.1	12.2
Oleic	C18:1 (n-9)	36.1	32.5	32.8
Linoleic	C18:2 (n-6)	2.4	2.7	14.2
α-Linolenic	C18:3 (n-3)	0.70	1.37	0.95
Arachidonic	C20:4 (n-6)	0.63	0.64	2.21
Eicosapentaenoic	C20:5 (n-3)	0.28	0.45	0.31
Dodecosahexanoic	C22:6 (n-3)	0.05	0.15	0.39

the fatty acid composition of the fats from cattle, sheep and pigs are compared, this difference is seen to be largely due to the higher content of linoleic acid in pig fat; and a higher content of polyunsaturated fatty acids.

Dietary fat has relatively little influence on the depot fat of all species of ruminant since ingested fatty acids are hydrogenated by rumen microorganisms (which may also effect a change in the length of the fatty acid chains). Although ruminant muscle thus has normally a low ratio of polyunsaturated to saturated fatty acids, it does contain various C_{20} and C_{22} polyunsaturated fatty acids of the n-6 and n-3 series which are valuable nutritionally to human consumers. Enser *et al.* (1998) found that the percentages of all n-3 polyunsaturated fatty acids were higher in the muscles of steers which had been fed on grass than in those of bulls which had been fed concentrates; whereas the content of the n-6 polyunsaturated fatty acids was higher in the latter.

The potential benefits for human health of conjugated linoleic acid (CLA), especially the isomers cis-9, trans-10 linoleic acid and trans-10, cis-12 linoleic acid have been investigated in recent years. Hydrogenation by rumen microorganisms converts ingested linoleic to stearic acid and produces CLA as intermediates. Supplementation of ruminant diets with fat sources and in C18:2, n-6 leads to an increase in the production of cis-9, trans-10 linoleic acid in the flesh (Mir et al., 2003; Noci et al., 2005). Nevertheless, there are seasonal alterations in the degree of unsaturation of the depot fats of ruminants, probably because of the ingestion of octadecatrienoic acid during pasture feeding; and there are distinct differences due to breed (Callow and Searle, 1956). Cattle regulate the degree of unsaturation of their fat by interchange between stearic and hexadecenoic acids whereas pigs do so by an exchange between stearic and oleic acids. As opposed to cattle, pigs and other non-ruminants such as the horse tend to deposit unchanged dietary fat. Indeed, pigs which have been fed large quantities of whale oil may have rancid fat in vivo (J. K. Walley, personal communication). With pigs the ratio of n-3 to n-6 polyunsaturated fatty acids is readily amenable to dietary manipulation; and, indeed, this has a greater influence on the ratio than genetic factors (Cameron et al., 2000 a, b). The nutritional value of pork, as represented by the ratio, can be increased by feeding linseed at a level which has no adverse effect on eating quality (Sheard et al., 2000).

Rabbits, eating the same grass as horses, lay down much more linoleic acid (Shorland, 1953). The fats of rabbit flesh are comparatively richer in palmitic, linoleic and myristic acids, and poorer in stearic acid, than the meat of other domestic species (Cambero et al., 1991). In Table 4.6, more recent data on the fatty acids present in the muscles of cattle, sheep and pigs are shown. It is evident, that as with the fats of these species, the muscles of pigs contain a markedly higher content of linoleic acid than those of cattle and sheep and more arachidonic acid. The muscles of cattle contain more hexadecenoic acid than those of the other two species and the muscles of sheep have a relatively high content of stearic acid.

Apart from static chemical differences due to species, in a given muscle, differences of a more dynamic kind are found. Thus, for example, the activity of cytochrome oxidase varies considerably in the psoas major muscles of different species (Table 4.7). Cytochrome oxidase is the enzyme principally for linking oxygen with the chain of electron carriers by which most substances in muscle are eventually oxidized in providing energy.

It is therefore not surprising to find that the activity of this enzyme should be particularly high in the muscles of the horse, which are obviously powerful, and low in those of the rabbit; and that the staying power of the hare and the relatively easy exhaustion of the rabbit are reflected in the capacity of their psoas muscles to gain energy by oxidation (Table 4.7). Another interspecies factor affects the issue, however, in the opposite sense, insofar as there is an inverse relationship between body size and basal metabolic rate (Benedict, 1958). Thus, it has been shown that the oxidative capacity of the diaphragm muscle in different species – as reflected by the number of red, myoglobin-rich fibres – is inversely related to body size (Gauthier and Padykula, 1986).

Some of the species differences in enzymic activity are reflected post-mortem. Thus at a given temperature (37 °C) and in a given muscle (l. dorsi) the characteristics of the onset of rigor mortis are different in horse, ox, pig and sheep (Table 4.8, Fig. 4.4a). The mean chain lengths of glycogens from the l. dorsi muscles of horse

Table 4.7 Activity (Q_{O_2}) of cytochrome oxidase in preparations from *psoas* muscles of various mammals (Lawrie, 1953b)

Species	Activity
Horse	1600
Ox	1200
Pig	1000
Sheep	950
Hare	650
Blue Whale	600
Rabbit	250

Table 4.8 Mean data on post-mortem glycolysis and the onset of rigor mortis in the *l. dorsi* of different species

Species	Time to onset of fast phase of rigor mortis (min/37°C/N₂)	Initial[a]	ATP/P[b] pH (at onset)	CP/P[b] pH (ultimate)	(as % TSP) at onset	(as % TSP) initial
Horse[c]	238	6.95	5 97	5.51	8.3	18.9
Ox[c]	163	6.74	6.07	5.50	13.2	13.2
Pig[c]	50	6.74	6.51	5.57	21.0	7.2
Lamb[d]	60	6.95	6.54	5.60	–	–

[a] i.e. at 1 h post-mortem. [b] ATP/P, CP/P = phosphorus due to adenosine triphosphate and creatine phosphate respectively (TSP = total soluble phosphorus). [c] Lawrie (1953c). [d] Marsh and Thompson (1958). Time to onset in sheep ca. 80 min.

and ox, pig and rabbit are 17, 15 and 13 glucose residues respectively (Kjølberg *et al.*, 1963).

It will be seen that while the ultimate pH is the same in all four species, the initial pH levels are high in horse and lamb and relatively low in ox and pig and that the pH at the onset of the fast phase of rigor mortis is low in horse and ox and high in lamb and pig. The level of ATP/P at the onset of rigor mortis is particularly high in the *l. dorsi* of the pig. In the *psoas major* of the rabbit at 37 °C, the fast phase of the onset of rigor mortis begins when the ATP/P has fallen to about 15 per cent soluble phosphorus (Bendall, 1951). It is also interesting to observe that the reserve of creatine phosphate – which, as mentioned in Chapter 3, is the most immediate mechanism for resynthesis of ATP – is highest in the horse *l. dorsi* and lowest in that of the pig, apparently reflecting their relative capacity for energy production. The ratio of the proteolytic enzyme, calpain II, to the inhibitory protein, calpastatin, is considerably higher in the *l. dorsi* of the pig than in this muscle in lamb or cattle (Ouali and Talmant, 1990). This difference is reflected by the fact that the tenderizing changes of post-mortem conditioning proceed fastest in pork, intermediate in lamb and slowest in beef.

Another post-mortem difference in enzymic constitution between the corresponding muscles of different species is shown by the activity of α-amylase. This

Table 4.9 Post-mortem activity of α-amylase in muscles of different species

Species	Rate glucose accumulation (mg/h/g at 20 °C)
Sheep	0.08
Horse and ox	0.04
Rabbit	0.50
Pig	0.90

Table 4.10 Myoglobin concentrations in muscles of different breeds of horse (Lawrie, 1950)

Breed	(% wet weight)	
	L. dorsi	*Psoas*
Draught horse	0.46	0.82
Thoroughbred	0.77	0.88

enzyme converts muscle glycogen to glucose and competes with the system whereby glycogen is converted to lactic acid. Comparative data are given in Table 4.9 from which can be seen the high rate of glucose accumulation in the muscles of pigs and rabbits in comparison with sheep and oxen (Sharp, 1958).

4.3.2 Breed

After species, breed exerts the most general intrinsic influence on the biochemistry and constitution of muscle. A particularly striking effect of breed is found in the horse. The percentage of myoglobin in the *l. dorsi* of thoroughbreds, which arch their backs strongly in running, is considerably higher than in this same muscle of the draught horse in which the *l. dorsi* is moved relatively little. On the other hand, the *psoas* muscles show no such difference (Table 4.10).

Among cattle, there are differences between those breeds which are primarily used for milk production and those which are more suitable for meat. Thus, the percentage of intramuscular fat in the *l. dorsi* muscle at the level of the 4th, 5th and 6th lumbar vertebrae tends to be markedly greater in beef cattle than in dairytype animals after the age of 18 months (Callow, 1947; Lawrie, 1961). The percentage of intramuscular fat in specially fattened beef animals (e.g. West Highland show cattle) may reach 17 per cent. Mean data for *l. dorsi* and *psoas* muscles from Herefords (a beef breed) and Friesians (predominantly a dairy breed) are given in Table 4.11. Calpastatin activity in muscles (cf. § 5.4.2) is genetically determined in different breeds of cattle (Shackelford *et al.*, 1994), the calpastatin gene correlating with toughness (Bishop *et al.*, 1993). *Bos indicus* cattle have a greater relative content of calpastatin (cf. § 4.3.5.1) in their muscles than do *Bos taurus* breeds. The consequently greater inhibition of calpain action in the former probably contributes to their toughness as meat (Shackelford *et al.*, 1991).

Table 4.11 Intramuscular fat and its iodine number in muscles of cattle (from Callow, 1962)

Breed	L. dorsi Intramuscular fat (%)	I.N.	Psoas Intramuscular fat (%)	I.N.
Hereford	7.1	54.6	5.6	50.5
Friesian	6.4	56.4	5.1	53.9

Table 4.12 Breed differences in fatty acid composition of neutral lipids and phospholipids of porcine *l. dorsi* muscles (after Wood *et al.*, 2003)

	Berkshire	Duroc	Large White	Tamworth
Neutral lipid fatty acids				
wt. (as % of muscle)	2.29	1.98	0.94	0.91
linoleic*	6.78	10.14	10.49	8.09
linolenic*	0.58	0.87	0.78	0.67
Phospholipid fatty acids				
wt. (as % of muscle)	0.45	0.45	0.39	0.40
linoleic*	29.10	30.10	29.35	31.40
linolenic*	0.70	0.67	0.61	0.75

* As % neutral or phospholipid content.

In sheep the percentage of intramuscular fat is considerably greater among improved breeds such as Hampshire and Suffolk than in semi-wild types, such as Soay and Shetland (Hammond, 1932a). In a study of the back and kidney fats of pigs, it was found that introduction of the Hampshire breed led to an increase in the number of pigs producing abnormally soft and highly unsaturated fats (Lea *et al.*, 1969). They concluded that, when the cause of softness was not ingested fat, there was a strong correlation between the ratio of the monoene to stearic and palmitic acids, and the melting point.

By using halothane-negative British Landrace pigs and those of the resistant Duroc breed, Cameron and Enser (1991) were able to assess the effect of breed on the intramuscular fat of *l. dorsi* muscles without confounding the issue by stress susceptibility. The intramuscular fat of Duroc pigs had a higher content of saturated and monounsaturated fatty acids, and a lower concentration of polyunsaturated fatty acids, than the British Landrace; and there was a higher level of fat in the former. From an investigation of the back fat from eleven breeds of pig, Warriss *et al.* (1990b) found that the characteristics of the fat (e.g. firmness) were largely determined by the level of fatness rather than by inherent breed factors. In a subsequent study of the fatty acid characteristics of porcine *l. dorsi*, Wood *et al.* (2003), found highly significant differences between breeds in respect of the contents of linoleic and linolenic acids in both neutral lipids and phospholipids (cf. Table 4.12).

Because of interest in pale exudative pig musculature (§§ 3.4.3 and 5.4.1), the effect of breed on muscle composition in this species has been investigated. At all locations from the 5th thoracic to the 6th lumbar vertebrae, the *l. dorsi* muscle of pigs of the Large White breed has more myoglobin and a higher ultimate pH than the corresponding muscle from pigs of Landrace breed (Lawrie and Gatherum, 1962). Again, for a given percentage of intramuscular fat in the *l. dorsi* muscle (lumbar), its iodine number and the area of cross-section of the muscle as a whole are greater in pigs of the Welsh breed than in Landrace, and in the latter than in Large White animals (Anon., 1962; Lawrie and Gatherum, 1964).

As determined by the analysis of *l. dorsi* muscles, the meat of the Piétrain has less water than that of other pig breeds, whereas *l. dorsi* of the Hampshire breed has less nitrogen (Sellier, 1988).

The rate of post-mortem glycolysis at 37 °C in the *l. dorsi* muscles of Large White pigs in the United Kingdom is only one-third to one-half that in the Danish Landrace (Bendall *et al.*, 1963). According to Bendall (1966), however, such breed differences disappear when pigs are curarized (or injected with myanesin) before slaughter. In these circumstances the rate of post-mortem glycolysis in the muscles of Danish Landrace pigs is as slow as that in those of Large White pigs in the United Kingdom. He suggested that electrical stunning can produce, in the pig only, a long-term stimulation of post-mortem glycolysis which would thus lead to that combination of low pH and high temperature responsible for the PSE condition (Hallund and Bendall, 1965). Were this the only reason for the superficial manifestations of the condition, however, there would remain unexplained the great variability in susceptibility shown by individual pigs. Moreover, Lister (1959) found that curare had no effect in diminishing the tendency for a swift production of lactic acid in stress-susceptible Poland China and Chester White pigs. Sayre *et al.* (1963a) in the USA found that the external chain length of glycogen from the *l. dorsi* muscle of Chester White pigs decreases to a greater extent – but at a slower rate during post-mortem glycolysis than that in pigs of the Hampshire and Poland China breeds; and the same workers (1963b) found that, when subjected to a temperature of 45 °C for 29–60 min preslaughter, the *l. dorsi* muscles of the latter two breeds became pale and exudative during subsequent post-mortem glycolysis, whereas those of the Chester White pigs remained firm and dark. The activities of phosphorylase and phosphofructokinase were markedly higher in *l. dorsi* muscles of Hampshire pigs than in those of Poland China or Chester White breeds (Sayre *et al.*, 1963c).

4.3.3 Sex

In general, males have less intramuscular fat than females, whereas the castrated members of each sex have more intramuscular fat than the corresponding sexually entire animals (Hammond, 1932a; Wierbicki *et al.*, 1956). Table 4.13 gives comparative data on the *l. dorsi* from a 12-month-old steer and bull of Ayrshire–Red Poll breed. It will be seen that with the exception of the percentage of intramuscular fat (which is relatively high) and the moisture (which is correspondingly low) in the steer in comparison with the bull, there are no striking differences in composition.

It will be noted, however, that despite the difference in the percentage of intramuscular fat their iodine numbers are the same; and it is found that, in the lumbar region of *l. dorsi*, the iodine number of intramuscular fat from bulls is markedly less than that from steers at a given level of fatness (Lawrie, 1961). In comparing

Table 4.13 Comparative composition of *l. dorsi* (lumbar) of Ayrshire–red poll steer and bull

	Steer	Bull
Intramuscular fat (%)	3.03	1.00
Intramuscular fat (I.N.)	51.39	51.06
Moisture (%)	74.09	77.30
Myoglobin (%)	0.20	0.19
Ash (%)	0.99	1.16
Total nitrogen (%)	3.61	3.50
Non-protein nitrogen (%)	0.38	0.41
Sarcoplasmic nitrogen (%)	0.94	1.08
Stroma nitrogen (%)*	0.36	0.28
Myofibrillar nitrogen (%)	1.93	1.81

* i.e. from insoluble protein, mainly originating from connective tissue.

Table 4.14 Comparative composition of *l. dorsi* muscles from control and hexoestrol implanted steers (Lawrie 1960a)

	(Mean of 6)	
	Control	Implanted
Intramuscular fat (%)	3.37	2.42
Intramuscular fat (I.N.)	57.45	59.31
Moisture (%)	74.12	74.88
Ash (%)	1.01	1.02
Total nitrogen (%)	3.51	3.53
Non-protein nitrogen (%)	0.43	0.44
Sarcoplasmic nitrogen (%)	0.88	0.89
Myofibrillar nitrogen 1%)	1.83	1.77
Stroma nitrogen (%)	0.36	0.44

castrated members of each sex, it has been found that the depot fats of steers have a more saturated fat than those of heifers (Terrell *et al.*, 1969). Fats from the latter have a higher percentage of oleic acid. No differences have been noted in the total cholesterol content of the depot fats between steers and heifers. Such sex differences can be further exemplified by comparing the composition of the *l. dorsi* muscle of control steers with those of the same age (36–40 months) and breed (Friesian) subjected to the masculizing effect of hexoestrol implants (Table 4.14), there being a somewhat lower content of intramuscular fat in the treated animals, but no other difference.

In pig, also, the major differences in muscle composition due to sex are found in the contents of intramuscular fat. The *l. dorsi* (lumbar) muscles of hogs (i.e. castrated males) contain about 30 per cent more intramuscular fat than those in gilts (i.e. mature females before pregnancy) of the same age (Lawrie and Gatherum, 1964), and this is reflected by its lower content of unsaturated fatty acids. The concentration of polyunsaturated fatty acids in the intramuscular lipids of *l. dorsi* is significantly higher in boars than in gilts (Cameron and Enser, 1991).

As assessed from a representative sample of UK animals, entire pigs were found to have a greater concentration of haem pigment (myoglobin plus haemoglobin) in their *l. dorsi* muscles than castrates (Warriss *et al.*, 1990a). Because the meat of mature boars can be associated with an unpleasant odour ('boar taint': cf. §§ 2.5.2.1 and 10.4.4) it is important to be able to detect the presence of meat from male pigs in various products. Meer and Eddinger (1996) employed a polymerase chain reaction on a male-specific fragment from the Y-chromosome for this purpose.

4.3.4 Age

Irrespective of species, breed or sex the composition of muscles varies with increasing animal age, there being a general increase in most parameters other than water, although the rates of increment are by no means identical in all muscles. Moreover, different components reach adult values at different times (Lawrie, 1961). Thus, in bovine *l. dorsi* those nitrogen fractions representing myofibrillar and sarcoplasmic proteins have reached 70–80 per cent of their mean adult value by birth, and their subsequent rates of increase become asymptotic at about 5 months of age. Non-protein nitrogen, however, does not attain its characteristic adult value until about 12 months of age; and the concentration of myoglobin increases rapidly until about 24 months of age. Intramuscular fat appears to increase and moisture content on a whole tissue basis to decrease, up to and beyond 40 months of age. Before birth, the moisture content of muscle is very high, being over 90 per cent for a considerable portion of the gestation period (Needham, 1931). On a fat-free basis, however, the moisture content remains fairly constant after 24 months of age. Some idea of the difference in composition of a given muscle in the bovine at two different ages is given in Table 4.15. The great increase in intramuscular fat and in myoglobin content, the lesser increase in total and sarcoplasmic nitrogen and the decrease in moisture and in stroma with age are evident. These trends are also evident in pig *l. dorsi* (Table 4.16) in which the composition is compared at 5, 6 and 7 months of age (Lawrie *et al.*, 1963b). Much of the increased saturation of intramuscular lipid in heavier pigs is due to an increase in the ratio of C_{18} to $C_{18.1}$ fatty acids in the neutral lipid fraction (Allen *et al.*, 1967a). It is interesting to note that odd-numbered fatty acids (C_{11}, C_{13}, C_{15}, C_{17}) are quite prevalent in the phospholipid fraction of porcine

Table 4.15 Comparative composition of *l. dorsi* muscle in calf and steer

	12-day-old calf	3-year-old steer
Intramuscular fat (%)	0.55	3.69
Intramuscular fat (I.N.)	82.41	56.50
Moisture (%)	77.96	74.11
Myoglobin (%)	0.07	0.46
Total nitrogen (%)	3.30	3.52
Non-protein nitrogen (%)	0.36	0.39
Sarcoplasmic nitrogen (%)	0.62	0.87
Myofibrillar nitrogen (%)	1.52	1.61
Stroma nitrogen (%)	0.80	0.65

Table 4.16 Comparative composition of *l. dorsi* muscles
from pigs at three ages

	(Mean of 10)		
	5 months	6 months	7 months
Intramuscular fat (%)	2.85	3.28	3.96
Intramuscular fat (I.N.)	57.4	55.8	55.5
Moisture (%)	76.72	76.37	75.90
Myoglobin (%)	0.030	0.038	0.044
Total nitrogen (%)	3.74	3.74	3.87

Table 4.17 Comparative rates of increase of oxygen-linked
factors in *psoas* muscles of immature and adult horses. The
rates of increase are expressed as percentages per year of
the average adult values for these factors

Age (years)	Myoglobin concentration	Cytochrome oxidase activity	Capacity for energy-rich phosphate resynthesis
0–2	42.8	37.9	43.8
2–12	2.1	0.6	1.0

intramuscular fat. In cattle the iodine number of the intramuscular fat falls very markedly with increasing age (Lawrie, 1961). In respect of the lipids of bovine *fatty* tissues, however, whilst the iodine number has been shown to decrease with increasing animal age and level of fatness (Callow and Searle, 1956), the ratio of linoleic to stearic acids, and the softness of the fat, increases. It is feasible that this apparent contradiction may reflect age-related changes in branched-chain fatty acids. In this species, one of the important factors involved is the development of the rumen microflora, which hydrogenate dietary fats. Calves, initially, tend to deposit dietary fat relatively unchanged (Hoflund *et al.*, 1956).

Myoglobin concentration appears to increase in a two-phase manner, an initial swift rate of increment being followed by one which is more gradual. The fast phase lasts about 1, 2 and 3 years in pigs, horses and cattle respectively. Reflecting the increases in myoglobin with age, there is a concomitant two-phase increment in the activity of the enzymes which govern respiration and, thereby, in energy production potential (Table 4.17; Lawrie, 1953a, b).

The activity of the respiratory enzymes increases in a not dissimilar fashion in various muscles. Typical values at birth and in the adult are given in Table 4.18 for horse muscles.

These activities are based on the rate of oxygen uptake (Q_o) of mitochondrial membrane preparations from the muscles concerned and are thus relative. They are proportional, however, to the absolute values for the muscles themselves.

It will be noted that the cytochrome oxidase activity of the heart has attained 33 per cent of its final value by birth, whereas that in *l. dorsi* is then only 8 per cent of

Table 4.18 Effect of animal age on enzymic activity in preparations from horse muscles

Muscle	Birth	Adult
(a) Cytochrome oxidase (Q_{o_2})		
Heart	1000	2700
Diaphragm	200	1700
Psoas	200	1600
L. dorsi	70	900
(b) Succinic oxidase (Q_{o_2})		
Heart	100	510
Diaphragm	20	260
Psoas	20	240
L. dorsi	10	130

Table 4.19 Collagen and elastin content of *l. dorsi* muscles in cattle (Wilson *et al.*, 1954)

Cattle	Collagen (%)	Elastin (%)
Calves	0.67	0.23
Steers	0.42	0.12
Old cows	0.41	0.10

the adult value: values for diaphragm and *psoas* are intermediate. This order of difference reflects the fact that the heart has been contracting powerfully for a considerable time during uterine life, whereas the other muscles have not: the latter develop preferentially in relation to the animal's need for increasing power to move about. A similar impression is given by the data on succinic oxidase activity. It is of interest that, in foetal and early post-natal skeletal muscle, all cells can synthesize the fast (≈anaerobic) forms of skeletal troponins C, T and I. Certain cells, in addition, have the capacity to synthesize the slow (≈aerobic) forms of these proteins (Dhoot and Perry, 1980). The stimulus for the activation of the genes controlling synthesis of the slow forms, and the suppression of those responsible for the fast forms, could be the changing pattern of innervation with development; but other factors must also be involved.

The connective tissue content of muscle is greater in young animals than in older ones (Bate-Smith, 1948; Wilson *et al.*, 1954). The concentrations of both collagen and elastin diminish with increasing animal age (Table 4.19). As the tenderness of veal and the toughness of older animals testifies, however, the *nature* of the connective tissue must be different at different ages.

There is a higher concentration of 'salt-soluble' collagen (a precursor of insoluble collagen) in young – or actively growing – muscle (Gross, 1958). The degree of intra- and inter-molecular cross-linking (cf. § 3.2.1) between the polypeptide chains in collagen increases with increasing animal age (Carmichael and Lawrie, 1967a, b; Bailey, 1968). The investigations of Bailey and his colleagues have provided detailed information on the age-related changes in the collagen of tendon, muscle and the other tissues. Whilst, in young animals, most of the cross-links are reducible, heat- and acid- labile and increase up to 2 years of age, thereafter they are gradually replaced by linkages which are thermally stable (Shimokomaki *et al.*, 1972; Bailey and Light, 1989; and cf. § 3.2.1). The change with maturation is greatest in the epimyseal connective tissue, which in young animals, consists mainly of thermally labile cross-links, whereas the endomysium already consists of thermally stable cross-links (Bailey and Light, 1989).

Even in mature animals collagen is known to have a significant turnover, and the rate differs according to the nature of the collagen. Thus, type III precursors turn over at a slower rate than those of type I (Bailey and Light, 1989). A high rate of collagen turnover, as in rapidly growing pigs, can have detrimental effects on the strength of the collagen laid down and can lead to lameness in the live pig, to a separation of fat from lean in the fresh meat and to 'lacy' bacon in the cured product.

4.3.5   Anatomical location
4.3.5.1   Muscles
The most complex of the intrinsic differentiating factors is anatomical. As we have already mentioned, muscles may be broadly classified as 'red' (slow-twitch) or 'white' (fast-twitch) according to whether they carry out sustained action or operate in short bursts. But the variation in shape, size, composition and function of the 300 muscles in the mammalian body obviously reflects a diversity of activity and development the details of which are still largely unknown.

In assessing the significance for meat quality of differences in the constitution of anatomically defined muscles, it should be appreciated that, in the past, both wholesale and retail cuts of beef were relatively large and thus represented aggregates of a number of specific muscles (or portions thereof). With the increasing tendency for the centralized preparation and prepackaging of portions of meat for the individual consumer, these may be derived from a single muscle or part thereof. With pork, the individual joints are frequently processed (e.g. cured) after separation from the carcass. Their specific composition is thus of greater interest, and nitrogen factors for the different joints have been published (Table 4.20: Anon., 1986).

Ramsbottom and Strandine (1948) analysed fifty muscles of adult beef animals. Moisture and fat contents ranged from 62.5 and 18.1 per cent respectively in the intercostal muscles to 76 and 1.5 per cent respectively in *extensor carpi radials*. The ultimate pH ranged from 5.4 in *semimembranosus* to 6.0 in *sternocephalicus*. Comparative data on chemical parameters of various beef and pork muscles are given in Table 4.21 (Lawrie *et al.*, 1963b, 1964). Particularly striking is the high iodine

Table 4.20   Mean nitrogen content (percentage fat-free) for (i) lean and (ii) lean, with associated fat and rind, in pork joints

		Leg	Rump loin	Joint rump belly	Rib loin	Rib belly	Collar	Hand
(i)	Lean	3.44	3.62	3.43	3.63	3.47	3.33	3.36
(ii)	Lean with fat & rind	3.60	3.81	3.67	3.83	3.72	3.50	3.59

Table 4.21   Some chemical parameters of various muscles

Muscle	Moisture (%)	Intramuscular fat (%)	Intramuscular fat (I.N.)	Total nitrogen (% fat free)	Hydroxyproline (µg/g)
(i) Beef					
L. dorsi (lumbar)	76.51	0.56	54.2	3.54	520
L. dorsi (thoracic)	77.10	0.90	56.6	3.47	610
Psoas major	77.34	1.46	52.9	3.30	350
Rectus femoris	78.07	1.49	67.8	3.40	550
Triceps (lateral head)	77.23	0.73	62.2	3.45	1000
Superficial digital flexor	78.67	0.40	81.4	3.27	1430
Sartorius	77.95	0.58	64.0	3.33	870
Extensor carpi radialis	74.83	0.60	68.1	3.29	1160
(ii) Pork					
L. dorsi (lumbar)	76.33	3.36	56.3	3.77	670
L. dorsi (thoracic)	76.94	3.26	55 5	3.69	527
Psoas major	77.98	1.66	62.8	3.58	426
Rectus femoris	78.46	0.99	71.5	3.41	795
Triceps (lateral head)	78.68	1.84	67.0	3.46	1680
Superficial digital flexor	78.87	1.90	65.3	3.35	1890
Sartorius	78.71	0.87	–	3.41	850
Extensor carpi radialis	79.04	1.39	69.7	3.36	2470

number and the low nitrogen content in *digital flexor superficialis* and its relatively high concentration of hydroxyproline (as an indicator of connective tissue). Using hydroxyproline as criterion, the lowest content of connective tissue may be seen to be in *psoas major* (350 µg/g); in *extensor carpi radialis,* however, it can be about 2500 µg/g.

In an assessment of commercial joints, Casey *et al.* (1985) found that the collagen content of the beef forequarter was significantly greater than that of the hindquarter, the shin having the highest value (4.8 per cent, fat-free basis). In pork, collagen was significantly highest in the hand (i.e. the lower forequarter). Bendall (1967) measured the elastin content of various beef muscles using a technique whereby the last traces of collagen and myofibrillar proteins were removed. In most of the muscles from the choice cuts of the hind quarter and loin the elastin content was generally less than 5 per cent of the total connective tissue. In *semitendinosus,* however, it constituted 40 per cent of the total. Of the muscle of the forequarter only *l. dorsi* had a comparable elastin content. In a detailed study of these two muscles, Rowe (1986) demonstrated that, while the general organization of the collagen and elastin fibres was similar in both (cf. § 3.2.1), the content of elastin fibres was markedly greater in *semitendinosus*. Bendall (1967) had concluded that elastin contributed to the toughness of cooked meat to about the same extent as denatured collagen and Bailey and Light (1989) suggested that the significance of elastin had been overlooked.

In a series of studies over 10 years, Bailey and his co-workers (Bailey *et al.*, 1979; Light and Bailey, 1983; Light *et al.*, 1985; Bailey and Light, 1989) demonstrated that muscles differ not only in their total content of connective tissue, but also in respect of the types of collagen molecule present (cf. § 3.2.1), in the ratio of heat-stable (oxo-imino) to heat-labile (aldimine) cross-links in these collagens, and in their histological distribution (Tables 4.22–24). Thus, Table 4.24 shows that the collagen of bovine *sternomandibularis* contains a greater proportion of heat-stable cross-links than *l. dorsi* at all locations – epimysium, perimysium and endomysium.

The collagen content and collagen fibre diameter of both perimysia and endomysia, and the ratio of heat-stable to heat-labile cross-links in epimysial, perimysial and endomysial connective tissue are each correlated positively with toughness, but the perimysium appears to be the most implicated in accounting for textural differences between muscles (Light *et al.*, 1985). Such findings have been confirmed by subsequent investigators. Thus, Torrescano *et al.* (2003), in a study of raw bovine muscles, found a strong correlation between toughness as assessed by Warner–Bratzler shear force and both total and insoluble collagen (Table 4.25). There was only a weak correlation with sarcomere length.

Table 4.22   Total collagen and its histological distribution in bovine muscles (after Bailey and Light, 1989)

Muscle	Total collagen (% dry wt.)	Percentage in epimysium	Percentage in perimysium	Percentage in endomysium
Psoas major	2.24	15	79	24
L. dorsi	2.76	13	80	34
Pectoralis profundus	4.96	22	98	42
Semitendinosus	4.75	29	54	41

Table 4.23   Connective tissue in various bovine muscles (after Light and Bailey, 1983; Bailey and Light, 1989)

Muscle	Texture score	Total collagen in perimysium (% dry wt)	Type III collagen in perimysium in endomysium (as % total collagen)		Elastin in perimysium (as % total collagen)
Psoas major	1	6.4	43.2	55.2	3.8
L. dorsi	2	6.5	24.5	53.3	2.9
Semitendinosus	3	6.7	41.7	54.4	37.0
Sternomandibularis	5	16.1	40.5	58.1	2.5

Table 4.24   Heat-labile (aldimine) and heat-stable (oxo-imine) cross-links in collagen regimes of bovine muscles (after Bailey and Light, 1989)

Muscle	Epimysium		Perimysium		Endomysium	
	aldimine (%)	oxo-imine (%)	aldimine (%)	oxo-imine (%)	aldimine (%)	oxo-imine (%)
L. dorsi	50	50	42	58	27	73
Sternomandibularis	21	79	28	72	13	87

Table 4.25   Shear force and insoluble collagen in bovine muscles (after Torrescano *et al.*, 2003)

Muscle	Warner–Bratzler shear force (kg)	Insoluble collagen (mg hydroxyproline/g wet wt)
Psoas major	2.11 ± 0.39	0.18 ± 0.01
Diaphragm	2.24 ± 0.90	0.39 ± 0.10
L. thoracis	2.29 ± 0.91	0.38 ± 0.03
L. lumborum	2.35 ± 0.59	0.42 ± 0.03
Gluteus medius	3.68 ± 0.51	0.45 ± 0.04
Semimembranosus	4.10 ± 0.79	0.49 ± 0.02
Infraspinatus	4.24 ± 0.83	0.76 ± 0.08
Quadriceps femoris	4.72 ± 0.89	0.52 ± 0.06
Semitendinosus	4.79 ± 0.58	0.64 ± 0.03
Triceps branchii	5.42 ± 0.82	0.78 ± 0.02
Biceps femoris	5.46 ± 0.84	0.60 ± 0.04
Flexor digitorum	5.93 ± 0.92	0.51 ± 0.06
Sternoimandibularis	6.32 ± 0.72	0.57 ± 0.06
Pectoralis profundus	6.66 ± 0.66	0.71 ± 0.04

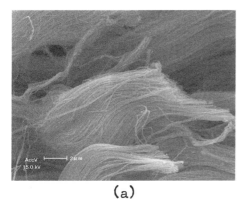
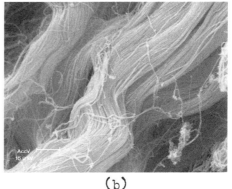

(a)   (b)

Fig. 4.8   High magnification scanning electron micrographs of the perimysia from (a) porcine *longissimus lumborum* and (b) porcine *pectoralis profundus* muscles. Bar = 2 μm. (Reproduced from *Meat Science*, Vol. 64, Y.-N. Nakamura, H. Iwamoto, Y. One, N. Shiba, S. Nishimura and S. Tabata, 'Relationships among collagen amount, distribution and architecture in the M. *longissimus thoracis* and M. *pectoralis profundus* from pigs', pp. 43–50. Copyright 2003, with permission from Elsevier Science and Prof. H. Iwamoto.)

In a study of porcine muscles by scanning electron microscopy, Nakamura *et al.* (2003) showed that the higher collagen content of *pectoralis profundus* in comparison with *longissimus lumborum* could be related to the more complex structure of the perimysium of the former, which involved longitudinal, circular and oblique collagen fibres crossing over one another in several bands; whereas the perimysium of *I. lumborum* was markedly simpler in its architecture (Fig. 4.8).

Although, within a given muscle, there is a hyperbolic relationship between the content of intramuscular fat and its iodine number (Callow and Searle, 1956) this clearly does not hold between different muscles. Certain differences between muscles in the iodine number or their fat have been attributed to corresponding differences in local temperature, a higher temperature being associated with a lower iodine number, and vice versa (Callow, 1958, 1962). There is obviously a wide range of values for these chemical parameters and this can be equally exemplified in corresponding muscles of beef and pork. In comparing those muscles which are common to both species (Table 4.21) it is seen that the muscles listed do not have the same relative composition in the two species. Thus while *extensor carpi radialis* has the lowest moisture content in beef, it has the highest value in the pig. Clearly both lumbar and thoracic *l. dorsi* muscles have the highest content of intramuscular fat in the pig; in beef, however, the percentage of intramuscular fat is relatively low in these muscles. Again, the hydroxyproline content of *digital flexor superficialis* is markedly greater in beef than in other muscles of this species. In the pig, however, the highest hydroxyproline content of this same group of muscles is found in *extensor carpi radialis*.

It is particularly interesting to note that, as judged by the hydroxyproline content, the connective tissue concentration is greater in most pig muscles than in corresponding beef muscles. Since pork is generally tender, this again emphasizes that the quality of connective tissue must be considered as a factor additional to its quantity.

Data on the ultimate pH of these muscles (not represented) show that as the percentage of moisture on a fat-free basis increases so also does the ultimate pH: there is, in fact, a significant positive correlation between these two variables in the pig. This obtains not only between muscles but within a given muscle also.

Studies of the fatty acid compositions of neutral lipids and phospholipids of different muscles have revealed a number of distinctions between them. For example, the phospholipid content and the ratio of C_{18} to C_{16} fatty acids in the total lipid are particularly high in the diaphragm (both in beef and pork) in comparison with such muscles as *l. dorsi* and *psoas* (Allen *et al.*, 1967b; Hornstein *et al.*, 1968).

In Table 4.26 comparative data are given on the fatty acid composition of intramuscular fats from nine muscles of an adult Large White × Landrace pig

Table 4.26   Relative fatty acid composition of total lipids from muscles of the adult pig

Muscle	Fatty acid (as per cent total fatty acids)									
	$C_{18:1}$	$C_{16:0}$	$C_{18:2}$	$C_{18:0}$	$C_{16:1}$	$C_{20:4}$	$C_{14:0}$	$C_{15:1}$	$C_{17:1}$	$C_{13:0}$
L. dorsi (lumbar 4–6)	40	24	11	10	5	2.5	1.5	1.5	0.8	–
L. dorsi (thoracic 13–15)	37	21	12.5	11	5	4.8	1.0	3.2	1.2	–
L. dorsi (thoracic 8–12)	34	30	11	10	3.5	4.3	1.0	2.5	0.5	–
L. dorsi (thoracic 5–7)	38	21	13.5	10	4	4.2	1.5	2.5	1.0	0.2
Semimembranosus	35	24	14	10	4.5	4.2	1.5	2.5	1.3	0.3
Rectus femoris	29	21	16	11	5	5.8	1.0	3.7	0.5	0.2
Psoas	33	22	17	12	5	3.8	1.2	2.3	1.5	–
Diaphragm	35	22	15	15	5	2.0	1.5	1.0	0.8	–
Supraspinatus	34	26	14	19	6	–	–	4.8	–	–

(Catchpole *et al.*, 1970). In respect of the major fatty acid components (oleic, palmitic, linoleic, stearic and palmitoleic) the pattern for the nine locations is not dissimilar. Nevertheless, the ratio of linoleic to oleic differs substantially between locations; and there are noteworthy distinctions in respect of arachidonic and pentadecenoic acids. The content of n-3 fatty acids is greater in red muscles than in white (Enser *et al.*, 1998). Differences in the fatty acid pattern between anatomical locations are *more* marked when muscles are fractionated into such functional components as mitochondria, and the phospholipids considered separately from the total and neutral lipids. Thus phosphatidylethanolamine concentration increases with the myoglobin content, i.e. as the metabolic type becomes more aerobic or 'red' (Leseigneur and Gandemer, 1991). Subsequently Alasnier *et al.* (1996) confirmed that the metabolic type of muscles (and of their constituent fibres) was systematically related to their lipid characteristics: white (glycolytic) muscles contained less total lipids, triglycerides and cholesterol than red muscles, and a lower content of polyunsaturated fatty acids than the latter.

Differences between muscles in other parameters are, of course, found; those in sodium, potassium and myoglobin are shown for five pig muscles in Table 4.27. In beef animals, in contrast to pigs, *extensor carpi radialis* does not have an abnormally low content of potassium. This may reflect differences in slaughtering procedures with the two species, as potassium is found at higher concentration in well-drained muscle (Pomeroy, 1971). The apparent variability in potassium content makes it difficult to accept that the radioactive emanations from K^{40} could accurately indicate the lean meat content of carcasses. Nitrogen distribution may also differ between muscles (Table 4.28), and this, together with differences in total nitrogen (Table 4.21), reflects both quantitative and qualitative differences in the proteins of different muscles. Indeed, starch gel electrophoresis reveals that the pattern of both

Table 4.27   Sodium, potassium and myoglobin in pig muscles (Lawrie and Pomeroy, 1963)

Muscle	Sodium (as % wet weight)	Potassium (as % wet weight)	Myoglobin (as % wet weight)
L. dorsi (lumbar)	0.05	0.35	0.044
Psoas major	0.05	0.37	0.082
Rectus femoris	0.05	0.38	0.086
Triceps (lateral head)	0.07	0.31	0.089
Extensor carpi radialis	0.08	0.29	0.099

Table 4.28   Distribution of nitrogen in bovine *psoas* and *l. dorsi* muscles (as per cent total nitrogen)

	Psoas	*L. dorsi*
Non-protein nitrogen	11.2	11.6
Sarcoplasmic nitrogen	21.3	26.0
Myofibrillar nitrogen	44.0	52.5
Stroma nitrogen	23.5	9.9

sarcoplasmic and myofibrillar proteins differs between muscles (cf. Table 4.32). Thus, the heart and, in the rabbit, the small red muscle deep in the proximal portion of the hind limb, differ from skeletal muscles generally in that the bands correspon- ding to the protein enzymes of the glycolytic pathway are virtually absent. More- over, such lactic dehydrogenase activity as is present is associated with a completely different protein to that in other muscles (R. K. Scopes, personal communication).

As a further reflection of intermuscular differences in proteins, it is found that the numbers of S–S bonds vary widely (from 0.5–2.0 moles/10^5 g protein), whereas the numbers of –SH bonds do not (averaging about 10.3 moles/10^5 g protein) (Hoffman and Hamm, 1978).

Quantitative differences in moisture and connective tissue (as indicated by hydroxyproline) have already been considered (Tables 4.21–24); many such exam- ples could be given. It is apparent, however, that there are also qualitative differ- ences in proteins between muscles within a given species and at a given age. Thus the proteins of beef *l. dorsi* and *psoas* muscles differ in their susceptibility to freez- ing damage. Even when the rate of post-mortem glycolysis and the ultimate pH are the same in each, the former exudes, on thawing, nearly twice as much fluid or 'drip'. Moreover, the solids content of the drip from *l. dorsi* is about 25 per cent higher than that from *psoas* (Howard *et al.*, 1960a), and the relationship of total solids to ash is quite different in the exudate from the two muscles. Differences in the sus- ceptibility to freeze-drying of the myosins prepared from porcine *psoas* and *l. dorsi* have been shown (Parsons *et al.*, 1969). While the pattern of electrophoretically separable components in both myosins is similar when prepared from fresh muscle, they differ considerably after freeze-drying, *l. dorsi* showing a greater degree of change. Tropomyosins from the two muscles, on the other hand, give similar pat- terns both before and after freezedrying. Herring (1968) has found differences in the actomyosins extracted from naturally tough and tender muscles and Dube *et al.* (1972) found that the myofibrillar proteins of bovine *l. dorsi* were less susceptible to oxidation of the sulphydryl groups (on heating to about 60 °C) than those of *psoas* muscle. Other aspects of differences in the types of protein found in various muscles are indicated by the moisture–protein ratio (Lockett *et al.*, 1962) and by the rela- tive susceptibility of pig muscles to become pale and exudative post-mortem, *l. dorsi* and *semimembranosus* being especially labile (Lawrie *et al.*, 1958; Wismer-Pedersen, 1959a). The susceptibility of myoglobin to oxidize during chill or frozen storage differs between different muscles (Ledward, 1971a; Owen and Lawrie, 1975). When the level of ultimate pH is high, the myofibrillar proteins of *l. dorsi* and *semi- membranosus* are altered less by freeze-drying than are those of *psoas* and *biceps femoris*. When the ultimate pH is normal, however, the myofibrillar proteins of *l. dorsi* are relatively the most susceptible to damage in freeze-drying (Table 4.29). The patterns obtained by starch gel electrophoresis of myofibrillar proteins from rabbit, ox and sheep indicate that, within each species, these differ between differ- ent muscles (Champion *et al.*, 1970).

Detailed studies of the myofibrillar proteins of 'red', 'white' and cardiac muscles have revealed various differences between specific components. Thus the elec- trophoretic patterns of the so-called myosin light chains (Perrie and Perry, 1970) and those of troponins (Cole and Perry, 1975) are different when these proteins are derived from the three types of muscle. This reflects differing degrees and mecha- nisms of phosphorylation of the proteins concerned (Perry *et al.*, 1975). Two sub-

Table 4.29 The percentage of myofibrillar protein from various pork muscles which is soluble in 0.92 M KCl at pH 6.0

Muscle	Ultimate pH 6.5–7.2 (frozen)	(freeze dried)	Ultimate pH 5.3–5.6 (frozen)	(freeze dried)
L. dorsi	91	85	53	41
Psoas	80	68	49	40
Biceps femoris	88	75	55	46
Semimembranosus	85	77	54	46

Table 4.30 Activity of certain enzymes in various beef muscles (after Berman, 1961)

Muscle	Lactic dehydrogenase[a]	Glutamic dehydrogenase[b]	Carbonic anhydrase[c]
L. dorsi	3.53	661	0.59
Semimembranosus	3.73	730	0.63
Serratus ventralis	1.08	960	0.63
Rectus abdominis	2.58	675	0.66
Semitendinosus	3.74	813	0.47
Trapezius	2.14	836	0.47

[a] Moles lactate oxidized per kg wet weight per 3 min.
[b] Micromoles glutamate oxidized per kg wet weight per h.
[c] Log pressure change per g wet weight per min.

units of tropomyosin, α and β, have been isolated (Cummins and Perry, 1973) and two of α-actinin (Robson *et al.*, 1970), the ratio of these subunits being different in 'red' and 'white' muscle. The myosin of 'red' muscle has a lower content of 3-methylhistidine than that of 'white' (Johnson and Perry, 1970), whereas the content in actin is apparently similar in both types. The concentration of 3-methylhistidine in the *masseter/malaris* muscles of the ruminant, which consist exclusively of type I ('red') fibres, is particularly low. The significance of this difference in using 3-methylhistidine as a robust, unequivocal index of lean meat in processed foods is considered in § 12.2.

In bovine *masseter* muscle only the slow type of myosin heavy chain is found whereas in the *cutaneous trunci*, a 'fast' muscle, only fast types of myosin heavy chain occur (Picard *et al.*, 1994). Differences between the myofibrillar proteins of different muscles are reflected in the nature of the gels they produce and which thus alter the emulsion products made from them (Young *et al.*, 1992) (cf. § 10.2.1.2).

In addition to purely chemical differences, muscles also vary in their enzymic constitution. Differences in the activity of the enzymes of the cytochrome system have already been exemplified (Table 4.18); comparative data for other specific enzyme activities are given in Table 4.30. The bovine *masseter* muscle has a particularly high capacity for respiratory metabolism and a very low titre of glycolytic enzymes (Talmant *et al.*, 1986). Thus, of 18 muscles studied by these workers, the *masseter*

Table 4.31 Characteristics of post-mortem glycolysis in horse muscles

Characteristic	L. dorsi	Psoas	Diaphragm	Heart
Time to onset fast phase rigor mortis				
(min/37 °C/N$_2$)[b]	214	173	148	50[a]
pH initial[b]	6.95	7.02	6.97	6.90[a]
at onset	5.97	6.24	6.28	6.30
ultimate	5.51	5.98	5.91	5.81
Glycogen:				
initial[b] (mg/100 g)	2249	1229	1715	584
residual (mg/100 g)	1411	606	1109	276
ATP/P at onset (as % TSP)	6	6	7	15
CP/P initial[b] (as % TSP)	19	10	3	1
Initial ~ P(γ/g)	700	450	520	100
Rate aerobic pH fall (pH/h)	0.13	0.06	0.07	0.0
Rate anaerobic pH fall (pH/h)	0.19	0.11	0.11	1.20[a]
Rel. capacity ~ P resynthesis aerobically				
(arbitrary units)	0.24	0.51	0.51	2.40
ATP-ase activity				
(i) unwashed (~ P/h/g)	2400	1800	1800	1800
(ii) washed (~ P/h/g)	9000	4800	5400	1800
Marsh-Bendall factor activity (ii)–(i)[c]	6600	3000	3600	0

[a] Because of the rate of pH fall in heart, special precautions are necessary to demonstrate the onset of rigor mortis; some data for heart are thus not strictly comparable with those from other muscles.
[b] Initial = 1 h post-mortem, except in heart.
ATP/P, CP/P = labile phosphorus of adenosine triphosphate and creatine phosphate, respectively.
TSP = total soluble phosphorus.
~ P = sum of ATP/P and CP/P.
[c] i.e. the capacity of the sarcoplasmic reticulum to pump Ca++ ions out of the sarcoplasm.

possessed the highest cytochrome oxidase activity (and the highest ultimate pH) and the *semimembranosus* the lowest; whereas the latter had the greatest buffering capacity, and the highest ATP-ase and phosphorylase activities, these features being lowest in *masseter*. The oxidative capacity of *masseter* was even greater than that of the diaphragm.

As between species (Table 4.8), so also between muscles there are broad differences in the pattern of post-mortem glycolysis and of the onset of rigor mortis. Although the nature of these differences is not yet understood in detail, they are most important in relation to the texture and appearance of meat post-mortem. Under controlled conditions muscles may differ in the times to the onset of rigor mortis, in their initial and ultimate pH, in their content of initial and residual glycogen, in the rates of pH fall aerobically and anaerobically, in their initial store of energy-rich phosphate (i.e. the labile phosphorus of ATP and CP), in their capacity for energy-rich phosphate resynthesis, and in their power to split ATP and to suppress such splitting (i.e. Marsh-Bendall factor activity*). Such differences are apparent from Table 4.31, in which data from four horse muscles are represented (Lawrie, 1952a, 1953a, b, c, 1955). Similar differences are obtained between muscles of beef animals (Howard and Lawrie, 1956, 1957a). In intact carcasses or cuts of meat, the

---

* Reflecting the relative capacity of the sarcoplasmic reticulum to remove Ca^{++} ions from the system.

environment of individual muscles clearly varies. Thus, temperature differences will affect the rates of post-mortem glycolysis; but, after correction for such circumstances, systematic intrinsic differences between muscles are still apparent (Bendall, 1978a). Post-mortem glycolysis will be considered again in Chapter 5.

The time to the onset of the fast phase of rigor mortis under anaerobic conditions at 37 °C tends to be proportionate to the initial store of energy-rich phosphorus and of glycogen. On the one hand *l. dorsi* has the characteristics of a 'white' muscle, capable of short bursts of activity, this being aided by the relatively large store of energy-rich phosphorus and a low capacity for the aerobic resynthesis of $\sim P$. On the other hand, the heart has a capacity for sustained activity, represented by its marked ability to resynthesize $\sim P$ aerobically, and a low $\sim P$ store. It is interesting to note that the power to suppress ATP-ase is least where – in the heart – there would be no lasting relaxation phase.

As a further reflection of the predominant respiratory metabolism of 'red' muscles, their store of initial glycogen tends to be less, their ultimate pH higher and their buffering capacity lower than those of 'white' muscles (Lawrie, 1952a, b, 1953c; and cf. Table 4.31). Rao and Gault (1989), in a detailed analysis of 12 beef muscles, found that the relative proportion of 'red' to 'white' fibres was correlated with their biochemical characteristics. Muscles which were composed predominantly of 'red' fibres had significantly higher ultimate pH, and lower acid buffering capacity, than those which were composed predominantly of 'white' fibres. The higher buffering capacity of the latter appeared to be due to their higher contents of inorganic phosphorus and of the dipeptide carnosine (cf. § 4.3.1). The content of various other chemical compounds differs between 'red' and 'white' muscles. Thus the 'red' (*masseter*) muscles in beef have ten times as much taurine, and three times as much coenzyme Q_{10} (*ubiquinone*), (but only a tenth of the carnosine content), as the relatively 'white' semitendinosus muscle (Purchas *et al.*, 2004a).

There are differences in glycogen metabolism between 'red' and 'white' muscles, the pathway from glycogen to glucose being more active in the former, as shown by the specific activity of residual, and trichloracetic acid-soluble, glycogens after the administration of C^{14}-glucose (Bocek *et al.*, 1966).

The activity of glucose–6-phosphate dehydrogenase, 6-phosphogluconic dehydrogenase and glycogen synthetase are also more marked in 'red' muscles. On the other hand, the enzymes in the pathway from glycogen to lactic acid are more active in 'white' muscle (e.g. glycogen-debranching enzyme* phosphorylase, phosphohexose-isomerase, phosphofructokinase, fructose–1:6-diphosphatase, aldolase, α-glycerophosphate dehydrogenase, pyruvate kinase, lactic dehydrogenase: Beatty and Bocek, 1970). There is a marked difference in the extent of binding of phosphofructokinase in rabbit 'white' muscle (95 per cent) and in rabbit 'red' muscle (45 per cent), signifying that the degree of phosphorylation of phosphofructokinase in fast 'white' fibres is more effective (Morton *et al.*, 1988). Enzymes involved in the complete oxidation of fat and carbohydrate are more prevalent in red muscles (e.g. β-hydroxy acyl CoA dehydrogenase, citrate synthetase, isocitric dehydrogenase, malic dehydrogenase, succinic dehydrogenase and cytochrome oxidase). Such

---

* The activity of glycogen-debranching enzyme diminishes – and thereby the rate of post-mortem glycolysis – by a rapid fall in temperature; but its activity is not depressed by high post-mortem temperatures (*ca.* 45 °C) and this may be a contributory factor in the development of PSE (Kylä-Puhju *et al.*, 2005).

differences are reflected in their relative post-mortem susceptibility to the benefits of incorporating antioxidants in the feed or by preslaughter injection (cf. § 7.1.1.3).

Red muscles tenderize less markedly than white muscles during conditioning and it might be supposed, therefore, that they contain a lower concentration of proteolytic enzymes (cf. § 5.4.2), including those activated by calcium ions (calcium activated sarcoplasmic factors CASF, calpains). Indeed this was reported by Goll et al. (1974). The differences between types of muscle in proteolytic capacity are complex. There are at least two types of calcium activated factors, referred to as calpains I and II (also referred to as µ- and m-calpains) which are respectively activated by micromolar or millimolar concentrations of Ca^{++} ions (Murachi et al., 1981), as well as an inhibitor, calpastatin. Whilst the *ratio* of calpain II/calpastatin is greater in white muscles than in red – an observation which accords with the greater susceptibility of the former to undergo conditioning changes – the *levels* of calpain II and of calpastatin are both greater in red muscles (Ouali and Talmant, 1990). On the other hand, the susceptibility of the myofibrillar proteins of red muscles to proteolysis is less than that of white muscles. In the latter, the post-mortem concentration of Ca^{++} ions rises more than in red – a factor favouring the action of the proteolytic enzymes (Ouali, 1992). These circumstances may dominate over the relatively low concentration of the calcium-activated proteases in white muscle and explain their greater tendency to tenderize in conditioning. Yet, in the *malaris* muscle, in which the fibres are exclusively red (Johnson et al., 1986), and which conforms in exhibiting little proteolytic post-mortem (Ouali, 1992) the calpain II/calpastatin ratio is high. The greater tenderness, during conditioning, in fast-glycolysing 'white' muscle has been attributed to proteolysis by released lysosomal cathepsins and calpain 1, uninhibited by calpastatin (O'Halloran et al., 1997), through the earlier attainment of acid pH levels. The complexity of the relationship between muscle type and the capacity for proteolysis post-mortem has been emphasized by Christiansen et al. (2004), in a study of porcine muscles. *L. dorsi* and *semimembranosus* muscles exhibited greater proteolytic capacity than *semitendinosus* (as measured by the degradation of desmin and T-troponin); and this accorded with their greater ratio of µ-calpain : calpastatin. Yet, judged by its relatively low ratio of type I ('red') to type II ('white') fibres, a high proteolytic capacity would have been anticipated in *semitendinosus*.

In vivo, greater activity of CASF in 'white' muscles, and their greater capacity to synthesize myofibrillar proteins (Iyengar and Goldspink, 1971), could suggest that the fast but intermittent contraction which they effect is more damaging to the muscles' structural integrity – and requires the removal and replenishment of its components more frequently – than the slower, continuous action of 'red' muscles.

In accordance with the lower power to suppress ATP-ase activity in 'red' muscle (i.e. a lower activity of the Marsh-Bendall factor), it is of interest that the grana of the sarcoplasmic reticulum isolated from 'red' muscle have less power to take up Ca^{++} compared with those from 'white' muscle (Gergely et al., 1965). This may help explain the greater susceptibility of red muscles to 'cold-shortening' (cf. §§ 10.3.3.1, 7.1.1.1). The vesicles of the sarcoplasmic reticulum isolated from heart muscle have a much lower capacity to accumulate Ca^{++} ions than those of either 'white' or 'red' skeletal muscle (Baskin and Deamer, 1969). The elements of the sarcoplasmic reticulum are also morphologically different in 'white', 'red' and cardiac muscles.

It is further of interest that it is in the grana of the sarcoplasmic reticulum that the adenyl cyclase system is found (Sutherland and Robinson, 1966) suggesting that the relative metabolism of 'red' and 'white' muscles may be mediated through cyclic 3'5'-AMP. Some hormones, such as catecholamines, appear to direct their control of tissues through the enzyme adenyl cyclase. This is said to be located on cell membranes and to govern the synthesis of cyclic 3'-5'-AMP from ATP.

In Table 4.31 *l. dorsi* may be regarded as a fairly typical muscle, having residual glycogen at an ultimate pH of about 5.5 (cf. §§ 4.2.2). It will be seen, however, that horse *psoas* and diaphragm have considerable residual glycogen at an ultimate pH of about 6. This phenomenon is also observed in beef *sternocephalicus* muscle. In the latter the residual glycogen has a shorter external chain length (9) than the initial glycogen (12) (Lawrie *et al.*, 1959), but this does not, in itself, appear to explain the cessation of glycolysis. It is conceivable that such qualitative differences in glycogen may represent some kind of specialization of muscle function, e.g. precision of movement, as a factor additional to the ability to contract quickly or in a sustained manner. It is thus of interest that the delicate muscles which operate the eye contain glycogen of unusual structure and that they are specialized for very fast contraction (Sartore *et al.*, 1987). Such extraocular muscles contain a specific type of 'fast' heavy myosin chain (Staton and Pette, 1990). On the other hand, variations in AMP-deaminase activity between different muscles could explain some of the characteristic differences found in ultimate pH (Scopes, 1970). Using reconstituted systems Scopes (1974) found that the *extent* of a post-mortem glycolysis depended significantly on the proportion of phosphorylase in the *a* form.

Even within a single muscle there may be systematic differences in composition and constitution. Figure 4.9 indicates how the ultimate pH and pigmentation vary along the *l. dorsi* muscle in Large White and Landrace pigs (Lawrie and Gatherum, 1962 and cf. § 4.3.5.2); and Lundström and Malmfors (1985) showed that this was reflected in corresponding variations in the light-scattering and water-holding properties of this muscle in Swedish Landrace × Yorkshire crossbred pigs. Similar findings, shewing locational differences in sensory quality (Van Oeckel and Warrants, 2003) and in water-binding capacity (Christensen, 2003) along the *longissimus dorsi et thoracis*, have been reported. Sensory analysis for tenderness across

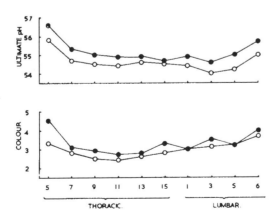

Fig. 4.9 The mean (20) ultimate pH and colour (arbitrary units) of *l. dorsi* muscles from Large White (●) and Landrace (○) pigs of similar weight, at the levels of the 5th, 7th, 9th, 13th and 15th thoracic and 1st, 3rd, 5th and 6th lumbar vertebrae. (After Lawrie and Gatherum, 1962.)

porcine *l. dorsi* revealed that there was a decrease from the caudal to the medial surface (Hansen *et al.*, 2003). On the other hand the tenderness of this muscle decreased significantly from the cranial to the caudal end. It is of interest that Rees *et al.* (2002a) found that the rate of post-mortem glycolysis was greatest at the caudal end of *l. dorsi*. In the *semimembranosus* muscles of pigs, areas only 19 cm apart may be pale and exudative and have an ultimate pH of 4.9 on the one hand or be pink and dry and have an ultimate pH of 5.6 on the other. In two-toned *semitendinosus* muscle in the pig, it has been shown that the concentrations of myoglobin, and of the cytochrome enzyme system, are higher in the darker (red) than in the lighter (white) portions; and the latter have a greater content of sarcoplasmic proteins (Beecher *et al.*, 1968). Variability within a given muscle may also be exemplified by bovine *l. dorsi*. The percentage of intramuscular fat is consistently higher in the region of the 4th, 5th and 6th lumbar vertebrae than in the region of the 8th, 9th and 10th thoracic vertebrae at all ages from birth to 36 months. But, while the iodine number of the intramuscular fat is initially lower at the former location, it eventually becomes significantly higher due to the elaboration of a greater proportion of C_{16} and C_{18} fatty acids in the lumbar region with increasing age. The physiological implications of such a difference and of the reversal in relative iodine number are unknown.

4.3.5.2 Myofibres

As indicated above, the differentiation of muscles overall as 'red' or 'white' reflects their relative content of individual 'red' and 'white' myofibres; but although, of course, commercial assessment of meat in terms of its fibre type would not be realistic, nevertheless it is increasingly important, in seeking control of the growth of muscles and, thereby, of meat quality, to consider interfibrillar differences in morphology and biochemistry. Thus 'white' (type II) fibres are wide in diameter and predominantly glycolytic in their metabolism: 'red' (type I) fibres are narrow, having a greater proportion of respiratory activity (George and Naik, 1958; Blanchaer and Van Wijhe, 1962; Hintz *et al.*, 1982) and being supplied by more blood capillaries (Romanul, 1964). The 'red' fibres are able to oxidize fat – indeed, in prolonged muscle activity, it is the principal fuel – and the inflow of lipids into muscle has been observed to supply such fibres (Vallyathan *et al.*, 1970). Lipids are stored mainly in type I fibres (Essen-Gustavsen *et al.*, 1994; Fernandez *et al.*, 1995).

Insofar as slow ('red') fibres are in more continuous action, depend more on respiration than anaerobic glycolysis to regenerate ATP, and produce considerable heat in doing so (which must be dissipated), they tend to be located closer to the blood supply of muscles (Suzuki and Tamate, 1988). The cheek muscles of the ruminant are unusual in that their constituent fibres are exclusively 'red' (type I, i.e. oxidative and myoglobin rich); whereas in porcine cheek muscles, and in the skeletal muscles of meat animals generally, only about 15–25 per cent of the constituent fibres are of this type (Johnson *et al.*, 1986), suggesting that the process of chewing in the ruminant involves especially slow, prolonged action. Subsequently it was demonstrated that all the fibres of bovine *masseter* muscle contain the protein 'myosin light chain 1 slow' (cf. Table 4.32), which they associated with active, albeit slow, movement.

It is a peculiarity of the pig that such 'red' fibres as are found in the muscles of this species are clustered together in a regular geometric pattern rather than randomly distributed. Giant fibres are prevalent in the muscles of exudative pork

(Cassens *et al.*, 1969) (§ 3.4.3). They have low phosphorylase and high ATP-ase activities and are thus intermediate between types I and II.

The glycogen of 'red' fibres resembles amylopectin, that of 'white' fibres resembles amylose in respect of the colour produced with iodine. The high glycogen content of horse muscles (cf. Table 4.31) and of the 'red' fibres of the guinea-pig (Gillespie *et al.*, 1970) emphasize that this criterion does not invariably signify a dependence on glycolytic, as opposed to respiratory, metabolism. Indeed histochemical studies have identified three types of muscle fibre. Type I is slow-acting, predominantly oxidative in metabolism and 'red'; but whereas type II fibres are generally fast-acting, predominantly glycolytic in metabolism and 'white', two subdivisions are recognized. Thus, type IIA has an appreciable capacity for oxidative metabolism: type IIB does not (Peters *et al.*, 1972). Each one of the three types is enervated by a distinct variant of motor unit (Close, 1967). Type I is also characterized by the presence of many large, spherical mitochondria, which form aggregates at the periphery of the fibres: in type IIB fibres, mitochondria are few and are mostly located in pairs at the I bands (Gauthier, 1970). These three types of fibre have each a characteristic variant of myosin heavy chain associated with them (Young and Davey, 1981) and a distinct level of myofibrillar ATP-ase activity (Young, 1984): the actins, however, are apparently identical in the three fibre types (Billeter *et al.*, 1982). The rate of protein turnover is two to five times faster in type I than in type II fibres (Citoler *et al.*, 1966). Reflecting more superficial differences between locations within a given muscle, Morita *et al.* (2000), in a detailed histochemical study of bovine *l. dorsi*, demonstrated that, whereas the relative percentages of fibre types was 37 per cent type I, 17 per cent type IIA and 46 per cent type IIB at the levels of both the 6th thoracic and 5th lumbar vertebrae, the percentages were 26 per cent type I, 15 per cent type IIA and 59 per cent type IIB in the region of the 11th thoracic vertebra. The further differentiation of fast-twitch fibres into three types, led to recognition of four isoforms of heavy chain myosin in adult skeletal muscle, one of each characterizing types I, IIA, IIX and IIB fibres. The contraction-rate and ATP-ase activity increased from type I to type IIB (Weiss and Leinward, 1996). Chikuni *et al.* (2001) determined the entire sequence of 1937–1939 amino acids of the heavy chain myosin isoforms from fibres of types IIA, IIX and IIB of porcine *l. dorsi* muscle. Although the sequence was highly conserved, the amino acid pattern of two regions of the heavy chain myosin isoform from type IIB fibres differed from those of the other type II fibres.

Pre-slaughter stress (cf. § 5.1) depletes glycogen from muscle fibres according to their nature and to its mode of inducement. Thus, mixing stress causes a greater depletion from fast fibres (types IIA and IIB) than from slow fibres (type I), whereas depletion of glycogen by adrenaline injection occurs more severely from slow fibres (Lacourt and Tarrant, 1985). Severe cold reduces the glycogen more in red fibres than in white since red muscles are those mainly responsible for shivering (Lupandin and Poleshchuk, 1979). Thus the susceptibility of different muscles to develop dark-cutting character will be affected by the relative proportion of fast and slow fibres they contain.

The problems of pale soft exudative (PSE) pork and of other undesirable quality conditions (cf. §§ 3.4.3, 8.3.2.1 and 10.1.2) have stimulated investigation of the relative susceptibilities to them of different muscles of the porcine carcass. Thus, Warner *et al.* (1993), compared the quality attributes of *longissimus lumborum* with those of muscles from the loin, ham and shoulder. They found that, when *l. lumborum* was classed as dark and non-exudative (DFD), the other muscles were also dark and their ultimate pH was high. When *l. lumborum* was pale and exudative, however, only the ham muscles (other than *rectus femoris*) were similarly defective.

Young and Bass (1984) showed that there was a markedly higher proportion of type IIB fibres in the muscles of steers than in those of bulls, and a higher proportion of type IIA fibres in the latter, suggesting that serum androgens exert a differential control on the growth of muscle. In support of this concept, Young *et al.* (1986) subsequently demonstrated that, in bulls implanted with zeranol, which effects chemical 'castration' as well as being an anabolic agent, the percentage of type IIB fibres increased at the expense of those of type IIA. Wild pigs have more oxidative type IIA fibres than domestic pigs, especially in fibres with less myoglobin (i.e. more glycolytic) and the cross-sectional area of types I, IIA and IIB is similar; whereas in domestic pigs the cross-sectional area of IIB fibres is much greater than those of types I and IIA (Ruusunen and Puolanne, 2004).

Electrophoretic analysis of individual fibres from bovine muscles has revealed a further degree of complexity in the myofibrillar proteins (Young and Davey, 1981). At least six types of fibre have now been identified by the patterns of the troponins and myosin light chains they possess. The classification is shown in Table 4.32.

Young and Cursons (1988) concluded that myosin light chain 1 slow (LCls) is specifically associated with movement as opposed to maintenance of posture; and it is absent from the fibres of muscles which have an obvious postural role and from the fibres of fast-acting muscles. Nucleotide sequences for three types of myosin heavy chain (MyHC) isoforms from bovine muscles were determined by Ohikuni *et al.* (2004). MyHC-2a, -2x and slow isoforms were present but not MyHC-2b. The latter isoform is the most glycolytic and is present in porcine muscles, whereas its absence from bovine muscles reflects their generally slower movement and greater fatigue resistance. Studies on ovine muscles (Sazili *et al.*, 2005) have shown that the slow myosin heavy chain isoform is correlated with calpastatin activity, a feature reflecting the fact that red muscles with predominantly oxidative fibres (type II)

Table 4.32 Semi-diagrammatic representation of electrophoretograms of myosin light chains and tropomins associated with six types of bovine muscle fibre (after Young and Davey, 1981)

Fast-twitch group (Type II)			Slow-twitch group (Type I)		
Fast 1	Fast 2	Fast 3	Slow 1	Slow 2	Slow 3
Troponin T_f	Troponin T_f (Myosin LCl_s?)	Troponin T_f (Myosin LCl_s?)	Troponin T_s Myosin LCl_s	Troponin T_s Myosin LCl_s	Troponin T_s
Myosin LCl_f	Myosin LCl_f Troponin $I_f(a)$		Troponin $I_s(a)$	Troponin $I_s(a)$ Troponin $I_s(b)$	Troponin $I_s(b)$
Troponin $I_f(b)$	Troponin $I_f(b)$	Troponin $I_f(b)$			
Troponin C_f	Troponin C_f	Troponin C_f			
			Troponin C_s Myosin $LC2_s$	Troponin C_s Myosin $LC2_s$	Troponin C_s Myosin $LC2_s$
Myosin $LC2_f$	Myosin $LC2_f$	Myosin $LC2_f$			
Myosin $LC3_f$	Myosin $LC3_f$	Myosin $LC3_f$			

are less effectively proteolysed by the calpain system; but this feature was not apparently correlated with tenderness.

In acknowledging that much of the biochemical differentiation of muscles can be related to their capacities for respiratory metabolism, and that the concentration of myoglobin in the muscle overall reflects this need, it is evident that it could be met either by a relatively small increase in the myoglobin content of fibres of types I, IIA and IIB, or by a relatively large increase in the myoglobin content of type I ('red') fibres, and in the percentage of the latter. (Thus, the myoglobin concentrations in bovine *psoas major* and *gluteus medius* are similar, but the former has twice the number of 'red' fibres: Hunt and Hedrick, 1977). Although the second mechanism no doubt predominates, the reason for the oxygen requirement – prolonged activity, tension development, high metabolic rate, diving (as in the whale) – may well determine how it is most effectively achieved.

4.3.6 Training and exercise

Implicit in the differences in constitution in a given muscle between active and inactive species and breeds, between young and old animals, and between 'red' and 'white' muscles in a given animal, is the concept that constant usage can cause a development of certain features and that, conversely, disuse can cause a reversion of such factors. Systematic usage over a period ('training') as opposed to fatiguing exercise immediately pre-slaughter, or cessation of activity in a previously active muscle, cause opposing changes in constitution. As to what training signifies, Müller (1957) has shown that hypertrophy – and presumably associated changes in muscle constitution – is not stimulated unless the exercise involves about two-thirds of the maximum resistance which the muscle can overcome. The most obvious alteration in constitution is the elaboration of myoglobin during systematic exercise and its failure to develop in inactive muscles (Lehmann, 1904; Hammond, 1932a). This is logical if myoglobin functions as a short-term oxygen store in muscle (Millikan, 1939) which facilitates its ability to develop power. There would appear to be a concomitant increase in the activity of the respiratory enzymes (Denny-Brown, 1961). Both training on a treadmill, and spontaneous activity, decreased the level of lactic dehydrogenase in porcine muscles, indicating an increase in the capacity for aerobic metabolism (Petersen *et al.*, 1997).

Another feature of training is the elaboration of increased stores of muscle glycogen, which, of course, leads to a lower ultimate pH post-mortem (Bate-Smith, 1948). This should not be confused, however, with the *depletion* of glycogen, and the high ultimate pH which exhausting exercise immediately prior to slaughter would cause (Chapter 5).

There is some suggestion that training may increase the equilibrium level of phosphorylase *a* (Cori, 1956) which is responsible for the first step in making glycogen available for energy production both in respiration and under the anaerobic conditions of glycolysis. Glutamine appears to enhance the deposition of glycogen in muscle. It is utilized at an increased rate during training and exercise (Keast *et al.*, 1995). As mentioned above, muscles, when working aerobically, i.e. when well trained, depend on fat rather than carbohydrate for their long-term energy requirements (Zebe, 1961). Regular exercise can also improve the status of ageing muscles (which tend to lose power because the muscle fibres become shorter and thinner).

The atrophy which arises when disuse is complete is invariably accompanied by a decrease in total nitrogen and in the percentages of sarcoplasmic and myofibrillar proteins; and by an increase in the amount of connective tissue proteins (Helander, 1957). Moderate inactivity causes a diminution in the sarcoplasmic and myofibrillar proteins only (Helander, 1958).

4.3.7   Plane of nutrition

The general effects of the level of nutrition on the growth of meat animals (Hammond, 1932a; McMeekan, 1940) are reflected in the composition of the individual muscles. As the percentage of fatty tissue in an animal increases, the percentage of intramuscular fat also tends to increase (Callow, 1948). This relationship is seen to hold, in general, with the data for cattle and sheep given in Table 4.33 (Callow and Searle, 1956; Callow, 1958).

The content of intramuscular fat also reflects the plane of nutrition when this is deliberately controlled. Data for the *l. dorsi* and *psoas* muscles of cattle are given in Table 4.34 (Callow, 1962).

Moreover, on a high plane of nutrition, a greater proportion of fat is synthesized from carbohydrate; and such fat has, consequently, a lower iodine number. With an increasing degree of emaciation, on the other hand, the relative percentage of linoleic acid increases and that of palmitic decreases in the phospholipids (Vickery, 1977). Attempts to reduce the fat content of porcine carcasses by genetic or nutritional manipulation can cause an increased softness and loss of cohesiveness in fatty tissue, leading to the separation of subcutaneous fat into layers. This undesirable

Table 4.33   Relationship between percentage of fatty tissue in carcass (FT/C) and percentage of intramuscular fat (F/MT) in *l. dorsi* and *psoas* muscles of cattle and sheep

	F/MT			
FT/C	*L. dorsi*		*Psoas*	
	Cattle	Sheep	Cattle	Sheep
20	4.5	5.8	4.7	2.4
26	6.1	8.9	5.3	4.6
34	7.0	8.8	5.6	5.6
39	11.1	8.7	8.2	5.1

Table 4.34   Percentage of intramuscular fat in *l. dorsi* and *psoas* muscles of cattle on four planes of nutrition

Plane of nutrition	*L. dorsi*	*Psoas*
High–high	8.3	7.1
Moderate–high	8.1	6.8
High–moderate	5.5	5.0
Moderate–moderate	7.7	6.6

Table 4.35 Effect of age and plane of nutrition on composition of *l. dorsi* muscles from ewes

	Plane of nutrition			
	High		Low	
	9 weeks	41 weeks	9 weeks	41 weeks
Intramuscular fat (%)	2.13	5.00	0.51	3.24
Intramuscular fat (I.N.)	65.67	55.85	99.65	51.3
Moisture (%)	77.59	73.59	81.03	74.13

Table 4.36 Effect of age and plane of nutrition on composition of *l. dorsi* muscle of pigs

	Plane of nutrition			
	High		Low	
	16 weeks	26 weeks	16 weeks	26 weeks
Intramuscular fat (%)	2.27	4.51	0.68	2.02
Intramuscular fat (I.N.)	62.90	59.20	95.40	66.80
Moisture (%)	74.39	71.78	78.09	73.74

effect is due to a decrease in the size of the fat cells and an increase in the ratio of linoleic to stearic acids in the fat (Wood, 1984b). As the contents of linoleic and α-linoleic acids increase, the fat becomes soft and more translucent (Maw *et al.*, 2003). Ingestion of high levels of unsaturated fat will cause the deposition of unsaturated intramuscular fat in pigs but not in ruminants unless the feed has been protected against the reducing action of rumen micro-organisms. Even in ruminants, however, when the feed is not so protected, the ingestion of different feeds can effect some changes in the fatty acid pattern of the fats laid down (Purchas *et al.*, 1986). The effects of plane of nutrition on the composition of muscles from ewes of two age groups are shown in Table 4.35 (Pállson and Vergés, 1952). Natural foraging, when pigs may ingest substantial quantities of nuts, acorns and other feeds which contain a high proportion of unsaturated fats, will cause the fat laid down to be soft and oily. (Although this is undesirable in general, such soft fat can be beneficial in certain products such as Iberian hams.)

It is obvious that a high plane of nutrition increases the percentage of intramuscular fat and decreases the percentage of moisture in sheep, both soon after birth and in older animals, the effect being superimposed upon trends due to the age. Similar data were obtained by McMeekan (1940) in pigs at 16 and 26 weeks (Table 4.36) and these show the same trends.

Janicki *et al.* (1963) have shown that, in addition to the above changes in the percentages of intramuscular fat and moisture, there is a progressive diminution in the percentage of myoglobin from 0.08 to 0.05 per cent in the *l. dorsi* as the plane of nutrition is increased in pigs. The moisture content of muscle may also be influenced by the *nature* of the diet (Lushbough and Urbin, 1963).

Undernutrition causes a marked increase in the water content of muscle. Thus, the muscles of pigs which had been kept severely undernourished from the age of 10 days to I year had a moisture content of 83 per cent in comparison with 74 per cent in the muscles of a corresponding 1-year-old, well-nourished pig (Widdowson *et al.*, 1960). Undernutrition is also associated with an increase in the percentage of intramuscular collagen and a decrease in its salt- and acid-soluble components – a factor conducive to greater toughness (Bailey and Light, 1989).

4.3.8 Inter-animal variability

The least understood of the intrinsic factors which affect the constitution of muscle is the variability between individual animals. Even between litter mates of the same sex considerable differences are found in the percentages of intramuscular fat, of moisture and of total nitrogen and in the distribution of nitrogen between sarcoplasmic, myofibrillar and stroma proteins (Lawrie and Gatherum, 1964). As has been indicated already (Chapter 2), such differences may be adventitiously determined by the position of the embryo in the uterus (McLaren and Michie, 1960) and, after birth, by the suckling order (Barber *et al.*, 1955; McBride, 1963); but the precise reasons have not been found so far.

Recessive genes no doubt account for apparently sporadic differences in the composition of muscles of animals within a given breed. Reference has already been made to the phenomenon of doppelender cattle, in which much of the musculature is hypertrophied and the ratio of muscle to bone and fat greatly increased (McKellar, 1960; Pomeroy and Williams, 1962; Dumont and Boccard, 1967). Although the musculature concerned has been reported as less developed than normal and to contain more moisture and less fat (Neuvy and Vissac, 1962), other investigations suggest, on the contrary, that the quality of the musculature generally is improved (in terms of eating quality), the content of total nitrogen being greater, the contents of moisture and of hydroxyproline being less than in normal muscle (Lawrie *et al.*, 1964) and the tenderness greater (Bouton *et al.*, 1978c). The intramuscular fat is also lower in the muscles of doppelender cattle. Moreover, hypertrophy does not simply involve the introduction of extra muscle protein into the same connective tissue framework (Lawrie *et al.*, 1964). The decrease in the hydroxyproline content of culard animals affects both hypertrophied and hypotrophied muscles (Boccard and Dumont, 1974). Selected data are given in Table 4.37. Such hypertrophied muscles have a lower succinic dehydrogenase activity than normal, the metabolism on the fibres being more of the glycolytic than oxidative type (Ashmore and Robinson, 1969; Boccard, 1981; Gagnière *et al.*, 1997). Since the former are thicker it has been suggested that this feature accounts for the muscular hypertrophy as superficially observed. The muscles of adult cattle which exhibit the double-muscling condition appear to contain an unusual form of myosin (Picard *et al.*, 1995).

Among more dynamic differences, interanimal variability in the rate of postmortem glycolysis in pigs may be considerable (Fig. 4.5). There is some evidence

Table 4.37 Composition of various muscles from normal (C) and doppelender (D) heifers

Muscle	Moisture (%)		Intramuscular fat (%)		Nitrogen (%)		Hydroxyproline mg/g	
	C	D	C	D	C	D	C	D
L. dorsi (lumbar)	76.51	75.63	0.56	0.27	3.54	3.70	520	350
Psoas major	77.34	77.41	1.46	0.43	3.30	3.35	350	265
Rectus femoris	78.07	76.81	1.49	0.30	3.40	3.51	550	330
Triceps (lat. head)	77.23	76.59	0.73	0.37	3.45	3.53	1000	770
Superficial digital flexor	78.67	77.79	0.40	0.42	3.27	3.40	1430	415
Sartorius	77.95	77.38	0.58	0.25	3.33	3.41	870	460

that a slow rate may be associated with the presence of the enzyme phosphogluco-mutase in a dephosphorylated form (R. K. Scopes, personal communication).

Random but significant anomalous differences in the composition of muscles are encountered. Thus the meat from identically reared and fed sibling heifers was found to differ six-fold in collagen content and greatly in toughness (R. A. Lawrie and P. C. B. Roberts, unpublished data). On the other hand, observed differences in toughness between the meat of pork-weight pigs showed no correlation with any difference in the content of either immature or mature cross-links in the perimysial collagen isolated from *longissimus lumborum* muscles (Avery *et al.*, 1996).

Clearly, substantial differences in the composition of muscle are caused by factors which are, as yet, unexplained: their elucidation will no doubt form the subject for a wide area of future research.

Chapter 5

The conversion of muscle to meat

In considering how meat animals grow and how their muscles develop and are differentiated, the distinction between the terms 'muscle' and 'meat' has not been emphasized. Meat, although largely reflecting the chemical and structural nature of the muscles of which it is the post-mortem aspect, differs from them because a series of biochemical and biophysical changes are initiated in muscle at the death of the animal. Some details of the conversion of muscle to meat will now be given.

5.1 Preslaughter handling

Although, at the most, only a few days elapse between the time when meat animals have attained the weight desired by the producer and the actual moment of slaughter, their condition may change appreciably in this period. This will happen to some extent irrespective of whether the animals are driven on the hoof or transported to the abattoir. There may be loss of weight, bruising and, if the animals are in road or rail trucks, suffocation due to inadequate ventilation. The lairage itself can have an appreciable effect on the level of bruising and it is recommended that those animals most vulnerable to bruising should occupy the quietest yards at the abattoir (Eldridge, 1988). In bruising, blood from damaged blood vessels visibly accumulates. (It can also occur in pigs through the use of markers to tattoo identification numbers on the skin). Both in pigs and cattle, such bruising can be caused through the animals slipping and falling especially when attempts are made to speed up their transit by using goads (Warriss, 2000). In 1971 bruising, oedema and emaciation accounted for 4 per cent, 65 per cent and 77 per cent of the total condemnations of pigs, sheep and cattle respectively, in Northern Ireland (Melrose and Gracey, 1975). In most cases of extensive bruising damage to muscles causes the release of enzymes into the blood stream; and the relative concentrations of creatine phosphokinase and aspartate transaminase permits an assessment of how long before slaughter bruising occurred (Shaw, 1973). Cows suffer more from bruising in transit than bullocks (Yeh et al., 1978). Nevertheless, although the incorporation of bruised beef in meat products is condemned by hygiene regulations in various countries, a comparison of its microbiological and technological properties with those of non-bruised tissue revealed no differences (Rogers and co-workers, 1992, 1993).

Death of pigs in transport increased greatly in the decade 1960–70 (Thornton, 1973). There was a marked seasonal effect, deaths both in transit and lairage being correlated with the environmental temperature (Allen et al., 1974). Above ca. 18 °C there is a very rapid increase in mortality. It was reported that the transport of pigs in double-decker road vehicles was particularly liable to affect the musculature adversely as meat (Williams, 1968). In unloading pigs from road vehicles, excessive steepness of the ramps and the absence of foot supports can be significant factors in increasing bruising (Warriss et al., 1991a).

During autumnal conditions (ambient temperature 16–24 °C), stocking densities of between 0.35 and 0.50 m^2/100 kg liveweight (the usual range of values in Europe) had little effect on the blood profile or meat quality of pigs when transported for up to three hours, although there was evidence of an increase of skin bruising, due to trampling and fighting with the lower allocation of space (Barton Gade and Christensen, 1998). Even more restriction of space, viz. less than 0.30 m^2/100 kg liveweight, is not acceptable, especially in journeys above three hours, since it is essential that the pigs have room to lie down (Warriss, et al., 1998a).

Sophisticated analytical techniques have revealed that there are differences in the overall pattern of proteins (the proteome: cf. §3.3.1) in pork meat from pigs slaughtered immediately and that in pork from pigs held for 12 hours before slaughter (Morzel et al., 2004).

It is increasingly recognized that thoughtless or rough handling of animals in the immediate preslaughter period will adversely affect the meat, quite apart from being inhumane. Recommendations on ante-mortem care (Thornton, 1973; Houthius, 1957; Grandin, 1993) will not be discussed at length in the present volume: but some aspects of the question should be considered.

It would appear that only recently have attempts been made to study the behaviour pattern of meat animals with a view to improving preslaughter handling. With pigs, for example, acknowledgement that these animals tend to fight when awaiting slaughter, especially if they come from different farms, led Danish workers to employ a halter which prevents biting and damage to the flesh (Jørgensen, 1963) and droving is so arranged that the more cautious animals follow the bolder ones. With particular reference to pigs, it is clear that responses to transport and handling depend not only on the stress susceptibility of the animal as a whole, but also on which muscles are being considered. Both the metabolic capability of individual muscles and the duration and severity of transport determine whether the PSE condition will develop or whether glycogen reserves will be depleted sufficiently to produce dark meat (Anon., 1971). Much of the stress sustained in transport and handling arises during loading and unloading.

In comparing various pig slaughtering premises, Warriss et al. (1994) detected a tendency for increased stress to be associated with the larger operations in which throughput was faster, restraining devices more prevalent and noise levels greater. Subjective assessments of stress correlated well with such objective measures as elevated levels of blood lactate and creatine phosphokinase. High levels of stress were associated with poorer quality meat. An analysis of pig meat quality in Denmark, Italy, the Netherlands, Portugal and the UK (Warriss et al., 1998b) indicated that preslaughter stress was more closely reflected by a high ultimate pH and dry, dark meat than by that of PSE characteristics. There is a greater potential for achieving optimal welfare in 'organic' systems of pig rearing than in intensive systems, but this requires more complex management (Barton Gade, 2002).

In reviewing the problems of preslaughter handling of meat animals, Gregory (1994) again emphasized the greater susceptibility of pigs to present difficulties than sheep or cattle. Over prolonged periods, however, sheep also suffer stress. Thus, in the seaboard transport of sheep from Australia to the Middle East, up to 6 per cent of the animals may die on the voyage, from lack of nourishment, salmonellosis or stress. Holding the sheep on board before unloading, as when there are no quayside lairage facilities available, also causes losses (Higgs *et al.*, 1991). Tranquillizing drugs (when permitted) may be effective in preventing fighting and struggling, and may reduce the incidence of exudative meat, injury and death.

Transport of animals under stressful conditions, especially when these are prolonged, is clearly inhumane and, indeed, can be lethal. It is now widely recognized that stressed animals are likely to have a subnormal content of glycogen in their muscles, whereby, post-mortem, the pH of their flesh fails to attain acidic values (cf. § 5.1.2 below); and the attributes of eating quality in the meat will be adversely affected (cf. Chapter 10). Both humane and organoleptic considerations may thus be expected to promote increasing legislation designed to ensure that the conditions to which animals are subjected in transport will be strictly controlled.

5.1.1 Moisture loss

The moisture content of pork muscle is especially liable to change because of even moderate fatigue or hunger in the immediate pre-slaughter period (Callow, 1938a, b). When fasted during transit, cattle lose weight less readily than sheep, and sheep less readily than pigs. With the latter species, the wasting can be about 3 lb/24 h (1.35 kg/24 h) in an animal weighing about 200 lb (90 kg) (Callow and Boaz, 1937). In one study it was found that pigs which had travelled for 8 h before slaughter yielded carcasses averaging 0.9 per cent less than corresponding animals which had travelled for only $\frac{1}{2}$ h pre-slaughter. This was regardless of whether they had been fed or not (Cuthbertson and Pomeroy, 1970). Moreover, pigs may lose further weight if they are given water on journeys over 36 h in duration (Callow, 1954). A loss of weight of this order cannot be accounted for entirely by the breakdown of fatty and muscular tissues to produce energy and heat for the fasting pig and may be due, in part, to a loss of water-holding capacity in the muscular tissues (Callow, 1938b). An animal killed immediately on arrival at a slaughterhouse, after a short journey, may provide both a heavier carcass and heavier offal than an animal which has been sent on a prolonged journey, then rested and fed for some days in lairage (Callow, 1955b). Although it is more difficult to cause carcass wastage in cattle, the extreme conditions in Australia, in which cattle may travel up to 1000 miles from feeding areas to abattoir, can cause losses of *ca.* 12 per cent of initial live weight. Dehydration is minimized when cattle have access to water for 3.5–7 h prior to a preslaughter fast (Wythes *et al.*, 1980). In a comparison of the effects of holding cattle for varying periods without feed or water before slaughter (Jones *et al.*, 1990), shrinkage varied from 31 g kg^{-1} in beef cattle slaughtered immediately to 106 g kg^{-1} for those held for 48 hours without feed or water. The weight of the liver and other offal, as a percentage of live weight, decreased progressively as the period of inanition increased; and there was some suggestion that the ultimate pH of the musculature was elevated (cf. § 5.1.2).

5.1.2   Glycogen loss

The influence of fasting in depleting the glycogen reserve of muscle has been known since the work of Bernard in 1877. Recognition of the importance of this fact in relation to the meat from domestic species is more recent. Callow (1936, 1938b) indicated that inadequate feeding in the period before slaughter could lower reserves of glycogen in the muscles of pigs. Bate-Smith and Bendall (1949) showed that fasting for only 48–72 h lowered the glycogen content of rabbit *psoas* muscle sufficiently to raise the ultimate pH from the normal (for the rabbit) value of 5.9 to 6.5. In contrast, when steers were fasted at normal ambient temperatures for periods up to 28 days, the ultimate pH was unaffected (Howard and Lawrie, 1956).

The importance of exhausting exercise as a factor in depleting glycogen reserves in muscle has also been recognized for a considerable period. The poor keeping quality of the meat from hunted wild cattle was referred to by Daniel Defoe in 1720. Mitchell and Hamilton (1933) showed that exhausting exercise immediately pre-slaughter could cause a high ultimate pH in the muscles of cattle; but Howard and Lawrie (1956) found it most difficult to deplete the glycogen reserves in this species, even when preslaughter exercise and fasting for 14 days were combined (Table 5.1). Yet such depletion occurred, without fasting, if enforced exercise took place immediately after train travel.

The glycogen reserve of pig muscle, however, is especially susceptible to depletion by even mild activity immediately preslaughter (Callow, 1938b, 1939); a walk of only a quarter of a mile may cause a small but significant elevation of ultimate pH. Bate-Smith (1937a) suggested that if an easily assimilable sugar were fed before death the reserves of muscle glycogen might be restored to a level high enough to permit the attainment of a normal, low ultimate pH – the latter being desirable to avoid microbial spoilage (Callow, 1935; Ingram, 1948). This principle was confirmed in commercial practice by Madsen (1942) and Wismer-Pedersen (1959b) in Denmark and by Gibbons and Rose (1950) in Canada (Table 5.2).

If pigs are rested for *prolonged* periods before slaughter, in an attempt to restore glycogen reserves naturally, there is danger that animals carrying undesirable

Table 5.1   Glycogen concentrations and ultimate pH in *psoas* and *l. dorsi* muscles of steers after enforced exercise and fasting

Treatment	L. dorsi		Psoas	
	Glycogen (mg %)	Ultimate pH	Glycogen (mg %)	Ultimate pH
Controls (fed and rested 14 days after train travel)	957	5.49	1017	5.48
Exercise (after train travel and 14 days fasting)	1028	5.55	508	5.55
Exercise 1½ hr (immediately after train travel)	628	5.72	352	6.15

Table 5.2   Effect of feeding sugar preslaughter on the
ultimate pH of pig muscles

Group treatment[a]	Muscle	Ultimate pH
(a) Held overnight without food	*psoas*	6.00
(b) Fed 3 lb (1.35 kg) sucrose at 22 h and 6 h preslaughter	*psoas*	5.54
(c) No food preslaughter	*psoas*	5.75
	biceps femoris	5.74
(d) 2 lb (0.9 kg) sugar fed 3–4 h preslaughter	*psoas*	5.56
	biceps femoris	5.57

[a] Groups (a) and (b), Gibbons and Rose (1950); groups (c) and (d),
Wismer-Pedersen (1959b).

bacteria may infect initially unaffected animals, e.g. with *Salmonella*, which can
endanger subsequent human consumers. This fact is reflected in current legislation
in the United Kingdom which does not permit holding for more than 48 h, since the
most recent outbreak of foot and mouth disease.

In the belief that the musculature of cattle might spare glycogen reserves during
activity, by metabolizing lipid stores more readily than monogastric animals,
Howard and Lawrie (1957a) injected cattle, after fasting, with neopyrithiamin to
inhibit thiamin pyrophosphate and, thereby, fat oxidation. Whereas fasting *per se*
had little effect in lowering the level of muscle glycogen in cattle, its combination
with neopyrithiamin lowered it markedly, leading to an average ultimate pH level
of above 6.1 throughout the musculature. Using similar reasoning, Lister and
Spencer (1981) induced 'dark-cutting' characteristics in the meat of sheep and cattle
by administering antilipolytic agents such as nicotinic acid and methyl pyrazole car-
boxylic acid, following exposure of the animals for some hours to isoprenaline (a β-
adrenergic agonist, which promotes lipolysis and glycolysis in muscle). The same
effect was induced in pigs (Spencer *et al.*, 1983) and annulled by the simultaneous
administration of caffeine which stimulates lipolysis (and spares glycogen). They
suggested that caffeine administration could be used prophylactically to conserve
muscle glycogen (and thus ensure a normal ultimate pH) in animals exposed to
stress.

The existence of some influence controlling the level of muscle glycogen other
than fatigue or inanition was suggested by the finding that certain steers which had
been well fed and rested, and would therefore have been expected to have ample
glycogen in their muscles, yielded meat of high ultimate pH (Howard and Lawrie,
1956). It appeared that these steers were of an excitable temperament. In such
animals, short range muscular tension, not manifested by external movement,
reduced glycogen reserves to a chronically low equilibrium level. Those animals in
a group that had shown the greatest resistance to handling, produced muscles with
the highest ultimate pH (Howard and Lawrie, 1956). Similar findings with reindeer
were reported by Petaja (1983).

Muscles differ in their susceptibility to preslaughter glycogen depletion by stress.
Thus, in the ox, Howard and Lawrie (1956) found that the ultimate pH of *psoas*

major tended to be elevated more than that of *l. dorsi*, whereas Warriss *et al.* (1984) reported the reverse. Insofar as red muscles are responsible for shivering, their glycogen stores are more readily lowered by severe cold than that of white muscles (Lupandin and Poleshchuk, 1979). Even individual fibres differ in their response to stress, and to its nature according to whether they are of 'fast' or 'slow' type (Lacourt and Tarrant, 1985; cf. § 4.3.5). The time required to restore glycogen levels in the muscles of young bulls, after their depletion by the stress of mixing animals of different origin, was shown to be not less than 48 h (Warriss *et al.*, 1984). Drugs given pre-slaughter to induce tremor considerably depleted glycogen reserves, causing a high ultimate pH, and confirmed the view that fear was an important factor in this context (Howard and Lawrie, 1957a). (The muscles of the cattle killed after excitement of train travel (Table 5.1) had a high ultimate pH, whereas those of cattle rested for 14 days after travel had normal values.)

A high ultimate pH, in the muscles of the cattle, causes the aesthetically unpleasant phenomenon of dark-cutting beef – known at least since 1774 (Kidwell, 1952) – and in those of pigs that of 'glazy' bacon (Callow, 1935). Apart from its poor appearance, the high pH of 'dark-cutting' beef enhances the growth of bacteria (Ingram, 1948).

Tarrant (1981) reported the results of a survey in which meat scientists in 20 countries were asked their views on 'dark-cutting' beef. The incidence was high in young bulls (reflecting the importance of excitability of temperament in causing the condition) and in cold, damp weather. Preslaughter stress generally was regarded as the prerequisite. Similar conclusions were reached by Brown *et al.* (1990) in a survey of 5000 cattle in the UK.

There is evidence that the relative susceptibility of individual cattle to develop 'dark-cutting' characteristics in their meat post-mortem is positively correlated with the number of 'slow' oxidative fibres in their muscles (Zerouala and Stickland, 1991).

Recognition that stress susceptibility is a factor in determining the condition of animals generally, and thereby the glycogen status of their muscles, has grown as the result of Selye's (1936) concept of the general adaptation syndrome. He noted that animals exposed to a variety of stress-producing factors such as emotional excitement, cold, fatigue, anoxia, etc., reacted by discharge of the same hormones from the adrenal gland irrespective of the nature of the stress – adrenaline from the adrenal medulla, 17-hydroxy- and 11-deoxycorticosterones from the adrenal cortex. These substances elicit a variety of typical responses in the animal. Adrenaline depletes muscle glycogen and potassium; 17-hydroxy-corticosterone and 11-deoxy-corticosterone, respectively, restore the equilibrium level of these substances in normal animals. The release of the latter two hormones is controlled by the secretion of ACTH by the pituitary; and ACTH production is controlled by a releasing factor produced in the hypothalamus (Harris *et al.*, 1966), the part of the brain which is reactive to external stimuli. As mentioned in § 2.5.2.1, an imbalance at various points in this complicated system can cause so-called diseases of adaptation (Selye, 1944, 1946). Such would be expected in individual animals which were stress-susceptible, and the imbalance could be manifested by low equilibrium levels of glycogen, disturbances in the rates of glycogen breakdown and so on (§ 3.4.3). In the plasma of pigs which yield pale, exudative flesh, for example, there is a deficiency of 17-hydroxy-corticosteroids (Topel *et al.*, 1967). Stress-susceptible pigs react to certain anaesthetics by a rise in body temperature of 1–4 C° (malignant hyperthermia: Gronert, 1980) and limb rigidity, and the development of rigor during light anaesthesia induced by halothane has been used to identify pigs which are prone to stress (Lister *et al.*, 1981). The glycolytic and oxidative pathways in the muscles of pigs of the genotype susceptible to hyperthermia under the influence of halothane, are significantly different from those of pigs which are insensitive to the anaesthetic: the former have a higher concentration of lactate and lower concentration of CP and ATP (Lundström *et al.*, 1989; cf. §§ 2.2 and 2.3).

Differences have been observed between species in their sensitivity to various stressors. Thus, pigs are more affected by sound than are sheep (Lister *et al.*, 1981). In humans it has been shown that the ratio of adrenaline to noradrenaline in the circulation, following stimulation of the hypothalamus, varies according to the nature of the stressor (e.g. heat, exercise, emotion) (Taggert *et al.*, 1972), and, since these two hormones affect different receptors, it may well be that the concept of the general adaptation syndrome, however useful, is an oversimplification of the response of animals to stress.

Tranquillizers known to offset stress susceptibility have been given to calm stock in transit, but they are not without danger as they may induce a state of relaxation so profound that the animals cannot stand and may be suffocated. The metabolic stresses which affect muscle were reviewed by Lawrie (1966).

At the cellular level, stress apparently induces the production of a group of proteins/polypeptides which possess cryoprotective activity (Lindquist and Craig, 1988). They exhibit a very high degree of conservation between species, suggesting their importance throughout evolution. Of these the polypeptide, ubiquitin, is an important member. It is a component of the filamentous inclusions which characterize neurodegenerative diseases, against which it appears to be mobilized as an aspect of immunological reactivity (Lowe and Mayer, 1990; cf. § 11.3).

5.2 Death of the animal

A major requirement for desirable eating and keeping qualities in meat is the removal of as much blood as possible from the carcass, since it can cause an unpleasant appearance and is an excellent medium for the growth of microorganisms. Despite reports that *delayed* bleeding has little effect on the eating quality of meat (Williams *et al.*, 1983), there is no suggestion that carcasses should remain completely unbled, of course.

Except in ritual slaughter, animals are anaesthetized before bleeding. The procedure at both stunning and bleeding is important. When special precautions are taken to ensure sterility, there is some evidence that unbled muscles undergo the tenderizing changes of conditioning to a greater extent than do those which are bled (Shestakov, 1962), but this could scarcely be regarded as a generally valid reason against bleeding.

5.2.1 Stunning and bleeding

Generally, cattle are stunned by a captive bolt pistol or by a blow from a pole-axe. In recent years the dressing of beef carcasses has been carried out more frequently as they hang vertically rather than when supine on the abattoir floor. These changing circumstances make it rather less important to ensure that the heart is still functioning as blood can drain quite effectively from the carcass even when heart action has ceased. Indeed it has been suggested that the vasoconstrictive effect of the stress of stunning will expel most of the blood from the musculature and that drainage is only necessary to remove blood from major blood vessels (Warriss, 1978; Warriss and Wotton, 1981). In certain countries cattle are stunned electrically.

Sheep and pigs are stunned electrically or anaesthetized by carbon dioxide. It has been observed that in sheep killed by a captive bolt pistol the epithelial lining of the intestines is shed, whereas it remains intact in anaesthetized animals (Badaway *et al.*, 1957); this could have microbial implications which will be referred to later. It is important to emphasize that drugs may not be used to induce unconsciousness in animals which are intended for human consumption since residue could remain in the meat.

In electrical stunning, the characteristics of the current must be carefully controlled, otherwise complete anaesthesia may not be attained and there may be convulsive muscular contractions. The siting of the electrodes is also important, since the current must pass through the brain. Variation in electrical resistance because of differing thicknesses in the skull can cause ineffective stunning. There are three phases in the animal's reaction: (i) as soon as the current is switched on there is violent contraction of all voluntary muscles and the animal falls over; respiration is arrested; (ii) after 10 s (the current being discontinued) the muscles relax and the animal lies flaccid; (iii) after a further 45–60 s the animal starts to make walking movements with its legs and respiration starts again. Usually, alternating current at 70–90 V and 0.3 A is used for 2–10 s (Cruft, 1957). Better relaxation and less internal bleeding is said to result if a high frequency current (2400–3000 Hz; Koledin, 1963) and a square wave form, instead of a sine wave (Blomquist, 1958), are employed.

There is some suggestion that electrical stunning may lower the glycogen reserves of the muscle slightly. The mean ultimate pH of *quadriceps femoris* from 518 electrically stunned pigs was 5.78; that in non-stunned controls was 5.67 (Blomquist, 1959). If the period between electrical stunning and bleeding is prolonged, the rather high pH may foster microbial spoilage (Warrington, 1974). In comparison with captive bolt stunning, electrical stunning has been shown to cause an elevation of amino acids in the plasma (especially of valine; the concentration of isoleucine falls somewhat) (Lynch *et al.*, 1966). It has been found that the level of corticosteroids in the blood of electrically stunned pigs is higher than that of those anaesthetized by carbon dioxide (Luyerink and Van Baal, 1969). There are said to be benefits in a diminished incidence of blood splash if a high-pressure water jet is combined with electrical immobilization in the stunning of pigs (Lambooij and Schatzmann, 1994). Repeated application of electrical stunning to pigs appears to cause no welfare problem but, of course, should normally be avoided (McKinstry and Anil, 2004).

Carbon dioxide anaesthesia is an effective alternative to electrical stunning provided the concentration of the gas is between 65–70 per cent. If the latter concentration is not exceeded, the musculature of the pigs is relaxed and the ultimate pH is slightly lower, and less variable, than with electrical stunning (Blomquist, 1957). One disadvantage of using a carbon dioxide chamber is that pigs differ somewhat in their susceptibility to anaesthesia by the gas, and that individual control of the animals is not feasible. Moreover, there is evidence that, prior to anaesthesia, animals suffer considerable stress; and, indeed, it has been suggested that carbon dioxide anaesthesia does not comply with the generally accepted definition of pre-slaughter stunning.

Von Mickwitz and Leach (1977) surveyed the various methods of stunning employed in the EEC (now EU). They rated concussion stunning of cattle as the most effective, followed by captive bolt stunning of sheep and electrical stunning of pigs. Electrical stunning of sheep and captive bolt stunning of calves were deemed ineffective procedures. They concluded that any attempt to standardize stunning methods must specify proper preslaughter treatment of animals as an integral part of the overall procedure. In a comparison of various modes of stunning in pigs, using ^{31}PNMR spectroscopy, Bertram et al. (2004a) found that captive bolt stunning caused the most rapid rate of post-mortem glycolysis and the greatest loss of fluid from the meat subsequently. CO_2 anaesthesia was associated with the slowest rate of post-mortem glycolysis and least loss of fluid; whereas electrical stunning was intermediate in these two respects. All three stunning procedures, however, were more stressful than general anaesthesia.

Insensibility has been associated with an electroencephalographic voltage of less than $10\,\mu V$. It has been demonstrated, however, that with electrical stunning before bleeding, the voltage takes longer to fall below $10\,\mu V$ than when throats are cut without prior stunning, suggesting that the electrical stunning of sheep and calves causes a prolonged increase in the electroencephalic voltage. The latter criterion is thus not a reliable index of insensibility with animals which have been electrically stunned (Devine et al., 1986).

Newhook and Blackmore (1982a, b) studied the efficacy of electrical stunning when applied to calves and lambs, using electroencephalography and electrocardiography to indicate the state of the animals. They defined 'death' as 'irreversible insensibility due to cardiac anoxia caused by complete severance of both common carotid arteries and jugular veins'. According to this criterion, lambs were technically dead by 10 s after exsanguination. Calves did not become insensible however until 90 s; and there were indications of recurrent sensibility for up to 5 min after bleeding. It was postulated that delayed death in calves was due to their brains being supplied more lavishly with blood via the vertebral arteries. In calves, following neck sticking, a considerable proportion of animals suffer carotid occlusion (ballooning), whereby large clots impede bleeding and may lead to sustained brain function (Anil et al., 1995). Anil et al. (2000) demonstrated that a relatively long cut in the thorax (chest sticking) in pigs (anaesthetized by the head-only procedure) was more humane insofar as it quickly stopped brain responsiveness. This problem does not appear to arise with chest sticking.

These findings related to conditions when the current was applied via two electrodes placed on the head; but when it was delivered via one electrode on the head and one on the back there was immediate cardiac dysfunction and permanent insensibility ensued. This mode of current application was thus regarded as more humane for the electrical stunning of calves (Blackmore and Newhook, 1982; Cook et al., 1995). Nevertheless Channon et al. (2002) reported that the latter procedure was associated with greater drip, paler muscle and a faster rate of post-mortem glycolysis than when the current was applied by electrodes on head and back; or with carbon dioxide anaesthesia.

Eike et al. (2005) used a computer model to study details of the current density in the head of electrically stunned pigs. The model confirmed that the placing of electrodes from eye to eye or from eye to ear was associated with high current density across the brain and effective stunning.

In cattle and sheep, bleeding is effected by severing the carotid artery and the jugular vein, and in pigs by severing the anterior vena cava. If the knife penetrates too far, blood may collect beneath the scapula and cause taint by early decomposition (Thornton, 1973). To avoid entry of micro-organisms, the cut made is minimal, especially with bacon pigs which are subsequently placed in a scalding tank. It has been said that bleeding after electrical stunning is more effective than after the use of the captive bolt pistol, but that it is less so than with carbon dioxide anaesthesia (Blomquist, 1957). Chrystall *et al.* (1980–81) could find no difference in the amounts of residual blood in lamb muscles (or in their microbiological status) whether the lambs had been electrically stunned by several procedures, bleeding had or had not been delayed or whether bleeding had been carried out without electrical stunning. Similarly, with head-only stunning of cattle (as required by halal slaughter), although the heart remains active, and its pumping action assists bleeding out, there appear to be no differences in residual blood in comparison with other stunning procedures (Anon., 1987–88). Even with effective bleeding only about 50 per cent of the total blood is removed (Thornton, 1973), different muscles retaining more or less blood according to their nature. In the horse, for example, 50 per cent of the total pigment left in the heart after bleeding is haemoglobin from the blood, whereas in *psoas* and *l. dorsi* the corresponding values are about 25 and 10 per cent respectively (Lawrie, 1950). It has been shown that electrical stimulation (§ 7.1.1.2) of the carcass after severance of the neck vessels increases the weight of blood which drains from the main veins and arteries and the organs, but does not affect its removal from the musculature (Graham and Husband, 1976).

Following the introduction of electrical stunning there was an increased frequency of 'blood splash', i.e. the appearance of numbers of small dark red areas in the muscles. These had previously been noted when pigs or lambs were shot. Blood splash is certainly more prevalent in electrically stunned lambs than in those stunned by captive bolt or percussion cap (Kirton *et al.*, 1981a). It is more frequently observed in *l. dorsi* and in various muscles of the hind limb. Microscopic examination has shown that blood splash arises where capillaries have ruptured through over-filling with blood (Anon., 1957a). When the current is applied there is a considerable rise in blood pressure, muscles are contracted and their capillaries are almost empty of blood. Subsequently the muscles relax and if the blood pressure is not released by external cutting, blood is forced into the capillaries again with sufficient force to rupture many of them and enter the muscle itself (Leet *et al.*, 1977). As indicated above, the positioning of the electrodes in stunning lambs is important. Thus, if both are placed on the head, blood pressure increases markedly and blood splash can arise. If, however, one electrode is placed on the head and the other on the back or leg, the blood pressure rises only transiently and there is little blood splash, but minute haemorrhages ('speckle') appear in the fat and connective tissue. These reflect the rupture of small vessels caused by the muscular spasms which occur (Gilbert and Devine, 1982). Emotional stress causes vasodilation of the blood vessels in skeletal muscle, and may enhance fibrinolytic activity in entrained blood. These effects, especially if combined with electrical stunning, could perhaps explain the higher incidence of 'blood splash' reported in excitable animals (Jansen, 1966). The remedy appears to be to bleed the pigs within 5 s of administering the anaesthetizing current (Blomquist, 1959).

Kirton *et al.* (1978) came to similar conclusions in respect of the electrical stunning of lambs. Bleeding should be performed as soon as possible after stunning whatever the method of stunning employed.

The extent of exsanguination is enhanced by vasoconstriction of the blood vessels (Warriss, 1978) which is induced by angiotensins produced from the plasma proteins by the enzyme rennin (Miller *et al.*, 1979). The latter is released by the kidneys when their blood supply is interrupted.

5.2.2   Dressing and cutting

Following bleeding, carcasses are 'dressed', i.e. the head, feet, hides (in the case of sheep and cattle), excess fat, viscera and offal (edible and inedible) are separated from the bones and edible muscular tissue. Cattle and pig carcasses, but not those of sheep, are split along the mid ventral axis into two sides. It is not appropriate here to detail dressing procedures; these are fully considered in other texts (Gerrard, 1951; Swatland, 1994; Warris, 2000). Nevertheless, recent developments in this field should be noted. As has been pointed out by Longdill (1989), the labour required to produce a dressed carcass is greater with small species such as sheep. Whereas beef requires *ca.* 22 man hours to produce 10,000 kg of carcasses, sheep requires *ca.* 80 man hours. In New Zealand, considerable advances have been made in the use of mechanical devices in slaughtering and dressing operations with sheep.

In the automated dressing line a series of mechanical devices stun the animal, remove the pelt (first from the brisket, then completely), eviscerate the carcass and process the head (Longdill, 1994). Other devices debone the loin and thoracic regions and have increased the yield of recovered lean meat. Overall hygiene is also improved. Automated slaughterhouses are also being developed for cattle in Australia and for pigs in the Netherlands (Longdill, 1994). These have reduced man hours by 40 per cent.

Considerable progress has been made in automating the entire dressing and cutting sequence, using portable television facilities to analyse these operations in detail; and the applicability to control these of the robotic equipment already available in industry is being assessed. In respect of dressed meat, video image analysis has been successfully applied to grading for speedy online determination of the fat/lean ratio (Newman, 1984) and fibre optic probes permit objective prediction of such textural defects in the meat as excessive paleness or darkness (MacDougall, 1984). In Denmark, a self-correcting computer system has been developed which determines the positioning of fibre optic probes in carcasses and has proved more accurate than grading by experts in calculating the percentage of lean in pigs.

Until recently it was commercial practice to chill dressed carcasses prior to preservation or processing (cf. Chapters 7, 8 and 9), and, after chilling (which signified after rigor mortis), to prepare primal, wholesale cuts (cf. Fig. 3.1; and Gerrard, 1951) from them. Traditionally, skeletal reference points and straight cutting lines have been used. These have contributed to variability in retail joints since the boneless, primal cuts are aggregates of several muscles rather than muscles isolated individually by 'seaming out' along the muscle fascia – as in certain continental practices (Strother, 1975).

Prior to the introduction of vacuum packaging, wholesale cuts were sold with bone intact to avoid evaporative losses and minimize contamination. Now, however, a considerable proportion of home killed beef in Britain is deboned centrally and delivered to retailers as boneless primal cuts. Not only is it likely that deboning of beef carcasses will become a standard abattoir operation in the future, such will be effected on hot carcasses immediately after slaughter.

It has been demonstrated that losses due to evaporation and exudation in vacuum-packed, hot deboned beef are markedly reduced. Moreover, it was reported that there was no significant difference between hot and conventionally deboned vacuum-packed beef in respect of the number of aerobic bacteria on the products, either initially or after storage at 0 °C (Sheridan and Sherrington, 1982), although the number of facultative anaerobes was significantly higher with hot deboned packs. Control of eating quality is greater since muscles can be 'seamed out' as anatomical entities. In the absence of bone, and because the meat cuts are less bulky, chilling and freezing can be more rapidly and economically effected. 'Cold-shortening' can be avoided by deboning in rooms at 5–15 °C and holding the vacuum packed cuts for at least 10 h at these temperatures (Schmidt and Gilbert, 1970; Follet et al., 1974).

Alternatively, 'cold-shortening' can be avoided by electrical stimulation of the carcass or side immediately after slaughter (Carse, 1973; Bendall, 1980) (cf. § 7.1.1.1). Clearly this procedure would be especially useful with relatively small, hot-deboned portions of meat. On the other hand, although the swift lowering of temperature which would thereafter be permissible would assist in controlling microbial spoilage, it would prevent the early accelerated conditioning to which the tenderness achieved by electrical stimulation can be partially attributed (Savell et al., 1977a,b; George et al., 1980). The principles have been extended to the hot cutting of lamb and mutton carcasses (McLeod et al., 1973), cuts being shrink-wrapped and held at 10 °C for 24 h before freezing. Tenderness, far from being diminished, was enhanced in some cuts over that of cuts from carcasses which had been chilled intact before cutting. Shrink wrapping, by moulding the warm fat and musculature, produces pleasing cuts from the initially untidy portions (Locker et al., 1975). It also eliminated weight losses. Similarly, the development of a hot-boning system for pork could increase processing efficiency and improve shelf-life (Reagan, 1983).

Weight losses, and in particular, exudation in deeper muscles where the combination of body temperature and low pH denature proteins, have been much reduced also by partial excision of beef muscles on the warm sides immediately after slaughter and exposing the 'seamed out' muscles to air at 0 °C (Follet, 1974). The bulk of the musculature, in these circumstances, prevents temperatures falling fast enough to cause 'cold-shortening'.

More detailed consideration will be given to 'cold-shortening', electrical stimulation and conditioning below (§§ 5.4, 7.1.1.2, 10.3.3.1 and 10.3.3.2).

5.3 General consequences of circulatory failure

Stoppage of the circulation of the blood at death initiates a complex series of changes in muscular tissue. The more important of these are outlined in Fig. 5.1. It will be appreciated, from what has been indicated in Chapter 4, that the speed and extent of these changes may be expected to differ in different muscles.

At the moment of death of the animal as a whole, its various tissues are continuing their particular types of metabolism under local control. Although muscle is not actively contracting at such a time, energy is being used to maintain its

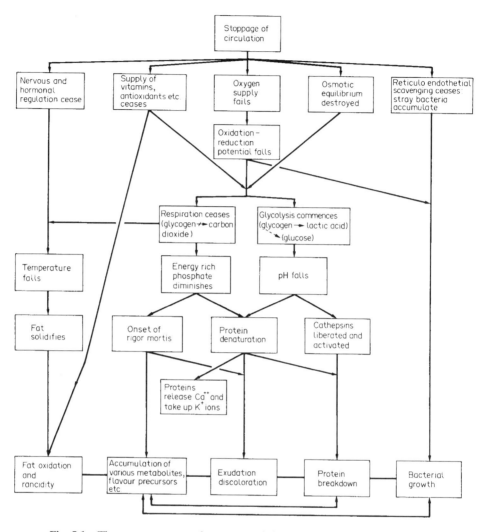

Fig. 5.1   The consequences of stoppage of the circulation in muscular tissue.

temperature and the organizational integrity of its cells against their spontaneous tendency to break down. The non-contractile ATP-ase of myosin, and not the contractile ATP-ase of actomyosin, is one of the enzymes involved in this context (Bendall, 1951). The most immediate change caused by bleeding is the elimination of the blood-borne oxygen supply to the muscles and the consequent fall in oxidation reduction potential. As a result the cytochrome enzyme system cannot operate, and the resynthesis of ATP from this source becomes impossible. The continuing operation of the noncontractile ATP-ase of myosin depletes the ATP level, simultaneously producing inorganic phosphate which stimulates the breakdown of glycogen to lactic acid. The ineffectual resynthesis of ATP by anaerobic glycolysis cannot maintain the ATP level and, as it drops, actomyosin forms and the inextensibility of rigor mortis ensues (as detailed in § 4.2.3). The lowered availability of ATP also increases the difficulty of maintaining the structural integrity of proteins. The lowered pH, caused by the accumulation of lactic acid, also makes them liable to denature. Denaturation is frequently accompanied by loss of the power to bind water and the falling pH causes the myofibrillar proteins to approach their isoelectric point. Both events cause exudation. Denaturation of the sarcoplasmic proteins also makes them liable to attack by the proteases or cathepsins of muscle, which are probably held inactive *in vivo* within particles known as lysosomes (De Duve and Beaufay, 1959) but are liberated and activated when the particle membranes are weakened by the falling pH.

The breakdown of proteins to peptides and amino acids, and the accumulation of various metabolises from the glycolytic process and from other sources, affords a rich medium for bacteria. Although growth of the latter is somewhat discouraged by the extent to which the pH falls, they are no longer subject to the scavenging action of 'white' blood corpuscles (since blood circulation has stopped).

A further aspect of the stoppage of the circulation is the cessation of long-term hormonal control of tissue metabolism. As it fails, the temperature falls and fat solidifies. The tendency for the fat to oxidize and become rancid is facilitated by failure of the blood to renew the supply of anti-oxidants, and by the accumulation of pro-oxidant molecules in the tissues.

5.4   Conditioning (ageing)

Although muscle is increasingly liable to suffer microbial spoilage in direct proportion to the time and temperature of holding post-mortem, hygienic abattoir operations will generally ensure satisfactory storage for a few days at room temperature and for about 6 weeks if the meat is held just above its freezing point ($-1.5\,°C$). Various processes applied to the commodity, such as curing, freezing, dehydration and irradiation, will vastly extend storage life, but, in so far as they are artificial, they are not relevant in this chapter. In the absence of microbial spoilage, the holding of unprocessed meat above the freezing point is known as 'conditioning' or 'ageing', and it has long been associated with an increase in tenderness and flavour (cf. Bouley, 1874). During the first 24–36 h post-mortem, the dominant circumstance is post-mortem glycolysis. This has already been considered in some detail in §§ 4.2.2 and 4.2.3. Even before the ultimate pH has been reached, however, other degradative changes have commenced. These continue until bacterial spoilage or gross denaturation and desiccation of the proteins have made the meat inedible. The extent of these changes, which affect the nature and amount of both proteins and small molecules, is generally limited, however, by the cooking and consumption of the meat.

5.4.1   Protein denaturation

Muscle, like all living tissues, represents a complexity of organization among molecules which is too improbable to have arisen from, or to be maintained by, their random orientation. The structure of the proteins which characterize contractile tissue can only be preserved against the tendency of the component atoms and molecules to become disorientated by the provision of energy (as ATP). Such energy is not available after death and the proteins will tend to denature. Denaturation may be defined as a physical or intramolecular rearrangement which does not involve hydrolysis of the chemical bonds linking the constituent amino acids of the proteins' polypeptide chains (Putnam, 1953). It is generally accompanied by an increase in the reactivity of various chemical groups, a loss of biological activity (in those proteins which are enzymes or hormones), a change in molecular shape or size and a decrease in solubility. Proteins are liable to denature if subjected, during post-mortem conditioning, to pH levels below those *in vivo*, to temperatures above 25 °C or below 0 °C, to desiccation and to non-physiological salt concentrations.

Of the proteins in muscle, it has been generally accepted that the collagen and elastin of connective tissue do not denature during conditioning (Ramsbottom and Strandine, 1949; Wierbicki et al., 1954). This view was supported by the concomitant absence of soluble, hydroxyproline-containing molecules, indicating that neither collagen nor elastin were proteolysed (Sharp, 1959); proteolysis would not normally precede denaturation. When, however, collagen is denatured, for example, by heating beef muscles to 60–70 °C for 20–25 min, there is a continuous nonenzymic, breakdown to hydroxyproline-containing derivatives (Sharp, 1963).*

During post-mortem conditioning, the proteins of the myofibril and of the sarcoplasm denature in varying degree. Immediately after death and before the onset of rigor mortis, muscles are pliable and tender when cooked. The principal proteins of the myofibril, actin and myosin, are dissociated and myosin is extractable at high ionic strength (Weber and Meyer, 1933; Bailey, 1954). With the onset of rigor mortis, as we have considered above, the muscle becomes inextensible and is tough when cooked (Marsh, 1964). As conditioning proceeds, the muscle becomes pliable once more (and increasingly tender on cooking); but this is *not* due to dissociation of actomyosin (Marsh, 1954); inextensibility remains. Although it was suggested (dos Remedios and Moens, 1995) that there are different levels of association between actin and myosin in rigor mortis which could change during conditioning, Hopkins and Thompson, (2001a, b) found no evidence that dissociation of actomyosin contributed to tenderness increments at this time. Increased lengthening under applied stress is observed, however, in post rigor muscle. This phenomenon appears to depend on changes in extracellular components (e.g. the mucopolysaccharide of the ground substance, the sarcolemma or collagen) which commence at the onset of rigor mortis (Dransfield et al., 1986). Busch et al. (1972b) followed the onset of rigor mortis by observing changes in isometric tension. This increases as extensibility decreases; but, whereas extensibility remains low, isometric tension diminishes again during the so-called resolution of rigor. Increase in isometric tension post-mortem is more marked in 'red' than in 'white' muscles. This difference appears to be related to dissimilarities in the sarcolemma of these two types of muscle.

The extractability at high ionic strength of total myofibrillar proteins decreases by about 75 per cent with the onset of rigor mortis, from the value immediately post-mortem, but on subsequent storage at 2 °C the extractability again rises – up to and even beyond the initial level (Locker, 1960a). It is significant that, in addition to a predominance of actomyosin, the myofibrillar proteins extractable at high ionic strength now include α-actinin, tropomyosin, and the troponin with which the latter is associated *in vivo* (Ebashi and Ebashi, 1964; Valin, 1968). This suggests that the

---

* It has become evident that collagen, together with an increasingly recognized number of other proteins, can exist in an unfolded form (i.e. not uniquely folded) under physiological conditions. Moreover, it seems likely that the unfolding of collagen induced by heat may well occur at a markedly lower temperature than has hitherto been believed (Gross, 2002).

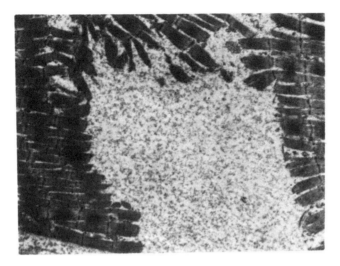

Fig. 5.2 Electron micrograph of a break across an aged, stretched muscle fibre. Each broken fibre has parted at the Z-line (×5000). (Courtesy Dr M. R. Dickson.)

process of conditioning detaches the actin filaments from the Z-line with which their union, probably via tropomyosin (Huxley, 1963), and zeugmatin (cf. Fig. 3.9) is weaker than with myosin. Figure 5.2 demonstrates that aged myofibrils break at the Z-lines on mild homogenizing (Davey and Dickson, 1970). The actin filaments collapse on to those of myosin, leading to lengthening of the A-bands (Davey and Gilbert, 1967), and there is increased weakness at the A–I junction, as shown by an increased gap between the A and I bands of the sarcomere (Davey and Graafhuis, 1976b).

During conditioning or ageing, the Z-lines in 'white' muscles appear to be more labile than those in the 'red' variety (Goll, 1970). Thus, they alter more rapidly in rabbit and porcine muscles than in those of the bovine (Henderson *et al.*, 1970), and more in bovine *semitendinosus* than in *psoas* (Goll *et al.*, 1974). Tenderness changes little in bovine *psoas* over 4 days ageing at 2 °C, whereas in *semitendinosus* it increases markedly during this period. It is significant that the latter has about three times the activity of CASF (calpains see below) as bovine *psoas*. It may be noted that 'white' muscles are less susceptible to 'cold shortening' than 'red', and that this has been attributed to their greater ability to control intramuscular concentrations of calcium ions because of a more effective sarcotubular system (§§ 4.3.5 and 10.3.3).

The extractability of myofibrillar proteins is affected by the ultimate pH of the muscle, a high ultimate pH tending towards greater extractability (cf. Table 4.29). The temperature post-mortem is also important, a high temperature being associated with lower extractability (Wierbicki *et al.*, 1956). This is partly due to the precipitation of sarcoplasmic proteins on to those of the myofibril (Bendall and Wismer-Pedersen, 1962). Some denaturation of the latter also occurs, however (cf. Table 5.3). This is implied by the greater difficulty of splitting muscle fibres into myofibrils after aseptic storage for 30 days at 37 °C than at 5 °C (Sharp, 1963), but in this case changes in the sarcoplasmic reticulum between each myofibril may be responsible (Lawrie and Voyle, 1962). Even at 35 °C denaturation of *isolated* myosin

Table 5.3 Percentage of sarcoplasmic protein precipitating from extracts of post-rigor beef *l. dorsi* (after Scopes, 1964)

Temp.:	0 °C	10 °C	15 °C	20 °C	25 °C	30 °C	37 °C	45 °C
pH								
4.5	4.4	3.8	4.2	4.0	4.4	4.7	5.1	8.1
4.8	6.6	5.7	4.8	6.4	6.6	7.4	5.3	15.0
5.2	5.0	4.9	5.5	6.2	7.8	10.5	18.5	35.0
5.7	3.1	2.9	3.2	3.1	4.2	6.2	12.2	34.0
6.0	2.1	1.9	2.2	2.5	3.2	6.3	8.5	29.0
6.5	0.6	0.8	0.7	2.1	2.8	3.4	6.6	24.5
7.1	0.4	0.4	0.6	0.5	1.2	2.1	5.2	22.0

Extracts exposed to temperature/pH conditions for 4 h.

is relatively speedy (Penny, 1967), and it can be presumed that some denaturation of actomyosin *in situ* occurs during post-mortem glycolysis.

An important aspect of changes in the myofibrillar proteins post-mortem, which is reflected in their extractability and tenderness, is the degree of shortening which occurs during the onset of rigor mortis (Locker, 1960a; Locker and Hagyard, 1963; Marsh, 1964). In muscles which go into rigor mortis in an extended condition, the filaments of actin and myosin overlap and cross-bond at fewer points, and the amount of actomyosin formed is small. Such meat is tender on cooking. On the other hand, when muscles go into rigor mortis in a contracted condition, there is considerable shortening since the actin and the myosin filaments interpenetrate extensively. There is much cross-bonding and the meat is relatively tough on cooking. Normally, muscle goes into rigor mortis in an intermediate condition wherein the overlapping of actin and myosin, the degree of cross-bonding and the toughness are somewhere between the two extremes (cf. Fig. 4.2). It has been shown that the rate of tenderizing during conditioning is minimal in muscles which have shortened substantially at onset of rigor mortis (Davey *et al.*, 1967). Lower activity of μ-calpain and increased calpastatin levels have been reported in cold-shortened muscles (Zanora *et al.*, 1998), and these circumstances must contribute to their toughness. The degree of shortening during rigor mortis is temperature dependent (cf. §§ 7.1.1 and 10.3.3).

By far the most labile proteins of muscle post-mortem are those of the sarcoplasm, the diversity of which is represented in Fig. 4.1 (p. 77). It has been realized for many years that proteins precipitate when muscle extracts of low ionic strength are allowed to stand at room temperature, the process being accelerated by raising the temperature and by the addition of salt and acid (Finn, 1932; Bate-Smith, 1937b). It has been shown, for example, that as the pH falls to acid levels during post-mortem glycolysis, glyceraldehyde phosphate dehydrogenase (cf. Fig. 4.3), which is a basic protein, denatures and combines with the more acidic myosin and actin, further promoting actomyosin formation and the predominance of actomyosin ATP-ase (rather than that of myosin) during subsequent conditioning (Matsuishi and Okitani, 2000). The behaviour of sarcoplasmic proteins in extracts from beef *l. dorsi* muscle under various temperature–pH combinations is shown in Table 5.3. It will be seen that an increase of temperature causes increasing precipitation of sarcoplasmic proteins at all pH values studied; that at all temperatures maximum

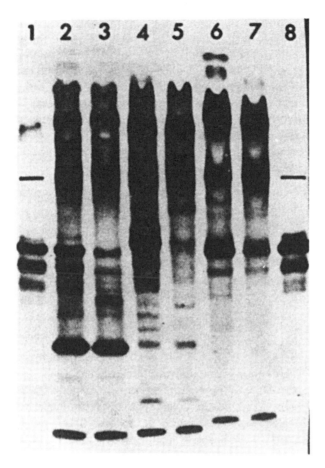

Fig. 5.3 Starch gel electrophoretograms showing relative stability of sarcoplasmic proteins in extracts from *l. dorsi* muscles of various species undergoing rigor mortis at 0 or 37 °C. (1) Purified beef creatine kinase. (2) Beef, 0 °C. (3) Beef, 37 °C. (4) Rabbit, 0 °C. (5) Rabbit, 37 °C. (6) Pig, 0 °C. (7) Pig, 37 °C. (8) Purified pig creatine kinase. (Courtesy Dr R. K. Scopes.)

precipitation occurs at a pH of 4.8–5.2; but that at some temperatures between 37 and 45 °C, a high ultimate pH no longer protects sarcoplasmic proteins against precipitation (Scopes, 1964). Even after heating at 60 °C for 10 h a proportion of the sarcoplasmic proteins is still soluble and will separate electrophoretically; but after 2 h at 80 °C almost all sarcoplasmic proteins except myoglobin have become insoluble (Laakkonen *et al.*, 1970). Obviously, even the attainment of a normal ultimate pH (about 5.5) during post-mortem glycolysis must be associated with the precipitation of some of the sarcoplasmic proteins (on the presumption that *in vivo* behaviour reflects that *in situ*). A high temperature during post-mortem glycolysis causes additional precipitation. This is exemplified in Fig. 5.3, from which it will be clear that one of the most labile of the sarcoplasmic proteins in the muscles of beef, rabbit and pig is the enzyme creatine kinase (Scopes, 1964). Particularly severe precipitation of sarcoplasmic proteins occurs during post-mortem glycolysis in the muscles of pigs which appear pale and are exudative post-mortem (cf. § 3.4.3; Scopes and Lawrie, 1963). In these there is a combination of low pH and high temperature

(Bendall and Wismer-Pedersen, 1962), and sarcoplasmic proteins precipitate on to those of the myofibril lowering their extractability and water-holding capacity. As observed, histologically, the precipitated sarcoplasmic proteins form bands across the muscle fibre (Fig. 3.11) and lower the extractability of the myofibrillar proteins, even although the latter may not be denatured themselves (Bendall and Wismer-Pedersen, 1962). The greater prevalence of such bands in affected musculature of ultimate pH 5.4 than in those of ultimate pH 4.7 and their presence in muscle of high ultimate pH (Lawrie *et al.*, 1963a) follows from the behaviour indicated in Table 5.3. There is clearly a critical temperature, between 37 and 45 °C, above which a high ultimate pH fails to keep sarcoplasmic proteins in solution. That pigs affected by the so-called PSE condition seem to have a higher temperature than normal immediately post-mortem (Bendall and Wismer-Pedersen, 1962) and that the condition is said to be induced artificially by holding pigs at 45 °C for a period before slaughter (Sayre *et al.*, 1963b) substantiates this view. It is interesting that, as already indicated, some adverse change should also occur in the myofibrillar proteins at about this temperature (Marsh, 1962).

After the ultimate pH has been reached, further changes occur in the sarcoplasmic proteins, there being a general alteration in the nature of the components (Deatherage and Fujimaki, 1964).

Denaturation of the principal muscle pigment, myoglobin, which is another of the sarcoplasmic proteins, accelerates the oxidation of its iron to the ferric form, the pigment turning brown (metmyoglobin). Although, considering the muscle as a whole, this is not an extensive process, it is, nevertheless, a very important one for it occurs preferentially near exposed surfaces or where the oxygen tension is about 4 mm (Brooks, 1935, 1938). Such factors as desiccation can initiate the denaturation and discoloration, especially where the ultimate pH is relatively low. It is also linked with still surviving activity in the oxygen utilizing enzymes (succinic dehydrogenase and cytochrome oxidase) which persists for some time at 0 °C. This matter will be referred to again below.

As far as meat quality is concerned, perhaps the most important manifestation of the post-mortem denaturation of the muscle proteins is their loss of water-holding capacity, because in practice it is a more universal phenomenon than discoloration. The point of minimum water-holding capacity of the principal proteins in muscle (i.e. the isoelectric point) is 5.4–5.5 (Weber and Meyer, 1933). Since, as we have seen in Chapter 4, the production of lactic acid from glycogen, at any given temperature and rate, will *generally* cause the pH to reach 5.5, normal meat will lose some fluid ('weep'). This will, obviously, be less if the ultimate pH is high, however (Empey, 1933).

The contribution by the sarcoplasmic proteins to overall water-holding capacity, once lost by precipitation during the attainment of even a normal ultimate pH, cannot be regained by applying a buffer of high ultimate pH to the muscle. Thus, the relatively low water-holding capacity of fibres prepared from muscle of low ultimate pH remains lower than that of fibres prepared from muscle of an intrinsically high ultimate pH, even when placed in a medium having the latter pH value (Penny *et al.*, 1963).

For a given muscle, water-holding capacity is at a minimum at the ultimate pH; thereafter, on subsequent conditioning of the meat, it tends to increase (Cook *et al.*, 1926). This may be due to an increased osmotic pressure, caused by the breakdown of protein molecules to smaller units (proteolysis will be discussed below); but much intramolecular rearrangement, not involving splitting but causing changes in the electrical charges on the protein, may also be responsible (Bendall, 1946). Insofar as Kristensen and Purslow (2001) found that vinculin, desmin and talin were proteolysed during conditioning, they suggested that the break-up of the cytoskeleton destroyed the force expressing water to the cell exterior; and its reabsorption could be responsible for the observed increase in water-holding capacity on ageing. There is concomitantly an increase in the pH of meat when it is held above the freezing point (Sair and Cook, 1938; Wierbicki et al., 1954; Bouton et al., 1958). The pH rise is more marked when the temperature of holding is high and is greater in pork than in beef (Lawrie et al., 1961).

These changes in pH are accompanied by changes in ion–protein relationships. Arnold et al. (1956) found that sodium and calcium ions are continuously released into the sarcoplasm by the muscle proteins, and potassium ions are absorbed after the first 24 h. Because of the large excess of potassium ions absorbed on to the muscle proteins, the net charge on the latter increases, and, thereby, the water-holding capacity.

5.4.2 Proteolysis

Denatured proteins are particularly liable to attack by proteolytic enzymes (Anson and Mirsky, 1932–1933; Lineweaver and Hoover, 1941). The increase in tenderness, observed on conditioning, was found many years ago to be associated with an increase in water-soluble nitrogen (Hoagland et al., 1917; Fearson and Foster, 1922), due to the production of peptides and amino acids from protein. There has been much controversy as to which proteins undergo proteolysis during the holding of meat at temperatures above the freezing point.

Although *extensive* proteolysis of the collagen and elastin of connective tissue might appear to be the most likely change causing increased tenderness, the proteins of connective tissue are not normally changed in this way during conditioning in skeletal muscle. This was conclusively shown by Sharp (1959). There is no increase in water-soluble hydroxyproline-containing derivatives, even after storage of sterile, fresh meat for one year at 37 °C. Despite the absence of massive proteolysis of native collagen during conditioning, such as would require the action of a true collagenase capable of cleaving all three chains in the helical region of tropocollagen (Gross, 1970), there are lysosomal enzymes (Valin, 1970) which can attack the cross-links in the non-helical telopeptide region of collagen. In 1974 Etherington isolated two collagenolytic cathepsins from bovine spleen which cleaved the non-helical, telopeptide region of native tropocollagen between the lysine-derived cross-links and the triple helix of the main body of the molecule. This resulted in longitudinal splitting and dissociation of the protofibrils. The enzymes operated at pH 4–5 (28 °C). Suzuki et al. (1985) isolated a collagenolytic enzyme from rabbit muscle which is bound to collagen; and Stanton and Light (1988) provided direct biochemical evidence for the action of proteolytic enzymes on the perimysial collagen during conditioning.

Collagen fibres appear to swell during conditioning.

The collagen of the endomysium, which opposes the swelling of the muscle fibre initially, is weakened during conditioning (Wilding et al., 1986). Electron micrography reveals that the sheaths of connective tissue become diffuse (Fig. 5.4; Nishimura et al., 1995). Stanton and Light (1990) demonstrated that it is the type III collagen which is preferentially attacked in the endomysium during conditioning rather than

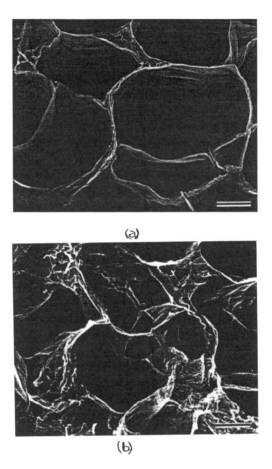

Fig. 5.4 Scanning electron micrographs of the endomysial sheaths of bovine *semitendinosus* muscle from which muscle fibres have been removed (a) immediately post-mortem and (b) after conditioning for 28 days at 4 °C, when the endomysial sheaths have become diffuse. The bar represents 25 μm. (Reprinted from Nishimura *et al.*, 1995, with kind permission from Elsevier Science Ltd. and courtesy of Prof. K. Takahashi.)

the type I component. Changes in the links between muscle fibres and their endomysial sheaths, whereby the latter are more readily removed as conditioning proceeds, may also be invoked in accounting for the swelling which is observed.

As tenderness increases, there is a concomitant increase in the titre of free β-glucuronidase (Dutson and Lawrie, 1974). This enzyme can attack the mucopolysaccharide of the ground substance or carbohydrate moieties in collagen itself. One of the points of attachment of carbohydrate to collagen is the ε-amino group of lysine; and the ε-aminoglycosylamines are probably involved in binding collagen to the ground substance (Robins and Bailey, 1972). Nishimura *et al.* (1996) clearly demonstrated that proteoglycans in both the basement membrane and the perimysium are degraded during conditioning. It may be, therefore, that splitting of both carbohydrate and peptide links contribute to increased tenderness in conditioning.

It is clear, however, that under some circumstances connective tissue proteins are much more labile than they appear to be post-mortem. Preceding the repair of damaged muscles, collagen and elastin are evidently removed *in vivo* (Partridge, 1962). There is a general increase in phagocytic (Rickenbacher, 1959) and in proteolytic activities – the latter due to the liberation of catheptic enzymes from lysosomes (Hamdy *et al.*, 1961). Anti-inflammatory (anti-rheumatic) drugs, such as cortisone, inhibit the formation of the acid mucopolysaccharides of the ground substance of connective tissue by suppressing sulphation (Whitehouse and Lash, 1961) and decrease the amount of free hydroxyproline (Kivirikko, 1963). Vitamin C deficiency interferes with collagen formation by inhibiting the hydroxylation of soluble proline (Stone and Meister, 1962) Again, during post-partum involution of the uterus, enzymes are elaborated which are capable of breaking down connective tissue proteins to their constituent amino acids (Woessner and Brewer, 1963). It has been postulated that the collagen fibril is first attacked extracellularly by a secreted neutral collagenase and that its subsequent digestion is intracellular (Etherington, 1973), by macrophages (Parakkal, 1969; Eisen *et al.*, 1971). These reactions imply that *in vivo* muscle is capable of elaborating enzymes which proteolyse connective tissue proteins, in abnormal circumstances, even if they are not present, or are inactive, during conditioning.

Notwithstanding the absence of massive proteolysis in the collagen and elastin of fresh sterile meat, even after 1 year at 37 °C, such breakdown does occur in sterile meat which has been heated. For example, in beef held at 37 °C, after heating for 15 min at 70 °C, soluble hydroxyproline rose from about 2 per cent to about 23 per cent of the total hydroxyproline during 97 days (Sharp, 1964). In corresponding beef which had been heated for 45 min at 100 °C (being thus cooked), the value rose from 12 per cent initially to 55 per cent over the same period of subsequent holding at 37 °C. Histological examination revealed that the connective tissues of the perimysium had been weakened, since fibre bundles were easily separated from one another. In view of the preceding heat treatment, however, the breakdown of collagen (or elastin) in these circumstances can scarcely have been due to enzymic action: progressive physical changes in the connective tissue proteins are probably involved (Gustavsen, 1956).

As mentioned in § 5.4.1, the absence of changes in the extensibility of muscle in conditioning – and subsequent to actomyosin formation during the onset of rigor mortis – despite the concomitant increase in tenderness, indicated that the latter phenomenon did not involve dissociation of actin from myosin (Marsh, 1954). A similar conclusion was reached by Locker (1960b). He applied Sanger's method of N-terminal analysis (1945) to the salt-soluble proteins of beef muscle during conditioning at low and high temperature and failed to detect any significant increase in the number of protein N groups. It must be appreciated, however, that significant changes in muscle proteins, which might alter the tenderness of meat, could occur without extensive proteolysis, if a few key bonds were broken, as indicated above.

At least some of the changes in the myofibrillar proteins during conditioning are apparently initiated by the release of Ca^{++} ions from the sarcoplasmic reticulum post-mortem (the capacity of which to accumulate Ca^{++} ions decreases during conditioning: Newbold and Tume, 1976) and operate through water-soluble enzymes. These enzymes are variously referred to as calcium-activated sarcoplasmic factors (CASF), calcium-activated neutral proteinases (CANP), calcium-dependent proteinases (CDP) or calpains. It is significant that the Ca^{++}-chelating agent, ethylenediamine tetraacetate, should prevent ageing changes (Penny, 1974; Koohmaraie *et al.*, 1988).

The calpains are now known to belong to a complex family of Ca^{++}-dependent proteinases (Sorimachi *et al.*, 1997). Whereas the calpain isoenzymes 1, 2 and 3 have been long recognized, recently other members, numbered 5, 7, 10, 12, 14 and 15 have been discovered in biological tissues (Dear *et al.*, 1997). Of these, Ilian *et al.* (2004) concluded that 10 was strongly correlated with the degree of tenderization of meat during ageing through its action on nebulin and desmin. The calpain proteolytic system in mammalian striated muscle comprises the ubiquitous enzymes, calpains 1, 2 and 10, and a tissue-specific isozyme, calpain 3 (Goll *et al.*, 2003).

The calpains are inhibited by the protein, calpastatin (cf. §4.3.5). The helical sequences of calpastatin prevent calpains from binding to membranes (Mellgren *et al.*, 1989). The calpains degrade desmin (Granger and Lazarides, 1978; Penny, 1980; Slinde and Kryvi, 1986) and weaken the binding of α-actinin to the Z-disc. Tropomyosin and *M*-line protein are also degraded (Penny, 1980). Penny and Dransfield (1979) and, later, Nishimura *et al.* (1996) showed that, during conditioning of beef muscles, troponin T is proteolysed, with concomitant production of four peptides, of which the principal member has a MW of 30,000. Earlier it had been demonstrated that CASF degrades the so-called 'gap filaments' (Locker, 1976) (connectin, titin; Maruyama *et al.*, 1979). Thus when *stretched* muscles are aged, subsequent cooking causes the 'gap filaments' to disappear (whilst the Z-lines appear intact; Davey and Graafhuis, 1976b) and Young *et al.*, 1980 detected increased solubility of connectin and aged beef muscle. In a detailed investigation of ageing changes in rabbit muscles, Mestre-Prates *et al.* (2002) proposed that tenderness was due to specific cleavage of titin and nebulin in the vicinity of the N_2 line by calpains; and these enzymes have been shown to degrade the integrin complex by which the cell proteins are attached to the cell membrane (Lawson, 2004; cf. §§ 3.2.2. and 10.2).

It is worth noting that the calpain system has no action on actin or myosin *per se* (Penny, 1974; Robson *et al.*, 1974) and that it is, in fact, located at the Z-line (Goll *et al.*, 1992). Koohmaraie *et al.* (1987) demonstrated that during storage of *l. dorsi* muscles over 14 days at 0 °C, there was a concomitant increase in myofibrillar fragmentation (as an index of conditioning changes) and a decrease in the activity of that calcium-activated neutral proteinase which depends on a *low* concentration of Ca^{++} ions (calpain I, μ-calpain), whereas the CANP which depends on a relatively high concentration (calpain II m-calpain) remained unchanged in activity. About 50 per cent of the total change in these parameters occurred during the first 12 hours, when the temperature was falling from *ca.* 37 to 10 °C, indicating that the tenderness increment during conditioning may well commence before the ultimate pH has been reached, especially if the temperature is slow to fall.

Because of the difficulty of attributing conditioning changes precisely to individual proteases and their inhibitors *in situ*, models of the system have been studied. Dransfield (1992, 1993) and Dransfield *et al.* (1992) examined extracts of beef muscle *in vitro* when these were stored at a range of temperatures from 0 to 30 °C. They postulated that the low levels of free Ca^{++} ions in the immediate post-mortem period would be insufficient to activate calpain I, but that, when the pH had fallen to *ca.* 6.1, the Ca^{++} level would have become high enough to activate this enzyme (calpain II would act similarly, but at a greater Ca^{++} level). At this pH calpains are bound to the inhibitor, calpastatin, but this inhibitory action falls as the pH drops further, from 6 to 5.5 and the activated calpains proteolyse the calpastatin. Tenderizing is initially due to the action of calpain I. Subsequently calpain II is responsible: it ceases as the calpains self-destruct by autolysis (Dransfield *et al.*, 1992; Dransfield, 1993). Subsequently, Dransfield (1994), in a detailed assessment of the data provided by modelling, concluded that variability in the post-mortem activity of calpains *per se* can be adduced to account for toughness, irrespective of such factors as sarcomere length. It must be acknowledged, however, that such models omit consideration of the contribution of the lysosomal cathepsins, of their activators and inhibitors and of the 'multicatalytic proteinase complex' (Orlowski, 1990) to post-mortem tenderizing (cf. below). Since the activation energy for the autolysis of calpain I is higher than for its proteolytic activity, less intense tenderizing occurs at higher temperatures. In addition to the solubilizing action of the calpains, however, it has been suggested that Ca^{++} ions *per se* cause non-enzymic 'salting-in' changes in certain myofibrillar proteins (Taylor and Etherington, 1991). Moreover, Tatsumi and Takahashi (1992) and Takahashi (1992) were able to demonstrate that *in vitro* 0.1 mM calcium chloride caused fragmentation of nebulin and the release of paratropomyosin (from the A-I junction region) which then binds to actin filaments. These non-enzymic factors lead to disruption of the latter and to lengthening of the sarcomeres. They thus concluded that the increasing level of Ca^{++} ions, as they are released from the sarcoplasmic reticulum post-mortem, contribute directly to the tenderizing changes during ageing. Although Whipple *et al.* (1994) postulated that most of the increase in tenderness was due to stimulation of the calpains by the Ca^{++} ions investigations by Hopkins and Thompson (2001b) could not confirm that they had a rôle in the process independently of the calpains.

The proteolysis of troponin T and increase of tenderness correlate well when conditioning takes place between 3 and 15 °C (Penny and Dransfield, 1979). At higher temperatures (25–35 °C), when protein denaturation is a contributory factor, and at 0 °C, when 'cold-shortening' may be anticipated, toughness is greater than the degree of proteolysis of troponin T would predict. On the other hand, the tenderness of electrically stimulated muscles is greater, for a given degree of troponin T breakdown, than that of controls (George *et al.*, 1980), suggesting that, although the proteolysis of troponin T is a useful indicator of change during conditioning, other factors must be considered.

There is at least one other enzyme system involved in meat conditioning, namely that of the lysosomes already mentioned. In contrast to the calcium-activated sarcoplasmic factors (calpains) which have pH optima above 6, the cathepsins (B, D, H and L) of the lysosomes represent a series of proteolytic enzymes with pH optima below 6 (Penny and Dransfield, 1979; Etherington, 1984). Of these, cathepsin H cannot degrade native myofibrillar proteins and, although cathepsin D can degrade myofibrillar proteins below pH 5, its action in post-mortem conditioning at the normal ultimate pH of 5.5 is minor (Ouali *et al.*, 1987).* On the other hand, both cathepsins B and L can degrade these proteins in post-mortem muscle (Bird *et al.*, 1977; Matsukura *et al.*, 1981). Cathepsin L is probably the most important lysosomal proteinase in conditioning (Mikami *et al.*, 1987). It degrades troponins T and I, and C-protein rapidly, and titin (connectin), nebulin, α-actinin, tropomyosin, actin and the light and heavy chains of myosin slowly. Its action at pH 5.5 is faster than

---

* *In vitro* examination has shown that cathepsin D can degrade (bovine) F-actin at numerous locations, bonds containing at least one hydrophobic amino acid residue being preferentially cleaved (Hughes *et al.*, 2000).

at pH 6, but slower than at pH 5 (Mikami *et al.*, 1987). The breakdown of myosin is not marked unless circumstances are exceptional (e.g. pH below 5, over 24 h at 25 °C: Penny and Ferguson-Pryce, 1979), or when meat is stored at 35 °C for some days (Penny and Dransfield, 1979).

The structurally important sites of enzyme actions by calpains and lysosomal enzymes (including cathepsins B, D and L) during conditioning can be summarized:

(a)   *CASF (calpains)*
    (i)   troponin T (above pH 6)
    (ii)   Z-lines (desmin)
    (iii)   connectin, 'gap filaments'
    (iv)   *M*-line proteins and tropomyosin
(b)   *Lysosomal enzymes (including cathepsins B, D and L)*
    (i)   troponins T and I (below pH 6) and C-protein: relatively rapidly
    (ii)   myosin (heavy and light chains), actin, tropomyosin, α-actinin, nebulin and titin ('gap filament'): relatively slowly above pH 5 or below 35 °C
    (iii)   cross-links of non-helical telopeptides of collagen
    (iv)   mucopolysaccharides of ground substance.

A third source of proteolytic activity has been discovered. This resides in a large complex of molecular weight *ca.* 700 kdalton, which consists of at least three distinct sub-units, each having different proteolytic activity against hydrophobic, basic and acidic amino acid sequences (Wilk *et al.*, 1979; Wilk and Orlowski, 1983). Since there is evidence that these different enzymic components cooperate in their proteolytic action (Wilk and Orlowski, 1983) and that disruption of the complex causes complete loss of proteolytic activity, it appears to be a functional unit and to justify the name 'multicatalytic protease complex' (Orlowski, 1990). Also referred to as the proteasome, its 20S component rapidly hydrolyses myofibrillar proteins *in vitro* (Robert *et al.*, 1999). Proteasome activity persists post-mortem and, after the action of μ and m-calpains, participates in a second wave of proteolysis, attacking denatured proteins (Lamare *et al.*, 2002). The complex has been identified in many tissues. Its presence in muscle suggests that it must be considered in elucidating proteolysis in this tissue.

Since most of the proteins of connective tissue and the myofibrils are not subjected to *extensive* proteolysis during conditioning, the considerable increments in the soluble products of protein breakdown must arise from the sarcoplasmic proteins. As we have seen, these denature in varying degrees during post-mortem glycolysis (§ 5.4.1); and chromatography of extracts prepared from muscle after increasing periods of storage show a gradual diminution of various components (Deatherage and Fujimaki, 1964).

During storage at 37 °C of sterile *l. dorsi* muscles of beef, the total soluble *protein* nitrogen was found to fall from 28 to 29 per cent of the total nitrogen to 13, 11 and 6 per cent after 20, 46 and 172 days respectively (Sharp, 1963). The lowered concentration of sarcoplasmic proteins was due rather to their proteolysis to amino acids and not to precipitation, which could only account for a small amount of the diminution: the nitrogen soluble in trichloroacetic acid rose from 11 per cent of the total protein to 17, 23 and 31.5 per cent respectively in these same periods. Moreover, in terms of a specific amino acid, the percentage of total tyrosine soluble in trichloroacetic acid rose from 11 per cent initially to 13, 17 and 35 per cent over 20, 46 and 172 days respectively (Sharp, 1963). Comparable changes

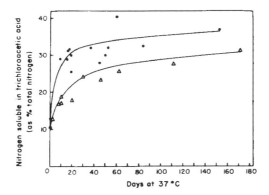

Fig. 5.5 Production of nitrogen soluble in trichloroacetic acid during aseptic storage of muscle at 37 °C. ●–●, rabbit muscle, △–△, beef muscle. (Courtesy of the late Dr J. G. Sharp.)

were found in rabbit *l. dorsi* although the rates of proteolysis are different in the two species (Fig. 5.5), and even between different muscles (J. G. Sharp, personal communication).

Ouali and Talmant (1990) and Monin and Ouali (1991) have confirmed such differences in extensive studies and established the basis for them. Variability in the rate of ageing reflects the contents of calcium-dependent proteinases (calpains I and II and their inhibitor, calpastatin), of lysosomal cathepsins B and L (and their inhibitors), the relative resistance of the muscle proteins to proteolysis and the intramuscular osmotic pressure. In turn, these variations are systematically related to the proportions of 'red', 'white' and intermediate type fibres which characterize the muscles concerned.

According to Radouco-Thomas *et al.* (1959), proteolysis is less marked in the muscles of pigs and sheep than in those of lamb and rabbit under comparable conditions. In a comparative study of conditioning changes in *l. dorsi*, Dransfield *et al.* (1981) confirmed these findings. Rates of tenderness increments in beef, lamb, rabbit and pork were, respectively, 0.17, 0.21, 0.25 and 0.33% per day. Subsequently, Etherington *et al.* (1987) showed that the relative rates were reflected by the relative concentrations of cathepsins B and L in the muscles; and Ouali and Talmant (1990) demonstrated that the rates were also correlated with the ratio of calpain II: calpastatin in the muscles of different species.

It is most important to note that these observations refer to a normal ultimate pH (i.e. about 5.5). At a higher ultimate pH the extent of proteolysis is less (Radouco-Thomas *et al.*, 1959). Thus, in rabbit *l. dorsi* after storage for 16 days at 37 °C, 17 per cent of the total tyrosine was soluble when the ultimate pH was 5.8. The corresponding value was only about 9 per cent, however, when the ultimate pH was 6.8 (Sharp, 1963); and there was a smaller degree of disintegration of muscle fibres during homogenizing.

Dransfield (1993) confirmed that, although the *extent* of proteolysis is relatively less at high ultimate pH, its *rate* is increased. The relationship between the rate of proteolysis and the ultimate pH was shown to be complex by Watanabe *et al.* (1996) in an extensive study of ovine *l. dorsi* of ultimate pH ranging from 5.5 to 7.0, the rate being minimal at *ca.* pH 6.0 and greater below and (especially) above this value. On the other hand myofibrillar fragmentation was least at pH 6.4.

Table 5.4 Dimensions of fibres present in greatest number in low-speed homogenates of sterile stored beef *l. dorsi* muscle of normal ultimate pH

Storage characteristics	Length (μm)	Diameter (μm)
Control (2 days at −20 °C)	650–1300	200–600
30 days at 37 °C	250–430	43–170
30 days at 5 °C	50–170	14–86

The extent of proteolysis is also temperature dependent, being greater at 37 °C than at 5 °C, although the degree of histological breakdown, as shown by the cohesiveness of fibres after homogenizing, is much greater at 5 than at 37 °C. This is, presumably, because there is a greater degree of denaturation of the myofibrillar proteins at the higher temperature (Table 5.4, Sharp, 1963) which would oppose breaking up of the tissue.

Low ultimate pH may enhance proteolytic activity in another way. The lysosomes, which contain enzymes having proteolytic activity and acid pH optima (De Duve, 1959a), have lipoprotein membranes which, whilst intact at *in vivo* pH levels under normal conditions, rupture when the pH falls post-mortem, or when there has been extensive tissue damage (Hamdy *et al.*, 1961), and liberate the proteolytic enzymes. The permeability of these membranes appears to be controlled by the vitamin A status of the tissue, hypervitaminosis A being associated with undue fragility (Fell and Dingle, 1963). It is also lowered following tissue breakdown (De Duve, 1959b) as in the dystrophies due to recessive genes or to vitamin E deficiency (Tappel *et al.*, 1962); in such dystrophies the activity of the lysosomal proteolytic enzymes is increased.

It is evident that some proteolytic activity may be due to residual blood in the muscle (Shestakov, 1962), and in 1974 Bailey and Kim showed that the proteinases from the porcine leucocyte lysosomes can degrade myofibrillar proteins. The question of whether the proteolytic activity observed in muscle post-mortem is a property of lysosomes intrinsic to the tissue or of those belonging to entrained phagocytes was resolved by Canonico and Bird (1970) who demonstrated that the former had relatively greater contents of acid phosphatases than of cathepsins. Venugopal and Bailey (1978) compared the lysosomal proteinases of the muscular tissue and leucocytes of beef and pork. They found that cathepsins D and E,[*] which had pH optima of 4.0 and 2.5, respectively, were the most active proteolytic enzymes found in both tissues, and that all the enzymes from the lysosomal leucocytes were more active than their counterparts in the lysosomes of the muscular tissue (cf. Table 5.5).

It has always been difficult to accept that proteolytic enzymes have only catabolic functions in muscle. It is now evident, however, that the calpain/calpastatin enzyme system is involved in protein turnover (§ 2.5.2.1), in the control of muscular excitation (§§ 3.2.2 and 4.2.1) and, indeed, in a number of other intracellular processes which are mediated by Ca^{++} ions.

---

[*] Terminology of Barrett (1977).

Table 5.5 Specific activity of bovine lysosomal proteinases (after Venugopal and Bailey, 1978) (pH of measurement in brackets)

Source	Carboxypeptidases			Cathepsins		Collagenase	Dipeptidyl-aminopeptidase I
	A (5.0)	B (6.0)	B (7.8)	D (4.0)	E (2.5)	(7.0)	(6.8)
Leucocytes	30	23	110	2751	1482	0.75	140
Muscle (diaphragm: 10,000g fraction)	12	8	24	1878	1132	0.15	28

5.4.3 Other chemical changes

By the time the ultimate pH has been reached, ATP has been largely broken down to inosinic acid, inorganic phosphate and ammonia (§ 4.2.3). Although some degradation of inosinic acid to phosphate, ribose and hypoxanthine will have occurred at this stage, the latter process is substantially a function of time, temperature and pH after the attainment of the ultimate pH (Solov'vev, 1952; Lee and Webster, 1963). According to Howard et al. (1960b), conditioning is organoleptically at an optimum when the hypoxanthine level has reached 1.5–2.0 μmoles/g. In beef this is attained after 10–13 days at 0 °C, 4–5 days at 10 °C, 30–40 h at 20 °C and 10–11 h at 30 °C (Lee and Webster, 1963). The rate of hypoxanthine formation is increased by a high ultimate pH, however, and this circumstance must be considered when assessing the time–temperature history of meat.

In view of the development of flavour which accompanies conditioning it is of interest that many years ago hypoxanthine, or its precursor inosinic acid, was reported to enhance flavour when added to meats (Kodama, 1913). It has been shown that inosinic acid (or inosine and inorganic phosphate) when heated with a glycoprotein containing alanine and glucose (also isolated from the water-soluble extracts of beef) produces a basic meat flavour and odour (Batzer et al., 1962). The breakdown of protein and fat during conditioning also contributes to flavour by producing hydrogen sulphide, ammonia, acetaldehyde, acetone and diacetyl (Yueh and Strong, 1960); but prolonged conditioning, e.g. 40–80 days at 0 °C is associated with loss of flavour (Hoagland et al., 1917). And, of course, where oxidative rancidity occurs in fat, the products affect flavour in a highly adverse manner (Lea, 1939). Oxidative rancidity in fat is retarded by a high ultimate pH as also is the oxidation of myoglobin (Watts, 1954) with which it is frequently linked. These phenomena will be considered in more detail in a later chapter.

Apart from the increase in free amino acids arising from proteolysis, their concentration is also augmented by the breakdown of various peptides. During conditioning, for example, the dipeptides carnosine and anserine are progressively hydrolysed to β-alanine and histidine (Bouton et al., 1958). The accumulation of free amino acids, and of soluble carbohydrates, such as glucose (by the action of α-amylase on glycogen; Sharp, 1958), glucose–6-phosphate (one of the intermediaries in the glycolytic pathway), ribose (from nucleotide breakdown) and other sugars in traces, is potentially undesirable. During the preparation of dehydrated meat, for example, the carbonyl groups of the carbohydrates will combine with the amino nitrogen of amino acids non-enzymically to form unsightly brown compounds which are also troublesome in having a bitter taste. The Maillard reaction, as it is known, may also take place between the sugars and intact protein (Lea and Hannan, 1950).

Although conditioning enhances the water-holding capacity of proteins to some extent, the loss due to denaturation changes and to post-mortem pH fall predominates, and meat exudes fluid post-mortem.

Chapter 6

The spoilage of meat by infecting organisms

Changes which take place during the conversion of muscle to meat, both immediately post-mortem and later, on keeping the commodity above the freezing point, were described in Chapter 5. It was emphasized that these occurred in meat irrespective of the presence or absence of extraneous organisms. But meat, like all human foodstuffs, is acceptable to other organisms and is susceptible to invasion by them. Their invasion of the meat (infection), the consequent production of unattractive changes (spoilage), the factors controlling their growth and the question of prophylaxis will now be separately considered.

6.1 Infection

The organisms which spoil meat may gain access through infection of the living animal (endogenous disease) or by contamination of the meat post-mortem (exogenous disease). The consumer is more likely to encounter the latter. Nevertheless, both aspects are important. It should be noted that serious human infections can be acquired from apparently healthy animals (Dolman, 1957).

6.1.1 Endogenous infections

Before considering diseases caused in people by consumption of meat from infected animals, brief reference should be made to those which are transmitted by contact, namely anthrax, bovine tuberculosis and brucellosis. These are caused by the microbes *Bacillus anthracis*, *Mycobacterium tuberculosis* and *Brucella* spp. respectively. Anthrax is mainly contracted by contact with the hides and hair; but consumption of underdone meat from infected animals can also cause human disease and death. Although the main vehicle for infection with bovine tuberculosis is raw milk, contact with affected carcasses is also a serious source (Dolman, 1957). Reflecting aspects of the intensive urbanization of the Industrial Revolution, it has been shown that 20–30 per cent of all cattle in Britain were affected by tuberculosis towards the end of the nineteenth century (Rixson, 2000). Although human beings are not apparently susceptible to the virus causing foot and mouth disease, it has long been a scourge of cattle, sheep and pigs. It was first reported in Britain in 1839 (Stratton and Brown, 1978). The skin (and the mucous membranes) is also the route of infection from carcasses carrying *Brucella* spp. These diseases tend to be localized in certain areas of the world.

Most other diseases arising from *in vivo* infection in meat animals are acquired by consumption of infected carcasses. The infections may be caused by bacteria or by parasitic worms. Perhaps the most important bacterial diseases in this category are those caused by members of the genus *Salmonella*. The consumption of inadequately cooked meat is the usual method of infection. *S. typhimurium* is found in lambs, calves and adult bovines, the principal source of their infection being on the farm (Nottingham and Urselmann, 1961). *S. cholerae suis* is mainly confined to pigs. Even healthy carriers may transfer *Salmonella* to normal animals whilst these are being held awaiting slaughter (Galton *et al.*, 1954). In an investigation in Australia nearly half of the animals passed as healthy had *Salmonella* spp. in the rumen liquor (Green and Bronlee, 1965). A very marked increase between 1961 and 1965 in the outbreaks of salmonellosis in calves, involving *S. typhimurium*, was attributed to the use of heavily infected premises for holding and sale (Melrose and Gracey, 1975). It is now evident that *Listeria monocytogenes* also can cause serious illness and death in people through the consumption of products such as milk and cheese from infected animals (Fleming *et al.*, 1985; James *et al.*, 1985); and ready-to-eat meat products if insufficiently processed. *M. paratuberculosis* ingested from beef which had been derived from animals grazing pasture contaminated with rabbit droppings, is believed to be the cause of increased incidence of Crohn's disease in UK consumers.

More prevalent are the infections acquired by ingestion of meat infested with parasitic worms. Stoll (1947) calculated that there may be a world total of 27 million cases of trichinosis and 42 million of taeniasis (caused by the roundworms and tapeworms of beef and pork). There are also about 100,000 cases of echinococcosis (disease caused by certain other tapeworms).

Trichinella spiralis is a small nematode worm with many potential hosts including humans. Infection may prove fatal (Zenker, 1860). Human trichinosis is a serious public health problem in the USA due to the practice there of feeding uncooked garbage to hogs in certain areas. The disease is found only where raw, inadequately cooked or improperly cured meat (especially pork) is eaten. Even in remote areas of the world the disease may be endemic, as in the Arctic where the flesh of whale, polar bear and walrus may be implicated. Fresh pork may be rendered harmless by exposure to sufficient heat, cold, salt, smoke or ionizing radiations. For example, exposure to $-38\,°C$ for 2 min will kill the larvae (Gould *et al.*, 1953). The effect of salt has been elucidated (Gammon *et al.*, 1968). It appears that *Trichinella spiralis* can survive the phase of salt equilibration during curing processes, but it begins to die off after 1 week, and none survives 1 month.

For the development of the adult beef tapeworm, *Taenia saginata*, which may grow to many feet in length, it is essential that a bovine should eat grass contaminated with the ova of the organism derived from the human intestine and that parts of the bovine containing the subsequent larvae of the organism (*Cysticercus bovis*) should be eaten raw by people. The predilection of the meat consumer for underdone beef may explain the greater prevalence of this kind of parasitic infection. Maintenance of the life cycle of the pork tapeworm (*T. solium*) requires an analogous relationship between pig and person. The corresponding larval stage in porcine muscle is known as *C. cellulosae*. Infestation with pork tapeworm is the more serious because the organism may develop into larvae in the human brain. Control of both diseases can be achieved by avoiding insanitary disposal of human faeces near cattle or swine feeding areas and by proper cooking. Exposure for 6 days to a temperature of −9.5 °C will destroy the cysts (Dolman, 1957). 'Measly' beef or pork, as the infested meat is called, is generally detected by thorough meat inspection.

Although human diseases caused by other tapeworms are not acquired directly by consuming infested meat, the domestic association of man with dogs, the usual host of *Echinococcus granulosis* (the parasite responsible), leads to human infection, as in sheep-rearing areas where sheep, humans and dog are in close contact.

Trematodes (especially *Fasciola hepatica*, the liver fluke) are communicable to people, but not through infested meat. Two recently identified protozoan parasites, which can cause disease in human consumers of raw or underdone meat, are *Cryptosporidium parvum* and *Toxoplasma gondii*. The latter is inactivated by heating to 70 °C.

6.1.2   Exogenous infections

Whereas the infections mentioned in § 6.1.1 arise from established disease in the live animal and may involve both parasitic worms and bacteria, meat spoilage and associated food poisoning reflect infection of the meat by bacteria (or fungi) after death. Where there is proper meat inspection to eliminate infected carcasses from distribution, the predominant mode of meat deterioration by invading organisms will be by exogenous infection.

6.1.2.1   Bacteraemia

Notwithstanding the predominantly post-mortem aspect of exogenous infection, the condition of the animal's blood immediately before and at slaughter is also appropriately considered in this context. In the large intestine there may be 33×10^{12} viable bacteria (Haines, 1937). Invasion of the tissues and organs of the body from the gut via the blood stream (bacteraemia) is opposed, however, by the mucous lining of the intestinal tract (Kohlbrugge, 1901), through agglutination of bacteria by circulating antibodies (formed from the gamma globulins of the blood in response to some previous minor exposure to the organisms concerned) and through phagocytosis of bacteria by the cells of the reticulo-endothelial system (in lymph nodes, in the blood and, possibly, in the tissues themselves). There is an equilibrium between invasion of the tissues and removal of the invading organisms such that the tissues of healthy animals are normally free from bacteria (Haines, 1937). Using bacteria labelled with C^{14} it has been confirmed that bacteria entering the lymphatic system from the intestines up to 24 hours post-mortem are destroyed by surviving action of the reticulo-endothelial system (Gill *et al.*, 1976). This explains the possibility of rearing disease-free pigs after sterile hysterectomy (Betts, 1961) referred to in § 2.5.2.3. In some species it would appear that the reticulo-endothelial system is more effective than in others, since venison, for example, can be hung for a considerable period at room temperature without undue precautions. Phagocytic activity, together with the gamma globulin content of the blood, can be enhanced by the administration of oestrogens (Charles and Nicol, 1961) and of certain other substances (Nicol *et al.*, 1961).

Invasion of the blood stream by organisms from the gut may be increased by fatigue in the animal (Haines, 1937; Burn and Burket, 1938; Robinson *et al.*, 1953), *prolonged* starvation (Ficker, 1905) or even feeding (Desoubry and Porcher, 1895; Gulbrandsen, 1935). Nevertheless it has been reported that feed withdrawal may increase levels of such pathogens as *Escherichia coli* O157:H7 in the gut of cattle (Small *et al.*, 2002). The mode of slaughter can also be implicated since breakdown of the intestinal mucosa was observed in sheep which had been shot (Badaway *et al.*, 1957). The bolt of a captive bolt pistol may carry a bacterial load of the order of 4×10^5 organisms/cm^2 of metal (Ingram, 1971). These observations explain the traditional reluctance to give animals food later than 24 h pre-slaughter and why the flesh of fatigued animals does not keep so well – although, in the latter, a high ultimate pH is a contributory factor. The organisms from the gut which can be distributed to the muscles by the blood include various streptococci (*Strep. bovis* in cattle and sheep; *Strep. faecalis*, *Strep. faecium* and *Strep. durans* mainly in sheep: Medrek and Barnes, 1962), *Clostridium welchii* and *Salmonella* spp.

If an infected knife is used, or organisms are inadvertently introduced from the skin whilst the main blood vessels are being severed, bleeding can itself lead to bacteraemia and to the infection of the animal's tissues (Empey and Scott, 1939; Jensen and Hess, 1941). It has been shown that meat can be contaminated by bacteria that persist on powered tools (e.g. air-driven knives) used in carcass dressing (Gill and McGinnis, 2004).

6.1.2.2   Sources and nature of external contamination
External contamination of the meat is a continuing possibility from the moment of bleeding until consumption. In the abattoir itself there are a large number of potential sources of infection by micro-organisms. These include the hide, soil adhering thereto, the contents of the gastrointestinal tract (if inadvertently released during dressing operations), airborne contamination, aqueous sources (the water used for washing the carcass, or for cleaning the floors), the instruments used in dressing (knives, saws, cleavers and hooks), various vessels and receptacles, and, finally, the personnel (Empey and Scott, 1939). Various pathogens in the intestinal contents may be derived from the feed and its mode of preparation (Wray and Sojka, 1977). In the lower parts of pasture, which are shielded from sunlight and resist drying, pathogens from infected slurry can persist and be ingested (Linton and Hinton, 1984). Antibiotic-resistant strains of *E. coli* may become established in the environment and colonize the gut of intensively reared veal calves (Hinton *et al.*, 1984). Some idea of the microbial loads which could be expected in an Australian slaughterhouse killing beef pre-1939 is given in Table 6.1. It is particularly important to avoid dirt from hides or fleece settling on exposed meat surfaces (Bryce-Jones, 1969). The fleece of sheep is a significant source of *Salmonella* contamination of the carcass: it can become a reservoir of these organisms after a day's holding at the abattoir lairage (Grau and Smith, 1974). There seems little difference, microbiologically, between flaying on the rail or in a 'cradle' (Nottingham *et al.*, 1973). By dosing sheep with cyclopropamide up to 10 days before slaughter, manual removal of the fleece can be readily carried out off the slaughter floor (Leach, 1971). Since legislation has now prohibited the use of wiping cloths, alternative ways of removing excess blood, etc., from the surface of the carcass have had to be considered

Table 6.1 Typical microbial counts in sources of microbial contamination in an abattoir (after Empey and Scott, 1939)

Sources and method of calculation	Temp. of incubation (°C)	Bacteria	Yeasts	Moulds
Hides (no. cm²-surface)	20	3.3×10^6	580	850
	-1	1.5×10^4	89	89
Surface soils (no. g dry wt)	20	1.1×10^5	5×10^4	1.2×10^5
	-1	2.8×10^6	1.4×10^4	1.0×10^4
Gastrointestinal contents:	20	9.0×10^7	2.0×10^5	6.0×10^4
Faeces (no. g dry wt)	-1	2.0×10^5	70	1700
Gastrointestinal contents:	20	5.3×10^7	1.8×10^5	1600
Rumen (no. g dry wt)	-1	5.2×10^4	50	60
Airborne contamination				
(no. deposited from air/	20	140	–	2
(cm²/hr)	-1	8	–	0.1
Water used on slaughter	20	1.6×10^5	30	480
floors (max. no./ml)	-1	1000	10	50
Water present in receptacles				
from immersion	20		1.4×10^5	
cloths (no./ml)	-1		40	

(Akers, 1969; Bryce-Jones, 1969). Pressure hosing is said to lead to loss of desirable surface appearance ('bloom') through uptake of water by the connective tissue. This latter can be avoided by applying warm water (about 43 °C) as a spray followed by *rapid* drying. The 'bloom' reappears after the carcasses have been at chill temperature for a few hours subsequently. The application of markedly higher temperatures to sheep carcasses – immersion in water at 80 °C for 10 s – has been shown to destroy about 99 per cent of contaminating coliforms initially present on surface tissues (Smith and Graham, 1978). Bacteria are liable to multiply rapidly in lukewarm water, and care must be taken to ensure that no pockets of moisture remain (if necessary by wiping with approved disposable, absorbent paper). Wetting carcasses does not spoil them provided the moisture is removed rapidly, and that the dirt is removed and not merely redistributed. In general it is preferable, of course, to avoid surface contamination by strict hygiene than to remove it.

It will be seen (Table 6.1) that the initial contamination acquired by beef surfaces during dressing operations under earlier conditions included more than 99 per cent of bacteria amongst those organisms viable at ordinary temperature (20 °C). These populations contained less than 1 per cent of organisms viable at –1 °C, although the percentage of yeast and moulds was greater in the populations viable at –1 °C than at 20 °C. The chief sources of the superficial microflora were the hide of the slaughtered animals and surface soils; the types of organisms in both localities were the same. Of the organisms viable at –1 °C, four principal bacterial genera were represented, namely *Achromobacter* (90 per cent), *Micrococcus* (about 7 per cent), *Flavobacterium* (about 3 per cent) and *Pseudomonas* (less than 1 per cent). The composition of the *Pseudomonas* population on the surface of stored beef, lamb and pork is substantially determined by the local, environmental microflora at each abattoir (Shaw and Latty, 1984). Of the mould genera the most common were *Penicillium*, *Mucor*, *Cladosporium*, *Alternaria*, *Sporotrichium* and *Thamnidium*. In Fig. 6.1

Fig. 6.1   Effect of incubating meat surface after contamination through contact with a hand
bearing mould spores. (Courtesy J. Barlow.)

the effect of incubating a piece of meat touched by the hand of a worker carrying
mould spores is shown.

Since the subcutaneous fat of beef may have a high initial microbial load and has
the capacity to support extensive growth, it could be a significant source of contamination of lean and of manufactured meat produced (Lasta *et al.*, 1995).

Inadequate cooling of the carcass at the abattoir will permit the proliferation of
putrefactive organisms, the most common type causing bone taint (Savage, 1918;
Acevado and Romat, 1929). These are external in origin and gain access to the
animal by cuts and abrasions on the skin (Cosnett *et al.*, 1956). Bone taint was
defined by Haines (1937), as 'the development of putrid or sour odours in the deep-
seated parts of meat, usually near the bone'. The bacterial flora of lymph nodes and
of tainted meat are similar (Nottingham, 1960), consisting of Gram-positive rods. It
seems likely that when the pH is relatively high and the temperature falls insufficiently quickly, bacteria proliferate within the lymph nodes (ischiatic and popliteal)
and spread into the surrounding meat (Cosnett *et al.*, 1956).

Although some sources of contamination are obviously removed when the carcasses leave the slaughter floor, contamination by contact with unhygienic surfaces,
by personnel and by airborne organisms will remain as a possibility in all opera-

Table 6.2 Some characteristics of bacterial meat poisoning

Causal organism	Time from ingest to onset symptoms	Reservoir of infecting organisms	Characteristic symptoms
Salmonella	8–72 h	Gut of animals	Abdominal pain; diarrhoea; nausea; pyrexia; prostration
Staphylococcus	1–6 h, often 2–4 h	Skin, nose, cuts in man, animals	As above; plus salivation and vomiting but sub-normal temperature
Enterococcus Cl. welchii (Cl. perfringens) Strep. faecalis	2–18 h	Gut of animals	Abdominal cramp; diarrhoea; no pyrexia or prostration
Cl. botulinum	2 h–8 days, often 12–48 h	Soil	Difficulty in swallowing; double vision; no pyrexia; respiratory paralysis

tions during the subsequent history of the meat – chilling, freezing, processing, cutting, packaging, transport, sale and domestic handling. The organisms derived from infected personnel or healthy carriers include *Salmonella* spp., *Shigella* spp., *E. coli*, *B. proteus*, *Staph. albus* and *Staph. aureus*, *Cl. welchii*, *B. cereus* and faecal streptococci: those from soil include *Cl. botulinum*. Together with those distributed to the tissue by the blood (§ 6.1.2.1), they constitute the source not only of much meat spoilage but also of food poisoning since they are conditioned to grow preferentially at body temperature. Inadequate cooking of contaminated meat or meat products which have been kept warm beforehand or reheating of partly cooked meat can be responsible. Since some of the organisms form spores or toxins, however, even cooking may fail to prevent infection, poisoning and even death (e.g. with *Cl. botulinum*, the toxin of which is one of the most potent poisons known). Fortunately, the microflora of the human gut normally provides an effective barrier against invasion by pathogens (Coates, 1987). Some reactions to food poisoning organisms are summarized in Table 6.2 (Anon., 1957b). There are six serologically distinguishable types of *Cl. botulinum* – A, B, C, D, E and F. Spores of A and B survive boiling for several hours but the organisms will only grow slowly below 10 °C. Spores of type E are killed by heating to 80 °C for 30 min, but they will grow at 3.3 °C in a beef-stew medium (Schmidt *et al.*, 1961). Fortunately, *Cl. botulinum* E, if present, would tend to be outgrown by psychrophiles in chilled meat. Types C and D are rarely implicated in human botulism: type F is so far known to have been involved in only one outbreak, caused by home-prepared liver paste. The geographical distribution of the various types is probably related to temperature, soil type and other unknown ecological factors. The toxins produced by *Cl. botulinum* are quite resistant to heat and to the enzymes in the digestive tract; in fact type E toxin is activated by trypsin (Duff *et al.*, 1956).

Of 164 food poisoning outbreaks in England and Wales in 1953 (Anon., 1957b), 44 were due to *Salmonellae*, 47 to *Staphylococci*, 19 to *Cl. welchii* and 54 to other organisms, mostly unidentified. It may be noted that, despite the provisions of the 1990 Food Act in the UK, individual notifications of clinically diagnosed food poisoning increased from 52,145 in 1990 to 69,990 in 1993 (Phillips *et al.*, 1995).

As diagnostic methods have improved, however, the implication of other organisms in food poisoning has been indicated. Thus, new strains of *Salmonella* are arising, some of which can be carried by animals which show no symptoms and cannot be detected by ante-mortem inspection. *Salmonella* Typhimurium DT 104 is an emerging strain that is pathogenic to both animals and human consumers. The mortality rate in humans appears to be greater than that from other *S.* Typhimurium strains (Hogue *et al.*, 1998). It has a multiple resistance pattern to a wide range of antibiotics, the use of which may foster growth by suppression of competing organisms.

Enteritis due to *Campylobacter jejuni* is increasingly identified. A study in the USA revealed that this organism can be isolated from more human faecal specimens than *Shigella* and *Salmonella* spp. combined. Bolton *et al.* (1982) isolated *Campylobacter jejuni* from the carcass surfaces of 32 per cent, 56 per cent and 70 per cent, respectively, of the cattle, pigs and sheep examined. The incidence on meat at retail level was much less, possibly due to the organism's susceptibility to chilling, drying and exposure to oxygen (Turnbull and Rose, 1982). Nevertheless, although chilling the carcasses of adult bovines significantly reduces the incidence of *C. jejuni* and *C. hyointestinalis*, these organisms survive more readily on the chilled carcasses of calves (Grau, 1988), possibly because of the greater surface moisture on the latter.

Although *E. coli* is frequently found in the intestinal tract of man and animals and occurs as many strains, some of which are potentially pathogenic, it is only recently that a virulent strain which produces several virotoxins has been isolated from beef and pork (0157:H7). It can cause haemorrhagic colitis (Boyle, 1987), is unusually tolerant of acids and has high infectivity (Doyle, 1998). Its acid tolerance may reflect its selection through exposure to factors in the environment currently causing increasing acidity. The toxins can be destroyed by adequate heating of the meat (viz. 70 °C for 2 min). A number of other recently identified organisms have been associated with poisoning outbreaks (e.g. *Giardia lamblia*, a microparasite, and *Yersinia enterocolitica*), some deriving from animal or poultry sources (Galbraith *et al.*, 1987). As the prevalence of the latter organism can be relatively high in pigs, and since pork is a major constituent of many meat products, it is interesting to note that several strains of lactobacillus (e.g. *L. sake*) produce sufficient lactic acid to inhibit the growth of *Y. enterocolitica* in fermented sausages (Rodriguez *et al.*, 1994).

Listeria monocytogenes, long recognized in animals, has now been acknowledged as a food-borne pathogen. Its presence in meat is of particular concern since it is capable of growing at refrigeration temperatures (Mossel *et al.*, 1975) and is relatively resistant to heat and curing salts (Shahamat *et al.*, 1980). The organism has been found in 65 per cent and 30 per cent respectively, of environmental samples from a lamb and beef chilling plant (Lowry and Tiong, 1988). The development of primary plating media which permit direct enumeration of *Listeria monocytogenes* from foodstuffs will enable a more accurate assessment of the organism's status in relation to food poisoning to be made (Buchanan *et al.*, 1987). The slaughterhouse environment is more strongly implicated as a source of *l. monocytogenes* than the faeces and skin of animals (Nesbakken and Skjerve, 1996). Not surprisingly, increasing awareness of the existence of such pathogens has prompted surveys to ascertain the extent of their occurrence in carcasses. Findings, at least in some areas, have been reassuring (Madden *et al.*, 2001; Guyon *et al.*, 2001).

Although testing procedures for micro-organisms have long been used by industry, the increasing incidence of food-borne disease clearly indicates that such methods are not sufficiently effective. A logical and systematic approach to prevention is essential. This involves the identification of hazards, understanding the risks they present, identifying the points where control must be exercised (critical control points), selecting the options for such control and for subsequent monitoring and implementing the degree of control needed. This approach is referred to as 'Hazard Analysis Critical Control Points' (HACCP) (Baird-Parker, 1987). Compendia of the views of various experts on the application of HACCP in the processing of meat products and fish (Pearson and Dutson, 1995) and in the meat industry generally (Brown, 2000) have been published.

To be effective, the HACCP approach requires methods for microbial identification and enumeration which are much more rapid than traditional procedures. These include phage and serological typing and direct epifluorescence filter techniques (DEFT). Plasmid profiling is particularly useful since it permits microbial strains to be firmly and swiftly identified. The technique depends on the isolation of plasmid (non-chromosomal) DNA and its separation by gel electrophoresis. Its success has been demonstrated, for example, in implicating the defeathering machines in poultry processing plants as the source of *Staph. aureus* contamination (Dodd *et al.*, 1988a, b). Oligonucleotide probes for the identification of five different enterotoxins (viz. SEA, SEB, SEC, SED, SEE) from strains of *Staph. aureus* have been developed (Jaulhac *et al.*, 1992), but the technique involved in their use is laborious and not sufficiently specific. Using an improved method of DNA purification, and a mixture of nucleotide primers – one being unique for each enterotoxin and one for the conserved region common to each – in a polymerase chain reaction, the five strains of *Staph. aureus* can be unequivocally detected in a single rapid operation (Sharma *et al.*, 2000). This approach should enhance the efficiency of microbial screening and control, and be extended to detect recently identified *SEG*, *SEH* and *SEI* genes. Another promising approach is to incorporate the *lux* genes from *Vibrio fischeri* into bacteriophages. When the amended bacteriophage DNA is injected into a bacterial cell light is produced. The technique permits as few as 10 cells to be detected within an hour and the metabolic status of the bacterial cells to be assessed (dormant cells are dark).

6.2   Symptoms of spoilage

In satisfying their requirements for nourishment and survival, invading organisms alter meat in a variety of ways. Some of these are not deleterious: a few are beneficial, but the vast majority are not and, indeed, may even be lethal, as we have seen.

The superficially recognizable effects of invasion by parasites are sometimes striking (Thornton, 1973) but, as they are generally detected by public health inspectors and are rarely recognizable or encountered by the consuming public, they will not be considered here. The types of spoilage caused by micro-organisms broadly depends on the availability of oxygen, although, as will be considered below, many other factors are involved (Table 6.3; Haines, 1937).

The nature, range and sequence of the changes in meat caused by the biochemical activities of a single species of invading organism can be exemplified by the behaviour of *Cl. welchii*, an anaerobe (Gale, 1947; Wilson and Miles, 1955). First, the

Table 6.3 Superficially recognizable symptoms of microbial spoilage of meat

Oxygen status	Type of micro-organism	Symptoms of spoilage
Present	Bacteria	Slime on meat surface: discoloration by destruction of meat pigments or growth of colonies of coloured organisms; gas production; off-odours and taints; fat decomposition
Present	Yeasts	Yeast slime; discoloration; off-odours and tastes; fat decomposition
Present	Moulds	Surface 'stickiness' and 'whiskers'; discoloration; odours and taints; fat decomposition
Absent	Bacteria	Putrefaction accompanied by foul odours; gas production; souring

meat liquifies because the organism excretes a collagenase which hydrolyses the connective tissue between the fibre bundles, causing them to disintegrate. This is followed by gas production. The free amino acids present are attacked by deaminases with the production of hydrogen, carbon dioxide and ammonia; and glycogen, if present, is fermented to give acetic and butyric acids. These activities cause foul smells and unpleasant tastes. Another enzyme produced by *Cl. welchii* decarboxylates histidine to histamine, which affects membrane permeability. Certain highly invasive strains of *Cl. welchii* produce hyaluronidase, which attacks mucopolysaccharides in the ground substance between cells and permits further penetration by the micro-organisms. In addition to all these actions, which are relatively harmless to the consumer, *Cl. welchii* produces toxins in the meat. On being ingested, these have various biological actions including haemolysis of the blood and destruction of tissue cells and, in severe infections, death. But the changes caused by individual organisms – and the corresponding symptoms of spoilage – are usually somewhat more limited in scope.

Surface slime is the superficially observable effect of the coalescence of a sufficiently large number of individual colonies of micro-organisms: the further apart these colonies are in the first place, i.e. the lower the initial infection, the longer the time will be until slime forms (Fig. 6.2; Haines, 1937). Slime formation signifies a general suitability of the temperature and moisture of meat surfaces and the adjacent air for growth; but the chemical nature of these two phases will select the type of organism found. Thus, there will be substantial representation of the genus *Achromobacter* on chilled beef (Empey and Scott, 1939), of *Micrococcus* on sides of pork and on matured bacon (Brooks *et al.*, 1940), and of *Lactobacillus* on vacuum packed, sliced bacon, if stored between 5 and 30 °C (Kitchell and Ingram, 1963). Slime formation on sausages, however, can be due to a white yeast (Haines, 1937).

Discoloration may be due to alteration or destruction of meat pigments. Myoglobin may be oxidized to brown metmyoglobin; it may combine with H_2S, produced by bacteria, to form sulphmyoglobin (Jensen, 1945); or be broken down to form yellow or green bile pigments by microbially produced hydrogen peroxide (Niven,

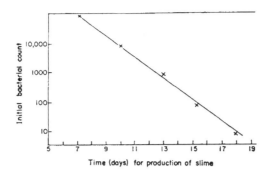

Fig. 6.2   The relationship between initial microbial load and the time for slime to develop on the surface of meat (Haines, 1937).

1951; Watts, 1954). Yeasts, growing on the fat surfaces of vacuum-packed chilled beef, have been found to cause the development of brown spots after six weeks' storage at 0 °C (Shay and Egan, 1976), due to their action on haem. Discoloration may also be due to the elaboration of foreign pigments of *Pseudomonas*, pink pigments of various types of micrococci, sarcinae and yeasts, and the red colour of *B. prodigiosus* (Haines, 1937). Moulds of the genera *Cladosporium*, *Sporotrichium* and *Penicillium* produce black, white and blue-green colours respectively. Black or red discoloration can be produced in salt meat and meat products by certain halophilic pseudomonads (Gibbons, 1958). Green cores can be formed in sausages by *L. viridescens* (Niven and Evans, 1957). Green, blue and silver surface lumines- cence may be caused by the activity of harmless bacteria belonging to many genera – a phenomenon known since ancient times (Jensen, 1949).

Putrid odours are produced mainly by anaerobes through the decomposition of proteins and amino acids (yielding indole, methylamine and H_2S) and sour odours through the decomposition of sugars and other small molecules (Haines, 1937). Such may be encountered in the interior of improperly cured hams, the organism respon- sible in this case being *B. putrifaciens* (McBryde, 1911): it may be caused by various Gram-positive rods in inadequately cooled beef carcasses (bone taint; § 6.1.2.2). The growth of anaerobes is associated with a more offensive decomposition than that of aerobes. There are several reasons for this (Haines, 1937). For example, the low energy yield of anaerobic processes compared with aerobic ones makes it necessary for anaerobes to break down a proportionately greater quantity of material than aerobes for a given degree of multiplication. Again, evil-smelling substances tend to be liberated particularly under reducing conditions. In stored, vacuum-packed beef, cadaverine and putrescine can produce foul odours. They appear to be derived from the growth of *Enterobacteriaceae* strains (Edwards *et al.*, 1985).

If *Pseudomonas fragi* is present during the storage of pork at 2–10 °C, some pro- teolysis of myofibrillar proteins occurs, and this will raise the emulsifying capacity of the meat (Borton *et al.*, 1970). Neither in their experiments nor in those of Dainty *et al.* (1975), who studied beef surfaces at 5 °C, was there detectable proteolysis before spoilage odours and slime had developed.

Many types of micro-organisms cause spoilage by producing free fatty acids and yellow or green pigments (Jensen, 1949) from the superficial fat in meat, and, indeed, such changes are frequently the limiting factors in storage.

Although, quite apart from microbial techniques, there are a large number of objective chemical tests which can be used to assess the degree of microbial spoilage in meat (Thornton, 1973; Rubashkina, 1953; Herschdoerfer and Dyett, 1959), human sensory evaluation of superficial symptoms has not lost its value.

6.3   Factors affecting the growth of meat-spoilage micro-organisms

Implicitly, the micro-organisms spoiling meat can gain therefrom their basic require-ments for growth – sources of carbon, nitrogen, bacterial vitamins, etc. – although the degree of accessibility of these nutrients will vary. A suitable temperature, mois-ture availability, osmotic pressure, pH, oxidation–reduction potential and atmos-phere are also essential; but these factors are interrelated and their individual importance varies with the particular circumstances being considered. Other less fundamental factors affect microbial growth. These include ionizing radiations, although their effect is more logically considered under prophylaxis, and some so far unidentified agencies. It was observed, for instance, that the growth of psy-chrophiles at 0 °C in extracts of beef *l. dorsi* muscles was inversely related to-the time elapsing between death of the animal and the onset of rigor mortis (Brown *et al.*, 1957). Muscle fibres shrink, however, as they go into rigor mortis (Pearson *et al.*, 1974), and it is thought that the spaces so formed between the fibres and the endomysium permit penetration of bacteria into the muscle tissue (Gill *et al.*, 1984). Since the shrinkage of actomyosin and the exudation from muscle bear an inverse relationship to the rate of onset of rigor mortis (Bendall, 1960; Lawrie, 1960b), the delayed growth observed may reflect the relative difficulty of access when fibre shrinkage has been minimal.

It has been observed that the application to food of very high pressures delays putrefaction. It may initiate the germination of bacterial spores or inactivate germinated forms (Clouston and Wills, 1969).

It is important to emphasize that the observed status of these factors, in or on meat at a given time, does not necessarily define whether or not the meat is (or is likely to be) spoiled. The products of past microbial activity will remain after the death of the organisms responsible and, in the case of toxins, may prove dangerous as well as distasteful. Moreover, against immediately unfavourable conditions, certain organisms form resistant spores which survive cooking and canning. In these, metabolic activities are largely suspended; but viability may remain and the rapid growth of the organism recommences when favourable circumstances again arise. Infection can thus be latent. Spores of *B. anthracis* have been known to survive as potential centres of infection for 60 years (Umeno and Nobata, 1938). Authentic evidence that spores of *Thermoactinomyces* spp. can survive for 1900 years has been presented (Seaward *et al.*, 1976).

Computer technology has made it possible to prepare models which can show how the growth and survival of specific micro-organisms will proceed in relation to the various factors which influence them and to predict, with considerable accuracy, the shelf life of meat commodities in which the organisms are present (Baranyi and Roberts, 1993). Growth and survival models for many pathogens, such as *L. mono-cytogenes, E. coli* 0157:H7, *Y. enterocolitica* and salmonellae, and for recovery fol-lowing thermal shock of such organisms as *Cl. botulinum* type B, salmonellae and *L. monocytogenes* are now available. Industry has access to these predictive models.

Combined physico-chemical and mathematical approaches to develop models for the prediction of how microenvironments within complex food systems affect micro-organisms have been developed (Anon., 1993c; Mackey *et al.*, 1994).

It has been discovered that bacteria can communicate with one another through a series of signalling molecules which are relatively simple in structure, some consisting of homoserine linked to an acyl group (Williams *et al.*, 1992). Such inter-bacterial communication is now referred to as 'quorum sensing'. Some of these signalling molecules appear to be unique to one bacterial species, others are common to several species and a number of different molecules may be produced by a single species. These signalling molecules can coordinate the behaviour of groups of bacteria, targeting specific genes, to induce growth, colony size, patho-genicity, the production of antibiotics and other features. Protein receptors of the signal-bearing molecules appear to encapsulate them and to enhance transcription of the appropriate DNA (Zhang, 2002). Moreover, by such signals, one bacterial species can control another whereby, for example, the presence of harmless bacte-ria in a community could induce virulence in a pathogen such as *Salmonella* if present. The efficacy of signalling molecules is related to their ability to diffuse through the medium in which they are present and this may explain why the various factors which affect bacterial growth and viability may operate more or less strongly than would be expected from *in vitro* observations.

The emergence of new, toxin-producing strains of micro-organisms has empha-sized the need for detailed knowledge of the mechanisms involved. Sheridan and McDowall (1998) reviewed the relationship between environmental stress and the development of pathogenicity. Pathogenicity and virulence reflect the adaptation of bacteria to counter hostile environments, mutations that foster survival being par-ticularly prevalent when cells are in the stationary phase. (Thus the transfer of genetic information between organisms could occur readily at this stage.) It has been suggested that a strategy to control the emergence of pathogens could be based on the induction of false signals to activate bacterial genes whose action was inap-propriate for survival at that time (Sheridan and McDowall, 1998).

It appears that exposure of pathogens (e.g. *S. typhimurium*) to successive cycles of stress, followed by growth under non-stress conditions, increases their resistance. This is associated with the *rpoS* gene, which is activated when the bacterial cells enter a stationary phase from one of exponential growth. Concomitantly, the pattern of protein synthesis by the bacterial cells alters. The changed profile of proteins can be detected and can indicate the nature of the stresses to which the bacteria have reacted.

6.3.1 Temperature

The most important single factor governing microbial growth is temperature (Haines, 1937). Broadly, the higher the temperature the greater is the rate of growth (Fig. 6.3; Haines, 1934). Many meat micro-organisms will grow to some extent at all temperatures from below 0 °C to above 65 °C, but, for a given organism, vigorous growth occurs in a more limited temperature range. It is customary to classify meat spoilage organisms in three categories. Psychrophiles (psychrotrophs) have tem-perature optima between –2 °C and 7 °C, mesophiles between 10 °C and 40 °C and thermophiles from 43 to 66 °C (Jensen, 1945). The distinction is by no means absolute, however, as certain Gram-negative rods, which are generally regarded as

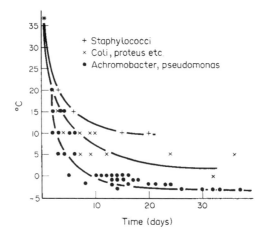

Fig. 6.3 The effect of temperature on the rate of growth of various bacteria (Haines, 1934).

mesophiles, will grow at –1.5 °C (Eddy and Kitchell, 1959). At chill temperatures under aerobic conditions, the spoilage flora of meat is dominated by pseudomonads and, under anaerobic conditions, by lactobacilli. In both cases, according to Gill and Newton (1978), the organisms attack glucose initially and amino acids subsequently.

An interesting aspect of temperature optima is the finding that the beef prepared from cattle in tropical areas will carry a relatively small percentage of organisms capable of growing when the meat is kept at chill temperatures (–1.5 °C) and will thus keep better than beef prepared from temperate zone cattle. The predominant microflora on beef surfaces is derived from the soil and is adjusted to grow at soil temperatures, which are high in the tropics (Empey and Scott, 1939).

Another reflection of temperature optima is the difference in behaviour of ham and gammon, despite that they are both produced from the leg of pork (Ingram, 1952). Hams are only lightly injected with curing salts. When taint occurs it does so at relatively high storage temperatures and is caused by mesophiles (faecal clostridia and streptococci) which are probably intrinsic to the animal body. Gammons, however, are injected with brine, the microbial flora of which consists of psychrophilic, salt-tolerant organisms (mainly micrococci). They will thus grow and cause taint even at chill temperatures. On the other hand, if curing is ineffective, spoilage of gammons by faecal streptococci can occur (Ingram, 1952). Similarly, bone taint in beef, as we have already mentioned in § 6.1.2.2, represents the development of mesophiles under conditions where the temperature is favourable for growth (through faulty cooling), and there exists a reservoir of infection in the animal (Cosnett et al., 1956).

Although the types of micro-organisms growing on prepackaged fresh beef, lamb or pork are the same at 3 and 7 °C (mainly *Achromobacter* and *Pseudomonas fluorescens*; Halleck et al., 1958), an apparent species difference has been reported in that both anaerobic or aerobic packs of lambs at 5 °C were spoiled mainly by *B. thermosphacta*, whereas, under similar conditions, beef was spoiled mainly by Gram-negative bacteria (Barlow and Kitchell, 1966). These differences may be partly attributed to the fact that the residual rate of oxygen utilization is considerably higher in lamb than in beef. The temperature of storage has marked influence on the micro-organisms growing in prepackaged cured meats. If prepackaged sliced bacon, for example, is kept at 37 °C its normal psychrophilic flora is replaced by mesophilic organisms, which include pathogenic staphylococci (Ingram, 1960). At temperatures of storage from 5 to 30 °C the halophilic micrococci and lactic acid bacteria, which constitute the greater part of the flora of packaged sliced bacon, both increase in numbers, but the lactic bacteria do so rapidly and they come to constitute a greater proportion of the population. On continued storage, the lactic acid bacteria reach maximum numbers and stop growing, but the micrococci continue to increase. Above 20 °C, however, the type of *Micrococcus* changes and organisms related to *Staph. aureus* predominate (Ingram, 1960). There is a marked, concomitant effect on bacon flavour – at 20 °C the spoilage odour is sour and at 30 °C putrid (Cavett, 1962) – and on the relative quantities of nitrate and nitrite present (Eddy and Gatherum, unpublished data). It is conceivable that micrococci reduce nitrate and that Gram-negative rods destroy nitrite, as in bacon curing brines (Eddy, 1958). Since temperatures alter the rate of bacterial respiration and the system is closed, the atmosphere within a pack, being of restricted volume, should contain more oxygen and less carbon dioxide if kept at low temperatures (Ingram, 1962). The reaction of within-pack atmospheres with temperature has to be considered as an additional determinant of bacterial flora and growth in such products; and it is of course, related to the nature of the wrapping medium and its relative permeability to various gases.

Micro-organisms may remain viable well outside the ranges quoted. They have been reported to survive for 10 h at –252 °C (MacFadyen and Rowlands, 1900). The effect of freezing must be distinguished from that of low temperature itself, however, since although –12 °C will stop all microbial growth in frozen carcasses, some organisms will grow in a super-cooled liquid at –20 °C (Richardson and Scherubel, 1909).* In respect of high temperature, spores can survive for at least $5\frac{1}{2}$ hours at 100 °C in the presence of moisture, and for $2\frac{1}{2}$ hours at 200 °C in the dry state (Tanner, 1944). In more recent times the existence of hyperthermophilic micro-organisms has been recognized. Thus, *Pyrolobus* survives at 113 °C and, indeed, is unable to grow below 90 °C (Stetter, 1998). The finding of such micro-organisms in various extreme environments in the present day Earth suggests that they may have been the channels along which life developed during the conditions prevailing 3000 million years ago.

The hyperthermophilic organisms form a group, *archaca*, of which the cell membrane lipids contain a high proportion of saturated fatty acids. The membranes are consequently more thermostable than those of other bacterial cells and of eukaryotes. The thermal stability, and high optimum temperature, of the enzymes in archacan bacteria, is due to relatively minor differences in their amino acid sequences whereby the enzyme proteins fold in a more robust configuration (Wharton, 2002).

By extrapolation of the genetic characteristics of present day bacteria, it has been shown that bacteria had a common ancestry in Precambrian times (*ca.* 1000 million years ago) (Brenner, 2002). The ancient enzymes they produced appear to have had

---

* The extreme robustness of microbial spores has been shown. Micro-organisms have been revived after their spores had been preserved for *ca.* 100 million years, preserved in amber.

temperature optima between 55 and 65 °C, being thus more heat-adapted than their modern counterparts.

In general, a reduction in the number of micro-organisms occurs when meat is frozen, but yeasts and moulds will grow at –5 °C although not at –10 °C (Haines, 1931). Carefully controlled experiments have failed to substantiate the prevalent view that thawed meat is *intrinsically* more perishable than meat which has not been frozen (Sulzbacher, 1952; Kitchell and Ingram, 1956). Even so, under commercial handling conditions, the moister surface of thawed meat would tend to pick up greater numbers of bacteria and hence be potentially more liable to spoil (Kitchell, 1959). Even with storage at 0 °C spoilage could occur through the activities of psychrophiles, e.g. *Ps. fluorescens* (Petersen and Gunderson, 1960).

The effect of temperature on microbial growth may differ according to the nature of the nutrients available. Thus, *Lactobacillus arabinosus* needs phenylalanine, tyrosine and aspartic acid for growth at 39 °C, phenylalanine and tyrosine at 37 °C and none of these amino acids at 26 °C (Borek and Waelsch, 1951). It is important to appreciate that, if there has been heavy microbial growth before freezing, a high concentration of microbial enzymes (e.g. lipases) may have been produced. Thus, even if microbial growth is arrested by the process of freezing, the enzymes may continue to produce deleterious quality changes even down to about –30 °C (Sulzbacher and Gaddis, 1968).

6.3.2 Moisture and osmotic pressure

After temperature, the availability of moisture is perhaps the most important requirement for microbial growth on meat, although some types of bacteria may remain dormant for lengthy periods at low moisture levels, and spores resist destruction by dry heat more than by moist heat (Tanner, 1944). Typical data from an experiment with cuts of lamb, showing the growth-promoting effects of moisture and temperature, are given in Table 6.4. The organisms present belong to several genera – *Pseudomonas, Achromobacter, Proteus* and *Micrococcus* (Jensen, 1945). The availability of moisture is complementary to that of osmotic pressure, which is a function of the concentration of soluble, dializable substances (salts, carbohydrates, etc.) in the aqueous medium. High solute concentrations tend to inhibit growth; desiccation of the substrate and not low temperature as such generally restricts microbial growth on frozen-meat products. Nevertheless, there is great variation between species and, although most of the organisms which will grow on meat are inhibited

Table 6.4 The effect of surface moisture and temperature on microbial growth on lamb cuts (after Jensen, 1945)

| Time (hr) | Aerobic bacteria/g | | | |
| | 2–3 °C | | 7–10 °C | |
	Wet surface	Dry surface	Wet surface	Dry surface
24	400,000	40,000	1,000,000	200,000
72	760,000	42,000	Putrid	4,000,000

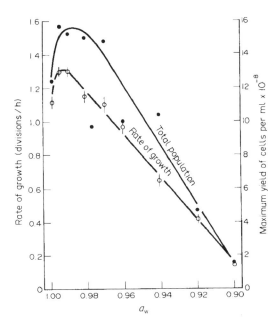

Fig. 6.4   Relationship between the mean rate of growth ($\bigcirc$), maximum yield of cells ($\bullet$) and a_w for 14 strains of *Staph. aureus* under controlled conditions (Scott, 1953). (Courtesy the late Dr W. J. Scott.)

by salt, there are many salt-tolerant organisms which grow successfully on bacon brines.

The water relations of meat-spoilage organisms were studied in detail by Scott, who used the term 'water activity' (a_w) in this context. The a_w of a solution is the ratio of its vapour pressure to that of pure water at the same temperature: it is inversely proportional to the number of solute molecules present (Scott, 1957). In general, moulds and yeasts tolerate higher osmotic pressures than bacteria (Haines, 1937), bacteria growing from an a_w of just under 1.0 down to an a_w of 0.75, and yeasts and moulds growing slowly at an a_w of 0.62 (Scott, 1957).

Scott (1936) showed that decreasing the a_w decreased the rates of growth of moulds, yeasts and bacteria on meat surfaces. The optimum a_w for several food poisoning strains of *Staph. aureus* was shown to be about 0.995 (Scott, 1953). Below this value, the rate of growth markedly diminished (Fig. 6.4). At 38 °C the level of a_w at which this organism would grow on dried meat was 0.88, corresponding to 23 per cent of available water. A water activity of 0.70–0.75, as in charqui, appears to effectively inhibit the growth of *Staph. aureus* and *Cl. botulinum* (Lara *et al.*, 2003). A similar study of the water requirements of *Salmonella* was made by Christian and Scott (1953), and of *Vibrio metschnikovi*, the latter having a very well-defined optimum a_w which was critical for growth (cf. Fig. 6.4). Temperature and pH are among various factors which affect the a_w (Scott, 1957).

Fresh meats have an a_w which is frequently about 0.99, and they are thus liable to spoil through the growth of a wide range of organisms (Scott, 1957). The importance of surface drying in restricting microbial growth on beef was demonstrated by Scott and Vickery (1939).

Because of the interaction of oxygen availability and a_w, it is feasible that in prepackaged meats, where the oxygen tension is low, growth of micro-organisms might be inhibited at a higher a_w (Ingram, 1962). Scott (1953) found, for example, that the growth of *Staph. aureus* would proceed at an a_w of 0.86 aerobically, but only above 0.90 anaerobically. On the other hand, when meat is prepackaged and sliced, surface growth will be possible between the pieces, and the surface drying which would normally inhibit growth on stored meat would not operate (Ingram, 1962). Whenever the mode of packaging limits evaporation and weight loss there will be an increased humidity within the pack and the danger of bacterial growth will be greater. Nevertheless, the microbiological undesirability of a wrap which is impermeable to water vapour may be offset if it is relatively permeable to oxygen and carbon dioxide (Shrimpton and Barnes, 1960). With bacon, a high salt/moisture ratio extends its storage life (Hankins *et al.*, 1950).

The effect of salt concentration in altering the types of micro-organisms capable of growing on meat products can be exemplified by comparing the behaviour of those isolated from bacon brines and pork sides (Table 6.5; after Kitchell, 1958). It has been pointed out, that in addition to the effect of salt in lowering a_w, the Na$^+$ ion has a specific inhibitory action on certain micro-organisms (Varnam and Sutherland, 1995).

It is obvious that the micro-organisms of bacon brine are predominantly of types capable of growing in the presence of 10 per cent salt, and relatively incapable of growing in the presence of 1 per cent salt, whereas, the converse is true for the micro-organisms of pork sides. High salt concentration shifts the balance of the microbial population towards halophilic organisms. Nevertheless, a certain affinity exists between the micro-organisms of raw pork and those of matured bacon (Shaw *et al.*, 1951), but both differ from those of brine. The ability to reduce nitrate and/or nitrite is one of the metabolic activities found preferentially among the bacteria capable of growing at high salt concentration (Table 6.6).

Table 6.5   Plate counts, at two salt concentrations, of micro-organisms from bacon brines and pork sides

Source	Counts	
	1% NaCl	10% NaCl
Bacon brines	1.43×10^6 per ml	11.14×10^6 per ml
Pork sides	15.8×10^3 per cm^2	8.4×10^3 per cm^2

Table 6.6   Proportions of the strains of micrococci, isolated from various sources, which reduce either nitrate or nitrite (after Kitchell, 1958)

Source	No. of strains examined	Per cent of reducing strains
Bacon brines	59	90
Immature bacon	22	83
Pork sides	31	68

Bacilli are known which will tolerate 15 per cent NaCl and in canned hams will vigorously reduce nitrate and nitrite with the production of sufficient nitrous oxide to 'blow' the cans (Eddy and Ingram, 1956).

In the maturation of bacon, colour fixation is due to the reaction of the pigment of fresh meat (myoglobin) with nitric oxide. The curing brine contains nitrate which is reduced to nitrite, and the latter to nitric oxide, by microbial action. An increase in salt concentration in the brine, however, suppresses these processes by lowering the metabolic rate of the halophilic micro-organisms responsible and vice versa: the effect is more marked at 5 than at $10\,^{\circ}C$ (Eddy and Kitchell, 1961). The microbial reduction of nitrate and nitrite also occurs in the bacon itself during maturation (Eddy *et al.*, 1960). The reduction of nitrite, in contrast to that of nitrate, may occur in bacon through the activity of tissue enzymes and in the absence of micro-organisms (Walters and Taylor, 1963).

Salt itself, depending on its origin, may harbour various bacteria, including halophiles which produce red colonies and thus spoil the product being preserved. These may be troublesome in the fish industry (Bain *et al.*, 1958) and with imported gut casings. Although, they have not appeared in the spoilage of bacon as such, they may well be responsible for a discoloration of bacon brines which has been spo-radically reported (D. P. Gatherum, personal communication). According to Yesair (1930), certain salts may harbour proteolytic anaerobes and should thus be avoided for meat curing.

Certain species of *Lactobacillus* will tolerate the high sugar concentrations used in ham curing brines in the USA, and will grow on cured unprocessed hams pro-ducing polysaccharides with concomitant deterioration in flavour and appearance (Deibel and Niven, 1959).

6.3.3   pH

As we have already considered, the post-mortem pH of meat will be determined by the amount of lactic acid produced from glycogen during anaerobic glycolysis (§ 5.1.2), and this will be curtailed if glycogen is depleted by fatigue, inanition or fear in the animal before slaughter. Since pH is an important determinant of micro-bial growth, it will be obvious that the ultimate pH of meat is significant for its resistance to spoilage. Most bacteria grow optimally at about pH 7 and not well below pH 4 or above pH 9 (Fig. 6.5; Cohen and Clark, 1919), but the pH of maximal growth is determined by the simultaneous operation of variables other than the degree of acidity or alkalinity itself. Some of the bacteriological enzymes which cause spoilage may have different optima from that of the organism itself. Thus, whereas bacterial proteolytic enzymes operate best near neutrality, the enzymes which attack carbohydrates tend to have optima below 6; and organisms such as lactic acid bacteria, of which the predominant activity is carbohydrate breakdown, have optima between pH 5.5 and 6.

In fresh meat, the encouragement given to bacteria by a high ultimate pH, espe-cially in the deeper areas of the carcass which are slow to cool, causes 'bone taint' (§§ 6.1.2.2 and 6.3.1). Muscle which has a high ultimate pH because of a deficiency of glycogen at death, also lacks the glucose which is produced by amylolysis post-mortem, albeit in much smaller quantity than lactic acid by glycolysis (cf. Table 4.9). In the absence of a readily available carbohydrate substrate, micro-organisms attack amino acids immediately, causing early spoilage, including off-odours (Newton and

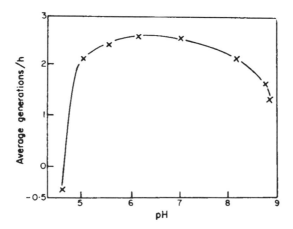

Fig. 6.5   The effect of pH of the medium on the rate of growth of *E. coli.* (Cohen and Clark, 1919.)

Table 6.7   Relation between ultimate pH of *psoas* muscle and subsequent spoilage in cured hams (Callow, 1937)

Degree of spoilage	No. of hams	Mean ultimate pH of *psoas*
None	148	5.65
Slight	47	5.71
Severe	18	5.84

Gill, 1978) and discoloration (§ 10.1.2). The use of a carbon dioxide atmosphere inside packaged beef is effective in limiting microbial growth, even with dark-cutting beef of high pH (Gill and Penney, 1986). High (90 per cent) concentrations of carbon dioxide are also effective in considerably extending the storage life of packaged lamb and pork of high ultimate pH at chill temperatures (Shay and Egan, 1986). Another effective alternative is to dip the meat surfaces in dilute acetic acid for 10 seconds at 55 °C prior to vacuum-packaging (Eustace, 1984). Attempts to farm red deer tend to cause stress which does not occur when wild deer are shot on open moorland, and this leads to a high ultimate pH in the venison (MacDougall *et al.*, 1979). This circumstance, however, in contrast to the result in beef and pork, seems to have no effect on the microbiological status of venison.

A particularly striking example of the adverse effects of high ultimate pH in spoiling cured meats was described by Callow (1937). In Northern Ireland, during re-organization of the pig industry in the early 1930s, it was required that pigs should be slaughtered at factories rather than on the farms of origin. An outbreak of soured hams resulted which was attributable to the greater glycogen depletion in the muscles of the factory-killed pigs during transit or in the excitement of slaughter (Table 6.7).

It will be apparent that an elevation in mean ultimate pH of only 0.2 units was critical. Of 46 micro-organisms which were isolated from tainted hams, 18 were

Table 6.8 Optimal pH values and salt concentration for ham spoilage organisms (after Ingram, 1948)

	3.1–4.0	4.1–5.0	5.1–6.0	6.1–7.0
A. pH *range*				
No. species having optimum pH within zone	1	7	31	7
Mean optimum salt concentration of these species (%)	5.0	6.9	8.2	11
B. *Range salt concentration* (%)	0–5	5–9	10–15	15
No. species having optimum salt concentration within zone	11	14	14	7
Mean optimum pH of these species	5.1	5.4	5.6	6.0

Table 6.9 Effect of pH on the aerobic nitrite tolerance of *Staph. aureus*

pH	Nitrite concentration (ppm)		Undissociated nitrous acid (ppm)	
	Growth	No growth	Growth	No growth
6.90	3500	4000	1.12	1.28
6.52	1800	2000	1.37	1.52
6.03	600	700	1.38	1.61
5.80	300	400	1.20	1.60
5.68	250	400	1.32	2.12
5.45	140	180	1.25	1.50
5.20	80	150	1.12	2.10
5.05	40	80	0.92	1.84

substantially inhibited below pH 5.7 and a further 14 below pH 5.4. Sugar feeding has been advocated to increase the muscle glycogen reserves immediately before death; and although the procedure does have some effect in lowering the ultimate pH to an inhibitory range (cf. Table 5.2), the benefits on storage life are not great.

The outbreak of spoilage referred to by Callow in 1937 concerned hams – a mildly cured product. Where the salt content in the product is higher, as in gammons, the effect of ultimate pH is not so critical (Table 6.8), and this may be why sugar feeding of bacon pigs has not been universally adopted so far. It will be seen that, in general, the micro-organisms which are less resistant to acid are more resistant to salt (mostly micrococci) and that the species which are not salt tolerant will not be inhibited by acid (mainly Gram-negative rods).

It might be thought that fresh meat, where the latter organisms would be preferentially found, would tend to spoil rather readily since salt is absent, but the muscles of beef and lamb, which are generally not cured, usually attain an ultimate pH of 5.5 or below. In cured products pH has another effect on microbial growth. It determines the proportion of the nitrite which is present as undissociated nitrous acid and thus inhibitory to bacterial growth (Table 6.9; Castellani and Niven, 1955).

It will be clear that an increase of 1 pH unit requires a tenfold increase in nitrite concentration to prevent growth, but that the undissociated nitrous acid is fairly constant. Nitrite happens to have a specific action in inhibiting *Cl. botulinum* (Roberts, 1971). The importance of nitrite in minimizing the risk from *Cl. botulinum* was reviewed by Roberts and Gibson (1986). It is likely to play an important part in the stability of uncooked cured meats and in the stability and safety of cooked, cured meats (Greenberg, 1972). In regard to the latter, however, it is present at too low a concentration to inhibit *Clostridia* at the pH values found in such products, suggesting that some additional factor is involved. Perigo *et al.* (1967) believe that when nitrite is sufficiently heated, it is involved in a reaction with the medium which causes the production of some inhibitory substance, which differs from nitrite *per se* in that, for example, its inhibitory action is only slightly pH-dependent. An inorganic compound derived from nitrite is possible (Spencer, 1971). Nitrite (at 20 ppm) lowers the concentration of salt required to stabilize mild-cured, vacuum-packed, green bacon for three weeks at 5 °C from 4 to 2.5 per cent (Wood and Evans, 1973). *Nitrate*, in the concentrations found in cured meats, is without *direct* effect on bacteria. Roberts *et al.* (1981a, b) have investigated the factors which control the growth of *Cl. botulinum* in pasteurized, cured meats. Both at low (5.5–6.3) and high (6.3–6.8) pH, increasing concentrations of nitrite, salt, isoascorbate or nitrate in the medium, and increased heat, significantly decreased toxin production by the organism. Although polyphosphate *per se* increased toxin production, it enhanced the action of isoascorbate in decreasing toxin production.

pH is also likely to be important with canned meat products since the resistance of faecal streptococci to heat at 60 °C for 1 h is apparently enhanced by a high value (Bagger, 1926).

6.3.4   Oxidation–reduction potential

Immediately after death, whilst temperature and pH are still high, it would be expected that the dangers of proliferation of, and spoilage by, anaerobes would be great. That such does not generally occur appears to be due to the level of the oxidation–reduction potential (E_h; Knight and Fildes, 1930) which usually does not fall for some time (Fig. 6.6; Barnes and Ingram, 1956). Growth of *Clostridia* on horse muscle does not take place until the E_h has fallen from about +150 mV to about –50 mV. The greatest rate of change in E_h occurs immediately after death and is probably due to the removal of the last traces of oxygen by the still surviving activities of the tissues' oxygen-utilizing enzyme systems. In whale meat, however, there is usually a greater reserve of oxygen because of the higher myoglobin content and of the low enzymic activity in the muscles of this species (Lawrie, 1953a, b) and the E_h may remain high for some time post-mortem (Robinson *et al.*, 1953; Ingram, 1959, unpublished data).

The effect of E_h on microbial growth is to prolong the initial lag phase: the eventual growth rate is not affected (Barnes and Ingram, 1956), since once the organisms become adjusted to a high E_h the rate is the same as at a low E_h. The E_h in meat is not, of course, a function of oxygen tension only; the concentration of molecules having a marked electropositive character is also important. Thus, the presence of nitrate in cured meats probably exerts an indirect antibacterial effect through raising the E_h of the system. Under anaerobic conditions at low pH (5.5) nitrate protects bacteria against nitrite (Eddy and Ingram, 1956).

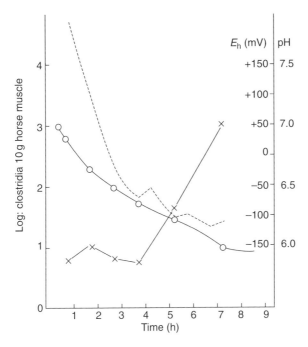

Fig. 6.6 Changes in pH (○), E_h (– – –) and numbers of viable *Clostridia* (×) in horse muscle after death (Barnes and Ingram, 1956). (Courtesy Dr E. M. Barnes.)

6.3.5 Atmosphere
Implicit in the concept of the oxidation–reduction potential, and in the contribution of oxygen tension to maintaining it at a high level, is the importance of the latter in determining the growth of surface spoilage organisms on meat and meat products. A major grouping of micro-organisms can be made on the basis of the oxygen tension which they need or can be made to tolerate – aerobes, anaerobes and fac-ultative anaerobes – although it is not possible to unequivocally correlate the growth of anaerobes with the amount of oxygen present.

The exposed surfaces of fresh meat at chill temperatures would normally support the growth of aerobes such as members of the genera *Achromobacter* and *Pseudomonas*.

In a computer-assisted classification of surface spoilage organisms isolated from fresh beef, lamb and pork (in which about 800 *Pseudomonas* strains were identi-fied) using 18 tests based on utilization of carbon sources, Shaw and Latty (1984) showed that 80 per cent of the isolates were strains of *Ps. fragi* and 4 per cent of *Ps. fluorescens*. The relative distribution of strains was similar for the three types of meat. Shaw and Latty concluded that the composition of the *Pseudomonas* popu-lation on stored beef, lamb and pork was determined principally by that of the initial contaminating microflora and that the particular environment at each abattoir was thus a major factor.

If the meat is wrapped, however, in a material which is impermeable, or only par-tially permeable, to the atmosphere a different situation obtains. Both the pressure and composition of the initial atmosphere within the pack can change. In an imper-meable wrap, where the oxygen supply is restricted, the growth of pseudomonads is hindered, and they are temporarily overgrown by organisms which tolerate low oxygen tension, whereas in permeable wraps spoilage tends to follow the same course as that on exposed surfaces (Ingram, 1962). Nevertheless, the partial pressure necessary to maintain aerobic bacteria is frequently much lower than that attained in an oxygen-impermeable film; and much of the inhibition to aerobic micro-organisms (including yeasts and moulds as well as bacteria) which arises may be due to the accumulation of carbon dioxide (Ingram, 1962). Haines (1933) showed that this gas specifically inhibited the growth of aerobes. With poultry, the delay in spoilage was found to increase regularly with increase in carbon dioxide concentration up to 25 per cent (Ogilvy and Ayres, 1951). But the inhibitory action of carbon dioxide is a selective one, moulds being sensitive and yeasts comparatively resistant (Ingram, 1958). Among bacteria, lactobacilli are relatively resistant and they occur in vacuum-packed fresh meat (Halleck et al., 1958) and bacon (Hansen, 1960), whereas Pseudomonas and Achromobacter spp. are susceptible (Haines, 1933). Although lactic acid bacteria predominate in vacuum-packed beef, lamb, pork and bacon, many of the strains involved have proved difficult to identify. Using numerical taxonomic methods, however, Shaw and Harding (1984) demonstrated that 90 per cent of 100 strains of lactobacilli isolated from such meats belonged to one or other of two clearly defined groups, both of which consisted of streptobacteria. A third group consisted of Leuconostoc strains. In 1986, Shaw and Harding classified 20 atypical strains of these organisms and identified a new species which they named Lactobacillus carnis. This organism is unusual in that lactobacillic acid is absent from the cellular lipids. It is distinct from the hitherto recognized L. divergens. Both the organisms appear to be important components of the microflora of vacuum-packed meats. It has been shown that surviving respiration of muscle in vacuum packs can produce 3–5 per cent of carbon dioxide within 3 h of packaging fresh pork (Gardner et al., 1967). In these circumstances the pork-spoilage bacteria Kurthia zopfi, pseudomonads and enterobacteria are inhibited, but lactobacilli and B. thermosphacta are not suppressed. With lamb carcasses, however, stored at 0 °C in packs flushed with carbon dioxide, the growth of lactic acid bacteria is suppressed (Grau et al., 1985). Atmospheres of 100 per cent carbon dioxide can be used to inhibit the growth of both lactobacilli and enterbacteria and thus extend the storage life of meat before it is prepared for display (Gill and Penney, 1986). Packaging in this gas has also been shown to limit the growth of micro-organisms on the surface of high pH, dark-cutting beef (Gill and Penney, 1986), lamb and pork (Shay and Egan, 1986). Thermosphacta is rarely found in cured meats, although such factors as carbohydrate encourage its presence in sausage. Comminution and packaging create conditions favourable for its multiplication. Gardner and Patton (1969) believe that routine examination for B. thermosphacta gives a good guide as to the presence of potential spoilage organisms. Since it is the psychrophilic organisms which are inhibited by carbon dioxide, an impermeable wrap is only likely to delay spoilage of meat products at chill temperatures. In general, the storage life of bacon could not be increased much by vacuum packaging since carbon dioxide has little effect on the micrococci and lactobacilli which are mainly responsible for its spoilage (Ingram, 1962).

The interaction of atmosphere and micro-organisms within packs tends to upset the usual correlation between spoilage and bacterial count: spoilage often becomes evident only after the number of the micro-organisms has been maximal for some time (Hansen, 1960). Wrapped meat in display cabinets can reach temperatures several degrees higher than the corresponding packaged meat, possibly because of a greater retention of radiant energy. This helps to account for cases of microbial spoilage of packaged meat even when good hygiene, and strict temperature control in the cabinet, were observed (Malton, 1971). The current interest in maintaining the bright red colour of oxymyoglobin on the surface of fresh, prepackaged meats has led to the successful use of in-pack atmospheres containing up to 80 per cent of oxygen. The incorporation of 20 per cent of carbon dioxide in such circumstances controls the microbial spoilage which would otherwise arise at high oxygen tensions. Its efficacy is markedly better at 0 than at 5 °C (Kitchell, 1971). Because meat surfaces are aerobic, post-mortem glycolysis would be expected to be delayed, and the pH higher, than in the anaerobic interior (cf. § 4.2.3). Rises in pH of the intact surfaces of beef and lamb carcasses were observed by Carse and Locker (1974); but differences in the surface pH did not appear to effect any corresponding differences in microbial growth.

6.3.6   High pressure

The effect of high pressure on micro-organisms has been reviewed by Cheftel (1995) and its specific application to meat by Cheftel and Culioli (1997). The degree of pressure inactivation is affected by the type of micro-organism, the pressure applied, the duration and temperature of the applied process and the pH of the medium. The inactivation of micro-organisms by high pressure appears to be due to intracellular damage (Simpson and Gilmore, 1997). The permeability of the cell membranes is impaired and numerous metabolic functions are affected, including ribosome morphology and the stability of DNA (Hugas *et al.*, 2002). Gram-negative bacteria appear to be more susceptible than Gram-positive organisms. Not surprisingly, bacterial spores are resistant to such pressure (unless above 100 °C), whereas vegetative micro-organisms may be inactivated by high pressure at temperatures as low as 0 °C. There is considerable variation in susceptibility to pressure between the types of micro-organism found in meat. Thus, when 400 MPa was applied for 10 min at 25 °C to pork homogenates, the population of such organisms as *E. coli*, *Campylobacter jejuni*, *Salmonella Typhimurium* and *Yersinia enterocolitica* was reduced by 6 log cycles, whereas *Staph. aureus* and *Str. faecalis* required a pressure of 500–600 MPa and spores of *B. cereus* were unaffected (Shigehisa *et al.*, 1991). Certain strains of micro-organisms are more resistant than others, e.g. *E. coli* O157:H7 (NCTC 12079). A 6 log reduction in the numbers of the latter organism, however, can be achieved by exposure to 500–600 MPa for 45 seconds (Ting, 1999).

There is evidence that a very small proportion of bacterial cells may be stressed, but not inactivated permanently, by high pressure and capable of growth after a resting period of some days at chill temperature (Carlez *et al.*, 1994), a feature requiring further investigation.

It has become clear that high-pressure treatment could be employed in conjunction with relatively low temperature to 'pasteurize' various heat-susceptible products (Zuber, 1993), for example prepackaged sliced ham. For such 'pressure pasteurization', pressures in the range 400–600 MPa, at 0–70 °C, over 1–10 min, have been suggested as the most useful procedures (Cheftel and Culioli, 1997). To date the full implications of pressure sterilization have not been determined (cf. § 12.1.2).

6.4 Prophylaxis

6.4.1 Hygiene

Veterinary and public health inspection must be relied upon to ensure that most of the meat which is endogenously infected with spoilage organisms, because of disease in the live animal, will never reach the consumer. Control of exogenous contamination requires the exercise of hygiene in slaughter-houses, meat stores, during transport, in wholesale and retail distribution and in the home. Such control is obviously difficult and, indeed, it is largely because of exogenous contamination that consumers encounter meat spoilage and meat poisoning. In recognition of this fact, most countries have statutory public health regulations designed to discourage unhygienic practice in handling meat and foods generally. Such regulations promulgate recommended procedures in the cleaning and maintenance of equipment, and in regard to personnel, at all points from pre-slaughter care of the animal to consumption of the meat (Anon., 1957b). These will not be given here.

An important aspect of meat hygiene is the design of abattoirs (Anderson, 1955). In general, whatever contributes to a smooth flow of animals will implicitly eliminate potential sources of contamination. In particular, inadequate lairages should be avoided, since overcrowding of animals increases the risk of infection by contact.

The spread of infection between animals can be substantially diminished by the establishment of specific pathogen-free herds (Skovgaard, 1987), and by avoiding over-high grazing density and the mixing of animals of different origin (Teugel, 1987).

It is especially important to make provision for the efficient removal from the carcass meat of blood, hides, guts and other portions of the animal which are likely to be heavily contaminated by micro-organisms (Anon., 1938, 1957b; Nottingham, 1963). In many large abattoirs in Australia and New Zealand this is achieved by killing above ground level and having chutes leading from the slaughter hall through which non-edible portions of the carcass proceed by gravity to the processing plant below. All equipment in the slaughter hall should be thoroughly and frequently cleansed by scrubbing, steaming and flushing with a solution of an approved antimicrobial agent. Until recently it was believed that methodical wiping of the surface of the carcass after dressing operations with a clean cloth was valuable in lowering contamination (Anon., 1938). Legislation has now prohibited the use of such cloths, and the application of a spray of warm water, followed by rapid drying, or pressure hosing, is employed for this purpose (Akers, 1969).

Provision for speedy and adequate chilling of the carcasses is essential. In this context, although it is microbiologically desirable to place these into mechanically operated chill rooms as soon as possible, the size of the chillers available and the rate of loading may dictate the accumulation of a group of carcasses in an external area before they are loaded. The space concerned must be designed to permit a through current of air to pass over the warm carcasses (Anderson, 1955). Walls and floors should be frequently washed with warm water, scrubbed with hot cleaning solution, rinsed, steamed and fumigated. It has been customary to use formaldehyde vapour for fumigation (e.g. in Australia; Anon., 1938), but it is sometimes necessary to leave the chill rooms vacant for 24–36h thereafter. Lactic acid has been successfully used. It achieves the same reduction in micro-organisms in 3h when sprayed as preheated lactic acid at a concentration of $300\,mg/cm^3$ of air (Shaw, 1963). Moreover, it is non-toxic. It is, however, expensive and the production of lactic acid through controlled fermentation by *L. plantarum*, inoculated into the meat surface, has been shown to be effective in reducing the growth of *pseudomonas* spp. under semi-tropical environments (Guerrero *et al.*, 1995). Ultraviolet light, ozone and carbon dioxide have been employed in discouraging the growth of micro-organisms on chilled meat products (Haines, 1937). The two former tend to cause oxidative rancidity in the fat and to accelerate the formation of brown metmyoglobin in the lean. The fact that ozone is toxic and that ultraviolet light fails to reach crevices limits their usefulness. Moreover, whilst ultraviolet light is effective against various psychrophilic bacteria and moulds, yeasts are relatively unaffected (Kaess and Weidemann, 1973).

The desirability of developing microbiological standards for meat has been considered from time to time. Clearly, whilst they would be feasible with packaged and comminuted products, it would be difficult to apply them to carcass meat. Such specifications might be workable if they served to verify good manufacturing and distribution practices: they could not, *per se*, achieve high quality (Mossel *et al.*, 1975). Standards should be derived from the results of surveys of products taken from production lines which have been examined for good manufacturing practice. Mandatory for the establishment of microbial standards would be precisely defined sampling procedures for assessing the numerical values specified (Mossel *et al.*, 1975).

Notwithstanding the importance of veterinary and public health inspection and of the assessment of the microbiological status of the end product in protecting consumers, food poisoning outbreaks still occur. The need for systematic and detailed control at all stages in the chain, from live animal to the consumer, by the identification of points where contamination is likely to occur or where micro-organisms can develop and, in particular, the need to eliminate specific pathogens as well as spoilage organisms, is now appreciated.

The concept of HACCP (cf. § 6.1.2.2) is being increasingly recognised as essential for the improvement of the microbial status of meat and meat products. In a detailed study of the beef supplied to supermarkets, and of the retail cuts sold by them, Nortjé *et al.* (1989) concluded that effective refrigeration *per se* guaranteed the preservation of quality which depends on a complex of interacting factors including the microbial status of the carcass meat, the maintenance of the cold chain, the sanitary condition of premises, equipment and personnel, and good management practice.

Procedures now employed to minimize microbial contamination of meat, at all stages in production, have also been reviewed by Huffman (2002). The season of the year, the type of animal, the location on the carcass and the stage in the processing operations, are all factors affecting the degree of initial contamination (Sofos *et al.*, 1999). Decontamination procedures, designed to improve hygiene in beef production, now include chemical dehiding (to remove hair, mud, manure, etc.) (Bowling and Clayton, 1992), knife-trimming, washing or spraying with various chemicals (Cutter *et al.*, 2000) and thermal treatment, including the use of pressurized hot water (Nutsch *et al.*, 1998) and steam pasteurization followed by vacuuming (Dorsa, 1996). The application of several of these 'hurdles' (Leistner, 1995) in sequence is synergistically effective (Sofos *et al.*, 1999; Huffman, 2002). The possibility of feeding probiotics to animals to limit subsequent infection by *E. coli* O157:H7 ('competitive exclusion') has been considered as a promising approach (Zhao *et al.*, 1998).

6.4.2 Biological control

Given the operation of efficient public health inspection and the exercise of hygiene, microbial spoilage can be controlled biologically, for the factors influencing growth, which we have considered above, can be manipulated to inhibit it. This fact is reflected in the processes used for preservation. The need by micro-organisms of a favourable temperature range for growth explains the possibility of preserving meat products by the imposition of sub-optimal temperatures (chilling, freezing) or of super-optimal temperatures (pasteurizing, sterilizing, cooking). Their requirement for moisture permits preservation by dehydration, freeze drying and by pickling in salt or sugar. The preferential development of anaerobes at a high ultimate pH can be prevented to some extent by ensuring that there is adequate glycogen in the muscles at the moment of death (§ 5.1.2). This involves avoiding fatigue, hunger and fear, which not only lower muscle glycogen but may cause bacteraemia (§ 6.1.2.1). The spoilage of meat surfaces by aerobic psychrophiles can be minimized by keeping the relative humidity of the atmosphere low and by incorporating in it about 10 per cent of carbon dioxide, to which such organisms are susceptible (as in the carriage of chilled beef from Australia: Empey and Vickery, 1933). Similarly, the storage life of prepackaged meat can be enhanced by incorporating carbon dioxide into the pack. Since the rationale for biological control is implicit in § 6.3, it will not be considered further at this point.

6.4.3 Antibiotics

It might be thought that the growth of micro-organisms on meat could be prevented easily by chemical treatment. Until the late 1940s, however, almost all chemicals which might have been used to kill or to prevent the growth of micro-organisms were also toxic to the human consumer. A new approach to food preservation became possible through the discovery of antibiotics. Appreciation of the fact that certain micro-organisms produced substances inimical to the growth of others – antibiotics – is attributable to Fleming (1929), who observed the production of what is now known to be penicillin by the fungus *Penicillium notatum*. Many other antibiotics, produced by bacteria, fungi or actinomycetes, have become known since then. Antibiotics are 100 to 1000 times more effective than permitted chemical preservatives. The low toxicity of many antibiotics to human beings, despite their power against micro-organisms, was firmly established in 1941 (Abraham *et al.*), but it was not until 10 years later that the possiblity of preserving foodstuffs by antibiotics was clearly demonstrated (Tarr *et al.*, 1952). The efficacy of the process has been reviewed by Wrenshall (1959).

Although many antibiotics are selective in their action, those which have been applied most frequently in the food industry are the so-called broad spectrum antibiotics, such as the tetracyclines, which are inhibitory to a wide range of both Gram-negative and Gram-positive bacteria:* few are active against yeasts and moulds. It is important to emphasize that, at practical concentrations, these antibiotics slow bacterial growth and hence delay spoilage: they do not sterilize. Aggregation into colonies or 'biofilms' enhances the resistance to antibiotics of bacteria and provides a reservoir of infective cells.

---

* Bacteria can be broadly classified according to whether they stain permanently with methyl violet (Gram-positive) or not (Gram-negative).

Widespread adoption of antibiotics in a non-medical context has been delayed – in particular to avoid the establishment by their indiscriminate use in human infection caused by organisms which have developed a resistance against them. Numerous antibiotic-resistant strains of micro-organisms are now known (Goldberg, 1962). Against this hazard the use, for food preservaton, of antibiotics which are not prescribed in medicine would seem appropriate. Such a substance is nisin, which is prepared by controlled fermentation of sterilized skimmed milk by the cheese starter organism *Strep. lactis* and in the United Kingdom is permitted in canned meats, provided these have been heated sufficiently to destroy *Cl. botulinium* (Taylor, 1963). The possiblity of 'R-factor' transfer must also be carefully considered (cf. § 2.5.2.2).

Another aspect of the use of antibiotics as preservatives requiring caution is the question of toxicity – due to residues of still active material in the treated products. Although one of the antibiotics most frequently used with meats, aureomycin, has been shown to be non-toxic when taken orally over long periods, its use in foods, at any rate in the USA, has been sanctioned only on the assumption that it is completely destroyed in cooking. Nevertheless, it has been detected in chops after frying (Barnes, 1956). The need for discretion is therefore obvious. It is also conceivable that antibiotics might eliminate the organisms normally responsible for creating the symptoms commonly associated with spoilage; and the meat could be contaminated with dangerous pathogens which did not indicate their presence by producing, for example, off-odours. When considering the use of tetracyclines for preserving meat in tropical or sub-tropical areas where refrigeration is not available, it is essential to establish whether the pathogenic hazard is increased or diminished by the treatment. This will depend on the numbers of resisting pathogens likely to be present. These will be increased if the animals have been fed the same antibiotic for growth stimulation (Barnes and Kampelmacher, 1966).

Antibiotics should not be used to replace good hygiene, but when employed with an appreciation of the dangers indicated above, and in conjunction with mild refrigeration (or some other process such as treatment with pasteurizing doses of ionizing radiations), they afford a means of preservation which does not materially alter the product.

The term antibiotic has come to signify more or less complex chemical products of micro-organisms, such as the penicillins and aureomyocins; but numerous simpler molecules, such as organic acids and alcohols, are also produced by microbial fermentation in foods. When derived from non-pathogenic species they can be encouraged to grow and to inhibit the development of other organisms which could cause spoilage or toxicity. Such fermentations have been employed inadvertently in preserving meats (e.g. as sausages) for many thousands of years, before the concepts of antibiotics or even of micro-organisms existed.

6.4.4 Ionizing radiations

In relatively recent times, another approach to the problem of preventing the growth of contaminating organisms on meat developed. This involves treatment of the commodity with ionizing radiations from electron or X-ray generators, or from a source of radioactivity (Brasch and Huber, 1947). Of the many types of ionizing radiations, only high-energy cathode rays, soft X-rays and γ-rays find application with foodstuffs (Hannan, 1955). The standard unit of dosage is the radiation which

Table 6.10   Doses of ionizing radiation required for various
biological effects (after Brynjolfsson, 1980)

Effect	Dose (10^3 Gy or 10^5 rad)
Death in higher animals	0.1–10
Inhibition of sprouting in vegetables	0.03–0.1
Death of insects, parasites	0.5–5
10^6 reduction in nos. fungi	1–10
10^6 reduction in nos. vegetative bacteria	0.5–19
10^6 reduction in nos. *Salmonella* spp.	2–7
10^6 reduction in nos. spores and dried bacteria	8–25
10^6 reduction in nos. viruses	10–40
Commercial sterility	45
10^6 reduction in nos. *Micrococcus radiodurans*	60
Inhibition of enzymes	1000

corresponds to the absorption of 100 ergs of radiation energy per gram of the absorbing substance (rad). It is more usual to measure dosage in terms of million rad (Mrad) or thousand gray (kGy).* The characteristic chemical property of these radiations is their ability to ionize receptor molecules, forming free radicals which cause other chemical changes in the area affected. Biologically, these effects can kill micro-organisms and parasitic life without raising the temperature of the product by more than a few degrees. The first significant publications on this effect were by Wyckoff (1930a, b). In general, the larger and more complex the organism, the more sensitive it is to radiation damage (Table 6.10; after Brynjolfsson, 1980).

Among bacteria, there is a 50-fold difference in the dose required for 90 per cent inactivation. It is feasible that radiation resistance in micro-organisms is associated with cystine-rich compounds, which may act by affording protection against the initial molecular changes induced by the radiation (Thornley, 1963).

Lea *et al.* (1936, 1937) found that the fraction of a population of microorganisms inactivated by a given dose was independent of the number of organisms originally present and that the degree of inactivation was determined by the total dose received and not by the rate at which it was delivered. They suggested that each micro-organism has a small 'target' area in which several ionizations had to occur to prove lethal: presumably the nucleic acid of the nucleus is the critical area (Weiss, 1952).

To destroy all micro-organisms, including viruses, would require 6 Mrad (cf. Table 6.10). But viruses are not at present regarded as a serious food poisoning hazard. Three to five Mrad produces the degree of sterility achieved in canning practice (Ingram and Rhodes, 1962): the spores of *Cl. botulinum* need 4.5 Mrad (Schmidt, 1961). Nevertheless, there is at least one type of non-sporeing organism – a red *Micrococcus* isolated from ground beef – which requires a dose of 6 Mrad (Anderson *et al.*, 1956) (cf. Table 6.10). Certain strains of *Moraxella acinetobacter*

---

* In 1975 the rad (100 erg g^{-1}) was replaced by the gray (Gy) as the unit of absorbed energy. 1 Gy = J kg^{-1} = 100 rad.

are also radio-resistant, and sufficiently heat-resistant to survive pre-irradiation heat treatment designed to inactivate enzymes (Maxcy and Rowley, 1978). Pasteurizing doses (about 0.5 Mrad) will destroy food-poisoning organisms (such as *salmonellae*, *staphylococci* and *Campylobacter jejuni*), and spoilage organisms, and thus prolong storage life, as with chilled beef (Drake *et al.*, 1961) and sausages (Coleby *et al.*, 1962). Parasites in meat, such as *Cysticercus bovis* and *Trichinella spiralis*, are killed by 0.01–0.1 Mrad (Tayler and Parfitt, 1959). Among psychrophilic food-spoilage organisms there are further minor differences in susceptibility to radiation, such that, even after only 0.25 Mrad, the spoilage microflora which, in poultry at 0 °C, is initially dominated by *Pseudomonas* and *Achromobacter* spp., is replaced by yeasts. When spoilage eventually occurs, however, and the bacteria have resumed dominance of the flora, non-pigmented pseudomonads are most numerous in both control and irradiated surfaces (Thornley *et al.*, 1960).

Although there are exceptions, one of the most important factors affecting the radiation resistance of micro-organisms is the presence of oxygen during irradiation. For example, the complete removal of oxygen, or the presence of reducing substances, may increase the radiation resistance of *E. coli* threefold (Niven, 1958). The removal of water by freezing or drying also protects microorganisms against radiation damage, probably by decreasing the possibility of free radical formation.

It would seem more practicable to combine an irradiation dosage which is lower than normally necessary to achieve sterility with another process (e.g. refrigeration, vacuum packaging, antibiotics, curing, heating). The bactericidal effects of combined processes are more than additive, for one treatment generally increases sensitivity to another (Ingram, 1959). Thus, for example, the heat resistance of spores of *Cl. botulinum* is only about one-third as great after treatment with 0.9 Mrad of γ-radiation, and as the radiation itself kills a proportion of the spores, the heat treatment needed to sterilize is reduced to about a quarter of that needed without irradiation (Kempe, 1955; Ingram and Rhodes, 1962). It must be remembered that radiation pasteurization *per se* does not inactivate any performed botulinum toxin whereas heat pasteurization does so (Gordon and Murrell, 1967). Again, chilling seems necessary to supplement a pasteurizing dose of irradiation, which might allow food-poisoning organisms to survive. Refrigeration is also a necessary adjunct with antibiotics and vacuum packaging to inhibit pathogens (Ingram, 1959). Nevertheless, the relationship between radiation resistance and product temperature is complex. Thus, the radiation resistance of *Yersinia enterolitica* was found to increase from 0.2 to 0.55 kGy as the temperature of the target product decreased from +5 °C to –76 °C (Sommers *et al.*, 2002).

Since complete sterilization by ionizing radiation cannot be achieved by any commercially practicable means, several terms have been developed to classify attainable objectives. Thus, *radappertization* is the application of doses of ionizing radiation sufficient to reduce the number of viable organisms (excluding viruses) to such an extent that few (if any) are detectable in the treated food by any recognized method, and no spoilage or toxicity of microbial origin is subsequently detectable, irrespective of the duration or conditions of storage (provided no recontamination occurs). *Radicidation* involves doses of ionizing radiation sufficient to reduce the number of viable, specific, non-sporeforming pathogenic micro-organisms (other than viruses) so that none are detectable in the treated food when it is examined by any standard method. *Radurization* is the term applied to a dose of ionizing radiation sufficient to enhance keeping quality by causing a substantial reduction in the numbers of viable, specific spoilage organisms. It is equivalent to pasteurization and it is thus essential to employ refrigeration during subsequent storage.

The physiological nature of the microorganisms which infect meat is reflected by the methods which can be used to discourage their growth, and these, in turn, are reflected by the processes which are used in practice to effect meat preservation. These will now be considered.

Chapter 7

The storage and preservation of meat: I Temperature control

The processes used in meat preservation are principally concerned with inhibiting microbial spoilage, although modes of preservation are sought which minimize concomitant depreciation of the quality of the commodity. The extent to which this secondary aim can be achieved is largely determined by the time of storage envisaged. The intrinsic changes which muscles undergo in becoming meat (Chapter 5) have not generally been considered in devising preservative processes, although it is increasingly recognized that their nature and extent may well determine the behaviour of the meat. Methods of meat preservation, however different superficially, are alike in that they employ environmental conditions which discourage the growth of micro-organisms (cf. § 6.3). They may be grouped in three broad categories based on control by temperature, by moisture and, more directly, by lethal agencies (bactericidal, bacteriostatic, fungicidal and fungistatic), although a particular method of preservation may involve several antimicrobial principles. Each principle may be regarded as a 'hurdle' against microbial proliferation and combinations of processes (so-called hurdle technology) can be devised to achieve particular objectives in terms of both microbial and organoleptic quality (Leistner, 1995). Thus a progressive reduction in the numbers of total micro-organisms, total coliforms and *E. coli* can be observed as beef progresses from its initial microbial load, through dehiding, washing and chilling and various decontamination methods are applied in sequence (Bacon *et al.*, 2000). Such control is becoming rather more deliberate and specific with increased scientific knowledge, but it was remarkably effective even when based on empirical observation and applied in making available to consumers a large variety of manufactured meats. A survey of the technology of currently available meat products was recently published by Varnam and Sutherland (1995).

It will have been obvious from the considerations in Chapter 6 that temperatures below or above the optimum range for microbial growth will have a preventative action on the latter. Meat and meat products may thus be preserved by refrigeration on the one hand or by heat treatment on the other.

7.1   Refrigeration

7.1.1   Storage above the freezing point
7.1.1.1   Fresh and chilled meat

That meat altered adversely sooner when kept during warm weather than under cooler conditions must have impressed early man. Appreciation of this fact led to the storage of meat in natural caves where the temperature was relatively low even in the warm season of the year. Later, as dwellings were built, cellars were constructed for food storage. These cellars were frequently quite sophisticated (Rixson, 2000). Much more recently ice, gathered from frozen ponds and lakes in winter, was used to keep cellar temperatures low (Leighton and Douglas, 1910). The principles of artificial ice formation and of mechanical refrigeration date from about 1750 (Raymond, 1929). Commercial scale operations based on mechanical refrigeration were in use 100 years later. Even after it became the practice to chill meat carcasses before either wholesale distribution or subsequent freezing, the idea persisted in the meat industry that carcasses should be placed in an area at ambient temperature to permit the escape of 'animal heat' before chilling in a mechanically refrigerated chamber. In seeking the reason for this view it is feasible that low surface temperatures on the meat in the chillers had been erroneously regarded as representing those throughout the carcasses and that refrigeration had been discontinued whilst deep temperatures were still high – thus giving rise to bone taint (Haines, 1937; Cosnett *et al.*, 1956) and other symptoms of microbial spoilage. *From the viewpoint of discouraging microbial growth*, the weight of evidence suggests that carcasses should be cooled as quickly as possible (Tamm, 1930; Scott and Vickery, 1939) provided that the temperature of the deepest portion of the carcass is used as an indicator of the efficacy of the process (Kuprianoff, 1956). Since, however, the surfaces of the meat carcasses are initially considerably higher than those of the chill room, evaporation of substantial quantities of water may occur. Even where this does not cause surface desiccation, it is obviously economically undesirable. The problem of cooling carcasses with the minimum of desiccation has been extensively studied (Scott and Vickery, 1939), and various combinations of air speeds and humidities have been considered. The achievement of rapid cooling requires a high air speed in the chill rooms or large air circulation volumes (60–100 changes of air/h; Kuprianoff, 1956). Although a higher air velocity will tend to cause a greater weight loss, it permits the use of a high relative humidity. At the beginning of chilling, the air temperature can be as low as $-10\,°C$ for pigs and sheep and air speeds as high as 180 m/min (600 ft/min). For beef carcasses air speeds of 120 m/min (400 ft/min) and an air temperature of $-1\,°C$ have been recommended (Kuprianoff, 1956). Once the difference in temperature between the meat surface and the air becomes small, the air speed must be reduced to avoid desiccation. Air, supersaturated with water vapour and moving at high speed, has been used to minimize evaporation from hot sides whilst providing a high capacity for heat removal (Hagan, 1954). Although there are undoubted advantages to be gained from fast chilling in reducing weight losses, the application of ultra-rapid chilling (air at $-30\,°C$ and 4 m/s) can cause 'cold-shortening' and toughening of pork (Van der Wal *et al.*, 1995). Spray chilling the surfaces of pig carcasses has been reported to enhance the oxygenation of myoglobin without any increase in metmyoglobin (Feldhusen *et al.*, 1995). In a comparison with air chilling, Greer and Jones (1997) showed that spray-chilling of beef carcasses with water mist at $1\,°C$, in four cycles during the first 4–16 hours, significantly reduced

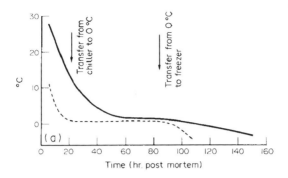

Fig. 7.1 Typical time–temperature curves for two grades of beef during chilling at 0 °C and freezing at –10 °C (Bouton *et al.*, 1957). ——— = grade 1, about 160 kg (360 lb) side weight. – – – = canner grade, about 90 kg (200 lb) side weight.

carcass shrinkage by 0.08 g/100 g per hour. They calculated that a large abattoir processing beef could diminish shrinkage by 2000 kg daily. It has been shown (Powell and Griffiths, 1988) that the spraying of beef, sheep and pork carcasses with a dilute (<1 per cent) aqueous emulsion of cetyl alcohol reduces the weight losses during chilling by 40–80 per cent and during holding at refrigeration temperatures by 22–50 per cent. Carcasses so treated retain a bright appearance and there is no difference in the microbial growth on the surface. No flavour changes were noted unless the concentration of cetyl alcohol was 3 per cent or above.

Obviously, the greater the bulk of the carcass and the greater its fat cover the longer it will take to cool with a given air speed and temperature (cf. Fig. 7.1; Bouton *et al.*, 1957). Workers at the Meat Research Institute, Bristol, obtained valuable comparative data on the rates of fall of temperature in the deep leg locations of beef sides under carefully controlled conditions (Bailey and Cox, 1976). The sides averaged 100, 180 and 260 kg. Air temperatures of 0, 4 and 8 °C were employed at air speeds of 0.5 to 3 metres per second. Whereas it took 80 hours for the deep leg temperature of 260 kg sides to attain 10 °C, in air at 8 °C travelling at 0.5 metres per second, this temperature was reached in only 16 hours with 100 kg sides in air at 0 °C travelling at 3 metres per second. For any given conditions in chilling, the weight losses from small, poorly covered sides will be greater than those from large sides having a good fat cover (Table 7.1; Bouton *et al.*, 1957).

It is important to point out that over-effective chilling of hot carcasses can lead to toughness. If the relationship between the refrigeration system and the bulk of meat exposed to it are such that the temperature of the muscles can be reduced below about 10–15 °C, whilst they are still in the early pre-rigor condition (pH about 6.0–6.4), there is a tendency for shortening and, thereby, toughness on subsequent cooking (Locker and Hagyard, 1963; Cook and Langsworth, 1966). This is the phenomenon referred to as 'cold-shortening'. The tendency is greater the closer the temperature attained by the pre-rigor muscle approaches the freezing point. At 2 °C the shortening is as great as that which occurs when muscles go into rigor mortis at 37 °C.

Locker and Hagyard (1963) failed to observe the cold-shortening effect in the *psoas* and *l. dorsi* muscles of rabbits (although these can be made to exhibit thaw-rigor: cf. § 7.1.2.2). Bendall (personal communication) found, however, that the

Table 7.1 Mean weight losses from good and poor quality beef sides under various chiller conditions

Conditions	% Weight loss	
	Good-quality sides	Poor-quality sides
1 day at 0 °C+		
2 days at 20 °C	1.7	3.8
3 days at 0 °C	0.7	0.9
14 days at 0 °C	1.6	3.2

soleus muscle of the rabbit *did* show the cold-shortening phenomenon. The latter is a so-called 'red' muscle, resembling in this respect the beef muscles studied by the New Zealand workers, whereas the *l. dorsi* of the rabbit is a so-called 'white' muscle. In studying porcine muscles, Bendall (1975) confirmed that 'red' muscles were susceptible to 'cold-shortening', whereas those with less myoglobin were much less so. The intensity of cold-shortening is correlated with the proportion of 'red' fibres in the muscle (Cena *et al.*, 1992). Similar considerations may explain the observation that the power developed in 'cold-shortening' by bovine *sternomandibularis* muscles increases with animal age (Davey and Gilbert, 1975a), i.e. as they change to a 'redder' type (Lawrie, 1952a).

Two mechanisms have been suggested for 'cold-shortening'. Cassens and Newbold (1967) believed that the sarcotubular system of pre-rigor muscle is stimulated to release calcium ions by the attainment of temperatures below about 15 °C, whereby the contractile actomyosin ATP-ase is much enhanced (Newbold and Scopes, 1967). Indeed a 30- to 40-fold increase of the concentration of calcium ions in the vicinity of the myofibrils, as the temperature falls from 15 to 0 °C, has been reported (Davey and Gilbert, 1974). It may be postulated that reabsorption of these ions could be more readily effected by a type of muscle wherein the sarcotubular system was relatively well developed; and Fawcett and Revell (1961) demonstrated that the system was more extensively elaborated in 'white' muscles than in 'red'. Indeed, *in vitro* studies have shown that the rate of Ca^{++} uptake by the sarcotubular system of 'white' muscle is greater than that of 'red', that inorganic phosphate ions substantially enhance Ca^{++} uptake by both types (Table 7.2: Newbold, 1980) and that the cold-induced release of Ca^{++} ions from the sarcotubular system is largely suppressed by relatively small concentrations of inorganic phosphate (Newbold and Tume, 1977).

Since the rate of production of inorganic phosphate during post-mortem glycolysis is considerably greater in 'white' muscle than in 'red' (Lawrie, 1952a), it may be that a major factor in explaining the absence of 'cold-shortening' in the former type is the attainment therein of a relatively high concentration of inorganic phosphate early post-mortem (Newbold, 1980). It would be expected, therefore, that the enhancement of actomyosin ATP-ase would be less easily suppressed in 'red' muscles and would more readily lead to marked interdigitation of myosin and actin filaments, i.e. shortening, as is observed.

Buegge and Marsh (1975), however, showed that pre-rigor 'red' muscles do not 'cold-shorten' when cooled below 15 °C if they are adequately supplied with oxygen.

Table 7.2 Effect of inorganic phosphate on the relative amounts of Ca^{++} accumulated by sarcoplasmic reticulum and mitochondria (in presence of ATP) (after Newbold, 1980)

Membrane source	Ca^{++} accumulation (mmoles/mg protein)	
	No phosphate	5–10 mM phosphate
Sarcoplasmic reticulum		
white muscle (rabbit)	240	8000
red muscle (rabbit)	40	2000
red muscle (ox)	80	3000
Mitochondria		
red muscle (ox)	2000	2000

This resistance is annulled by reagents which uncouple the uptake of oxygen from resynthesis of ATP. They thus inferred that cold-shortening is due to the discharge of calcium ions from the muscle mitochondria under anaerobic conditions post-mortem and to failure by the sarcotubular system to reabsorb them effectively at low temperatures. Since 'red' muscles contain a greater number of mitochondria than 'white' (Paul and Sperling, 1952: Lawrie, 1952a), this feature also could have significance in relation to the relative 'cold-shortening' susceptibility of the two types of muscle. On the other hand, muscles can be induced to 'cold-shorten' many hours post-mortem (Locker and Hagyard, 1963) when they would long have been anaerobic. Subsequently Mickelson (1983), in a study of beef muscle mitochondria (under conditions when the sarcoplasmic reticulum had been excluded), demon-strated that these would not leak Ca^{++} ions in the pre-rigor period and thus that anoxia *per se* would not cause such release provided there was still sufficient ATP available to support calcium uptake by the mitochondria. The cold-induced release of Ca^{++} from mitochondria is much more limited than that from the sarcotubular system (Newbold, 1980) and inorganic phosphate has no effect in enhancing Ca^{++} uptake by these organelles (Table 7.2). Again, since the concentration of calcium-carrier protein in mitochondria is relatively very low in comparison with the calcium-binding mechanism in the sarcoplasmic reticulum, the involvement of mitochondrial Ca^{++} release in 'cold-shortening' seems likely to be less significant than exchanges of Ca^{++} between the sarcoplasm and the sarcotubular system (Greaser, 1977).

There is clearly evidence for the involvement of both mechanisms. The situation is not yet completely resolved, however, since porcine *l. dorsi*, although a 'white' muscle, undergoes vigorous 'cold-shortening' (Bendall, 1975; Dransfield, 1983). Moreover the muscles of the ostrich, which are red (containing more myoglobin on average than those of beef), are apparently not susceptible to 'cold shortening' (Sales and Mellett, 1996).

Cold-shortening can be avoided by cooling the meat swiftly to about 15 °C and holding it at this temperature to allow the onset of rigor mortis. The temperature can then be lowered as quickly as is compatible with minimal surface desiccation. Moreover, not all the muscles of carcasses, even if their temperature falls below about 15 °C whilst they are in the early pre-rigor condition, are free to shorten. Their attachments on the skeleton may restrain them sufficiently to prevent it. In the conventional mode of carcass suspension by the Achilles tendon, certain muscles are more liable than others to shorten; and, by altering the posture of the suspended carcass in various ways, the pattern of muscles which shorten may be changed. Thus pelvic hanging of lamb and beef carcasses (field-stretching), by preventing shortening in muscles which are normally able to respond to cold stimulus, can bring about a degree of tenderness which would otherwise require more than a week's conditioning (Herring et al., 1965; Davey et al., 1971; Bouton et al., 1974). Such altered carcass postures, however, while permitting control of cold-shortening during the fast chilling of muscles early pre-rigor, could be inconvenient in abattoir operations; and an alternative method of minimizing cold-shortening was devised by workers in New Zealand, involving electrical stimulation of the carcass in the immediate post-mortem period. This technique will be considered in § 7.1.1.2: it has developed rapidly in recent years.

Because of the relatively small bulk of lamb carcasses in comparison with those of beef, their post-mortem temperature can be more speedily and accurately controlled by the refrigeration regime; and, in 1980, Davey and Garnett suggested it might be possible to chill pre-rigor lamb carcasses without electrical stimulation and still avoid the toughening of 'cold-shortening'. This was confirmed by Sheridan (1990), who demonstrated that, by cooling pre-rigor lamb carcasses in air at –20 °C and at 1.5 metres per second, the meat was as tender as that which had been conventionally chilled at 4 °C for 24 hours when assessed after 7 days of subsequent chill storage. He attributed the absence of 'cold-shortening' during such rapid refrigeration to the skeletal restraint of the carcass and to 'hardening' (but not freezing) of its surface – a factor which Davey and Garnett (1980) had envisaged as being sufficient to avoid toughening.

Since 1990 a number of alternative fat chilling procedures, including spraying the carcasses with cold water, have been developed. Such procedures substantially reduce evaporative losses and with careful control, avoid the toughening of 'cold-shortening' (Brown et al., 1993; Jones et al., 1993).

Currently there is considerable interest within the European Union in devising a commercial procedure for very fast chilling (defined as the attainment of a temperature of –1 °C within 5 h post-mortem), which would ensure tenderness through precise control of the factors which promote or inhibit it (Joseph, 1996).

7.1.1.2   Electrical stimulation
(a)   General

In 1749, Benjamin Franklin observed that electrical stimulation of turkeys immediately post-mortem tenderized their flesh. Much later, Harsham and Deatherage (1951) filed a patent for the tenderizing of meat by this device, but its application to commercial practice was not considered seriously until 1973. Harsham and Deatherage (1951) had shown that one effect of the procedure was a considerable acceleration of post-mortem glycolysis; and the primary biochemical purpose of renewed interest in electrical stimulation was to swiftly lower the pH of the musculature, thereby, to below 6 (when muscles are no longer susceptible to cold shock: see above and, incidentally, no longer respond to electrical stimulation; Davey, 1980), so that speedy refrigeration could be employed without risk of cold-shortening (Carse, 1973).

Electrical stimulation shortens the time to the onset of rigor mortis through two phases of acceleration of glycolysis, the first during stimulation and the second, less precipitate phase, following stimulation. During electrical stimulation pH falls of the order of 0.7 units can occur in two minutes. This represents a 100- to 150-fold increase in the rate (Chrystall and Hagyard, 1976). Whilst in lamb and rabbit (Bendall, 1976) and pig (Hallund and Bendall, 1965) the post-stimulation rate of pH fall appears to be about twice as fast as normal (for a given temperature range), the reaction of bovine muscles, in this phase, depends upon the orientation of current-flow to fibre direction and the current strength (Bendall, 1977). When stimulation is across the fibres the rate of pH fall does not increase until the current is about 60-fold greater, when partial breakdown of muscle membranes may occur.

(b)   Mode of application

The scientific literature on electrical stimulation indicates that the electrode system, the type of current (voltage, frequency of pulse and duration), the pathways (via nerve or direct) and the time post-mortem have varied considerably between investigators. Bendall (1980) comprehensively reviewed the position. Most appear to have applied the current via the thoracic region of the carcass and to have used the Achilles tendon for return to earth. Although low voltages (<100 V) are intrinsically safer in operation, they are less consistent in effect than voltages of 500–1000 V or more (Bouton et al., 1980a; Bendall, 1980). High voltages are effective in accelerating post-mortem glycolysis when applied for 1.5–2 min, whereas longer times (~4 min) are required with voltages of the order of 100 V.

Because of differences in their intrinsic electrical resistance, intact carcasses will allow more current flow for the application of a given voltage. Thus, for example, a peak voltage of 680 V, between electrodes 200 cm apart, gave a peak current of 5.2 amps with intact beef carcasses (wherein conductivity is high because of the relatively large cross-sectional area and the presence of the wet gut contents), 3.3 amps with dressed carcasses and only 2.4 amps with dressed sides (Bendall, 1980). The latter tend to jerk outwards since there are no contralateral intracostal or *l. dorsi* muscles to oppose movement (Bendall, 1978a).

In respect of other aspects of the current, optimum pulse rate appears to be between 15–25 pps (higher frequencies tend to be relatively ineffective since they fall within the latency period of the muscles concerned), and the optimum pulse width is about 20–40 ms (shorter widths may fail to activate all the muscle fibres).

The response of beef carcasses to electrical stimulation falls off quickly after about 50 min post-mortem (Bendall et al., 1976) and that of lamb carcasses even sooner. It is thus desirable to apply the current within about 30 min of slaughter.* Since *isolated muscles*, on the other hand, will respond as well at 3 h post-mortem as at 20 min (Bendall, 1977) it appears that in electrical stimulation of the carcass or side the musculature is usually activated via the nervous pathways. If the applied voltage is sufficiently high, direct stimulation of the muscles can occur at a later time (Chrystall et al., 1980) when decay of the nervous pathways has made low voltages ineffective. Such observations are useful insofar as electrical stimulation may

---

* If electrical stimulation, whether with high or low voltage, is applied at *ca.* 3 min post-mortem, its effect is relatively less, in tenderizing, than at *ca.* 40 min post-mortem, owing to an early reduction in the activity of calpain I (Wang and Thompson, 2001).

not be feasible in certain abattoirs until after carcass dressing (Hagyard and Hand, 1976–77).

The importance of a still-functioning nervous system in making low voltages effective was re-emphasized by Swedish (Rudérhus, 1980) and Australian (Morton and Newbold, 1982) work. By placing one electrode in the nerve centre of the muzzle (and earthing via the Achilles tendon) during the first 10 min post-mortem, a typical acceleration of post-mortem glysolysis was achieved by applying various stimulation procedures lasting for 1 min and having a peak voltage of 80 V. A current of 14 pps, applied either in 1 s pulses or continuously, was effective. Although the peak voltage was 80 V, the duration of each peak was only 5 ms, and the actual voltage very low, being thus relatively safe in operation. Australian regulations now require that the peak voltage should be 45 V. With such a system, in order to ensure that at least 95 per cent of the stimulated carcasses have a minimum standard tenderness (not more than 8 kg shear force), stimulation must be applied within 4 min of slaughter and the high resistance of the upper leg and shoulder must be bypassed by earthing via the anus or leg (Powell et al., 1983).

Another low voltage procedure is to push a plastic rod (with an electrode at its tip) down the spinal cord immediately after bleeding. The current used with this device is pulsed at 25 pps for 2 min, the peak voltage being 140 V (Bendall, 1980). Such procedures have obvious economic advantages, but not all abattoirs would be in the position to stimulate carcasses so early post-mortem.

Although the duration of the current required to achieve anaesthesia in the electrical stunning of calves and sheep (1–3 s) is much less than that involved in post-mortem electrical stimulation (ca. 2–4 min) there is evidence that electrical stunning may also accelerate post-mortem glycolysis (Lambooy, 1981). The very fast cooling which electrical stimulation permits immediately post-mortem, especially with the relatively small portions of meat now produced by hot-deboning, not only avoids 'cold-shortening' but also greatly reduces microbial problems from residual blood. In such changed circumstances, electrical stunning could be used both to kill the animal and to accelerate post-mortem glycolysis in its musculature. There would be no need to have a period for anaesthesia and bleeding, and abattoir operations could be even more efficient. Indeed Geesink et al. (2001) demonstrated that post-mortem electrical stimulation of beef carcasses could be omitted, without sacrificing tenderness, when animals are electrically stunned or immobilized with a relatively low electrical input (75 V, 20 s). On the other hand, prolongation of the current for 80 s can be associated with tougher beef, partly because of sarcomere shortening in such circumstances.

(c)   Mechanism and effect on muscle

The vast acceleration of post-mortem glycolysis caused by electrical stimulation whilst the current is flowing signifies a concomitantly high rate of ATP breakdown (Bendall et al., 1976) – which, in turn, reflects marked activation of the contractile actomyosin ATP-ase by released Ca^{++} ions. The latter also enhance the titre of phosphorylase a, which is an additional factor accounting for the increased rate of post-mortem glycolysis (Newbold and Small, 1985). Yet such are precisely the circumstances which cause the toughening in 'cold-shortening', the avoidance of which is the principal reason for applying electrical stimulation in the first place. It appears that the membrane of the sarcoplasmic reticulum is altered by electrical stimulation, and that the calcium-binding protein, calsequestrin, becomes more exposed (Tume, 1980). Possibly this feature enhances the capability of the system to retain Ca^{++} ions, but this seems unable to explain the anomaly fully.

Moreover, the attainment of a relatively low pH, whilst temperatures are at *in vivo* levels, denatures muscle proteins, thus causing loss of water-holding capacity. This can be exemplified *in vitro* (Scopes, 1964) by the musculature of pigs which produce pale, soft exudative (PSE) pork (Bendall and Wismer-Pedersen, 1962; Penny, 1969) and by the slow-cooling, deep musculature of beef hindquarters (Follet *et al.*, 1974). Yet electrically stimulated muscles do not overtly lose drip fluid (at least not initially). There is some eventual loss of fluid however. This has been demonstrated (Taylor *et al.*, 1981). Thus, whereas the expected benefits of diminished drip were found in hot-deboned, vacuum-packed primal cuts (whether these had been removed from electrically stimulated sides or not), in comparison with cold-deboned primal cuts after 5 days storage, the electrically stimulated, hot-boned joints had lost relatively more fluid after 21 days storage – although the quantity was still less than that of corresponding joints from cold deboned sides.

These seeming biochemical paradoxes have not been fully explained, but recent findings resolve them partially.

That electrical stimulation is not associated with permanently shortened sarcomeres (and toughening) may reflect the fact that the current is short-lived and that, when it is discontinued, the ATP level is still relatively high and the temperature has fallen little from its *in vivo* value. In these circumstances the sarcotubular system can presumably recapture Ca^{++} ions readily, thus suppressing ATP-ase activity whilst the ATP level is sufficient to effect muscular relaxation and the restoration of resting sarcomere length. In 'cold-shortening', on the other hand, the low temperature prevents effective operation of the ATP-fuelled Ca^{++} pumps of the sarcotubular system, and stimulation of ATP breakdown is thus not arrested. The situation may be similar to that in pre-rigor frozen muscle fibres when these are thawed very rapidly (cf. § 7.1.2.2). Thaw rigor is avoided in these (unusual) circumstances because Ca^{++} ions are reabsorbed by the sarcotubular system before much ATP can be broken down, and the muscle relaxes after brief shortening.

There has been disagreement between research workers on the reason for the association of post-mortem electrical stimulation with tenderness. Some (Bouton *et al.*, 1980a) attribute this benefit to the avoidance of 'cold-shortening', and they have published data showing the retention of rest length in the sarcomeres of muscles which were electrically stimulated, whereas control sarcomeres shortened, when both were exposed to environments causing 'cold-shortening' that is, in the absence of high-temperature conditioning. Other workers have attributed the benefit to early and extensive conditioning changes arising from the combination of low pH with *in vivo* temperatures (Bendall, 1980; Dutson *et al.*, 1982). Dutson (1977) and Savell *et al.* (1977a) postulated that the release of Ca^{++} ions from the sarcotubular system on electrical stimulation may enhance proteolysis by the calpains and as the pH falls further, whereby lysosomal membranes are damaged, cathepains may be liberated (Dutson *et al.*, 1980; Locker, 1989). Indeed various workers have shown that the activity of calpain I (μ-calpain is enhanced by electrical stimulation (Wang and Thompson, 2001). The acceleration of such proteolytic activity by the high temperatures (near *in vivo*) could contribute to the observed tenderness in muscles which have been stimulated electrically in the immediate post-mortem period as a factor additional to the avoidance of 'cold-shortening'. Indeed, under certain conditions of electrical stimulation no differences in sarcomere length can be detected between electrically stimulated and control muscles, although the former are more tender as meat (Dutson, 1977; Savell *et al.*, 1977b).

Marsh and his co-workers (Marsh *et al.*, 1981; Takahashi *et al.*, 1984) also endeavoured to distinguish between the tenderizing effects of electrical stimulation in preventing 'cold-shortening' and in promoting other textural changes. They demonstrated (Takahashi *et al.*, 1984) that, when the frequency of the electrical current employed was only 2 Hz instead of *ca.* 50–60 Hz (which is more usual in commercial practice), stimulated sides were less tender than unstimulated controls, both sets having been held at 37 °C for three hours under circumstances *when muscle shortening was prevented.** Yet the stimulated sides exhibited the much accelerated post-mortem glycolysis expected. It was inferred that, because of the relatively slow pH fall in the unstimulated sides, proteolytic enzymes with pH optima near neutrality had enhanced their tenderness above that of the faster-acidifying stimulated ones, especially at the high temperature of holding used, and that the tenderizing changes of conditioning must commence before the ultimate pH is attained (cf. §§ 5.4.2 and 10.3.3.1; and Koohmaraie *et al.*, 1987).

Marsh *et al.* (1981) found no histological evidence of tissue disruption in muscles which had been electrically stimulated by current of frequency 2 Hz, whereas there was both severe contraction and breaking of sarcomeres (as revealed by electron microscopy) when the current frequency was 50–60 Hz (Takahashi *et al.*, 1984, 1987). They thus inferred that the additional effect of electrical stimulation as normally used in commercial operations was due to such tissue disruption and not to the early attainment of low pH, although they acknowledged that both low- and high-frequency electrical stimulation are effective in preventing the toughening of 'cold-shortening' when fast cooling prevails. Other workers, however, demonstrated that electrical stimulation at 50–60 Hz had no effect in tenderizing muscles of high ultimate pH (Dutson *et al.*, 1982) despite the extensive contraction and tearing of sarcomeres it caused (Fabiansson *et al.*, 1985). This conflicting evidence suggests that the tissue disruption induced by electrical stimulation at 50–60 Hz requires a concomitant low pH for it to effect tenderness and supports the view that acidophilic proteolytic enzymes are involved in conditioning as well as those which operate at more neutral pH values (and cf. § 12.1.2).

During further detailed studies of the relationship between electrical stimulation and beef tenderness, in which various combinations of rates of post-mortem cooling and modes of electrical stimulation were employed, Marsh *et al.* (1987) showed that tenderness was optimal when the rate of post-mortem glycolysis, which was a function of the temperature (cf. § 4.2.2) and of the type of electrical stimulation, was of intermediate value. The rate of glycolysis was assessed by the pH attained by 3 h post-mortem (pH_3). Tenderness was greatest at pH_3 values of 5.9–6.0. Marsh *et al.* (1987) pointed out that those combinations of temperature and modes of electrical stimulation which yielded a pH_3 above or below 5.9–6.0 would be associated with relatively greater toughness and that deliberate induction of a rapid rate of pH

---

* Not surprisingly, if *excised*, electrically stimulated muscles are exposed to temperatures *ca.* 39 °C in the immediate post-mortem period (i.e. under circumstances when they are free to shorten), they become tougher than corresponding, non-stimulated controls since the shortening of high-temperature rigor mortis will occur and will be accelerated by the stimulation (Harris and Shorthose, 1987). In severely shortened muscle, conditioning is minimal (Davey *et al.*, 1967).

fall by electrical stimulation may not *per se* ensure tenderness even when 'cold-shortening' was avoided.

Insofar as heavy carcasses will cool more slowly than those of lighter weight, when exposed to given chilling conditions, the temperature of their muscles will tend to remain for longer above the range at which 'cold-shortening' is indicated (indeed this may be an additional explanation for the traditional belief that marbled meat tends to be more tender than lean meat).

Electrical stimulation of such carcasses may cause exudative meat through the acceleration of post-mortem glycolysis and the attainment thereby of acid pH values whilst the temperature is still relatively high. It is evident that the temperature of the meat could modify the effect of electrical stimulation and must be considered when applying the technique in commercial practice (cf. § 10.3.3.1).

Conditioning changes – as assessed by the decrease in shear force – have a high-temperature coefficient (Davey and Gilbert, 1976a). By comparing the temperature/pH history of control and electrically stimulated muscle, George *et al.* (1980) calculated that conditioning changes proceed at about twice the rate in the latter during the first 24–30 h post-mortem – in circumstances when 'cold-shortening' was avoided by not exposing the muscles to environmental temperatures below 10 °C until 8 h post-mortem. After 21 days storage at 0 °C, the tenderness of control muscles had increased substantially through the operation of conditioning processes at the normal rate, but it was still slightly less than that of the electrically stimulated muscles at this time. The efficacy of electrical stimulation in enhancing tenderness under circumstances when the avoidance of cold-shortening is not an issue, diminishes in relation to non-stimulated controls the further the subsequent period of conditioning proceeds (Harris and Shorthose, 1987), presumably because tenderness approaches an asymptotic value and cannot be improved beyond it.

George *et al.* (1980) also presented histological evidence which clearly showed that, in electrically stimulated bovine muscles, in which near *in vivo* temperature and pH had prevailed, bands of denatured protein formed gradually within the fibres. These were similar to those observed in the musculature of PSE pork (Bendall and Wismer-Pedersen, 1962). Insofar as the proteins of electrically stimulated muscles conform to expectation and denature, therefore, one apparent anomaly of the procedure has been resolved. On the other hand, Savell *et al.* (1977b), although they found no evidence of denaturation, demonstrated histologically that massive contracture bands are a feature of the muscles from electrically stimulated carcasses. Voyle (1981) found such bands of intensive contraction in bovine *l. dorsi* (but not *semitendinosus*) muscles, following stimulation, although they developed only after 24 h at 15 °C. He suggested this may be a reflection of damage to the membrane of the sarcoplasmic reticulum whereby, when muscles relax after stimulation has ceased, local leaks of calcium ions arise and cause contraction zones. There is evidence that electrical stimulation lowers the shrink temperature of collagen (Judge *et al.*, 1980).

There remains to be explained the absence of marked exudation in electrically stimulated bovine muscle. As remarked above, such loss of water-holding capacity is striking in PSE pork, in the deep musculature of beef hindquarters and when *in vivo* temperature and low pH are combined *in vitro*. It is feasible that the acceleration of conditioning changes, which is one effect of the fast pH fall achieved by such stimulation, enhances intracellular osmotic pressure sufficiently to accommodate the loss of water-holding capacity by the muscle proteins. Certainly normal slow conditioning of beef at $0\,^{\circ}C$ has this effect (Cook *et al.*, 1926; Bouton *et al.*, 1958). On the other hand pork undergoes conditioning changes to a greater extent than beef (Henderson *et al.*, 1970) and this might have been expected to raise intra-cellular osmotic pressure even more than with the former type of meat. Possibly the proteins of pork ($\approx$'white' muscle) are intrinsically more labile than those of beef. There is some evidence that the proteins of so-called 'white' muscle generally are more readily denatured post-mortem than those of the so-called 'red' type (Howard *et al.*, 1960a). Again, the sarcolemma of porcine muscle is more permeable to water than that of beef (George *et al.*, 1980).*

In most cases, carcasses can be frozen swiftly, after electrical stimulation, without pre-rigor freezing and resultant 'thaw-rigor' development. Nevertheless, there is a phase in post-mortem glycolysis when the muscles, although no longer reactive to cold-shortening, are still pre-rigor and when, therefore, 'thaw-rigor' is still a pos-sibility. Pelvic hanging appears necessary for maximum tenderness in electrically stimulated carcasses which are placed into blast freezers within 30 min of stunning (Shaw *et al.*, 1976). When such rapid freezing is intended this should be delayed until 6 h after stimulation (Bendall, 1980).

Apart from the avoidance of toughening, electrical stimulation has been associ-ated with an improved flavour and an enhanced brightness of the red colour on cut meat surfaces (Savell *et al.*, 1978; Smith *et al.*, 1980). The latter effect possibly arises because the process depletes the metabolises of surviving oxidative pathways in the muscle, or because the fast fall in pH causes the muscle proteins from treated car-casses to approach their isolectric point much sooner, thereby 'opening up' the struc-ture and easing oxygenation of myoglobin. The latter explanation seems at least partially responsible because, although the cut surface from electrically stimulated sides develops a brighter red than those from control sides when exposed to the air at *ca.* 12–20 h post-mortem, this difference is not apparent when the muscles are exposed later, e.g. 48–72 h post-mortem (Dr T. R. Dutson: personal communication). By this time both electrically stimulated and control muscles would have attained their ultimate pH, the structure in both would be 'open' and residual oxygen uti-lization of the same order. On the other hand, in certain muscles (e.g. *semimem-branosus* of beef) electrical stimulation is associated with increased metmyoglobin formation and thus loss of colour (Ledward *et al.*, 1986). With low voltage stimula-tion of beef the combination of pH values below 6 and near *in vivo* temperature causes loss of colour which is partly due to denaturation of myosin (Hector *et al.*, 1992).

(d)   Practical application

In endeavouring to assess the practical implications of electrical stimulation it is important to appreciate that, concomitantly with the desire to enhance the effi-ciency of abattoir operations by speeding throughout, there has been an increased tendency for abattoirs to undertake the centralized preparation of prepackaged cuts, both of commercial joints and of portions for the individual consumer. This has reflected changing patterns of consumption – smaller families, canteen meals, convenience foods and domestic freezer storage. Moreover, deboning of the still hot carcass and vacuum-packaging of the warm cuts is a further extension of

* It is thus of interest that electrical stimulation causes a 4–5-fold increase in exudation from porcine muscle in comparison with non-stimulated controls (Gigiel and James, 1984).

these tendencies, which have been shown to diminish evaporative and exudative losses.

With such relatively small portions of meat, electrical stimulation of the carcass or side could prove especially useful in avoiding 'cold-shortening', since the latter would otherwise be readily induced under the very rapid rates of cooling which would occur with them. Moreover, the rapid lowering of temperature markedly lessens microbial growth (Raccach and Henrickson, 1980), a factor of importance with portions of meat having a large surface of volume ratio. Unfortunately, however, although the toughening of 'cold-shortening' would be easily avoided, these very circumstances would prevent the positive contribution to tenderness of electrical stimulation through its creating favourable circumstances for early and rapid conditioning: i.e. a combination of low pH with *in vivo* temperatures, as alluded to above.

It may be that the solution to this dilemma lies in subjecting the various joints and portions to different cooling regimes. Muscles differ biochemically. Neither in their response to 'cold-shortening' nor in their conditioning behaviour are they identical. Indeed, it has been shown that a pulse frequency of 14 pps is associated with greater tenderness in calf *l. dorsi* than 40 pps; whereas for *semimembranosus* the latter frequency is apparently superior (Bouton et al., 1980a).

This approach may well be expedited by the electrical stimulation of individual muscles or groups in a manner appropriate for each following their partial or complete removal from the hot carcass. The issue is likely to be rather complex, however, since Devine *et al.* (1984) showed that so-called 'red' muscles, which are relatively susceptible to 'cold-shortening', are little affected by electrical stimulation; whereas, in so-called 'white' muscles, which are little affected by the conditions which cause 'cold-shortening', the process is very effective in lowering the initial pH and in speeding the subsequent rate of pH fall.

Insofar as 'cold-shortening' is particularly marked in so-called 'red' muscles, its incidence in pork, the musculature of which is generally pale, would be expected to be minimal. Nevertheless, with the increasing possibility of very rapid chilling, its effect on pork sides was investigated by Gigiel and James (1984). Their results indicated that electrical stimulation was associated with somewhat increased tenderness in comparison with sides which had also been chilled very rapidly, but not stimulated. Though it appeared that this effect was probably due more to the promotion of early conditioning than to the removal of susceptibility to 'cold-shortening', for which there was only slight evidence in the non-stimulated sides, Dransfield (1983) had reported that when excised porcine *l. dorsi* muscles were chilled rapidly using air at temperatures well below the freezing point, some 'cold-shortening' was observed. Moreover, Møller and Vestergaard (1986) demonstrated that pelvic suspension, which would tend to oppose 'cold-shortening' of *l. dorsi* muscles, was associated with longer sarcomeres and greater tenderness on cooking than suspension by the Achilles tendon in rapidly chilled pork carcasses. Subsequently Dransfield and Lockyer (1985) found that excised *l. dorsi* muscles 'cold-shortened' more and were tougher on cooking as the *rate* of cooling down to 10 °C increased. In this context, Brown *et al.* (1988) confirmed that speedy refrigeration of vacuum-packaged primal cuts of pork in brine at 0 °C caused a significant increase in the toughness of the meat, but reduced evaporative losses in comparison with conventionally chilled packs in the air at 0 °C. Dransfield *et al.* (1991) later demonstrated that meat from the *l. dorsi* but not that from *semimembranosus*, in sides of pork which had been chilled rapidly to 0 °C, was tougher than that from slowly chilled controls – and that the presumed cause ('cold-shortening') could be avoided either by pelvic suspension or by electrical stimulation before chilling. Pelvic suspension, however, was also associated with a higher water-holding capacity whereas electrical stimulation, in the sides chilled at slow rate, was associated with increased exudation (as measured at 48 h post-mortem) in the absence of 'cold-shortening' conditions. Taylor (1992) showed that whereas high voltage (700 V) electrical stimulation of rapidly chilled sides of pork was associated with greater exudation (from *l. dorsi*) than that of controls if applied at 5 min post-mortem, it was associated with markedly less exudation if applied at 20 min post-mortem. He attributed this desirable effect to the less pronounced acceleration of post-mortem glycolysis caused by the latter. Both modes of electrical stimulation were effective in preventing 'cold-shortening' in rapidly chilled sides. Subsequent studies confirmed these findings (Taylor *et al.*, 1995) and again demonstrated that electrical stimulation has a beneficial effect on tenderness in addition to its prevention of 'cold-shortening'.

Thus, despite their markedly diminished susceptibility to 'cold-shortening' in comparison with those of beef and lamb, it is evident that certain porcine muscles might well suffer toughening when subjected to the extemely rapid rates of chilling now available commercially. It is also clear, however, that electrical stimulation, if judicially applied, can ensure that quality is retained.*

Because of substantial development of red deer for meat in New Zealand, the effects of electrical stimulation on meat quality were studied by Wiklund *et al.* (2001). Although the procedure, insofar as it avoided 'cold-shortening', was initially associated with increased tenderness in comparison with non-stimulated venison, this benefit became inapparent during subsequent ageing at −1.5 °C, suggesting that electrical stimulation offers no advantage for products intended for long-term storage at chill temperature.

Since hot-deboning in certain tropical countries may involve exposure of meat to ambient temperatures of 30–40 °C, it has been of interest to assess the effects of electrical stimulation in such circumstances, where early excision of muscles could conceivably cause contraction and toughening during the onset of rigor mortis (Busch *et al.*, 1967; Locker *et al.*, 1975). Electrical stimulation (100 V, 25 pps, 4 min) of hot-deboned bovine *l. dorsi* muscles was found to achieve a marked tenderizing effect over control muscles, either held on the carcass or excised, when the meat was subsequently exposed to temperatures of 30 or 40 °C for 5 h (Babiker and Lawrie, 1983). At the latter temperature, however, and in contrast with the findings at 30 °C, there was appreciably greater loss of water-holding capacity (and subsequently greater development of microbial numbers); moreover, the tenderness of the electrically stimulated muscles was proportionately rather less. These findings suggest that the beneficial effects of high temperature in accelerating proteolysis were being overridden by increased protein denaturation of the enzymes responsible for protein breakdown. It has been demonstrated that the heat-lability of cathepsins is marked above 40 °C (Okitani *et al.*, 1980). The susceptibility of muscle proteins generally to denaturation at temperatures above 35 °C was earlier indicated by the studies of Hamm (1960), Sharp (1963) and Scopes (1964), and more recently by those of Penny and Dransfield (1979).

---

* It is surprising that the cryogenic chilling of pork, by the immersion of warm sides in liquid nitrogen for 1–3 min was not found to produce the toughening of 'cold-shortening' (Jones *et al.*, 1991).

It should be recollected that the musculature from electrically stimulated carcasses, notwithstanding its advantages in terms of colour and tenderness, might not be suitable for all purposes. Thus, it would be less useful than non-stimulated muscle for the production of those cured and freeze-dried products which depend upon the retention of near *in vivo* levels of ATP at time of processing for their high water-holding capacity (Honikel and Hamm, 1978).

Electrical stimulation was the subject of a comprehensive review edited by Pearson and Dutson (1985).

7.1.1.3   Storage changes: prepackaging effects

Given adequate cooling of the hot carcasses and cuts, the deterioration of fresh chilled meat is due to surface changes. The natural surface consists of fat and connective tissue, and during cooling the consistency of the latter changes, so that further loss of water by evaporation is restricted. On the other hand, muscle surfaces continue to lose water at a fairly fast rate, and this desiccation leads to an increased concentration of salts at the surface which causes oxidation of the muscle pigment to brown or greyish metmyoglobin (Brooks, 1931) and a darkening of colour due to optical changes in the tissue (Brooks, 1938). Different muscles show differing susceptibility to such browning due to differing tendencies to desiccation (Ledward, 1971a). If surfaces are more moist, moulds of various colours will tend to grow (cf. Chapter 6), and these may affect the fat, causing rancidity and off-odours due to other changes. If the surfaces are moister still, bacteria can grow and, in sufficient numbers, produce off-odours and aggregate in visible colonies (slime). Apart from moisture, these features are a function of time and temperature (cf. Table 7.3).

When meat is removed from chill storage, moisture tends to condense on the cool surfaces, especially when the relative humidity of the atmosphere is high. This phenomenon is known as 'sweating'. Apart from its potential effect in encouraging microbial growth, it causes the collagen fibres of connective tissue to swell and become white and opaque. This change is reversible, however, and there is no evidence that 'sweating' causes a permanent loss of 'bloom' (Moran and Smith, 1929), the term given by the trade to a pleasing superficial appearance in the meat. The most exacting requirements for meat storage above the freezing point were those encountered in the shipment of chilled beef from Australia and New Zealand to the United Kingdom, when temperatures had to be maintained at −1.4 °C (Scott and Vickery, 1939; Law and Vere-Jones, 1955). The essential difference between so-called

Table 7.3   Time for bacterial slime to develop on meat surfaces exposed under moist conditions (after Haines and Smith, 1933)

Temperature (°C)	Time (days)
0	10
1	7
3	4
5	3
10	2
16	1

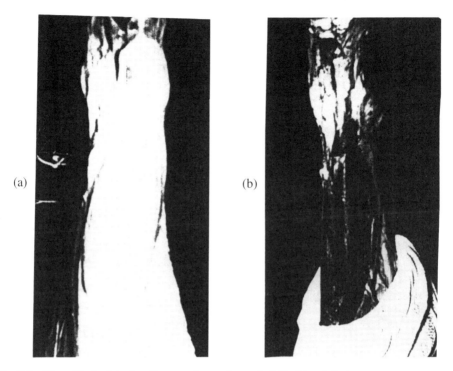

Fig. 7.2 The effect of desiccation on the surface of chilled beef during a 45-day period at 0 °C. In (a) the hindquarter was covered by a moisture-impermeable wrap. In (b) the meat was exposed to the atmosphere. (Courtesy the late K. C. Hales.)

chilled meat and the fresh commodity is the length of time during which it is expected to resist substantial change. By far the most important consideration is the minimizing of microbial contamination during preparation of the commodity (§ 6.1.2.2) and the strictest hygiene is essential (§ 6.4.1). On the one hand, a high relative humidity in the storage chambers will prevent desiccation and loss of bloom; on the other hand, it will encourage microbial growth. A balance has to be established between these two extremes over storage times which may extend up to 60 days. During this time beef quarters will lose about 1½–2 per cent of their weight by evaporation (Hicks *et al.*, 1956). Although some degree of desiccation is desirable, excessive drying, of course, must be avoided, especially because of its effects on the layer of connective tissue separating the muscles from the exterior. Although this layer is very thin, it imparts a pleasing, translucent appearance to the surface of the carcass, even when there is little subcutaneous fat. When the layer becomes desiccated the superficial appearance deteriorates remarkably (Fig. 7.2).

The rate of evaporation from different parts of a carcass may vary by a factor of 10 (Hicks *et al.*, 1956). With lambs held 10 days at 3 °C, the connective tissue over the fat on the loin reached equilibrium with the atmosphere in the cold store at an a_w of 0.862, but the a_w in the *panniculus* muscle was still 0.906.

With chilled beef carcasses, the control of microbial growth on the surface of the neck muscles is the most difficult to achieve (Scott and Vickery, 1939). This portion of the carcase had been observed to spoil relatively rapidly (Anon., 1816) long before it was appreciated that micro-organisms were responsible. Successful storage is related to conditions prevailing whilst the hot carcass is cooling, when there should be fast rates of temperature fall combined with a high speed of air movement over the beef, provided the possibilities of 'cold-shortening' and of desiccation are carefully considered. This will have little effect on areas of exposed muscle where a high rate of moisture diffusion from within can be maintained during storage. In regions such as the *panniculus*, however, where rates of moisture diffusion to the surface are low (Scott and Vickery, 1939), a high rate of moisture removal, whilst the carcass is cooling, will deplete surface moisture to an extent incompatible with the retention of bloom during storage.

Callow (1955a) reviewed the various means which were employed to discourage microbial growth on the surface of chilled meat during prolonged storage at relative humidities high enough to prevent undue desiccation. The use of formaldehyde in this context, whilst successful, was prohibited on the grounds of toxicity in 1925. In due course it was shown that replacement of air by 100 per cent carbon dioxide in the chillers would prevent microbial growth but cause brown discoloration because of metmyoglobin.* Nevertheless, concentrations up to 20 per cent had a negligible effect on colour (Brooks, 1933), and 10 per cent severely inhibited the growth of the two most common types of bacteria found on chilled meat (Haines, 1933), and, thus, indirectly, doubled the effective storage life by slowing down the rate of fat oxidation (Lea, 1931). In 1933 the first shipment of chilled beef in 10 per cent carbon dioxide was carried from New Zealand to the United Kingdom without trace of microbial spoilage. Much of the success of chilled beef carriage under carbon dioxide – by 1938, 26 per cent of the beef from Australia and 60 per cent of that from New Zealand was so carried to the United Kingdom – was attributable to the efforts of shipping companies to construct gas-tight holds. Leakage rates were reduced from about 30 to 0.25 ft^3 gas/ton of beef (about 0.84 to 0.007 m^3 gas/tonne) (Empey and Vickery, 1933).

Although oxidation of fat and the formation of brown metmyoglobin proceed slowly on the exposed surface of carcasses from the moment such exposure commences, the changes are generally negligible by the time the meat is sold. These changes began to become apparent, however, when carcasses could be kept above the freezing point and free from microbial spoilage for upwards of 40–50 days – as with chilled beef under carbon dioxide. The development of prepackaged methods of sale, in which large areas of surface are exposed and the superficial appearance of the meat is especially important, has created conditions favouring and revealing undesirable changes in the commodity even over relatively short periods of time. Improved packaging procedures, however, can substantially diminish spoilage problems and exploit the antimicrobial action of inpack atmospheres. An increase in saleable life is particularly important to countries such as New Zealand in relation to the distance of major markets, and an effective procedure based on carbon dioxide has been developed. When chilled lamb is packaged in a foil-laminate container, which is completely impermeable to oxygen, under 100 per cent carbon dioxide, the inpack atmosphere can be maintained irrespective of the cut, size or shape of the meat portions (Gill, 1987). A master pack of foil-laminate is employed. The carbon dioxide not only inhibits microbial growth but also prevents

* In the 1920s and 1930s '100% CO_2' was difficult to ensure and traces of oxygen could remain, oxidation of myoglobin to metmyoglobin being maximal at 4 mm O_2 pressure (cf. Fig. 7.3).

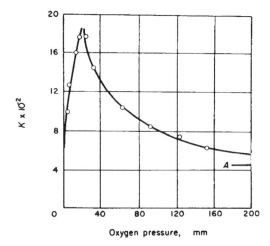

Fig. 7.3   The relationship between oxygen pressure (mm) and rate of oxidation of muscle pigment (K) (Brooks, 1935). A = value at atmospheric oxygen tension.

discoloration due to metmyoglobin formation from traces of oxygen (cf. Fig. 7.3). The meat pieces within the master pack can be packaged in various ways so long as the material in which they are wrapped allows free access of carbon dioxide. The storage life is at least 16 weeks and markedly longer than that achieved by vacuum packaging. The appearance of lamb in such conditions does not begin to deteriorate until the master pack is opened, the meat in the consumer pack having a display life at chill temperature equal to that of corresponding freshly prepared lamb. The 100 per cent carbon dioxide atmosphere ensures that myoglobin remains in the reduced form, any metmyoglobin present at time of packaging being readily reconverted to myoglobin by surviving enzymic activity of the muscles. The purplish-red colour of the myoglobin quickly changes to bright red oxymyoglobin on exposure to the air. The off-odours sometimes encountered on opening vacuum packs are not found, although, on extended storage, an aged (but acceptable) flavour may develop.

Gill and McGinnis (1995) have pointed out, however, that in controlled atmosphere packaging, transient metmyoglobin formation can arise due to traces of oxygen, and it is thus important to ensure that the meat is not exposed to the atmosphere prematurely (i.e. before the muscles' enzyme systems have reduced any metmyoglobin initially formed).

Provided strict hygiene is applied to the meat to minimize initial microbial loading, the storage of consumer packs under controlled atmospheres within master packs can provide a useful display life (in terms of fresh colour, absence of microbial spoilage and general acceptability) of beef steaks, after their removal from master packs, in which they have been stored for 2, 4 or 7 weeks under O_2/CO_2 (2:1), N_2 or CO_2, respectively (Gill and Jones, 1994). On the other hand, because of the (currently) poor hygiene condition of pork, its master packaging under such oxygen-depleted atmospheres limits its useful retail storage life to little more than one week (Gill and Jones, 1996).

In a comparison of vacuum and carbon dioxide packed, electrically stimulated hot-deboned beef loins (which were chilled to 7 °C within 24 h of leaving the slaughter floor and subsequently stored at –1 °C), Bell *et al.* (1996) found that beef subjected to both modes of packaging deteriorated because of darkness in the lean and greyish or greenish discoloration in the fat. They concluded that vacuum packaging, combined with cooling at 5 °C for 24 h and subsequent storage at –1 °C, was the procedure most likely to ensure the retention of eating quality over 70 days of chilled storage.

It is important to note that carbon dioxide is very soluble both in fatty and muscular tissues (Gill, 1988). This factor can cause the partial pressure of the gas to fall from its original value in the mixture applied unless sufficient excess has been included to saturate the meat. Moreover, with 100 per cent carbon dioxide, the packs will tend to collapse. It has been shown that the solubility of carbon dioxide in muscular tissue increases by *ca.* 35 per cent for each unit of pH increment in the meat; and it is similar for beef, pork and lamb (Gill, 1988). Its solubility in fatty tissue, however, varies with species. Because the cherry-red colour of carboxymyoglobin is more stable than the bright red of oxymyoglobin, the incorporation of a very low concentration of carbon monoxide (*ca.* 0.5%), in a within-pack atmosphere of 60% O_2/39.5% N_2, can extend the useful storage life of packed meat (Sørheim *et al.*, 1999). Luño *et al.* (2000) reported a similar finding using 0.1–1.0% carbon monoxide in an atmosphere of 24% O_2, 50% CO_2 and 25–25.5% N_2. The carbon monoxide also inhibited the growth of aerobes (e.g. *Brochothrix thermosphacta*) and caused a significant retardation of lipid oxidation and of metmyoglobin formation.

Vacuum-packaging provides an alternative to modified atmospheres for extending the storage life of fresh and chilled meat. The relative storage lives of different meats at 0°C are compared in Table 7.4. As Egan and Shay (1988) point out: 'The properties of the various groups of micro-organisms suggest strategies to be used in attempting to preserve meats by the use of packaging systems. The practical result . . . is the creation of conditions under which lactic acid bacteria become the only group which can grow readily.' Some of the visible defects in vacuum-packaged beef, pork and lamb are listed in Table 7.5. *Brochothrix thermosphacta* is resistant to inhibition by carbon dioxide and growth can occur at high concentrations of this gas. The organism does not grow if the pH of the meat is below 5.8, inhibition being due to the lactic acid produced during post-mortem glycolysis.

The sale of prepackaged meat reflects technical advances in the production of plastic films and the development, with changing economic circumstances, of self-service stores (Bryce-Jones, 1962; Paine and Paine, 1983). Some properties of the packaging materials, which are commonly used, are given in Table 7.6. They may be

Table 7.4 Comparative storage lives of vacuum-packaged primal cuts at 0 °C (after Egan and Shay, 1988)

Species	Meat pH	Storage life (weeks)	Spoilage defect
Beef	5.5–5.8	10–12	flavour (souring
Pork	5.5–5.8	6	flavour
	6.0–6.3	4–6	colour (greening)
Lamb	variable*	6–8	colour, fat appearance

* Since lamb cuts involve numbers of small muscles of differing ultimate pH.

Table 7.5 Visible defects in vacuum-packaged fresh meats (after Egan and Shay, 1988)

Defect	Cause
Greening	Sulphmyoglobin from bacterial H_2S
(a) high ult. pH	Gram-negative bacteria (e.g *Aeromonas*)
(b) normal ult. pH	Lactobacilli
Browning	Metmyoglobin formed by excess O_2 in pack
Brown spots	Haem pigments assoc. with growth of yeasts
Loose packs	Poor seals, punctured films
Blown packs	Excess growth of CO_2-producing bacteria

Table 7.6 Permeability of packaging films to gas and water vapour (after Paine and Paine, 1983)

Material	Permeability $(ml/m^2/MPa/day)$*			
	N_2 (at 30 °C)	O_2 (at 30 °C)	CO_2 (at 30 °C)	H_2O (at 25 °C & 90% RH)
Polyvinylidene chloride (Saran)	0.07	0.35	1.9	94
Polychloro-trifluorethylene	0.20	0.66	4.8	19
Polyester (Mylar A)	0.33	1.47	10	8700
Polyamide (Nylon 6)	0.67	2.5	10	47000
Polyvinylchloride (unplasticized)	2.7	8.0	6.7	10000
Cellulose acetate (P912)	19	52	450	500000
Polyethylene				
(*d* = 0.95–0.96)	18	71	230	860
(*d* = 0.92)	120	360	2300	5300
Polystyrene	19	73	590	80000
Polypropylene (*d* = 0.91)	–	150	610	4500

* All permeabilities calculated for a 25 µm thick film.

combined in laminates, or coated, to alter their individual characteristics for specific purposes. For each packaging material listed in Table 7.6 (Paine and Paine, 1983), it is evident that the permeability to nitrogen is less than that to oxygen and that the latter is less than that to carbon dioxide. Water vapour permeability is much more rapid than that of the other three gases. For many processed and fresh meats, bags made for copolymers of vinylidene chloride and vinyl chloride are heat-shrunk, so as to conform closely to the contours of the product, after evacuation of air. Such packs have an attractive appearance and good storage life, primarily because oxygen has been eliminated. A later development had been the overwrapping of complete pallet loads by a heat shrunk film, whereby the use of paper or board coverings is dispensed with.

The growth of trade in vacuum-packaged, deboned, primal cuts and portions of meat has emphasized the great variability in the degree of exudation and discoloration which can develop. In the deeper locations of beef carcasses, even with fast chilling, the problem is not so much cold-shortening but delay in temperature fall. As a result, *in vivo* temperatures persist, there is a consequently fast rate of post-mortem glycolysis (cf. § 4.2.2) and the pH attains acid values swiftly. These conditions, being similar to those in PSE pork, lead to loss of water-holding capacity by the proteins and to much exudation. By exposing the deeper locations of beef carcasses to the chilling environment, much faster rates of chilling are attainable. These slow the rate of post-mortem glycolysis, and they are reflected by substantial reductions in exudate from vacuum-packed primal cuts, by diminished discoloration and by some increase in tenderness (Follet, 1974; Follet *et al.*, 1974). Such exposure is achieved by 'seaming out' the fascia of the major muscles, leaving them attached to the carcass at one end. The bulk of the muscles prevents their temperature being lowered below 10°C until the pH has fallen beyond the point of susceptibility to cold-shortening. Microbial contamination is only marginally increased by the procedure.

With the advent of hot deboning of beef carcasses and of hot cutting of lamb, and their preparation as shrink-wrapped cuts (cf. § 4.2.3), much faster refrigeration, economies of space and substantial reduction of evaporative and exudative losses, in comparison with those which apply when carcasses are chilled before butchering, have been achieved (Locker *et al.*, 1975), and abattoir operations are likely to alter accordingly. Such procedures would clearly involve a risk of cold-shortening (or thaw-rigor). If, however, carcasses are electrically stimulated immediately after dressing (cf. § 7.1.1.2) fast chilling can be applied without the danger of toughening and without any increase in the microbial load. Unfortunately, however, although the toughening of cold-shortening would thus be easily avoided, these very circumstances would prevent the positive contribution to tenderness of electrical stimulation through its creating favourable circumstances for early and rapid conditioning, i.e. a combination of low pH and *in vivo* temperatures, as alluded to above.*

The impression is current that, as a means of preservation, prepackaging represents an advance over older methods of sale by diminishing microbial hazards. But this cannot be taken for granted (Ingram, 1962), since meat prepared in small pieces is subjected to much more initial handling than, for example, that cut from joints; and unless the operations during prepackaging are hygienically controlled, they may provide additional opportunities for contamination. With fresh meats, a bright red surface colour is particularly desirable. As a result, oxygen-permeable wraps tend to be used for this commodity to permit the formation and retention of a layer of oxymyoglobin on the surface. Obviously, oxygen is not so freely able to saturate the surface in such a pack as when the meat is completely exposed. Oxygen tension tends to become limiting not only because of difficulty of entry, but also

---

* Offer (1991) demonstrated a further complexity in the effect of temperature on post-mortem glycolysis. Thus, although increasing the intensity of pre-rigor chilling (in the absence of 'cold-shortening' circumstances) generally reduces the denaturation of myosin, it has little protective effect when the intrinsic rate of post-mortem glycolysis is high – and the muscle enters rigor mortis rapidly. Moreover, when muscles are not chilled, the denaturation of myosin is greater in more slowly-glycolysing muscles in which rigor mortis is relatively delayed. Offer (1991) explains these unexpected observations by postulating that actin markedly inhibits the denaturation of myosin when the latter combines with it in forming actomyosin.

because surviving activities in the oxygen-utilizing enzymes of the meat deplete the atmosphere within the pack of this gas and produce carbon dioxide (cf. § 4.3). The partial pressure of the gases constituting the internal atmosphere of the pack will influence the type of micro-organism which can grow on the meat, and, in turn, their metabolism will further change the internal atmosphere (Ingram, 1962). Moreover, the characteristics of the prepackaging films which are used are generally such as to prevent loss of water vapour. Whilst this would oppose weight loss, the absence of surface desiccation tends to encourage microbial growth, especially that of bacteria. Again, although prepackaging (once it is complete) may prevent further contamination of the meat from the exterior, it may encourage consumers to expose the meat to high temperatures and other conditions which they would not consider suitable for unwrapped meat, with consequent dangers from spoilage or food poisoning. It is important to emphasize that prepackaging of meat is an adjunct to preservation by control of temperature or of other factors involved in deterioration.

It will be clear that there are several general factors likely to cause differences between the microbial status of stored, prepackaged meat and that cut from carcasses or large joints. More specifically, inherent differences between the muscles of meat animals (Chapters 4 and 5) may become apparent. Although such differences exist in the intact carcasses, they are exaggerated in prepackaging for the reasons already indicated – namely, exposure of individual muscles under conditions when their different susceptibilities to microbial growth, fat oxidation, discoloration and exudation are emphasized. Much variability in the keeping quality of prepackaged fresh meat can be explained in this way. Thus, as indicated above, functional differentiation of muscles will be reflected in such features as moisture content, salt concentration, pH, degree of exudation, the surviving activity of oxidizing enzymes, the degree of protein denaturation, the content of amino acids and other micronutrients and, thereby, microbial growth.

As mentioned in § 7.1.1.1, the formation of brown metmyoglobin and the oxidation of fat are among other aspects of changes during storage which prepackaging emphasizes. It is particularly important that the reaction by which metmyoglobin is formed is maximal not at a high oxygen tension but when the partial pressure of the gas is 4 mm (Fig. 7.3; Brooks, 1935). Such a situation could well arise in fresh prepackaged meat, depending on the permeability of the wrap used. When fresh meat is packaged under vacuum the purplish-red colour of reduced myoglobin predominates; and, indeed, metmyoglobin, if initially formed, tends to be reduced (Ball, 1959). As indicated in § 6.3.5, an alternative method for retaining the bright red colour of oxymyoglobin on the surface of packaged meat is to seal it under high oxygen tension with sufficient carbon dioxide to inhibit aerobes (Georgala and Davidson, 1970; Gill and Tan, 1980).

Normally, fresh meat exudes fluid (weep) from cut surfaces post-mortem after attaining its ultimate pH (§§ 5.3, 5.4.1 and 10.1.2); but with the large areas of exposed surface in prepackaged cuts, the phenomenon is much more noticeable. It may be exaggerated if the contents of the pack are under tension from the wrapping material. Among procedures employed to make weep less apparent is the incorporation of an absorbent tray in the pack. This soaks up fluid as it exudes from the meat. The weight of meat offered for prepackaged sale, as we have mentioned above, will tend to emphasize the differences between individual muscles in their degree of exudation. Packaging systems that do not involve the application of a vacuum can considerably reduce fluid loss (Payne et al., 1998).

The weep derived from fresh meat on standing produces a similar electrophoretic pattern, in respect of its protein components, as muscle press juice or an extract of muscle in dilute salt solution: it appears to be mainly sarcoplasmic in origin (Howard *et al.*, 1960a). This view was confirmed by Savage *et al.* (1990). They made a detailed study of the exudate from fresh rabbit and porcine muscle (which they referred to as 'drip', although this term usually signifies the exudate from thawing muscle: cf. § 7.1.2.1). As the percentage of exudate increased, its protein concentration fell; but this was not due solely to dilution: denaturation of certain proteins and their insolubility appeared to be involved also. Creatine kinase was particularly labile (cf. Fig. 5.3; Scopes, 1964). Savage *et al.* (1990) speculated that fluid from the myofibrils might contribute.

Deterioration in meat lipids may be due to direct chemical action or through the intermediary activity of enzymes (either indigenous or derived from micro-organisms). As a rule, direct chemical deterioration is not so important in fresh meat. Two types of deterioration occur: hydrolysis and oxidation. Lipolytic enzymes split fatty acids from triglycerides (§ 4.1.2) leaving, ultimately, free glycerol. With phospholipids, inorganic phosphate is also produced. The fatty acids liberated in meat are generally not so offensive as those produced in milk (Lea, 1962). Since the rate of auto-oxidation of fatty acids increases with the number of double bonds they contain, and since acids with several double bonds tend to produce off-flavours, the content of unsaturated fatty acids has an important effect on the susceptibility of a given fat to oxidation. From the considerations in Chapter 4, the rate of oxidation of intramuscular fat would thus tend to be higher (a) in non-ruminants than in ruminants (e.g. in whale meat and pork in comparison with beef and mutton), (b) in the less improved breeds, (c) in young animals rather than in older ones, (d) in muscles with relatively low contents of intramuscular fat, (e) in the lumbar region of *l. dorsi* in the pig rather than the thoracic region, the reverse being true in beef animals, (f) in animals on a low plane of nutrition and (g) in animals receiving large proportions of unsaturated fat in their diet, particularly in non-ruminants.

A considerable number of such differences may operate simultaneously. The relative tendency of pork muscles to become rancid and discoloured exemplifies this complexity. Porcine *psoas* muscle has a higher proportion of unsaturated fatty acids, especially in the phospholipid fraction (Owen *et al.*, 1975), than the *l. dorsi* of this species; and more myoglobin, which, when present in the metform, can act as a pro-oxidant. Yet, during prolonged storage at –10 °C, minced porcine *l. dorsi* undergoes oxidative rancidity, and concomitant metmyoglobin formation, to a markedly greater extent than *psoas*. This anomalous behaviour appears to be related to the higher ultimate pH of the latter (Owen *et al.*, 1975). At high pH the activity of the cytochrome system of enzymes is much enhanced (Lawrie, 1952b) and this increases their metmyoglobin-reducing activity (Cheah, 1971). Moreover, such enzymes are found at higher concentration in *psoas* (Lawrie, 1953a). In porcine *psoas* muscle, therefore, the relatively high ultimate pH, by minimizing pro-oxidant conditions, more than offsets the inherently greater tendency of its lipids to oxidize.

On the other hand, the ultimate pH of both bovine *psoas* and *l. dorsi* muscles is generally normal (Howard and Lawrie, 1956) and this potentiates the effect of the higher proportion of polyunsaturated fatty acids in the former (Rhee *et al.*, 1988). Clearly, a considerable number of factors must be known before accurate prediction of the behaviour of a given muscle can be made.

The pro-oxidant effect of haematin compounds in fat oxidation is reciprocal since unsaturated fatty acids accelerate the oxidation of myoglobin (Niell and Hastings, 1925). Since myoglobin and fats are brought into intimate contact with one another in meats, their coupled reaction will contribute to rancidity and discoloration (Lea, 1937; Watts, 1954). During cooking, both haem-bound and nonhaem iron released from the latter accelerate lipid oxidation (Igene *et al.*, 1979). Nevertheless, it should be pointed out that the behaviour of haem pigments and unsaturated fats, when in juxtaposition, is not fully understood (D. A. Ledward, personal communication). Thus, Kendrick and Watts (1969) postulated that at low lipid : haem ratios, haem compounds can stabilize peroxides or free radicals, and exert an anti-oxidant effect. Ultraviolet light (and other ionizing radiations) and ozone, which have been used to discourage microbial growth on meat held above the freezing point, accelerate fat oxidation.

Various anti-oxidants such as polyhydroxy phenols (e.g. propylgallate, hydroxyanisole) have been incorporated successfully in meat products to retard fat oxidation (Barron and Lyman, 1938; Lineweaver *et al.*, 1952); but these have been reported to be associated with adverse health effects and the anti-oxidant potential of various natural plant extracts – such as fenugreek, rosemary, sage, soya protein and tea catechins – has been demonstrated in several pork products (McCarthy *et al.*, 2001). Of anti-oxidants which have been fed to animals in the hope of enhancing the post-mortem anti-oxidant potential of their tissues, only tocopherols, such as vitamin E, are stored to any extent (Barnes *et al.*, 1943; Major and Watts, 1948). By feeding pigs diets containing beef tallow or soya oil, with or without supplementation by α-tocopherol acetate, Monahan *et al.* (1992) found a significant protective action of the vitamin E on the susceptibility of the porcine muscle lipids to undergo oxidative rancidity. They found a similar beneficial effect of α-tocopherol acetate feeding on veal when it was administered to calves (Engeseth *et al.*, 1993). Jensen *et al.* (1997) reported that, in feeding α-tocopherol acetate to pigs, at levels of 100–700 mg/kg of diet, there was a linear relationship between the levels subsequently found in the muscles and the logarithm of the dose fed. Increasing the supplement of the vitamin from 100 to 200–700 mg/kg of feed significantly reduced lipid oxidation during chilled storage of the pork (but did not affect myoglobin oxidation).

The biochemical differentiation of muscles (cf. § 4.3.5.1) is reflected in their post-mortem susceptibility to the incorporation of anti-oxidants in the feed or by preslaughter injection. Thus, whereas the tendency for lipid oxidation and met-myoglobin formation in *psoas major* and *gluteus medius* was found to diminish following preslaughter injection of vitamin C in beef animals, the stability of *longissimus dorsi* was unaffected (Hood, 1975). On the other hand, Arnold *et al.* (1992) found that the colour stability of *longissimus dorsi* was increased more than that of *gluteus medius* when vitamin E was incorporated in the feed. Jensen *et al.* (1997) noted that in *psoas major* the level of vitamin E established by preslaughter feeding was markedly greater than that in *longissimus dorsi*. The formation of oxidized products from cholesterol (which are reputed to exacerbate the tendency to atherosclerosis), appears to be suppressed by vitamin E supplementation in the feed (Galvin *et al.*, 2000). Monosodium glutamate is said to be quite an effective anti-oxidant, particularly in frozen meat (Sulzbacher and Gaddis, 1968), although its use in foods has been criticized recently.

Consumer concerns have encouraged research on the use of protein-based edible films for meat packaging. Apart from the long-established use of collagen for sausage casings, the properties of a variety of protein films derived, for example, from corn, wheat, milk and meat itself, have been examined. Although such films are not impermeable to moisture, they are relatively efficient barriers against the entry of oxygen (Perez and Desoubry, 2002).

7.1.2   Storage below the freezing point

The efficacy of freezing in preserving meat has obviously been long understood by the Eskimoes and other peoples inhabiting regions with Arctic climates. An extreme example is the reported edibility of meat from mammoths adventitiously frozen for 15–20,000 years in northern Siberia under conditions preventing desiccation (Stenbock-Fermor, 1915; Tolmachoff, 1929) (Fig. 7.4). Bone marrow from a horse which had been frozen in Alaska for 50,000 years was served at a dinner in New York (R. C. S. Williams, personal communication, 1969). Modern views on the freezing of meat are based on an understanding of the changes caused by the process as well as on its preservative aspects.

7.1.2.1   Effects of freezing on muscular tissue

The advantages of temperatures below the freezing point in prolonging the useful storage life of meat, and in discouraging microbial and chemical changes, tend to be offset by the exudation of fluid ('drip') on thawing. Proteins, peptides, amino acids, lactic acid, purines, vitamins of the B complex and various salts are among the many constituents of drip fluid (Empey, 1933; Howard *et al.*, 1960a; Pearson *et al.*, 1959). The extent of drip is determined by factors of two kinds. In one category are the factors which determine the extent to which the fluid, once formed, will in fact drain from the meat. Among these are the size and shape of the pieces of meat (in particular the ratio of cut surface to volume), the orientation of cut surface with respect to muscle fibre axis, the prevalence of large blood vessels and the relative tendency for evaporation or condensation to occur in the thawing chamber. Factors of this kind are of greater importance with beef than with pork or mutton, for with the former more cutting is required to produce an easily handled quantity.

Factors in the second category are much more fundamental. They are concerned with the nature of the freezing process in muscular tissue and with the water-holding capacity of the muscle proteins, thus determining the volume of the fluid which forms on thawing. In general, the proportion of the total water in muscle which freezes increases rapidly at first as the temperature is lowered further below the freezing point; then more slowly, approaching an asymptote of about 98.2 per cent at $-20\,°C$ (Moran, 1930). Because not all the water in muscle freezes, the latent heat is lower than would be anticipated. The non-frozen portion appears to increase as the fat content of the muscles increases (Fleming, 1969).* However, as well as the *extent*, the *rate* at which the temperature of the meat falls is a most important consideration: the time taken to pass from $0\,°C$ to $-5\,°C$ is usually regarded as an indicator of the speed of freezing. The fastest times so far obtained are of the order of $1\,s$. These have been achieved by placing a single muscle fibre in isopentane at

---

* Love and Elerian (1963) showed that damage to proteins in fish muscle increases progressively as temperature is lowered to $-183\,°C$, because more and more structural water is irreversibly frozen out of them.

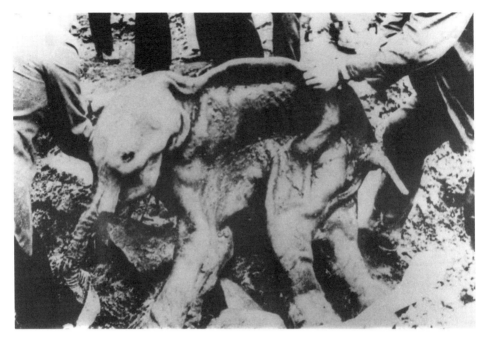

Fig. 7.4 Carcass of young mammoth recovered from frozen Siberian tundra after *ca.* 15,000 years. (Reproduced by courtesy of the Novosti Press Agency, London.)

–150°C. At such rapid rates water freezes between the actual filaments (≃molecules) of myosin and actin in aggregates so small that they do not distort the structure, even at the level of observation possible with the electron microscope (Menz and Luyet, 1961). These minute aggregates still appear to be crystalline and not amorphous and vitreous (Luyet, 1961). As the time to freeze increases, structural damage to the muscle also increases. To assess it, Love (1955) used the concentration of deoxyribose nucleic acid phosphorus (DNAP) in the expressible fluid of the muscle during a series of studies on fish. Since in undamaged muscle DNAP is found exclusively in the nucleus of the muscle cells, and thus inside the cell membrane (sarcolemma), its appearance in the extracellular fluid indicated a degree of damage sufficient to rupture the sarcolemma.

Some comparative data, compiled from the publications of Love (1958) and Love and Haraldsson (1961), are given in Table 7.7. It will be seen that the sarcolemma is still undamaged when the freezing time has been extended from 1 sec to 5 min (although with the latter time there would be considerable distortion of the myofibrils inside the muscle fibre: Menz and Luyet, 1961). It will also be seen that as the time to freeze increases beyond about 5 min, damage to the sarcolemma proceeds through a series of maxima and minima corresponding to different kinds of ice formation, firstly within and eventually outside the fibre. With those freezing times less than about 75 min and where there is little damage to the sarcolemma (e.g. 5 and 50 min), the muscle can be thawed with little formation of drip irrespective of ultimate pH – the water which separated as ice being completely reincorporated by the proteins. Not all investigators agree that fast rates of freezing are associated with less drip than slow rates. Thus, whereas Cook *et al.* (1926) found such a rela-

Table 7.7 Relation between freezing rate and DNAP (as an indicator of cell damage) in press juice from thawed fish muscle (after Love, 1958; Love and Haraldsson, 1961)

Approximate time to freeze (0 to –5 °C) (min)	DNAP concentration	Type of ice formation	Presumed explanation
5	Low	Many very small *intra*cellular crystals	Sarcolemma not damaged
25[a]	High	Four *intra*cellular ice columns	Unfrozen intracellular contents too thinly spread over surface between ice crystals and sarcolemma to avoid damaging latter
50	Low	Relatively small single *intra*cellular column	Layer of unfrozen intracellular contents sufficiently thick to separate ice from sarcolemma and protect it
75[b]	High	Very large single *intra*cellular column	Unfrozen intracellular contents too sparse to protect sarcolemma from ice damage
100	Low	Relatively small *extra*cellular crystals	Water osmotically drawn from cell through intact sarcolemma to form crystals outside, which are too small to distort fibres
200–500	High	Large *extra*cellular ice crystals	Extracellular crystals are sufficiently large to distort muscle fibres and damage sarcolemma
>750	Low	Very large *extra*cellular ice masses	Ice masses so large that fibres press together: the sarcolemma of interior members thus protected from damage

[a] Denotes slowest freezing rate at which more than one ice crystal forms within muscle cell. [b]Denotes slowest freezing rate at which ice forms intracellularly.

tionship, Bailey (1972) did not. More recently, Anon and Calvelo (1980) re-examined the question. In a study of beef, they found that, as the time to freeze (defined as the time to pass between –1 and –7 °C) decreased below 15–20 min, the extent of drip decreased rapidly, this being associated with increasing proportions of intracellular ice formation. With times of freezing between 20–200 min the amount of drip was appreciable and associated with extracellular ice formation but relatively constant. The relationship between the temperature of freezing (which is approximately inverse to the rates of freezing) and ice crystal formation in beef *l. dorsi* muscle was studied by Rahelic *et al.* (1985). Whereas water formed ice only intercellularly at –10 and –33 °C, and only *intra*cellularly at –78 °C and below, ice formed both inter- and *intra*cellularly at –22 °C, when the greatest structural damage was observed. At this temperature, also, the solubility of the myofibrillar proteins appears least (Petrovic, 1982). Ice crystals formed within the I-band, but not within the A-band at –22 °C, possibly because the water-holding capacity of the actin filaments is weaker than that of myosin filaments and the difference can be elicited in this temperature range (Rahelic *et al.*, 1985). The rates of freezing feasible commercially are much too slow to cause intracellular ice formation. With such rates of freezing, ice crystals tend to form first outside the fibre, since the extracellular osmotic pressure is less than that within the muscle cell (Chambers and Hale, 1932). As extracellular ice formation proceeds, the remaining unfrozen extracellular fluid increases in ionic strength and draws water osmotically from the super-cooled interior of the muscle cell. This freezes on to the existing ice crystals, causing them to grow, thus distorting and damaging the fibres. Moreover, the high ionic strength denatures some of the muscle proteins (Finn, 1932; Love, 1956; and § 5.4.1), and this factor, quite apart from the translocation of water, largely accounts for the loss of waterholding capacity of the muscle proteins and for the failure of the fibres to reabsorb, on thawing, all the water removed by freezing – this being manifested as drip (Moran, 1927). On electrophoresis and ultracentrifugation, changes in the pattern of separated proteins between fresh and frozen muscle are detectable (Howard *et al.*, 1960a). The sarcoplasm of fresh muscle shows only one isozyme of glutamate-oxalacetate transaminase. After freezing, the mitochondrial isozyme is liberated; and the presence of both in a sarcoplasmic extract is thus an indication that freezing has taken place (Hamm and Kormendy, 1966). It has also been reported that, in the drip from thawed pork, the concentration of α-glucosidase is significantly greater than in the exudate from fresh pork (Toldra *et al.*, 1991). Protein damage is generally a function of time and temperature of freezing (Meryman, 1956). At –4 °C, for example, increasing damage is caused to both sarcoplasmic and myofibrillar proteins as time of storage increases (Awad *et al.*, 1968). The number of protein components separated by gel electrophoresis of urea-treated bovine actomyosin decreased from six to three over 8 weeks' storage at –4 °C. Specific sarcoplasmic protein components also became insoluble. Nevertheless, the extractability of myofibrillar proteins has been reported to increase slightly during the early stages of storage at –20 °C, the affinity of actin for myosin being greatest after one week.

Although various cryoprotectants, such as glycerol and dimethyl sulphoxide, are now commonly used to depress the freezing point of solutions and thus minimize damage to cells, the high concentrations necessary for efficacy are toxic and cause other deleterious changes which prevent their application in the freezing of foods. The study of animals, plants and micro-organisms which survive freezing environments, however, has revealed the existence of a number of antifreeze proteins which lower the freezing point of water by adsorption–inhibition (DeVries, 1988) and also alter the pattern of ice crystal formation to yield needle-like forms instead of large crystal aggregates. In addition, they inhibit ice recrystallization (as occurs by sublimation when temperatures fluctuate), whereby cell damage is limited. Conservation of the native configuration of proteins against environmental damage appears to be a general phenomenon exerted by various organic solutes – osmolytes – which control the availability of water in cells (Yancey, 2003) (and cf. Fig. 7.7).

The possibility of utilizing antifreeze proteins to protect meat during freezing and frozen storage was investigated by Payne *et al.* (1994). They demonstrated that antifreeze glycoproteins of the Antarctic cod – which are α-helical repeat polymers of Ala-Ala-Thr, with galactose-*N*-acetylgalactosamine linked via the hydroxyl groups of threonine – when used to soak portions of meat before freezing at –20 °C, specifically reduced the size of ice crystals in comparison with controls. These glycoproteins exemplify the production, by a single gene, of a polyprotein (Hsiao *et al.*, 1990), which can be subsequently split into separate functional units. The mechanism is similar to that in viruses which translate their entire genome into a single polyprotein (Douglas *et al.*, 1984).

Subsequently, Payne and Young (1995) showed that, when the antifreeze glycoprotein from Antarctic cod was injected into lambs 24 h preslaughter, the ice crystals found subsequently in the frozen *l. dorsi* muscles were markedly smaller than those of non-injected controls and the drip loss was less on thawing (Fig. 7.5). It is thus feasible that future development of this approach could improve the quality of frozen meat.

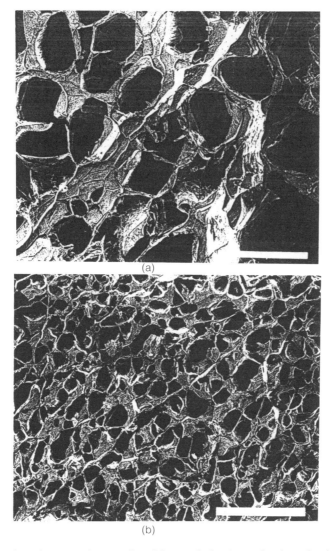

Fig. 7.5 Scanning electron micrographs of frozen *l. dorsi* muscles from lamb. (a) Control. (b) Lamb injected 24 h preslaughter with 0.01 µg/kg of antifreeze glycoprotein from Antarctic cod (bar = 100 µm). (Reprinted from Payne and Young, 1995, with kind permission of Elsevier Science Ltd and Drs S. R. Payne and O. A. Young.)

Despite the prolonged freezing times which are inevitable when dealing with meat in commercial handling – carcasses, cuts or steaks – the degree of drip may be greatly lessened by various procedures. These will be considered in § 7.1.2.2.

7.1.2.2   Frozen carcass meat

Large-scale preservation of meat by freezing dates from about 1880, as far as the United Kingdom is concerned, when the first frozen beef and mutton arrived from Australia. Increasing industrialization in Britain was accompanied by increasing population and decreasing livestock: there was a surplus of meat animals in the southern hemisphere, especially in New Zealand and Australia: and freezing offered a means of preserving meat during the long voyages involved between the two areas (Critchell and Raymond, 1912). By 1960 Britain was importing more than 500,000 tons per year of frozen beef, veal, mutton and lamb (Commonwealth Economic Committee, 1961).

The rate of freezing is dependent not only on the bulk of the meat (Fig. 7.1) and its thermal properties (e.g. specific heat and thermal conductivity), but on the temperature of the refrigerating environment, on the method of applying the refrigeration and, with smaller cuts of meat, on the nature of the wrapping material used. Table 7.8 gives the thermal conductivities, at various temperatures, of muscle, bone and fat of several meat species (Morley, 1966).

The low conductivity of fat at both ambient and freezing temperatures, and the greatly increased conductivity of meat when frozen, are apparent. Such data enable

Table 7.8   Thermal conductivity of meat

Sample	Temperature (°C)	Thermal conductivity 10^{-3} J sec^{-1} cm^{-1} °C^{-1}
Miscellaneous		
Gelatin gel	+20	6.2
Ice	−19	24.7
Muscle		
L. dorsi (lumbar), beef	+16	5.2
	−19	13.0
L. dorsi (lumbar), pork	+16	5.2
	−19	18.0
L. dorsi (lumbar), lamb	−19	17.7
Fat		
Sirloin, beef	+16	2.1
	−19	2.7
pork	+16	2.0
Kidney, beef	+16	1.9
(rendered)	+16	1.3
Bone (cancellated)	+16	2.6
Femur, beef	−19	3.3
Bone (compact)		
Femur, pork	+19	5.8
Rib, beef	+19	5.4
Radius, beef	+19	5.7

accurate cooling and freezing rates to be calculated for commercial operations (Earle and Fleming, 1967).

It is not appropriate here to discuss freezing techniques in detail: these have been considered in various publications (cf. Tressler and Evers, 1947; Calvalo, 1981); but some comparative data will be given.

It used to be normal commercial practice in Australia to chill beef quarters for 1–3 days at about 1 °C and then to freeze them for 3 to 5 days at –10 °C (Vickery, 1953), but rates of freezing have been increased either by using smaller cuts than quarters (Law and Vere-Jones, 1955) or by placing the hot meat directly after slaughtering into a blast tunnel freezer without prior chilling (Jasper, 1958; Howard and Lawrie, 1956). The latter procedure is now used extensively in preparing New Zealand lamb. Apart from rapidity of freezing, it involves biochemical factors which will be considered below. Some idea of the rate of freezing of beef quarters in a blast tunnel can be gained from the time–temperature curves in Fig. 7.6 (Howard and Lawrie, 1957b). Curve (a) records the temperature in the deepest part of a hindquarter of good quality beef when placed hot in a tunnel operating at 1000 ft/min (300 m/min) and –40 °C. The time for entry to completion of freezing is about 18 h compared with 24 h for beef of similar quality in a tunnel operating at 250 ft/min (75 m/min) and –35 °C (curve (b)). A thick section of sirloin under the latter conditions froze in about 2 h (curve (c)). The process obviously eliminates weight losses caused by evaporation during chilling and lessens those during

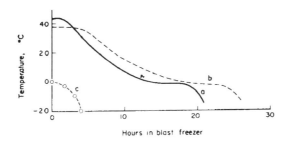

Hours in blast freezer

Fig. 7.6   Time–temperature curves for beef placed whilst warm into a blast freezer (Howard and Lawrie, 1957b). (a) Temperature in deepest portion of side: blast of 1000 ft/min (300 m/min); –40 °C. (b) Temperature in deepest portion of side: blast of 250 ft/min (75 m/min); –30 °C. (c) Temperature at centre of thick section of sirloin.

Table 7.9   Weight losses during chilling, freezing and frozen storage of beef quarters (Howard and Lawrie, 1956)

Process	Quarters blast frozen without prior chilling (as % hot weight)	Quarters normally frozen after chilling (as % hot weight)
Chilling	–	1.50
Freezing	0.60	1.33
Storage at –10 °C for 20 weeks	0.74	0.36
Total	1.34	3.19

freezing (Table 7.9); but apparently causes the frozen quarters to be more susceptible to loss during subsequent storage at –10 °C. Even so, the overall loss is less. The loss in weight of carcasses of New Zealand lamb over 90 days is generally about 4 per cent (Griffiths et al. 1932).

It is still customary in Britain to store frozen meat at –10 °C; yet weight losses by evaporation at –30 °C are only 20 per cent of those at –10 °C (Cutting and Malton, 1971). Moreover, wastage at the latter temperature increases the need to trim discoloured surfaces. Packing in polythene reduces evaporation at –10 °C to about the same level as that of meat at –30 °C without such packaging.

When unprotected meat surfaces are blast frozen, there is considerable freezer burn. Freezer burn is the name given to the whitish or amber-coloured patches seen on the surface of frozen meats. The patches are caused by the sublimation of ice crystals into the atmosphere of the cold store, thus creating small air pockets on the meat which scatter incident light (Brooks, 1929a). This happens because the vapour pressure of water over the coils of the refrigerating machinery is much less than that above the meat surface.

Kaess and Weidemann (1961, 1962, 1963) made an exhaustive study of freezer burn. The phenomenon involves the formation of a condensed layer of muscular tissue near the surface. This prevents access of water from below, thus enhancing surface desiccation. Freezer burn is maximal during storage when the meat has been frozen rapidly under conditions where evaporation has been prevented; and conversely.

The construction of cold stores which are entirely surrounded by an insulating air jacket has made feasible a substantial reduction in evaporative loss during the storage of frozen meat. In Moscow a store capable of holding 35,000 tons of meat had been so jacketed (Gindlin et al., 1958) many years ago; but the extension of pre-packaging methods of meat handling has largely invalidated this technical development. Currently there is growing interest in importing frozen meat from Australia and New Zealand in standardized insulated containers. Apart from speeding loading and unloading operations, this obviates any risks from exposure to ambient temperatures. Containers measuring 20 × 8 × 8 ft (6 × 2.5 × 2.5 m) are in service (Middlehurst et al., 1969). When used to hold 60-lb (27 kg) cartons of frozen boneless beef, such containers require additional refrigeration at varying times, dependent on the initial temperature of the frozen meat when loaded into the containers and the ambient temperature. 'Clipon' refrigeration units can be attached for this purpose. Temperatures should not be permitted to rise above –4 °C with frozen beef, lamb or mutton, or above –5 °C with offal, as otherwise damage and distortion due to softening occur (Haughey and Marer, 1971).

With pieces of meat smaller than quarters or joints considerably faster rates of freezing are naturally possible, the actual rate depending on the temperature of the refrigerating environment, on its nature (Table 7.10) and on the wrapping material which may be employed (Table 7.11).

It will be seen that such meat freezes fastest in an air blast and slowest in still air, direct contact with freezing plates giving intermediate values; and that certain types of wrap will double the time to freeze over that for corresponding unwrapped meat (Dunker and Hankins, 1955), depending on their insulating properties (cf. Table 7.6).

Even the rates of freezing in Table 7.10, being greater than 75 min, will cause extracellular ice formation (Table 7.7) and thus potential drip on thawing. Never-

Table 7.10 Rates of freezing in relation to temperature and type of freezing (after Dunker and Hankins, 1955) (6 inch (150 mm) cubes)

Ambient temperature (°C)	Type of freezing	Approximate time to freeze 3° to −12 °C (hr)
−17	Air blast	12
	Plate freezer	16
	Still air	19
−56	Air blast	2
	Plate freezer	3.5
	Still air	4

Table 7.11 Effect of wrapping on freezing rate of lean beef (6 × 6 × 23 in. (150 × 150 × 580 mm)) frozen in an air blast at −17 °C (after Dunker and Hankins, 1955)

Type of wrapping	Approximate time to freeze 3 °C to −15 °C (hr)
Nil	7
Greaseproof paper	9.5
Cellophane	10
Aluminium foil	12
White parchment	12
Polyethylene	14.5

theless, the extracellular crystals formed are much smaller and more finely distributed than those in normal commercial practice (Moran, 1932; Cook *et al.*, 1926). The consequently lower order of fibre distortion and of translocation of water is associated with a somewhat smaller degree of drip (Ramsbottom and Koonz, 1939). It should be pointed out that these benefits from the formation of small, *extracellular* ice crystals will be lost if the frozen meat is subsequently held at too high a temperature because of the phenomenon of recrystallization (Moran, 1932; Meryman, 1956). If the temperature of storage is not far below the freezing point, the smallest ice crystals will sublime and recrystallize on to those which are somewhat larger, thus increasing the size of the latter still further; and damage to the tissue, manifested by drip or thawing, will increase accordingly.

The great difficulty of freezing even small portions of muscle sufficiently rapidly to produce minute intracellular ice crystals, and thus diminish drip on thawing, has been alluded to above (§ 7.1.2.1). As is well known, the application of pressure lowers the freezing point of water. Martino *et al.* (1998) applied this principle to 0.4 kg portions of porcine *l. dorsi* muscle, under a pressure of 210 MPa, when they were immersed in ethylene glycol–water at −20 °C. On releasing the pressure to 0.1 MPa, numerous small ice nuclei formed uniformly throughout the muscle cells. Their size was less than that of those produced by freezing in liquid nitrogen, or by air blast, under normal pressure (cf. Table 7.12; Fig. 7.7). Development of the

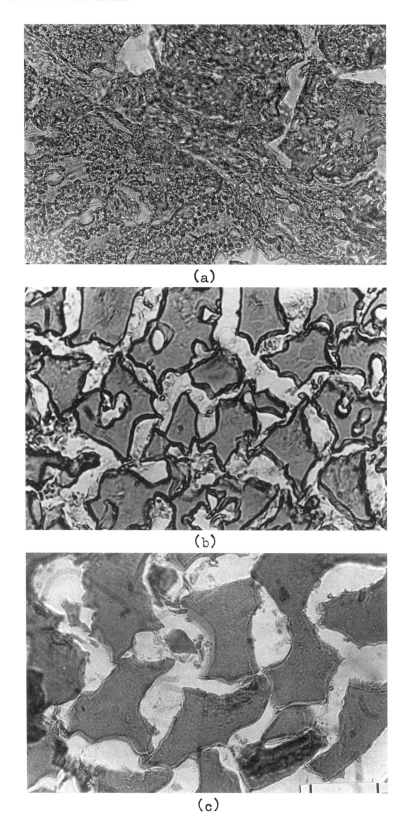

(a)

(b)

(c)

Table 7.12 Mean ice crystal diameters (μm) at the surface and centre of 0.4 kg samples of porcine *l. dorsi* muscle frozen by different methods

Freezing method	Surface	Centre
High pressure (210 MPa) in fluid at –20 °C	2.34 ± 1.08	4.13 ± 2.16
Liquid nitrogen	3.03 ± 1.76	18.81 ± 1.36
Air blast	20.10 ± 3.04	31.11 ± 7.24

technique for the freezing of commercial-sized cuts of meat, whereby small ice crystals would form intracellularly and drip on thawing would be markedly diminished, is envisaged (Sanz *et al.*, 1999).

It should also be noted that, during freezing at ambient pressure, water is converted into Ice I, which is of a lower density than water and thus occupies a larger volume, causing tissue damage. However, at 632 to 2216 MPa, water freezes to Ice VI and, as such, causes no disruption of the meat microstructure (Molina-Garcia *et al.*, 2004) since the volume change on freezing is negative. The meat can be kept under these conditions at ambient temperature.

The possibility of unlimited quality retention in frozen meat would depend on complete prevention of enzymic reactions, oxidative rancidity and ice recrystallization. This would involve complete immobilization of solutes by 'glass' formation. Determination of the glass transition temperature of frozen pork (T_g) has suggested that temperatures below –13 °C would be required. There is evidence, however, that some residual mobility of solutes persists down to a second glass transition temperature at –55 °C, suggesting that frozen storage below this level would be necessary to completely prevent quality changes in the frozen commodity (Hansen *et al.*, 2003) (cf. Table 7.14).

Apart from its effect on the size of extracellular ice crystals, the rapid freezing of meat without prior chilling influences the water-holding capacity in another way, since post-mortem glycolysis will still be proceeding in the muscles (§ 4.2.2). If the muscle should be frozen before the ATP level has fallen appreciably (i.e. pre-rigor), there will be a greatly enhanced ATP-ase activity on thawing which will cause marked shortening and the exudation of excessive quantities of drip which may amount to 30–40 per cent of the muscle weight ('thaw-rigor': Moran and Smith, 1929; Bendall and Marsh, 1951) unless the muscle is held taut (e.g. *in situ* on the carcass: Marsh and Thompson, 1958). In attempting to explain the phenomenon of thaw-rigor, Bendall (1973) pointed out that the rate of contraction depends mainly on the rate of thawing. To elucidate this problem he employed thin strips of muscle wherein the time of heat transfer was negligible. He found that during thaw-rigor

Fig. 7.7 Photomicrographs of central locations of porcine *l. dorsi* muscles from portions (5 × 7.5 × 11 cm³) frozen (a) under 210 MPa pressure in ethylene glycol–water at –20 °C, (b) in liquid nitrogen and (c) in an airblast at –35 °C and 5.5 m/s. Scale bar = 50 μm. (Reproduced from *Meat Science*, Vol. 50, M. N. Martino, L. Otero, P. D. Sanz and N. E. Zaritzky, 'Size and location of ice crystals in pork frozen by high pressure assisted freezing as compared to classical methods', pp. 303–313. Copyright 1998, with permission from Elsevier Science and Dr M. N. Martino.)

in such strips both creatine phosphate (CP) and ATP fell very swiftly from their initial level, post-mortem glycolysis being complete in a fraction of an hour. The rate of ATP breakdown was ten times greater than in normal rigor mortis at 37 °C and this could only be explained by presuming that the contractile ATP-ase had been stimulated by pre-rigor freezing and thawing: the non-contractile ATP-ase of myosin is responsible for ATP breakdown during the onset of rigor mortis normally. Bendall further observed that the shortening of the muscles was complete before the ATP level had declined significantly, whereas if shortening occurs at all in normal rigor it does so at the stage where the ATP level is falling swiftly. It can be shown that there is a characteristic notch in the shortening curve – indicative of relaxation. When even faster thawing of pre-rigor frozen muscle is achieved (by the use of quantities of muscle containing only a few fibres) contraction is almost instanta-neous (about 30 s) and develops considerable power, being then followed by almost complete relaxation. On the basis of these results, and investigations on model systems, Bendall concluded that the first effect of thawing is to cause an extensive salt flux whereby Ca^{++} ions are released by the sarcotubular system and stimulate the breakdown of the considerable level of ATP present in the pre-rigor frozen muscle. Where thawing is more or less instantaneous, calcium is quickly recaptured by the sarcotubular system. Since, thereby, the stimulation of actomyosin ATP-ase is so brief, ATP is still virtually at its pre-rigor level, H-meromyosin cannot cross bond with actin and relaxation ensues. Luyet and co-workers (1965) demonstrated the visual changes in this sequence of events on film.

The rate of ATP breakdown in thaw-rigor is also affected by the rate of pre-rigor freezing: fast frozen muscles metabolize ATP faster than slow frozen muscles. This is again a reflection of the degree of stimulation of the contractile ATP-ase of acto-myosin by calcium ions which appear to be released by the sarcotubular system to a greater extent in the former case (Scopes and Newbold, 1968). Both in thaw-rigor and cold shortening phosphorylase is markedly activated by the AMP produced (Newbold and Scopes, 1967; Scopes and Newbold, 1968).

According to Shikama (1963) the activity of isolated myosin ATP-ase is sub-stantially lost by exposure to temperatures below –20 °C for even 10 min, due to dis-ruption of the ordered array of water molecules adjacent to the protein surface. This suggests that the rate of ATP breakdown on thawing muscle which has been frozen pre-rigor might be somewhat less if the temperature attained on freezing was suf-ficiently low. Arrest of ATP breakdown sometimes fails to occur despite rates of freezing which are theoretically fast enough to achieve it (Howard and Lawrie, 1956). This is probably because the rate of ATP breakdown before freezing has been *enhanced* by cold shock (and caused 'cold-shortening') or where localized concen-trations of salt have arisen just before the tissue as a whole has frozen (Smith, 1929). The rate of onset of rigor mortis at –3 °C is as rapid as at 20 °C (Behnke *et al.*, 1973). Moreover, when pre-rigor frozen muscle is held for a few days at –2 °C (Marsh and Thompson, 1958) or for a few weeks at –12 °C (Davey and Gilbert, 1976a), there is a slow breakdown of the ATP, whilst the structure is still rigid whereby the prereq-uisites for the shortening and exudation are removed.

Where the rate of freezing is insufficient either to arrest ATP breakdown (as in pre-rigor freezing) or to speed it up (as by salt concentration in slower freezing or by 'cold-shock'), it may be sufficient, nevertheless, to slow it. This would be associ-ated with a greater water-holding capacity (§ 5.4.1) and less drip on thawing. The rate of ATP breakdown can be further slowed by preslaughter injection of relaxing

Table 7.13 Percentage drip from beef *psoas* muscles under different conditions

Treatment	Normal freezing after chilling	Blast freezing without prior chilling
Control	9.1	6.6
Pre-slaughter injection with magnesium sulphate	7.9	4.4

doses of magnesium sulphate (Table 7.13; Howard and Lawrie, 1957b), and drip on thawing is diminished even more. It is of interest that various studies in which magnesium sulphate has been included in the diet of pigs, have reported a reduced incidence of PSE, with enhanced water-holding capacity and colour – possibly because the rate of post-mortem glycolysis has been diminished thereby (Hamilton *et al.*, 2003) (cf. § 4.2.2).

Conversely, the pre-slaughter administration of calcium salts, by increasing ATP breakdown during post-mortem glycolysis, enhances drip formation (Howard and Lawrie, 1956).

There are other ways in which drip can be minimized despite the extracellular ice formation which is inevitable with carcass meat. Conditioning of meat before freezing (§ 5.4) diminishes drip – possibly by increasing intracellular osmotic pressure and thus opposing the egress of fluid to the extracellular ice crystals (Cook *et al.*, 1926; Bouton *et al.*, 1958). Part of the conditioning effect may be due to alterations in ion–protein relationships since, during holding, sodium and calcium ions are released and potassium ions absorbed by the myofibrillar proteins (Arnold *et al.*, 1956).

The work of Empey (1933) showed that one of the most important factors determining the availability of drip fluid was pH. Even with the relatively low rates of freezing in carcass meat, which produce extracellular ice formation, the induction of a high ultimate pH can virtually eliminate drip (Fig. 7.8). The necessary glycogen depletion may be brought about in various ways (cf. §§ 4.2.2 and 5.1.2). With quarters of meat, however, the diminution in drip caused by high ultimate pH is not so marked, because much of the fluid in these circumstances is derived from blood vessels and other extracellular spaces rather than from the muscular tissue as such (Howard *et al.*, 1960).

It is interesting to note that even with the same rate of post-mortem pH fall, and at the same ultimate pH, drip from the *l. dorsi* muscle is generally twice as great as that from *psoas*; and it has a different composition (Howard *et al.*, 1960a). Clearly, different muscles have different intrinsic susceptibilities to damage during freezing and thawing (Fig. 7.8). Especially with the advent of frozen cuts and individual portions of meat, there has been increasing interest in establishing optimum modes of thawing. Various investigations have been undertaken to clarify the issue. Thus, Bailey *et al.* (1974) found that, for a given temperature, frozen pork legs thaw faster in water than in air, although the appearance after the latter treatment was preferable. According to Vanichsensi *et al.* (1972) the thawing of frozen lamb shoulders in water (45 °C for 2–2½ hours) is suitable for short-term batch or continuous

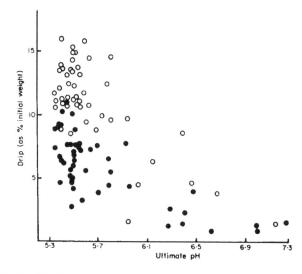

Fig. 7.8 The relationship between the percentage of drip on thawing and the ultimate pH
in *psoas* (●) and *l. dorsi* (○) muscles of the ox (Lawrie, 1959).

processing immediately prior to boning out. For a batch process with a cycle time
of 8–10 hours, however, air thawing at 2–18 °C and high humidity is better.

The long periods of storage which temperatures below the freezing point make
possible may lead to considerable oxidative rancidity in exposed fat. The fats of beef
and lamb are relatively resistant to such oxidation and may still be good after 18
months, storage at –10 °C (Lea, 1938). The more highly unsaturated fat of pork
becomes rancid rather more quickly, and that of rabbit, which may have an iodine
value of 180, turns yellow and oxidizes after only 4 months at –10 °C (Vickery, 1932).
As indicated above, the formation of brown metmyoglobin tends to develop in the
lean in parallel with fat oxidation and rancidity (Watts, 1954).

7.1.2.3 Prepackaging aspects
The sale of prepackaged frozen meat has developed in parallel with the fresh article
for several reasons. In the first place, it adds to the convenience of handling, since
there is a great extension of storage life, making it attractive for institutional and
other bulk catering. Secondly, and in relation to the increasing availability of freezing compartments in domestic refrigerators, it gives the consumers a new facility in
creating a reserve of quasi-fresh food against unexpected demands. Although temperatures below the freezing point are beneficial in permitting prolonged storage,
they may make more severe demands on the packaging material than do chill or
ambient temperatures (cf. Table 7.6). The meat is exposed for times long enough to
cause desiccation and for oxidation of the fat and the pigments, notwithstanding the
slowness of these changes because of the low temperature. The wrapping must effectively prevent – or disguise – them at temperatures for which many packing materials are unsuitable. Over prolonged storage periods, exposure to strong light, as in
display cabinets, will accelerate oxidation of fats (Lea, 1938) and cause discoloration
(especially at wavelengths between 5600–6300 Å: Townsend and Bratzler, 1958), and
this is one reason for the use of an opaque wrap for frozen prepackaged products.

Another reason is that frozen meat, by its hardness, is liable to pierce the films of polyethylene or polyvinylidene chloride which are often used. To avoid this the film can be prevented from making close contact with the meat surfaces by the walls of the carton base. Because of the space above the meat, sublimation of ice can occur, causing extensive desiccation on the surface. An opaque wrap – stamped with a photographic representation of the contents when cooked – offsets the superficially unpleasant impression this would create. Moreover, instructions are frequently given to cook without thawing, thus disguising drip – the major disadvantage of frozen meats. Even so, the pieces of meat used in prepackaging are sufficiently small to permit really fast freezing, with a consequently diminished tendency to drip on thawing (Bouton and Howard, 1956). If the pieces of meat are free of sharp edges and are thus not liable to rupture the packaging film, the latter may be shrunk on to the meat surfaces, thus eliminating the possibility of surface desiccation.

There is a definite relation between the ability of wrapping materials to retain moisture and a desirable flavour and odour in the cooked meat. For frozen prepackaged meat, aluminium foil, polyethylene and polyvinylidene chloride are particularly good in this respect (Emerson et al., 1955; Bouton and Howard, 1956). Palatability can be enhanced by packing in nitrogen (Steinberg et al., 1949) or in vacuum (Hiner et al., 1951) although it is also more expensive. Dahl recommends –30 °C as the desirable temperature of storage for frozen meat, but this might be difficult to achieve commercially on a large scale. Fluctuating temperatures are particularly damaging for discoloration, rancidity and desiccation (Emerson et al., 1955; Townsend and Bratzler, 1958).

Packaging materials (such as 'Sarlyn') are available which soften on heating and can be tightly wrapped over sharp areas under vacuum without causing puncturing. These are particularly useful with frozen cuts. There is no residual space in which frost can form. The films are transparent and have moderately high oxygen permeability. Their use permits the retention of the bright-red colour of oxymyoglobin on the surface of the frozen meat for up to a year, provide it is stored in the dark below –20 °C (Taylor, 1985). In a comparison of lamb cuts, either wrapped in film of high oxygen permeability or vacuum-packed in film of low oxygen permeability, and stored in the dark at –10, –20 or –35 °C for 0–20 weeks, it was found that the former had better colour retention during subsequent display at –20 °C (Moore, 1990). Cuts stored at –10 °C had no display life after 10 weeks, whereas those stored at –20 or –35 °C had acceptable display life after 20 weeks.

From what has already been indicated in relation to fresh meat (§ 7.1.1.2) the lower the temperature of frozen storage, the longer is the time before oxidative rancidity in fat, and the production of brown metmyoglobin in the lean, develop. These features, and the greater susceptibility to spoilage of pork than beef, have been emphasized in the work of many investigators (Emerson et al., 1955; Palmer et al., 1955; Dahl, 1958b): Table 7.14 is taken from the latter. The feeding of vitamin E to pigs was found to increase slightly the storage life of the vacuum-packaged pork when stored at –20 °C. Since the storage life of the commodity at this temperature is 9 months without such supplement, its use seemed superfluous unless even longer storage or storage in the presence of light were envisaged (Houben and Krol, 1985). Similar supplementation of cattle was found to improve the stability of both lipids and pigments when beef, prepackaged in film of both high and low oxygen permeability, was stored at –20 °C (Lanari et al., 1994). Kinetic studies showed that the rate of auto-oxidation of oxymyoglobin was decreased by the vitamin E (cf. § 7.1.1.3).

Table 7.14 Approximate times for appearance of distinct rancidity in fat or discoloration of lean in unwrapped meat

Meat	Storage temperature			
	–8 °C	–15 °C	–22 °C	–30 °C
Beef (rib roasts)	3 months	6 months	12 months	–
Pork (loin without rind and back fat)	–	3 months	6 months	12 months

Glazing the meat with water greatly retards these changes since it protects the meat itself from desiccation and by preventing access of oxygen to the surface retards oxidation (Dahl, 1958b). In the USA the wrapping of frozen cuts in edible coatings (such as acetylated monoglycerates) has been advocated. They can be applied to frozen meat (–20 °C) by dropping the latter in a bath of the coatings substance at 130 °C for 5 s. Even at –30 °C such coatings will not crack, and will retain their efficacy against desiccation.

Experiments with slices of *l. dorsi* (about 1 inch (25 mm) thick) and ground hamburger beef have shown that cryogenic freezing in liquid nitrogen gives initially lower weight losses than those with blast-freezing at –30 °C. Subsequently, however, cooking losses from the cryogenically frozen material are greater. It has been suggested that cryogenic freezing causes cracks to develop in the product from which moisture is readily lost on cooking (Jakobbson and Bergtsson, 1969).

Although the low temperatures involved in the storage of frozen prepackaged meat diminish the risk of microbial contamination, they do not eliminate it (§ 6.1.2.2). The same dangers of contamination during preparation, which apply to fresh prepackaged meat, apply here (§ 7.1.1.3). Spores and diseases caused by sporulating organisms or toxins may survive freezing. In the first place, bacteria would be able to multiply in the meat if it were to stay much above freezing temperatures before consumption. In the second case, sufficient toxin may form before bacterial growth is stopped by freezing to affect customers subsequently eating the meat after frozen storage.

The salient points on the freezing of meat may be summarized.

Post-rigor meat
1. The faster the rate of freezing the less will be the drip on thawing. (Not all research workers agree on this aspect.) A high speed of freezing will also tend to enhance tenderness.
2. A high ultimate pH (however it arises) will give greatly enhanced water-holding capacity and diminish drip on thawing if the meat is subsequently frozen; but it may lead to excessive tenderness and to loss of colour and flavour.
3. Increased tenderness and enhanced water-holding capacity (with diminished drip on thawing) will develop if the meat is 'conditioned' for some days after reaching the ultimate pH. (This applies particularly to beef.)

Pre-rigor meat
The consequences of freezing are determined by the freezing rate which is feasible in particular circumstances.

1. With intact sides of pork and beef the rate of heat removal is likely to be too slow to cause much 'cold-shortening': and there will be virtually none of the meat frozen pre-rigor.
2. When freezing is applied to pre-rigor meat of smaller dimensions there may be some diminution of drip insofar as the process of postmortem glycolysis is slowed; but, as the rate of temperature fall is increased still further, firstly 'cold-shortening' then 'thaw-rigor' (due to pre-rigor freezing) will ensue. When very fast freezing rates are possible, pre-rigor freezing will ensue before 'cold-shortening' can develop.
3. 'Cold-shortening' can be avoided by ensuring that the temperature does not fall below 10–15 °C before about 10 h post-mortem by altering the mode of suspension of the side or by electrical stimulation of the warm carcass.
4. The extreme disadvantages of 'thaw-rigor' can be avoided if the time and temperature of frozen storage are such as to allow the ATP level to fall before thawing.
5. If pre-rigor meat is salted, freezing is not associated with loss of water-holding capacity (cf. § 10.2.1.2).

7.2   Thermal processing

Preservation of meat by thermal processing dates from the beginning of the nineteenth century when Appert (1810), whilst not aware of the nature of the processes involved, found that meat would remain edible if it were heated in a sealed container and the seal maintained until the meat was to be eaten. This method of preservation has developed into the canning industry (although glass containers as well as metal cans may be employed). Canned meat and meat products may be subjected to heat at two levels: pasteurization, which is designed to stop microbial growth with minimum damage; and sterilization, in which all or most bacteria are killed, but which alters the meat to a considerably greater degree.

An important consideration in achieving sterility is the fact that certain microorganisms form spores which may be exceedingly heat-resistant (cf. § 6.3.1). To destroy the spores of certain thermophiles would require a degree of heating which would greatly lower the organoleptic attributes of the commodity. In canning practice 'commercial sterility' is achieved by giving a degree of heat treatment sufficient to kill non-sporing bacteria and all spores that might germinate and grow during storage without refrigeration. To avoid the growth of any thermophiles potentially present it is essential to cool the cans rapidly after processing and to prevent storage at high ambient temperatures. The eating quality of canned, cured meats can only be retained by using a low degree of heat treatment (pasteurizing) which is insufficient to kill spores; but in such products the curing ingredients reinforce the bactericidal and bacteriostatic effects of the heat.

7.2.1   Pasteurization

Pork is more susceptible to heat damage than other meats; moreover, the ingredients used in making certain products from pork (as in hams) tend to weaken the structure still further. As a result, the heat treatment given to such commodities as canned hams cannot be sufficient to kill all micro-organisms. The definition of pasteurized or semi-preserved meats indicates that such products do not remain unchanged, and consumable, for any length of time in temperate climates unless special precautions are taken during transport and storage (Maillet, 1955). Although containers of semi-preserved meats are labelled to recommend low-temperature storage, this was frequently disregarded when United Kingdom consumption of continental canned hams and other such products increased after the Second World War, since there was a current impression that all canned meats were sterile. The widespread spoilage of the product, manifested by 'blown' cans and otherwise, altered this view and has led to appropriate handling precautions (Hobbs, 1955). It is nevertheless a fact that pasteurized cured meats are usually remarkably free from microbial spoilage. Perigo et al. (1967) found evidence that, on heating during the pasteurizing process, nitrite reacts with some component of the medium to produce a substance which is strongly inhibitory to the growth of Clostridia. This may help to explain the fortunate – but unexpected – stability of this type of product. It is evident that the efficacy of a given pasteurization treatment, at constant levels of salt and nitrite, in controlling the growth of Cl. botulinum, varies between different portions of the pig carcass and between individual animals (Gibson et al., 1982). Breed appears to have no consistent effect.

Spoilage of canned hams may be attributed to various associations of bacteria, including faecal streptococci of porcine origin. The organisms responsible can be classified (Table 7.15: after Mossel, 1955). The frequent presence of micrococci (which have only normal heat tolerance) indicates that under-processing often occurs (Ingram and Hobbs, 1954).

Although it is not possible to sample the interior of a ham during heating to determine the bacterial count, the process has been followed in ham which has been minced under sterile conditions. Typical data are given in Table 7.16 (after Grever, 1955). No actual increase in bacterial numbers occurred; and after the centre temperature reached about 45–55 °C there was a definite decrease in numbers. Apart from ensuring that the meat used in the preparation of semi-preserved products is kept cool before canning, and has the minimum bacterial load deriving from the animal or from slaughter operations, it is important to ascer-

Table 7.15 Classification of bacteria causing spoilage in pasteurized canned meat products

Degree of resistance to pasteurization	Genera or species	Type of spoilage
(1) Occasional	Salmonella	Human disease – public health risk
	Lactobacillus	Greening
	Achromobacter pseudomonas	Putrefaction
	Coliforms	Swelling
(2) Frequent	Strept. faecalis	Off-flavours
	Micrococcus	Swelling
(3) Complete	Bacillus	Swelling and off-flavours
	Clostridium	Swelling

Table 7.16 Pasteurization of hams at 75 °C: aerobic counts
of nutrient agar

Pasteurizing (h)	Temperature of ham (°C)	Bacterial count
0	10	1.25×10^6
1	15	2.0×10^6
2	30	2.1×10^6
3	44	1.15×10^6
4	55	2.5×10^5
5	62	7.9×10^3
6	66	700
7	69	<10

tain that the minor ingredients which may be used – such as spices, condiments, curing salts, sugar, milk powder – are sterile: they frequently harbour micro-organisms which may resist canning. Some of these may be harmful, for example spore-bearing anaerobes in spices. Black pepper, pimento and mustard seed, in their natural state, are reservoirs for moulds of *Penicillium* and *Aspergillus* spp. The latter can include *Aspergillus flavus* (Hadlok, 1969). On the other hand, some spices, because of their essential oil content, have bactericidal properties. Thus the antibacterial action of cloves is due to eugenol and that of mustard seed to allyl isothiocyanate.

Many muscle enzymes are inactivated in the temperature range used in pasteurizing, especially the more complex ones such as hexokinase: others such as creatine kinase are not inactivated until a temperature of 60 °C is reached: but an enzyme such as adenylic kinase can stand a temperature of 100 °C at pH 1 and, clearly, would still be operative (Dixon and Webb, 1958). Fortunately, changes effected by the latter are of minor importance. Nevertheless, it is obvious that in semi-preserved meats, wherein the temperature is not raised much above 60 °C, there may still be residual enzymic activity. This could be undesirable even though the microbial status of the product was satisfactory.

From the point of view of minimizing damage to texture, it is preferable to administer the dose of heat required for stabilizing the microbial status of the product by a short period at high temperature, rather than by longer period at lower temperature (Ball, 1938). In the future, the severity of the heat may be minimized if it is combined with the use of irradiation, infrared treatment or high pressure.

7.2.2 Sterilization

The majority of canned meats are 'commercially' sterilized, i.e. they are processed to the point at which most micro-organisms and their spores have been killed: this permits more or less indefinite storage life in the can, at any ambient temperature, provided it is kept sealed; but the product is markedly different from fresh meat, and may alter chemically and physically in the course of time. Canned meat has remained edible for 114 years (Drummond and Macara, 1938). In the early days of canning, the meat products were heated in an open water bath. Under these conditions, the temperature of the cans failed to attain 100 °C and a long processing time was necessary to achieve commercial sterility. Increasing the boiling point of the water by adding salts such as calcium chloride made possible a great reduction in the processing time. By 1874 a controllable pressure steam retort had been invented; and between 1920 and 1930 information on the heat resistance of bacterial spores, and on heat penetration into cans, permitted the preparation of time–temperature processing schedules, to control the canning process instead of relying upon empiricism (Howard, 1949). The concept of thermal death time (TDT) proved most useful in evaluating the efficacy of thermal processes (Brigelow and Esty, 1920). It is also referred to as the Fahrenheit value (F), with a subscript to represent temperature, the unit F_0 being most frequent and representing the time, in minutes, needed to achieve sterility at a temperature of 250 °F (121 °C). The decimal reduction value (D) is the time in minutes required to kill 90 per cent of the bacteria cells at a given temperature. By plotting the log of F or D against heating temperature, a straight line is obtained. Its gradient is known as the z-value, and it may be defined as the number of Fahrenheit degrees by which the temperature must be raised to obtain a ten-fold increase of death rate of bacterial cells. For spores of *Cl. botulinum* and of some other sporulating *anaerobes* (in phosphate buffer) z has a value of 18 (Doty, 1960).

As in the case of semi-preserved meats, underprocessed or faulty cans are liable to microbial spoilage, which may take various forms. A frequent result is the production of gas in sufficient quantities to swell or 'blow' the can (cf. Table 7.15). Although its pH is generally on the acid side of neutrality, meat is regarded as a low acid food. Since the most lethal food-poisoning organism, *Cl. botulinum*, has a lower limit of growth at pH 4.5, all foods such as meat which support its growth are given heat treatment sufficient to destroy it. Its F_0 value is 2.8; and at 100 °C the toxin is destroyed in 10 min. The presence of curing ingredients in products such as canned hams makes them less liable to harbour *Cl. botulinum* and this permits pasteurization (Halvorson, 1955). *Cl. sporogenes*, another spore-forming organism capable of growth on meat, is more heat resistant than *Cl. botulinum* and is used to evaluate heat processing (Desrosier, 1959). Certain thermophilic bacteria capable of withstanding very severe heat treatment may also be present. To eliminate these would require a degree of processing which would seriously affect the meat and lower its nutritive value. These organisms are controlled, as far as possible, by avoiding initial contamination (Howard, 1949). The thermo-resistance of bacteria in meat products appears to bear some relation to the type of meat. Thus *S. faecalis* is more resistant to thermal inactivation in comminuted salt pork than in the corresponding beef product (Zakula, 1969). The effect has been attributed to differences in the a_w.

There is evidence that the thermal stability of bacterial enzymes is determined more by their intracellular environment than by their molecular structure. Thus, when the luciferase activity of *vibrio fischeri* was expressed in strains of *E. coli*, *S. Typhimurium*, *Listeria monocytogenes* and *Brochothrix thermosphacta* (following introduction of the *lux* AB genes via the plasmid pSP13), it was found that the thermostability of the enzyme was determined by the internal environment of the microorganism in which it was located (and not directly by the relative thermostability of that organism) (Mackey *et al.*, 1994).

Since most proteins are denatured by heat (Putnam, 1953), sterilized canned meats suffer considerable change in the process. There is an increase in free –SH groups (Bendall, 1946) and the proteins may coagulate and precipitate. The texture of canned meat after sterilization is thus more like the cooked than the fresh commodity. If heat treatment is excessive, marked deterioration in aesthetic appeal and eating quality occurs. Moreover, since meat (especially pork) contains appreciable quantities of thiamin (vitamin B_1) and ascorbic acid (vitamin C) and these are destroyed by heat, the nutritive value of the canned product will be lower than that of fresh meat. It must be remembered, however, that meat is not primarily eaten for its vitamin content, and that these vitamins would, in any case, be largely destroyed in cooking. The loss of such labile nutrients will be exaggerated if the cans are subsequently stored for long periods at high ambient temperatures. The colour of canned meats will also tend to resemble that of the cooked commodity, since the high temperatures will change the red pigment (myoglobin) to brown myo-haemochromogen (Lemberg and Legge, 1949). If the interior of the cans is not lacquered, there may be discoloration due to the reaction of H_2S (produced from the meat proteins) with the plate metals (Howard, 1949).

Except, of course, where there is microbial spoilage, flavour changes during canning are not generally a problem since, as has been mentioned, canning represents a degree of cooking and meat is generally cooked before consumption.

There is some suggestion that the actual biological value of the proteins of meat may be lowered if processing temperatures are mainained at 113 °C for periods longer than about 5 min (Beuk *et al.*, 1949).

A degree of heat adequate to kill micro-organisms, but insufficient to seriously damage texture, has been achieved by agitation of the cans during the canning process (Fischer *et al.*, 1954) and by flash-heating the meat to temperatures *ca.* 150 °C (Ball, 1938) *before* placing it in sterile cans (Martin, 1948). The latter two processes are not readily applicable to meats, however; but the former process has resulted in a 50 per cent reduction in the heating time required for corned beef and pork (in a loose mix with muscle juices: Jul, 1957).

A further advance in achieving commercial sterility with minimal concomitant damage to the product has been due to the development of heat-sterilizable flexible bags (retort pouches). These are produced from multiple laminates which are hermetically sealable. A typical 4-ply laminate for this purpose could consist of 12 μm polyester/12 μm Al foil/12 μm polyester/70 μm polyolefin (Paine and Paine, 1983). The retort pouch offers several advantages. Thus, the thermal process time tends to be significantly shorter than that for metal or glass containers (which minimizes loss of quality in the product); the shelf-life is comparable to that of a frozen product without the need for a frozen chain for storage and distribution; there is less container/product interaction; and, by permitting 'boil-in-the-bag' (where this operation is appropriate), the retort pouch speeds preparation and serving (Paine and Paine, 1983).

7.2.3   Novel thermal generating procedures

Several other procedures for the thermal processing of meat (and other foods) have become available, which achieve the required degree of microbial control with minimum damage to the nutritive and organoleptic properties of the meat (Hugas *et al.*, 2002). Thus, in ohmic heating, an elevated temperature is developed by passing an electric current through the meat (which has a high resistance). The process permits continuous production, without heat transfer, and pasteurization or sterilization at relatively low temperature. Ohmic heating, however, can be applied only to pumpable products and is costly.

In dielectric heating, high-frequency radiowaves (1–100 MHz) cause oscillation of water molecules in the meat and generate heat by friction. At even higher frequencies (300 MHz to 300 GHz) – microwave heating – the temperature required to decrease microbial loads, with minimum damage to the product, can be achieved in minutes rather than hours (cf. § 10.3.3.3).

Chapter 8

The storage and preservation of meat:
II Moisture control

From the considerations in § 6.3 it is apparent that deprivation of available moisture can not only prevent the growth of the micro-organisms found on meat but may also kill them. Water may be made unavailable by direct removal, as in dehydration and freeze dehydration, or by increasing the extracellular osmotic pressure, as in curing. With these processes the prevention of microbial change and the preservation of edibility involves the creation of a commodity which necessarily differs more from the fresh meat than does refrigerated meat, although on subsequent cooking these differences in nature are less apparent.

8.1   Dehydration

As with refrigeration, the efficacy of drying had been empirically recognized in meat preservation since very early times. The idea that dehydration could preserve muscular tissue was obviously known in Egypt 5000 years ago and utilized in preparing mummies. A variety of dried and cured meats was available 3000 years ago: indeed, muscle fibres can still be recognized in material from Jericho (A. McM. Taylor, personal communication).

Early drying procedures, which are still continued in remoter areas of the world, involved exposure of strips of lean meat to sunlight, as in the manufacture of pemmican by North American Indians, or a combination of light salting followed by air drying, as in the preparation of charqui (in South America) and biltong (in South Africa) (Sharp, 1953). Such products are considerably different from fresh meat – and to most tastes lower in eating quality. Their preservation may be enhanced by fermentation via halophilic micro-organisms.

The large-scale commercial production of dehydrated meat in a form which, when cooked, was similar in nutritive value and in palatability to the fresh commodity resulted from research carried out during the Second World War (Dunker et al., 1945; Sharp, 1953). To effect the necessary degree of moisture removal by the current of hot air used in the process, it was essential to have a high surface : volume

Table 8.1 Comparative density of different forms of beef (after Sharp, 1953)

Commodity	ft³/ton[a]	ft³/ton meat solids
Frozen quarters	95	264
Frozen quarters (boneless)	78	173
Canned corned beef	54	157
Dehydrated beef (compressed to specific gravity 1)	54	75

[a] $1 \text{ ft}^3/\text{ton} \approx 27 \times 10^{-6} \text{ m}^3/\text{kg}$.

ratio in the meat: minces were, therefore, used. The use of raw, minced meat proved unsatisfactory, since it quickly became case-hardened, thus opposing further moisture removal. By first cooking the meat in slices, however, mincing it and then drying it under carefully controlled conditions (the temperature being kept below 70°C: Dunker et al., 1945; Sharp, 1953) a product could be prepared which was almost indistinguishable in flavour and texture from raw minced meat, when comparisons of the fully cooked meats were made. Although minced and precooked dehydrated meat obviously cannot serve the same purposes as fresh meat, it was designed to conserve the essential nutrients and eating quality of the latter without refrigeration and in small bulk – against emergency conditions. It was never issued as such to the domestic consumer, although it was utilized for certain manufactured meat products; but it proved most useful for the Armed Forces.

The data in Table 8.1 illustrate the saving in space involved. It is important to note that although 1 ton of both canned corned beef and dehydrated beef occupies 54 ft³, the density of essential nutrients is more than twice as high in the latter, since canned beef contains about 53 per cent water and the dehydrated meat quoted contained a mean of 7.5 per cent water (Sharp, 1953).

Before outlining some of the considerations involved in preparing a dehydrated meat of satisfactory quality, it is desirable to mention briefly some of the implicit biochemical aspects.

8.1.1 Biochemical aspects

The difference between dehydrated and fresh meat will obviously be minimized if the water removed from the former can be incorporated again on rehydration. The degree of reincorporation depends on the surviving water-holding capacity of the muscle – in terms both of microscopic structure and of the chemical state of the muscle proteins.

Loss of water from both raw and precooked meat is accompanied by diminishing space between groups of muscle fibres and between the individual fibres and by a progressive reduction in muscle fibre diameter (Wang et al., 1953). The rate of moisture removal and of muscle fibre shrinkage is more rapid with precooked than with raw meat and proceeds further. Potassium was found to accumulate on the periphery of the dehydrated muscle fibres. Since this circumstance would denature the proteins in this location, thus obstructing re-entry of water on rehydration, experiments on muscle from which potassium has been removed by electrolysis

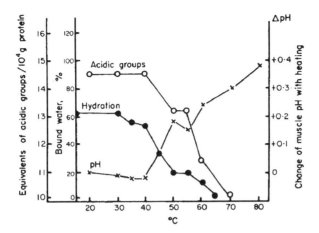

Fig. 8.1   The effect of temperature on degree of hydration, pH and acidic groups of beef muscle homogenates (Hamm, 1960). (Courtesy Prof. R. Hamm.)

were carried out (Wang *et al.*, 1954a). On rehydration, electrolysed muscle absorbed about twice as much water as control material: this was attributed to its pH (2.8) which was considerably on the acid side of the iso-electric point. This procedure could not be developed in practice. Ease of rehydration is said to be enhanced by the use of irradiated meat (M. C. Brockmann, personal communication).

With hot-air drying procedures, lack of rehydratability is largely due to changes similar to those occurring during heat denaturation. In the range 0–20 °C (as measured by the amount of so-called bound water; Grau and Hamm, 1952) the water-holding capacity of meat decreases with increasing temperature (Wierbicki and Deatherage, 1958) presumably by its effect upon the sarcoplasmic proteins (cf. Table 5.3). Between 20–30 °C there is no change in the degree of hydration (Fig. 8.1). Between 30 and 40 °C polypeptide chains in the muscle protein unfold and new electrostatic or hydrogen bonds form (Wierbicki and Deatherage, 1958), and there is a slight fall in the degree of hydration at the iso-electric point (about 5.5). Although the heat coagulation of isolated actomyosin *in vitro* starts at about 35 °C (Locker, 1956), muscle proteins are more stable *in situ* (Engelhardt, 1946), and marked changes in hydratability do not occur until the temperature rises above 40 °C (Fig. 8.1). Between 40 and 50 °C, however, there is a loss in water-holding capacity which is associated with a corresponding diminution in the titratable acidic groups. There is a further decrease in the water-holding capacity between 50 and 80 °C, but this is less marked (although loss of acidic groups continues). Above 80 °C, free H_2S begins to form and increases with increasing temperature (Tilgner, 1958). The loss of free acidic groups explains the considerable rise in the pH of the meat (cf. Fig. 8.1) and the iso-electric point of the muscle changes to higher pH values (Fig. 8.2), thus tending to offset the increase in water-holding capacity which the higher pH would normally cause (Hamm, 1960). It is important to note that relative differences in the water-holding capacity of fresh meat are retained after heating. For example, meat of high ultimate pH which has a high water-holding capacity when fresh, has also a higher water-holding capacity after heating than that of normal ultimate pH (Bendall, 1946; Hamm and Deatherage, 1960a). This circumstance is important in the production of frankfurters (Grau, 1952) and canned ham (Koeppe, 1954).

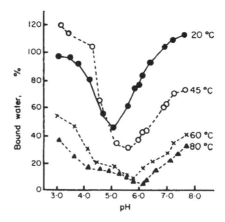

Fig. 8.2   The effect of temperature on the pH hydration curve of beef muscle of ultimate pH 5.5 (Hamm, 1960). (Courtesy Prof. R. Hamm.)

The collagen of associated connective tissue also changes as the temperature is raised. Thus, at about 60–65 °C collagen is transformed into more soluble form. At somewhat higher temperatures the latter swells and softens as it takes up water and finally disintegrates, forming gelatin. The latter does not occur appreciably below 100 °C. These changes in collagen with heating will tend to increase the water-holding capacity of the meat (Hamm, 1960), but the reactions of sarcoplasmic and myofibrillar proteins are quantitatively much more important and clearly the lower the heating temperature during drying the less will be the loss of water-holding capacity and the greater degree of reconstitutability.

8.1.2   Physical aspects

Various physical factors determine the efficacy of a hot-air draught in dehydrating precooked minced meat.

In the first place, while the size of the meat particles is not important in itself, within the range 0.3–0.8 cm diameter it influences the density of loading of the drying trays and thus has an effect on the time of drying. Thus the time taken for finely minced particles to reach a water content of 5 per cent was 3, 5 and $6\frac{1}{2}$ h with loading of 1, 2 and 4 lb/ft² (0.04, 0.08, and 0.16 kg/m²) respectively (Sharp, 1953).

As will be apparent from § 8.1.1, the temperature of drying is important. Heat damage is characterized by toughness, grittiness and burnt flavour. Whilst the water content is still about 77 per cent air temperatures of 80 °C can be tolerated for 2 h without loss of quality; but even 50 °C can produce some deterioration when the water content is low. Nevertheless, quality is maintained at a satisfactory level when the drying temperature is 70 °C throughout the process (Dunker et al., 1945; Sharp, 1953). A typical drying curve showing water content and meat temperature is given in Fig. 8.3 (Sharp, 1953). The operating conditions refer to an air speed of 600 ft/min (180 m/min), an air temperature of 60 °C, a relative humidity of 40 per cent and a tray loading of 2 lb/ft² (0.08 kg/m²). It will be observed that the temperature of the meat remains relatively low (at the web bulb temperature) until about half the moisture has been driven off. It is at this same point that a relatively high fat content

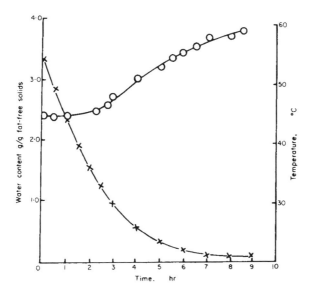

Fig. 8.3 Typical drying curve for meat. Drying temperature 60 °C (dry bulb), 44 °C (wet bulb, R.H. 0.40 (Sharp 1953). ×–× = water content. O–O = meat temperature. (Courtesy the late Dr J. G. Sharp.)

begins to retard the rate of drying, although it has little influence initially. This results in a serious increase in the time of dehydration if the fat content is above about 35 per cent of the dry weight; and where it is above 40 per cent of the dry weight the spongy texture of the dry meat can no longer hold the molten fat and it drips away (Sharp, 1953; Prater and Coote, 1962). On the other hand, when the fat content is below about 35 per cent of the dry weight, dehydrated meat of satisfactory water content can be obtained in a continuous hot-air drier and at a fixed drying time, despite considerable variability in the fat content. In general, however, a low-grade meat from a leaner type of carcass is preferred for the preparation of dehydrated meat in these circumstances.

The degree of precooking is an important factor. If the meat is overcooked its connective tissue framework will be changed to gelatin, and although it will give dry granules which reconstitute quickly, it will break down under compression (§ 8.1.3). An undercooked meat, however, will have a slow drying rate and a slow rate of reconstitution, yielding a dry and brittle texture. Since the aqueous liquor exuding from the meat during the precooking period contains various soluble substances, it must be returned to the cooked meat before dehydration commences to retain the full meat flavour and nutritive value of the fresh commodity. Any fat rendered out on cooking may be returned to the meat according to the fat content desired.

8.1.3 Organoleptic aspects

For long-term storage, such dehydrated meat must be compressed to exclude pockets of air or moisture and kept in an airtight and moisture-proof container, preferably a tin-plate can (Sharp, 1953), since its finely divided porous state makes it specially liable to attack by oxygen. Non-oxidative changes, whether enzymic or chemical, are of secondary importance except at high storage temperatures. Thus, while restriction of oxygen will maintain the flavour of dehydrated meat for 12 months or longer at 15 °C, non-oxidative deterioration may develop under nitrogen at 37 °C.

In the *absence* of oxygen the main changes during storage are caused by the Maillard reaction (Henrickson *et al.*, 1955; Sharp, 1957), wherein carbonyl groups of reducing sugars react with the amino groups of proteins and amino acids nonenzymically. A dark brown coloration and a bitter, burnt flavour develop. Glucose and glucose-6-phosphate are formed in muscle post-mortem, the rate of formation being much higher in pork and rabbit than in beef. In the former two species the breakdown of muscle glycogen by α-amylolysis is more intense (Sharp, 1958). The rate of nonenzymic browning increases with pH (Sharp, 1957) and is not inhibited above 37 °C unless the moisture content of the meat is 2 per cent or less. Dehydrated raw meat deteriorates to a greater extent than the cooked product, possibly because the residual amylolytic activity is greater. To keep the reactants at the lowest possible concentration, meat must be dehydrated immediately after the death of the animal or held for some time below –10 °C before dehydration. The removal of glucose by glucose oxidase is not practicable commercially. Because of the Maillard reaction, dehydrated meat can become unpalatable in 6 months when kept at high temperature (Sharp and Rolfe, 1958). Concomitant losses in the water-holding capacity of the proteins cause brittleness of texture. The storage life can be extended considerably, however, by drying to very low moisture contents.

In the *presence* of oxygen, the storage of dehydrated meat (of high moisture content) at high temperature causes it to become pale and yellow, due to the conversion of myoglobin to bile pigments. A mealy odour develops and fat oxidation occurs giving rise to paint-like odours. In dehydrated raw meat there is still considerable lipolytic action (Table 8.2). Precooking the meat before drying greatly reduces, but does not destroy, its lipolytic activity which is greater the higher the storage temperature and the residual moisture content. Fat rancidity does not develop when the moisture content is reduced to 1.5 per cent, but at such a level flavour and texture are likely to be seriously affected. The stability of the fat to oxygen is increased by the incorporation of antioxidants such as gum guaiac or ethylgallate in the meat during precooking (Anon., 1944). But other changes then become apparent. These are characterized by the mealy odours already mentioned (Tappel, 1956). Rancidity may also be avoided by gas-packing the dehydrated meat: only a slight odour, like that of crab meat, is then apparent and the latter disappears during reconstitution and cooking. Dehydrated cured meat is especially liable

Table 8.2 Free acidity of fat in air-dried raw beef after 12 months' storage (Sharp, 1953)

Moisture content	(As per cent oleic acid) Temperature of storage	
	20 °C	37 °C
7.5	17	36
5.0	12	24
3.2	6	13

Table 8.3 Loss of thiamine in dehydrated pork stored at various temperatures

Storage temperature (°C)	Per cent thiamin retained		
	7 days	14 days	21 days
−29	100	100	100
3	100	100	96
27	–	89	77
37	70	55	43
49	15	7	0
63	4	0	0

to undergo oxidative rancidity because of the production of a pro-oxidant during curing.

Generally the moisture content of dehydrated meat is too low to permit bacterial growth, but if it rises above 10 per cent mould growth may occur after some weeks (generally, *Penicillium* and *Aspergillus* spp.).

Ideally, foods undergoing dehydration should be heated at a temperature such that microbial growth cannot occur before the moisture content has dropped sufficiently low to prevent it. It appears, however, that even if *Cl. botulinum* were present, it could not form appreciable amounts of toxin during the period of dehydration (Dozier, 1924). Precautions must be taken, however, to prevent contamination during the period of reconstitution.

As with thermally processed meats, the thiamin content of dehydrated meat diminishes during storage, especially at high temperatures (Table 8.3: Rice *et al.*, 1944).

In general, when compressed into blocks of specific gravity 0.8 to 1.0 in sealed cans under nitrogen, dehydrated pork, mutton or beef will keep without deterioration in flavour and odour for 3 years or more at moderate storage temperatures (Sharp, 1953) although the degree of reconstitution decreases after about 12 months (Grau and Friess-Schultheiss, 1962).

8.2 Freeze dehydration

The necessity of using cooked mince to produce a satisfactory product with air drying procedures has been mentioned above: the possibility of vacuum drying, with advantages in quality resulting from the reduced heat treatment required, were realized in commercial operations at the end of the Second World War when vacuum contact-dehydration (VCD) plant was developed in Denmark (Hanson, 1961). Nevertheless, although the VCD method permitted technological advances, the meat so processed was difficult to reconstitute. The possibility of removing water from meat by sublimation from the frozen state rather than by evaporation of liquid had been apparent for some years, since freeze drying was a well-known procedure in the production of highly priced pharmaceuticals and biological materials in laboratory scale operations. It has been appreciated as the mildest method known for drying meat (Wang *et al.*, 1954b; Regier and Tappel, 1956). No satisfactory way of applying the process to production line operations was initially foreseen. Between 1955 and 1960, however, a large-scale process for freeze drying was developed in the United Kingdom. It was fully described by Rolfe (1958). Since it incorporated plates to enhance heat exchange during the initial phase of sublimation and to supply heat to them to aid drying during the second phase, the process was called accelerated freeze drying (AFD). AFD permitted the freeze drying of hundreds of kilograms of *raw* meat per run: in only 4 h the moisture content could be reduced to about 2 per cent. Moreover, whole steaks (about 1.5 cm thick) could be processed; and they rehydrated quickly and easily. In short, meat could be handled in portions similar to those available to the consumer of fresh meat – or of that prepared by refrigeration, curing, canning or irradiation. The low operating temperatures and speed of the process, the avoidance of translocation of salts, etc., during drying and the honeycomb texture created by the direct sublimation of ice from the minute interstices of the tissue, caused little damage to the meat proteins. The quality of the product thus resembled fresh meat. The use of spikes, to conduct the heat into the centre of the drying tissue, permitted the freeze drying of $2\frac{1}{2}$ lb (1.1 kg) roasts in 5–8 h (Brynko and Smithies, 1956).

8.2.1 Histological aspects

If operated at a low plate temperature, the freeze drying process causes relatively little gross histological change (Wang *et al.*, 1954b; Luyet, 1962). Whereas evaporation of water from the fluid phase involves the movement of salts with the water front towards the surface of the meat, with consequent distortion of the muscle, sublimation of water vapour occurs direct from each crystal nucleus of ice during freeze drying; and any distortion is limited to the areas around these nuclei (Luyet, 1961, 1962). Such distortion obviously depends on the size and number of the ice nuclei, which, in turn, is a function of the rate of freezing (§ 7.1.2.1). The smaller they are the less will be the histological damage, and the finer the honeycomb of air spaces left after sublimation. With a sufficient speed of freezing, ice crystals will form inside the myofibrils between the molecules of actin and myosin causing no alteration in structure, even at the level of observation given by the electron microscope (Menz and Luyet, 1961). It is obvious, of course, that such rates of freezing are impossible with even the smallest piece of meat encountered by the consumer. In any case, even with much slower rates, where ice crystal size and distribution is *relatively* gross, the honeycomb structure left on sublimation can be too fine to permit easy rehydration (Wang *et al.*, 1954b), because air bubbles are trapped within the structure. The pre-frozen steaks used in the AFD process have a relatively open honeycomb structure which facilitates drying (Rolfe, 1958; Hanson, 1961). They also rehydrate easily. If attempts are made to freeze-dry steaks which have been placed whilst unfrozen into the vacuum chamber, their water will tend to freeze by evaporative cooling, and, by causing local concentrations of salt on the surface, will create thereon a skin of denatured, protein (cf. § 7.1.2.1) which will oppose re-entry of water (Rolfe, 1958). On the other hand, if a small quantity of water forms in the meat by thawing between pre-freezing and placing in a vacuum chamber, it quickly re-freezes and sublimes satisfactorily. Nevertheless, if the rate of such re-freezing is exceptionally fast, very small intracellular ice crystals will form causing a texture which will oppose rehydration (Wang *et al.*, 1954b).

Even when operated under optimum conditions, the AFD process appears to cause some change in meat: when homogenized the myofibrils tend to adhere along their length as observed at the histological level (Voyle and Lawrie, 1964). It is thus interesting to note that the sarcoplasmic reticulum, which is located between the myofibrils (§ 3.2.2), will no longer stain with the Veratti reagent after freeze drying, although electron microscopy indicates that the structure is still present (Voyle and Lawrie, 1964).

If the plate temperature is below 60 °C the histological appearance of the reconstituted muscle is similar to that of fresh (Aitken *et al.*, 1962): above this temperature there is progressively greater histological change which is paralleled by difficulty in reconstitution.

8.2.2   Physical and biochemical aspects

Unlike the heating involved in hot air dehydration (cf. Fig. 8.2), freeze dehydration will not change the iso-electric point of muscle if operated under optimum conditions. Moreover, instead of lowering the water-holding capacity generally, it tends to do so only at the iso-electric point (Fig. 8.4). According to Hamm and Deatherage (1960b), this loss of water-holding capacity is not due to the freezing aspect of the freeze-drying process. The plate temperature during the final phase of drying is implicated. Although the latter has little effect on the total moisture content after reconstitution (Table 8.4), it affects the amount of moisture which is firmly bound (here defined as that remaining after centrifuging).

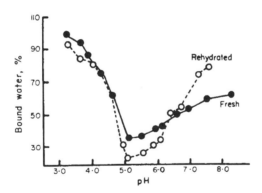

Fig. 8.4   The effect of freeze drying on the pH hydration curve of beef muscle (Hamm, 1960). ● = fresh muscle. ○ = freeze dried and rehydrated muscle. (Courtesy Prof. R. Hamm.)

Table 8.4   The effect of plate temperature on various characteristics of freeze-dried meat after rehydration (Aitken *et al.*, unpublished)

Characteristic	Control (frozen)	Temperature of plate (°C)				
		20	40	60	80	100
Total moisture (per cent)	77.2	75.4	75.9	75.5	76.8	75.3
'Bound' water (per cent)	65.3	63.8	63.9	63.1	62.8	53.8
Meat solids (per cent) in reconstitution water	12.3	10.1	9.9	10.9	10.3	5.4

Table 8.5   The percentage of myofibrillar protein soluble in
9.95 M KCl-glycerophosphate buffer, pH 6

Muscle	A. Control (low ultimate pH)		B. Adrenaline treated (high ultimate pH)	
	Frozen	Freeze dried	Frozen	Freeze dried
L. dorsi	53	41	91	85
Psoas	49	40	80	68
Biceps femoris	55	46	88	75
Semimembranosus	54	46	85	77

There is a distinct drop between 80 and 100 °C in the latter. Moreover, the percentage of total meat solids which is leached out by the reconstitution water is considerably less at a plate temperature of 100 °C, indicating that there has been heat denaturation of sarcoplasmic proteins causing their insolubility. The contribution of sarcoplasmic proteins to water-holding capacity is also indicated by the effect of the time of reconstitution of freeze-dried meat on its bound water content. This decreases appreciably at times of reconstitution greater than 1 h, due to the leaching out of sarcoplasmic proteins. When the plate temperatures are about 20–30 °C, the accelerated freeze-drying process does not affect the extractability of the myofibrillar proteins at any pH between 5 and 7 in comparison with frozen material (Scopes, 1964). On the other hand, where the plate temperature is about 60–70 °C (as is normal commercially), there is a drop in the solubility of the myofibrillar proteins, indicating some measure of denaturation (Table 8.5: Penny, Voyle and Lawrie, unpublished data), and a change in the electrophoretic pattern, indicating some alteration in the actomyosin complex (Thompson et al., 1962).

Clearly, in the AFD process under commercial conditions, there is some loss of water-holding capacity attributable to the effect of the plate temperature on the myofibrillar and sarcoplasmic proteins; and these changes may explain the woodiness in texture already referred to. Occasionally, even with optimum operating conditions, excessive woodiness and difficulty of reconstitution is encountered. This can be explained on the basis of exceptionally severe denaturation changes in the sarcoplasmic and myofibrillar proteins of animals, wherein post-mortem glycolysis has been very fast or where the ultimate pH has been unusually low, for example in the PSE condition of pigs (§§ 3.4.3, 4.2.2, 5.4.1). The induction of a high ultimate pH by pre-slaughter injection of adrenaline will enhance the water-holding capacity of these proteins. This is manifested, in both cattle and pigs (Penny et al., 1963, 1964), by the behaviour of the myofibrils from the fresh meat at various environmental pH levels; and the differences between muscles of originally high or low ultimate pH are retained after freeze drying (Fig. 8.5). It is also reflected in easy reconstitution of the dehydrated muscles (Table 8.6) and organoleptically by the virtual absence of woodiness of texture (cf. § 8.2.3).

It is important to note that rehydration of freeze-dried meat having originally a low ultimate pH in a fluid of high ultimate pH does not enhance its water-holding

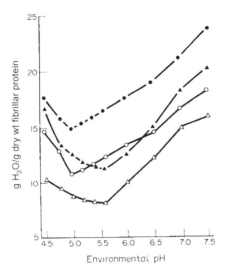

Fig. 8.5 The effect of environmental pH on water-holding capacity of myofibrils prepared from fresh or freeze-dried muscle of high or normal ultimate pH (Penny *et al.*, 1963). ● = fresh, ult. pH 6.7. ○ = rehydrated, ult. pH 6.7. ▲ = fresh, ult. pH 5.6. △ = rehydrated, ult. pH 5.6.

Table 8.6 Reconstitution ratios (g H_2O/g dry wt) of dehydrated muscle from adrenaline-treated (high ultimate pH) and control (low ultimate pH) steers

Muscle	High ultimate pH	Low ultimate pH
Semimembranosus	3.10	1.99
Biceps femoris	3.12	1.96
Psoas	3.58	2.08
L. dorsi (lumbar)	3.22	2.02
L. dorsi (thoracic)	3.34	1.98
Deep pectoral	3.65	1.95

capacity to the same extent as does the rehydration of freeze-dried meat having an originally high ultimate pH in water (Fig. 8.6). This may well reflect the fact that damage to sarcoplasmic proteins (and, to some extent, those of the myofibril) by the normal low ultimate pH attained during post-mortem glycolysis (Scopes and Lawrie, 1963; Bendall and Wismer-Pedersen, 1962; Scopes, 1964) cannot be completely overcome merely by raising the pH again. It may be that the protective action of high ultimate pH involves some change in the mode of chemical union between proteins and residual water. The high water-holding capacity of pre-rigor meat can also be retained by salting before the onset of rigor mortis and subsequently freeze-drying the commodity (Honikel and Hamm, 1978). The AFD process removes all the loosely held or capillary-condensed water in muscle (about 310 g/100 g dry protein), and much of the more tightly bound water (about 37 g/100 g dry protein) which is attached in two layers of differing bond strength to hydrophilic groups on the proteins (Hill, 1930; Hamm, 1960), since the residual moisture content

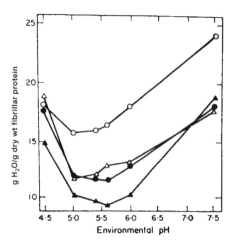

Fig. 8.6   The effect of adjusting the pH of homogenates of beef *biceps femoris* muscle, of high or normal ultimate pH, on water-holding capacity of myofibrils (Penny *et al.*, 1963). ○ = ult. pH 6.7. ● = ult. pH 6.7. ▲ = ult. pH 5.6. △ = ult. pH 5.6, adjusted to pH 6.7.

in the freeze-dried product is about 2.5 g water/100 g dry protein. In general, pork muscle is more susceptible to damage during freeze drying than in corresponding beef muscle.

That proteins may survive freeze dehydration in a substantially unaltered condition is indicated by the finding that the ATP-ase activity of actomyosin in rehydrated meat may still be 80 per cent of its initial value and that the muscle fibres (as observed histologically) will still contract on the addition of ATP (Hunt and Matheson, 1958). After isolation from muscle, the ATP-ase activity of the heavy microsomal fraction and its ability to accumulate Ca^{++} ions survive freeze drying (Diehl, 1966). On subsequent storage, however, calcium-binding power is lost, due to a structural deterioration of the sarcoplasmic reticulum membranes which comprise the heavy sarcosome fraction. This may explain the myofibrillar cohesion concomitantly observed in muscle samples which have been freeze-dried intact (cf. § 8.2.1). The mildness of the AFD process is further exemplified by the survival of oxymyoglobin. Myoglobin on the exposed surfaces of the steaks used in the AFD process is particularly susceptible to oxidation or denaturation, forming the brown pigments metmyoglobin or myohaemochromogen (§ 5.4.1). Yet spectroscopic study of aqueous extracts (Table 8.7) and of the meat surface (by reflectance measurements) shows that the spectrum is mainly that of oxymyoglobin, indicating that a considerable quantity of the myoglobin in the meat is not denatured by freeze drying. This is especially so at a low plate temperature (Penny, 1960a) suggesting that it is the temperature during the second phase of drying which is the important factor causing denaturation (Fig. 8.7).

8.2.3   Organoleptic aspects
It will be apparent from Table 8.7 that although the raw steaks processed by accelerated freeze drying retain much of the bright colour of fresh meat, there is a some-

Table 8.7 The effect of the AFD process on the proportion of myoglobin derivatives in aqueous meat extracts

	Metmyoglobin	Oxymyoglobin	Myoglobin
Fresh meat	5.2	91.8	3.0
Freeze-dried meat (60–70°C plate temp.)	30.7	69.3	0.0
Freeze-dried meat (30°C plate temp.)	14.7	85.3	0.0

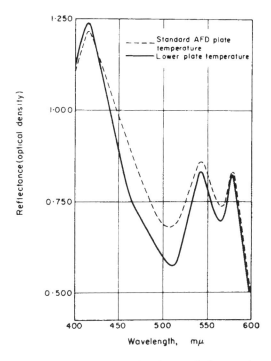

Fig. 8.7 The effect of plate temperature during freeze drying on the reflectance spectrum of dehydrated beef (Penny, 1960a). – – – – = standard plate temp. ―――― = low plate temp. (Courtesy Dr I. F. Penny.)

what higher concentration of brown metmyoglobin on the surface. Prolongation of storage exaggerates this discoloration, which becomes worse the higher the storage temperature and the longer the time of storage (Penny *et al.*, 1963). But, as in the case of the air-dried product (§ 8.1.3), non-enzymic Maillard browning (Sharp, 1957) is also involved. The reactants glucose and glucose-6-phosphate disappear even at a residual moisture level of 2 per cent, although the reaction is faster at higher moisture levels (Matheson, 1962). The non-enzymic nature of the changes will be evident in Table 8.8: cooking has little effect.

Other aspects of organoleptic deterioration can occur at extremely low residual moisture levels. These include fat oxidation following surviving lipolytic activity (Lea, 1938).

Table 8.8 The effect of freeze dehydration and storage of glucose and glucose-6-phosphate in raw and cooked pork

	Glucose (mg/g dry wt)		Glucose-6-phosphate (mg/g dry wt)	
	Raw	**Cooked**	**Raw**	**Cooked**
Fresh	4.9	3.6	3.0	1.5
Freeze-dried, initial	3.4	3.2	2.6	1.7
Freeze-dried, stored for 1 month at 37 °C				
(a) 2% residual moisture	1.0	1.4	1.5	1.0
(b) 10.5% residual moisture	0.4	0.4	0.0	0.0

Table 8.9 The sum of the tenderness[a] rankings given by eight tasters for control (low ultimate pH) and adrenaline-treated (high ultimate pH) beef before and after freeze dehydration (after Penny *et al.*, 1963)

	A. Control (low ultimate pH)		B. Adrenaline-treated (high ultimate pH)	
	Frozen	**Freeze dried**	**Frozen**	**Freeze dried**
Semimembranosus	23	24	12	16
Biceps femoris	26	25	11	15
Psoas	23	25	12	14
L. dorsi (lumbar)	19	21	19	17
L. dorsi (thoracic)	26	25	11	12
Deep pectoral	25	27	10	16
Total	142	147	75	90
No. times 'woodiness' detected	10	16	2	9

[a] i.e. low numbers in the table represent desirable degrees of tenderness and vice versa.

In comparison with fresh or frozen meat, freeze-dried meat is somewhat lower in tenderness and juiciness (Penny *et al.*, 1963). The data on tenderness are shown in Table 8.9 wherein this attribute is given for various beef muscles as assessed by taste panel: objective measurements of tenderness by tenderometer (Table 8.10) confirm taste-panel results.

Some of this difference is attributable to 'woodiness', although it will be seen that this characteristic is also noted in fresh (frozen) meat. The benefits of high ultimate pH (induced by pre-slaughter injection of adrenaline) in protecting muscle proteins and in enhancing their water-holding capacity are reflected in greatly enhanced tenderness and diminished 'woodiness' and these benefits are retained after freeze drying (Tables 8.9 and 8.10; Penny *et al.*, 1963). Results for pig muscles are similar (Penny *et al.*, 1964).

Table 8.10   Measurement of work done (ergs $\times$ 10^6/cm) by tenderometer in shearing beef

	A. Control (low ultimate pH)		B. Adrenaline-treated (high ultimate pH)	
	Frozen	Freeze dried	Frozen	Freeze dried
Semimembranosus	24.5	25.6	9.8	15.4
Biceps femoris	12.6	15.2	12.1	13.6
Psoas	12.9	12.4	9.2	9.5
L. dorsi (lumbar)	15.2	17.7	11.5	11.8
L. dorsi (thoracic)	12.9	17.2	6.9	9.1
Deep pectoral	18.6	19.2	14.8	18.3

Table 8.11   Effect of storage and freeze-dried beef *l. dorsi* muscle (after Penny *et al.*, 1963)

Storage temperature (°C)	Ultimate pH 5.6		Ultimate pH 6.7	
	−20	37	−20	37
Tenderness	3.4	2.0	4.3	4.1
Juiciness	4.3	3.1	4.4	4.4
Flavour	4.0	4.4	4.7	4.3
Shear force (ergs $\times$ 10^6/cm)	12.7	18.4	11.5	11.5
Reflectance (at 400 μm)	0.15	0.50	0.13	0.35

An interesting aspect of the adrenaline treatment was the differential effect it had on the various muscles in altering their resistance to shear, the *semimembranosus* showing the greatest increase in tenderness. The eating quality of freeze-dried meat falls both with increasing storage temperature and increasing moisture content (Thompson *et al.*, 1962). For instance, if held at 37 °C it is worse than at −20 °C – as manifested by tenderness, juiciness, flavour, shear force and non-enzymic browning (as measured by reflectance at 400 μm). Again, however, a high ultimate pH in the meat is beneficial not only in enhancing eating quality but in minimizing the storage changes (Table 8.11).

Apart from the question of non-enzymic browning, the percentage of metmyoglobin is increased by a higher temperature of storage; but this also is less marked when the ultimate pH is high (Penny *et al.*, unpublished data).

From the nutritive point of view, freeze drying does not alter the biological value of the meat proteins (Hanson, 1961) and, indeed, may enhance it (Adachi *et al.*, 1958). Although there is a loss of about 30 per cent of the thiamin content of the meat during freeze drying, this would occur on cooking in any case. The process causes a similar loss in riboflavin from mutton but not from beef or pork (Hanson, 1961).

Rehydration of freeze-dried meat with aqueous solutions of tenderizing enzymes, such as papain, helps to offset the somewhat adverse effect of freeze drying on tenderness (Wang and Maynard, 1955; Penny, 1960b). In the USA freeze-dried steaks are now available in a moisture-proof pack with a compartment containing dried proteolytic enzymes, the latter being added to the reconstitution water on rehydration.

Apart from its value under emergency conditions, freeze dehydration can provide a very palatable and nourishing meat which will resist spoilage for a considerable period without refrigeration in remote areas, and it has been successfully used during the ascent of Everest and in trans-Antarctic expeditions.* Because of its lightness and high protein content, it has proved valuable in space flights. The weight of most other forms of meat would seem to preclude their use in such a context. Although relatively high costs tended to depress the development of freeze drying, there has been renewed interest in the process (e.g. the production of freeze-dried beef for the Japanese market by Australia). As in the case of freeze-dried meat, the tenderness of osmotically dehydrated meat can be enhanced by its rehydration in solutions containing proteolytic enzymes (e.g. the proteases of *Asperigillus oryzae*) (Gerelt *et al.*, 2000).

8.3   Curing

The empirical observation that salting would preserve meat without refrigeration was made several thousand years ago. By 1000 BC salted and smoked meats were available (Jensen, 1949). The efficacy of the process, and of the many variants which have developed (including the use of sugar), arises primarily from the discouragement to microbial growth caused by the enhanced osmotic pressure in such products. As time has passed, cured meats have come to be valued for their organoleptic quality *per se* and there has thus been a tendency to lower the concentration of the curing ingredients. This has made these mildly cured or semi-preserved products more liable to spoilage and reintroduced the need for some degree of refrigeration. Recognition of the value of sodium nitrate in producing an attractive colour may well have been due to adventitious impurities in the sodium chloride employed. At the end of the nineteenth century it had become recognized that meat-curing brines contained nitrite, that this was the colour-fixing agent and that the nitrite was produced by a reduction of nitrate.

Preservation was originally effected by sprinkling salt on to the meat surfaces. In due course the meat was placed in brine; but both dry salt curing and tank curing have been practised until the present time. Vascular pumping, or multiple injection of salt solution, is now employed to hasten curing. Granulated salt was formerly called 'corn' and accounts for the term corned beef. It is of interest to note that the crystalline form and size of the salt used may affect its preservative properties. Smoking over a wood fire was originally employed to enhance the preservative action of curing, but it is now used mainly because people like the flavour of the smoked product.

---

* The rehydration of beef, dehydrated osmotically, in a solution of calcium chloride, enhances tenderness apparently due to its action in increasing myofibrillar fragmentation (Gerelt *et al.*, 2002).

As we have seen, the characteristics of the muscles of pork, beef and mutton are different (§ 4.3) and this is reflected in the organoleptic quality of the cured meats made from them. As a result, now that other methods of preservation are available, salted mutton and beef have fallen in popularity, whereas the various forms of salted pork, which are widely regarded as attractive, continue to be extensively prepared. Predominant consideration will, therefore, be given here to cured pork. Nevertheless, because of the limited supply and high price of pig carcasses in New Zealand, there has been renewed interest in the possibility of curing of mutton and lamb (Moore *et al.*, 1976–7). Lamb is far superior to mutton in this respect, and the product, although distinguishable from cured pork, is equally acceptable. (It is perhaps appropriate to mention that cured meats now face increasing competition from cured turkey, which has low fat content.)

The part of the pig's carcass which was most difficult to cure was the top of the hind limb where the depth of meat was greatest. Consequently prolonged curing was frequently required, sometimes up to 80 days (Jensen, 1949). Perhaps for this reason the methods employed for curing this area are particularly diversified. Where it is cut off from the side and separately cured it is referred to as ham. The names of the available types in the United Kingdom – e.g. Yorkshire, Suffolk, Cumberland, Bradenham, Belfast – indicate more or less subtle variations in the process. Where the area concerned is cured on the side it is referred to as gammon.

In Europe various dry-cured hams, which are eaten without cooking, are available, for example, Parma ham in Italy. During the long drying period, there is considerable proteolysis and lipolysis. These confer distinctive flavours on the products.

Before considering the biochemical factors involved in curing, it is desirable to outline the procedure used in the production of Wiltshire sides, since this underlies most other curing methods.

8.3.1   Wiltshire cure and variants

Much of the following description is based on that given by Callow (1934). Pigs, preferably well rested, are anaesthetized (by electrical shock or carbon dioxide: § 5.2) and bled. The carcass is placed into a scalding tank at about 63 °C to loosen the hair which is then generally removed mechanically.* The carcass is then singed to remove coarse hair, sprayed with cold water and cleaned. The backbone is chopped out during butchering operations and the carcass (now as sides) is cooled either at ambient temperatures or in a chiller. After cooling, the sides are trimmed. This involves removal of the *psoas* muscle, the scapula and the aitch bone (bones of pelvis). The trimmed sides are next chilled to the temperature of the curing cellar (3–7 °C) where curing takes place in four stages: (i) brine (pickle) is pumped into the sides, (ii) the sides are either sprinkled with dry salt or placed in a tank of brine, (iii) the sides are removed and stacked for some time in a maturing cellar (at 3–7 °C) and (iv) the sides may be smoked. Higher temperatures speed the curing and maturation processes but increase the danger of spoilage from the growth of undesirable bacteria and moulds. It has been suggested that the processing time for cured sides would be considerably shorter, and their microbial status better, if they

---

* Prescalding brushing of pig carcasses has been found to increase, not diminish, their microbial load (Rahkio *et al.*, 1992).

were stitch-pumped at body temperature immediately after slaughter, then placed in brine at $-10\,°C$ (Kassai and Kárpáti, 1963). Henrickson *et al.* (1969) compared the effects of curing pork with brine at $35\,°C$ (followed by swift chilling) with normal practice. The rate of salt penetration – and the extent of nitrosomyoglobin formation – were both somewhat greater in the former case.

For speed, the brine is introduced by pumping under a pressure of about $75\text{--}100\,lb/in^2$ ($500\text{--}700 \times 10^3\,N$). The concentration of sodium chloride in the injected brine (pump pickle) is about 25–30 per cent. It also contains 2.5–4 per cent of potassium or sodium nitrate, and in some cases 0.5–1 per cent sugar. About 18–25 injections are required, most being given to the gammon region. The total amount of brine injected is about 5 per cent of the weight of the side. The 'shoulder pockets' (scapula cavities) are filled with solid salt and the sides placed in curing tanks.

In a large tank the sides are stacked about 12 deep and lightly covered with sodium chloride and potassium nitrate in the ratio 10:1. They are battened down, brine is run in and the sides remain submerged in it for 4–5 days. The composition of this brine (tank pickle) is between 20 and 28 per cent with respect to sodium chloride and 3–4 per cent with respect to potassium nitrate when first prepared; but before it can be used in curing it must be seeded with specific, salt-tolerant micro-organisms. They are encouraged to grow by traces of protein leached from meat previously immersed in the brine and they are responsible for converting nitrate to nitrite. The latter is essential for colour fixation. The number of micro-organisms (and hence the nitrite content of the pickle) can be controlled by increasing the sodium chloride content of the brine if the nitrite is too high and by omitting the sprinkling of the sides if it is too low (Ingram *et al.*, 1947).

After removal from the brine tanks the sides are stacked in cellars for 7–14 days or longer. During this period, known as maturation, the sodium chloride, the nitrate and the nitrite become more evenly distributed throughout the musculature and the typical colour and flavour bacon develop.

The bacon thus produced may be consumed unsmoked (green), but probably the greater proportion is smoked for 2–3 days. Apart from adding flavour, the smoke contains phenols and phenanthrene derivatives. These have a preservative action and also delay rancidity in the fat on the surfaces of the bacon sides. If the temperature is allowed to rise too high during smoking, on the other hand, deep-seated bacterial growth may be encouraged.

The addition of nitrite obviates reliance on micro-organisms to produce it from nitrate, but the level must not rise above 0.05 per cent (cf. § 11.2). In this connection it may be mentioned that if all the nitrate added during traditional curing were converted to nitrite, the level of the latter would rise to 0.25 per cent (Eddy *et al.*, 1960). Fortunately, most of the nitrate appears to be changed in some other way or is not broken down (Eddy *et al.*, 1960). That bacon could be made by the direct addition of nitrite and without bacterial intervention was shown in the early 1940s (Brooks *et al.*, 1940).

Another development in the preparation of bacon is slice curing (Holmes, 1960). In this process slices of pork muscle 2–8mm thick are passed for 2–15min through a brine containing 8–10 per cent sodium chloride and 0.02 per cent sodium nitrite. Maturation occurs over a few hours. The process gives a uniform product in less than a day instead of the 10–21 days of the traditional Wiltshire process. The brines can be filtered under pressure to remove organisms which gradually accumulate, thus regenerating the brine and permitting a substantial saving in costs (Dyett, 1969).

Endeavours have been made to avoid the remote possibility of excess production of nitrite from nitrate in the traditional Wiltshire cure, by omitting the use of nitrate in the pickle and adding only about 200 ppm of nitrite. Such bacon is said to develop the flavour associated with the traditional cure provided it is permitted to mature for a reasonable period (e.g. upwards of 2 weeks). In such a cure the nitrite content clearly *cannot* exceed 200 ppm.

A more recent development in curing is the 'tumbling' or massaging of pork in rotating drums. In effect this procedure involves massaging the pieces of meat against one another in the presence of about 0.6% of their weight of salt. This draws out salt-soluble proteins (mainly actomyosin) to the meat surface and enhances the overall water-holding capacity when these gel on heating. NMR has elucidated the effect of tumbling on porcine muscle. The increase in water-holding capacity observed is related to the concomitant binding of brine (Dolata *et al.*, 2004). Tumbling can shorten the curing period to 24 h (US Pat. No. 3076713, 1960) partly by aiding the distribution of curing salts. If tumbling is carried out for more than 12 h, however, the binding strength of the ham tends to decline. The procedure has been increasingly applied to the production of ham (from the major muscles of the hindquarter) during the past 20–25 years. Most recently, tumblers which operate under vacuum have been used, whereby subsequent cooking losses are virtually eliminated (partly because dissolved air is minimized: R. S. Hannan, personal communication), and this makes it possible to produce ham with a greater brine content than in traditionally produced ham. The low fat content of the muscles now used for ham production also enhances the brine content of the muscular tissue.

In general, the composition of cured meat differs markedly from fresh (McCance and Widdowson, 1946). As would be expected, its ash content is high (about 5 per cent). There is also less water (about 45–55 per cent if tank cured and about 25 per cent if dry salt cured compared with about 75 per cent raw), less protein (about 14 per cent compared with 20 per cent) and considerably more fat. Moreover, although the modern tendency to use very lean meat counters the latter feature, the composition of cured products from tumbling procedures will include greater levels of salt.

8.3.2 Biochemical aspects

8.3.2.1 Curing

The biochemical mechanisms of curing were first extensively investigated by Callow (1932, 1933, 1936). During curing, the initial outward flow of water and soluble proteins from the muscle to the brine, by virtue of the higher osmotic pressure of the latter, is eventually reversed. This is because the salt, which diffuses inwards, forms a complex with the proteins of the meat which has a higher osmotic pressure than the brine. Normally, diffusion of sodium chloride into the muscles is rapid, equilibrium being established in about 48 h in 25 per cent brine (Callow, 1930). The slower the diffusion inwards, however, the longer is the period of outflow of water from the muscle. Slow inward diffusion is favoured by immersion of the meat in relatively weak salt solutions and by a close micro-structure in the tissue. The amounts of protein extracted are a function of salt concentration, the maximum being in

6–9 per cent brine. Smaller amounts are extracted by distilled water or brine of higher salt concentrations (Callow, 1931).

When meat is initially placed in curing brine, the exterior muscles will be exposed to a much higher concentration of salt than that established when equilibrium between meat and brine is subsequently attained. On the other hand, interior muscle locations will be subjected to a slow increase in salt concentration from physiological to equilibrium level. Since the behaviour of meat in curing brines would thus be expected to vary according to the relative position of its constituent muscles to the brine, Knight and Parsons (1988) undertook a detailed histological and biochemical study of the process. Their findings confirmed and extended those of Callow in the early 1930s. When isolated myofibrils were exposed to 1 M sodium chloride, they swelled maximally and A-band protein (mainly myosin) was extracted from them. Further increase in concentration to 5M was associated with inhibition of swelling and of protein extractability. Knight and Parsons (1988) also found that when myofibrils were exposed to solutions of increasing molarity in 1M steps, the extractability of protein was more complete when the muscles were exposed to 5M sodium chloride as Callow's more empirical observations had indicated (Callow, 1930, 1931). On the other hand, less protein was extracted from the A-band by 1M sodium chloride if the myofibrils had first been subjected to 5M brine, possibly because the latter had caused some denaturation of the myosin. Further aspects of water-holding capacity in salted meat are discussed in § 10.2.

The greater water-binding capacity of the salt protein complex is indicated by the fact that about 61 g out of the 73 g of water in 100 g lean pork could be expressed mechanically, whereas only about 26 g of the 63 g water in 100 g lean bacon could be so removed (Callow, 1927a). Studies on the electrical resistance of meat and bacon suggest that this difference arises because of the more swollen structure in the latter (Banfield and Callow, 1934, 1935). It is particularly interesting that the electrical resistance of the muscles of pork sides which have been cooled quickly and in which, therefore, post-mortem glycolysis is relatively slow (§ 4.2.2) is higher than that of those which have been cooled slowly and in which, therefore, post-mortem glycolysis is relatively fast. The high resistance of the muscles of rapidly cooled sides gradually falls to the level of that of slowly cooled sides over some 14 days' storage at $0\,°C$. It is conceivable that slow denaturation of the sarcoplasmic proteins (§ 5.4.1) changes in the binding of ions by the proteins (K^+ is taken up, Ca^{++} is slowly released on storage: Arnold et al., 1956) or the slow synaeresis of actomyosin filaments could be implicated. The degree of synaeresis or shortening of actomyosin filaments increases with the speed of onset of rigor mortis (§ 4.2.3). Its extent could be initially restricted by fast cooling; but shortening might slowly continue on storage to the point immediately attained in slowly cooled muscles. Any or all of these circumstances could lower the resistance of the fluid phase of the muscle; but the phenomenon has not so far been elucidated.

Quick chilling, whilst lowering shrinkage losses at this stage in processing, appears to be associated with an increase in the loss of fluid from the muscles during curing and maturation. As a result there is no overall gain (Gatherum, 1956, unpublished data; Jul et al., 1958).

Another circumstance causing high electrical resistance is a high ultimate pH (Callow, 1936). Partly because of the greater difficulty of salt penetration into such muscles, and of the direct effect of high ultimate pH in stimulating bacterial growth

(§ 6.3.3), bacon made therefrom tended to taint frequently when curing by the dry salt method was common. Indeed, tainted bacon was associated with factory killing but not with farm killing, for the journey from farm to factory caused sufficient fatigue and depletion of muscle glycogen (§ 5.1.2) to give a high ultimate pH (Callow, 1936). Taint arises less frequently with tank curing, because a high concentration of salt builds up quickly and discourages bacterial growth – even if the pH is high. It may be undesirable, nevertheless, because of its fiery colour and sticky consistency ('glazy' bacon). On the other hand, organoleptic advantages are claimed for bacon of relatively high ultimate pH, including increased tenderness, and decreased shrink on curing and cooking (Kauffman *et al.*, 1964). A high ultimate pH has been deliberately induced in bacon by pre-slaughter injection of adrenaline (Rongey *et al.*, 1959) and by incorporating phosphate in the curing brine (Hall, 1950; Brissey, 1952).

It may be mentioned that the salt flavour of bacon is not closely related to its salt content, since apparently only some of the salt has time to affect the palate before the bacon is swallowed (Ingram, 1949a). This effect is particularly marked in the swollen structure of bacon at high ultimate pH. Attempts have been made to cure pork with acid brines (about pH 4.5) in order to enhance antimicrobial action, but increased disappearance of nitrite, precipitation of protein in the brine and somewhat lowered eating quality in the bacon more than offset the moderate increase in storage life brought about thereby (Ingram, 1949b). The slower penetration of salt into pork muscle of high ultimate pH has been attributed to the physical effect of greater muscle fibre size rather than to a direct chemical change (Körmendy and Gantner, 1958).

Proper resting before slaughter, or the feeding of sugar (Bate-Smith, 1937a; Gibbons and Rose, 1950; Wismer-Pedersen, 1959b), builds up muscle glycogen, giving a lower ultimate pH and increasing storage life (although this latter point has been disputed). According to Wismer-Pedersen (1959b), the ultimate pH of the muscle of sugar-fed pigs was about 0.2 pH units lower than that of controls. They gained more weight during curing (5.65 per cent compared with 5.49 per cent) and shrunk less during maturation (2.75 per cent compared with 3.16 per cent). Wismer-Pedersen believed that the enhanced water-holding capacity of the meat of low ultimate pH (contrary to expectation) was due to the greater ease of salt penetration possible with its more open structure and the consequently greater formation of salt–protein complex. The higher concentration of reducing sugars in the muscles of the sugar-fed pigs may also have been partly responsible.

It would be anticipated that the lower waterholding capacity of the muscles of pigs affected by the so-called PSE condition (§§ 3.4.3 and 4.2.2) would be reflected in their reaction to curing. This was studied by Wismer-Pedersen (1960). Ground meat from muscles, in which post-mortem glycolysis was fast, and in which the structure was watery, had a poor absorption of salt compared with normal meat, the weight of brine absorbed being about 7 per cent in the former and 40 per cent in the latter. The weight gain of sides from watery pork during curing was about 3 per cent, whereas it was about 7 per cent in normal pork. It must be presumed that the greater penetrability of the watery pork to salt, which would have been expected to increase water binding, is more than offset by loss of water-holding capacity through damage to the proteins under the abnormally severe conditions of post-mortem glycolysis. In this context, more recently, in comparing the cured meat

Table 8.12 Salt content of unfrozen and thawed pork
muscle during curing

Time of immersion in 25% brine (min)	Salt (%)	
	Unfrozen	Frozen and thawed
5	1.5	2.2
10	2.1	2.7
20	3.0	3.6
60	4.5	5.6

products from the three halothane genotypes (NN, Nn and nn: cf. § 2.2), Fisher *et al.* (2000) demonstrated that there was a progressive lowering of bacon yield, and an increase in cooking loss, as the gene signifying halothane sensitivity (viz. nn) increased. Nevertheless, in curing meat from PSE pork, the incorporation of porcine collagen enhances the water-holding capacity of the product (Schilling *et al.*, 2003).

Freezing affects the structure of muscular tissue (§ 7.1.2.1): the penetration of salt into pork which has been frozen and thawed is about 20 per cent greater than into fresh meat (Table 8.12; Callow, 1939), and when dealing with frozen carcasses curers shorten the tanking period by 1 day in every 5.

Apart from the salt concentrations of the brine (and the time of contact with the meat) and the microscopic structure of the musculature, various other factors affect the penetration of salt during curing. An increased temperature will increase the velocity of penetration (Callow, 1934; Wistreich *et al.*, 1959; Holmes, 1960; Henrickson *et al.*, 1969), but this obviously requires the strictest hygiene since it increases the risk of microbial spoilage. Thus, if curing salts are added early postmortem, before the carcass has completely chilled, they diffuse more rapidly and cured colour develops faster. Moreover, the yield of raw product is improved and cooking loss is less (Wiener *et al.*, 1964). Such hot curing has generally not been applied to entire sides, as in the Wiltshire cure, because of chilling difficulties, but it can be applied readily in a modified Wiltshire process in which immersion in brine is replaced by dry salting (Taylor *et al.*, 1980), and sides are chilled quickly by being hung individually. Acceptable bacon can be produced in this way within 5 days of slaughter (Taylor *et al.*, 1982b).

Curing brines containing phosphates (especially polyphosphates) have been used to enhance the water-binding capacity of bacon and hams (Taylor, 1958). In some cases, the effect may be due to an elevated pH; but pyrophosphate is said to have a specific effect because it resembles ATP and interacts with actomyosin (Bendall, 1954). On the other hand, Hamm (1955) believes that the polyphosphates increase water-binding capacity by sequestering calcium ions.

8.3.2.2 Maturing

Relatively little is yet known about the reactions which cause the development of flavour when sides of pork, after immersion in curing brine, are removed and stacked in cellars to mature. But there has been much research on the development of colour, both desirable and undesirable during processing.

The attractive red (pink) colour of cured meats before cooking is essentially that of nitrosomyoglobin (nitric oxide myoglobin). *In vitro* nitric oxide can combine directly with myoglobin, and indeed a process for producing bacon by subjecting pork to a high pressure of nitric oxide gas has been devised in the USA. Normally, however, the sequence of events in cured meats is more complicated. In the latter, nitrite firstly reacts with oxymyoglobin (i.e. in the presence of oxygen) to form met-myoglobin (Greenberg *et al.*, 1943). Although in the absence of oxygen, nitrite reacts with haemoglobin to form

$$NO_2^- \quad MygO_2$$
$$+$$
$$(\text{Nitrite}) \quad (\text{Oxymyoglobin})$$
$$\rightarrow \quad NO_3^- \quad + \quad \text{Metmyog.} \quad + O_2$$
$$(\text{Nitrate}) \quad (\text{Metmyoglobin})$$

equimolar quantities of methaemoglobin and nitrosohaemoglobin, if substances capable of reducing methaemoglobin and nitrite are not present (Brooks, 1937), with myoglobin these conditions produce only metmyoglobin (C. L. Walters, personal communication). Subsequently, although brine micro-organisms can convert nitrite to nitric oxide (Eddy *et al.*, 1960), the reduction of both metmyoglobin to myoglobin and of nitrite to nitric oxide is probably brought about by surviving activity of enzyme systems of the muscle itself (Watts and Lehmann, 1952a; Walters and Taylor, 1963, 1964). Some years ago, Fox (1962) and Fox and Thomson (1963) showed that nitric oxide can react directly with metmyoglobin and that the complex could then be reduced to nitrosomyoglobin. The rate of formation of nitrosomyoglobin is proportional to the concentration of nitrite up to the point where the nitrite:metmyoglobin ratio is 5:1. Beyond this point nitrite appears to be inhibitory and this may explain why the conversion of myoglobin to the cured meat pigment is frequently incomplete despite apparently much more than adequate nitrite concentrations. Detailed studies have elucidated the details of nitrosomyoglobin formation. The mechanism is as follows: (i) nitrite oxidizes myoglobin to metmyoglobin; (ii) nitrite also oxidizes ferrocytochrome *c* to nitrosoferricytochrome *c* (catalysed by cytochrome oxidase); (iii) the nitroso group is transferred from nitrosoferricytochrome *c* to the metmyoglobin by NADH-cytochrome *c* reductase action, forming nitrosometmyoglobin; and (iv) the nitrosometmyoglobin is reduced to nitrosomyoglobin by enzyme systems of the muscle mitochondria (even in the presence of nitrite concentrations causing rapid oxidation of oxymyoglobin). Nitrosometmyoglobin also autoreduces to nitrosomyoglobin under anaerobic conditions, but aerobically it breaks down to give metmyoglobin (Walters and Taylor, 1965; Walters *et al.*, 1967). Nitrite also reacts with tryptophan residues in non-haem proteins and these are capable of transferring nitrite to metmyoglobin to form nitrosometmyoglobin which, as indicated, can be reduced to nitrosomyoglobin.

In processing sausages, emulsion curing has been employed in recent years. This involves cooking the meat emulsion, immediately after adding the curing ingredients, for 90 min at 75 °C. The formation of nitrosomyoglobin in these conditions is much better in the absence of air (Ando and Nagata, 1970). In comparison with entire porcine muscles, colour development was decreasingly satisfactory with fractions consisting of sarcoplasm, mitochondria, microsomes or myofibrils – a finding of interest in relation to the colour fixing mechanism.

Although reduction of nitrite to nitric oxide can be effected either by bacteria or by the muscles' own enzyme system, the reduction of nitrate can be effected only by the former. Hence the need for careful control of bacon curing brines to ensure that the necessary microbial reduction of nitrate will occur. Moreover, various contaminating organisms in the brine can exert a deleterious effect on the cured meat, e.g. souring, putrefaction and excessive sweetening caused, respectively, by *Lactobacillus*, *Vibrio* and *Bacillus* (Leistner, 1960). When nitrite is added as such (Brooks *et al.*, 1940; Holmes, 1960) colour fixation is a purely chemical process.

The stability of the red (pink) colour of bacon is enhanced if nitrosomyoglobin is converted to nitric oxide myohaemochromogen in which the globin portion of the molecule is denatured, e.g. by salts or heat, as in cooking. However, it is now believed that the stable red 'cooked' pigment is a nitrosohaemochromogen, which is a penta-coordinated ferrous haem, where nitric oxide is the fifth ligand and which is not bound to the protein (Killday *et al.*, 1988). Thus the same agents which cause discoloration in fresh meat are advantageous in retaining the colour of cured meats (Watts and Lehmann, 1952b).

Incorporation of ascorbic acid in the curing brine accelerates the reduction of metmyoglobin and probably the conversion of nitrite to nitric oxide. Yet nitrite alone converts oxymyoglobin to metmyoglobin, and ascorbic acid alone gives a mixture of metmyoglobin and choleglobin (Lemberg and Legg, 1950, cf. Table 10.1). Choleglobin may also form in meat when micro-organisms producing hydrogen peroxide are present. In fresh meat, hydrogen peroxide is quickly destroyed by catalase, but the latter is destroyed during curing and thus cured meats are more liable to become green. Another green pigment, which may appear during maturation, is associated with the production of hydroxylamine during the reduction of nitrate (Iskandaryan, 1958).

At an elevated pH the oxidation of nitrosomyoglobin to metmyoglobin is retarded (Urbain and Jensen, 1940). Moreover, the activity of the muscle enzyme systems which are capable of reducing metmyoglobin and nitrite is also increased (Lawrie, 1952b; Walters and Taylor, 1963). As mentioned above, a high pH may be induced in muscle by the preslaughter injection of adrenaline; a similar result may be achieved by the incorporation of an alkaline phosphate in the cure (Hall, 1950; Brissey, 1952). Unfortunately, a high ultimate pH enhances the bacterial conversion of nitrate (as added, together with nitrite, in the traditional Wiltshire cure) to nitrite; and unacceptably high levels of the latter can be produced during a few days' storage at chill temperature of bacon derived from muscle of ultimate pH above 6 (Jolley, 1979). On the other hand, when bacon is produced rapidly, in the absence of nitrate-reducing bacteria, as in slice curing, the levels of residual nitrite remain within accepted limits, irrespective of the pH of the pork processed. Metmyoglobin, formed by exposing cured bacon to the atmosphere, can be reduced by surviving enzyme systems in the cured musculature, and reconverted to nitrosomyoglobin, by repacking the bacon *in vacuo* and storing it at 5 °C for 1–2 weeks (Cheah, 1976). Lactic dehydrogenase and NADH are involved.

It is not the purpose of this book to give detailed consideration to the nature of the various processed meats that are available in many countries. Nevertheless there is much recent interest in relating the changes occurring during maturing (ripening) of such products as dry cured ham. The development of proteomic techniques, such as two-dimensional gel electrophoresis, has revealed the protein components produced by proteolysis at various stages in ripening (Luccia *et al.*, 2005); and such investigations can be expected to provide the knowledge necessary to deliberately enhance the expression of desired flavour components.

8.3.2.3 Smoking

Smoke, generally produced by the slow combustion of sawdust derived from hard woods (consisting of about 40–60 per cent cellulose, 20–30 per cent hemicellulose, and 20–30 per cent lignin), inhibits microbial growth, retards fat oxidation and imparts flavour to cured meat (Callow, 1927a, 1932). Traditionally, smoking was uncontrolled and consisted of burning the wood beneath the meat. The process can be more speedily carried out, and a product of consistent quality produced, by controlled smoking in a kiln and by electrostatic deposition of wood smoke particles (Cutting and Bannerman, 1951; Forster and Jason, 1954).

Part of the bactericidal action of smoke is due to formaldehyde (Callow, 1927b; Hess, 1928), but the composition of wood smoke is complex. According to Foster and Simpson (1961) it consists of two phases: a disperse, liquid phase containing smoke particles and a dispersing gas phase. Direct deposition of smoke particles makes a negligible contribution to the process: vapour absorption by surface and interstitial water is much more important. The vapour phase can be separated into acids, phenols, carbonyls, alcohols and polycyclic hydrocarbons (Hollenbeck and Marinelli, 1963). The major components include formic, acetic, butyric, caprylic, vanillic and syringic acids, dimethoxyphenol, methyl, glyoxal, furfural, methanol, ethanol, octanal, acetaldehyde, diacetyl, acetone and 3,4-benzpyrene; but there are said to be more than 200 components (Wilson, 1961). The various alcohols and acids are derived from the celluloses and hemicelluloses which decompose at a lower temperature than lignin. The latter decomposes above 310 °C, yielding phenolic substances and tars.

The detection of carcinogenic compounds such as 3,4-benzpyrene and 1,2,5,6-phenanthracene has led to studies of the effect of smoke-generating conditions on their production. They are formed from lignin above 350 °C and must be present, generally, since the temperature of the combustion zone is about 1000 °C (Miler, 1963). Although it is felt that the danger of carcinogenesis from smoked meat is extremely small (Keller and Heidtmann, 1955), there have been many attempts to produce carcinogen-free smoke, for example, by condensation, followed by fractional distillation. The selected fraction is diluted with water in which the benzpyrenes are insoluble (Lapshin, 1962). The use of such liquid smokes is increasing on the continent, and they may be supplemented by the addition of specific phenolic substances having a fruity flavour and odour (Wilson, 1961). The composition of liquid smoke has also been elucidated by gas liquid chromatography (Fiddler *et al.*, 1970).

The flavour imparted by smoking varies according to the conditions used to produce the smoke (Tilgner *et al.*, 1962). Moreover, the same smoke will produce different aromas with different meats. To some extent, therefore, the flavour of the smoked product depends on the reaction between the components of the smoke and the functional groups of the meat proteins. Thus phenols and polyphenols react with –SH groups and carbonyls with amino groups (Krylova *et al.*, 1962). Organoleptic evaluation of the phenolic substances present in wood smoke suggests that guaiacol is the most effective (Wasserman, 1966).

8.3.3 Organoleptic aspects

Unlike dehydration or freeze dehydration, curing is not designed to preserve meat in a condition resembling that of the fresh commodity. Indeed, cured meats are valued for the differences in organoleptic quality produced by curing. It may be mentioned that the biological value of the proteins is not lowered by curing (Dunker *et al.*, 1953) and the vitamins of the B group are almost unaffected (Schweigert *et al.*, 1944). During storage, cured meats deteriorate in the first instance because of discoloration, secondly because of oxidative rancidity in the fat and thirdly on account of microbial changes – the latter having become of somewhat greater importance since the advent of prepackaged methods of sale.

Although the pigments of cured meats (nitrosomyoglobin or nîtrosomy-ohaemochromogen) are stable in the absence of oxygen, or even under vacuum (Urbain and Jensen, 1940), oxidation, to metmyoglobin in the uncooked product, is very rapid when oxygen is present (Watts, 1954). Unlike myoglobin itself, where the rate of oxidation is maximal at 4 mm oxygen partial pressure, the rate of nitro-somyoglobin oxidation increases directly with increasing oxygen tension (Brooks, 1935). The only practical and effective anti-oxidant so far extensively used is ascorbic acid, either incorporated in the curing brine or sprayed on to the surface of the product after maturing (cf. § 8.3.2.2).

Nitrosomyoglobin and the cooked pigment are much more susceptible to light than myoglobin. Cured meats may fade in 1 h under display lighting conditions, whereas fresh meats will not alter over 3 days (Watts, 1954).

The latter are not affected by visible light, although they oxidize under ultraviolet radiation. The pigments of cured meats are equally affected by both. Since light accelerates oxidative changes only in the presence of oxygen, however, vacuum packaging, or packaging under nitrogen, can eliminate the effect – although, of course, adding to the cost of the product. Occasionally, a particularly swift fading of the red pigment of cured meat is observed. In such cases the labile form may be nitric oxide metmyoglobin and not nitrosomyoglobin or nitrosomyohaemochro-mogen, since *in vitro* the former is very easily dissociated by oxygen forming brown metmyoglobin. If this is so, it suggests that the enzyme systems in the muscle which reduce nitrite and metmyoglobin are not identical.

In uncooked bacon and hams the maximum formation of nitrosomyoglobin is attained with a high ultimate pH, whereas in the cooked products a low ultimate pH gives a greater proportion of cured pigment (Hornsey, 1959).

Salt has an accelerating effect on the oxidation of fat. As a result, cured meats are more liable than fresh to spoil through oxidative rancidity in the fat (Lea, 1931). For this reason it is preferable to import frozen pork and cure it than to import frozen bacon (Callow, 1931). The process of curing reduces the resistance of pork fat to oxidation to a much greater extent than would be expected if the direct influence of temperature were the sole factor involved. This is said to be because salt accelerates the action of a lipoxidase present in the muscle (Lea, 1937). Nevertheless, those pork muscles that have relatively high contents of glutathione peroxidase and catalase are less susceptible to salt-accelerated lipid oxidation (Hernandez *et al.*, 2002). Smoking decreases oxidative rancidity, partly on account of the phenolic anti-oxidants it contains (cf. § 8.3.2.3). Exclusion of oxygen by storage under carbon dioxide in the case of sides (Callow, 1932) effectively prevents fat oxidation in cold storage; but the concentration of the gas required is dangerous and also difficult to maintain. For packaging, as already indicated, gas packs employing nitrogen, or vacuum packs, are effective. Nevertheless, nitrite *per se* has an anti-oxidant effect, inhibiting fat oxidation in *cooked* cured meats (Zipster *et al.*, 1964). This appears to be due to its chelation of non-haem iron and to the effect of the cooked pigment in blocking the prooxidant action of haem iron (Morrissey and Tichivangana, 1985).

Unfortunately, ascorbic acid, notwithstanding that it preserves the colour of cured meats, can accelerate the oxidation of fat when the tocopherol content of the fat is low (Abrahamson, 1949; Scarborough and Watts, 1949) or in the absence of other specific fat anti-oxidants. Ascorbic acid has an inhibitory effect on fat oxidation, however, when a metal-chelating agent (e.g. polyphosphate) is present (Lehmann and Watts, 1951). As in the case of nitrosomyoglobin, light accelerates the oxidation of fat (Lea, 1939). In addition to the independent oxidation of unsaturated fatty acids and pigments in cured meats, each accelerates the oxidation of the other (Robinson, 1924; Niell and Hastings, 1925; Tappel, 1952). The effect is eliminated in cooked cured products, since it depends on the presence of relatively undenatured proteins.

Apart from the general preservative effect of salt, the storage life of cured meats is enhanced by the specific antimicrobial action of nitrite in the curing brines (§ 6.3.3). As indicated in § 7.1.1.3, however, the recent development of prepackaging has introduced new potential hazards of spoilage from microbial action. Thus, while the procedure obviously lowers the risk of contamination of the product after wrapping, it increases the possibility of contamination during preparation, especially as large areas of cut surface are frequently exposed (Ingram, 1962). The relatively high salt content of bacon, and its own halophilic microflora, tend to discourage the growth of the kinds of micro-organism likely to be introduced during handling: but with cooked, cured products and, especially, semi-preserved items, the nature and number of introduced contaminants might have a pronounced effect on the organoleptic behaviour of the product and, if they happened to be pathogens, on its safety.

Because vacuum packaging helps to prevent the oxidation of fat and pigments in cured meats, it is frequently employed. As a result the atmosphere within the pack may be altered by surviving microbial activity. For example, oxygen (residual) may be absorbed and carbon dioxide generated (Ingram, 1962). This circumstance would inhibit the normal microflora of cured meats, causing their replacement by other micro-organisms capable of changing flavour, odour and perhaps the safety of the products, e.g. lactic acid bacteria will grow, causing souring (Kitchell and Ingram, 1963). Higher temperatures of storage will, of course, increase the number of micro-organisms, and the eating quality decreases faster. It is of interest, however, that bacterial numbers reach their maximum many days *before* the eating quality is noticeably affected (Cavett, 1962; Kitchell and Ingram, 1963). Unpacked bacon generally goes off organoleptically when the bacterial load has attained only 10 per cent of its final maximum (Haines, 1933). The development of unpleasant odours may be related to the preferential growth of certain types of micrococci which split fat and protein. Above 20 °C organisms like *Staph. aureus* dominate the microflora (Cavett, 1962), when the spoilage odour is 'scented-sour', and at 30 °C it is putrid. The tendency for prepacked cured meats (like other prepacked products) to be subjected to temperature conditions which would not be considered suitable for meat in more traditional form, makes such spoilage a distinct possibility.

Although the potential nitrite content of bacon, represented by the nitrate employed, is not attained in normal practice because of destruction of both nitrite and nitrate by bacterial or tissue enzymes (Eddy *et al.*, 1960; Walters and Taylor, 1963), dangerously high concentrations of nitrite (0.27 per cent) can occur, if rarely, in vacuum-packaged bacon (Bardsley and Taylor, 1962). The necessary conditions are as yet unknown, but temperatures of storage about 15–20 °C cause maximum nitrite accumulation (Eddy and Ingram, 1965). It has been shown, also, that when the pH of bacon is above 6.5, bacteria will convert an exceptionally high proportion of nitrate into nitrite (Jolley, 1979).

The possibility that carcinogenic nitrosomines may be produced from nitrite in curing processes has been raised (Lijinsky and Epstein, 1970), but so far there has been no evidence to indicate that any public health hazard exists with cured meats. Nevertheless, there have been vigorous efforts to lower the residual nitrite contents of such products. Although the pigment nitrosomyoglobin requires only 5 ppm of nitrite for its formation, a stable pink colour – and the variability of myoglobin concentration between muscles – necessitates 50 ppm. For the development of bacon flavour, 5–100 ppm is needed; and, most importantly, a minimum of 100 ppm is required to inhibit *Cl. botulinum* (Anon., 1974a). *Cl. botulinum* has been demonstrated in pork in the United Kingdom; and in pasteurized, cured meat heat treatment, nitrite concentration and storage temperature are important factors in determining whether growth of *Cl. botulinum* types A and B will occur (Roberts and Ingram, 1976). The frequent production of toxin at 15 °C emphasizes the need for refrigerated storage of pasteurized cured meat products and the dangers which might arise if nitrite were to be markedly reduced in such products, without any compensatory safeguard. In terms of the destruction of *Cl. botulinum*, safe long-term storage life at ambient temperature can be as assured *without* using nitrite if the bacon is sterilized by 2–3 Mrad of ionizing radiation (Wierbicki and Heiligman, 1980), but 20–40 ppm of nitrite is necessary to ensure normal colour and flavour. In such circumstances no nitrosamines have been detected. They were also reported to be absent from cured pork which had been prepared without nitrite and in which colour was derived from preformed cooked cured meat pigment (Shahidi *et al.*, 1994).

8.3.4   Intermediate moisture meat

The need by expeditions and similar groups for palatable and nutritious food which would be stable under tropical conditions (Brockmann, 1970) and for astronauts (Klicka, 1969) has led to the development of intermediate moisture food technology. The intention is to lower water activity to the point at which bacteria will not grow, even at high ambient temperature, without lowering the water content to the point at which the product becomes unpalatable. As usually applied, the method involves the soaking of the food in an infusing solution of higher osmotic pressure so that, after equilibration, its water activity is lowered to the desired level. Equilibration can be accelerated by raising the temperature, as in the cook–soak–equilibration procedure (Hollis *et al.*, 1968).

As applied to meat, the lean is cut into portions about 1 cm³ in volume and immersed in about one and a half times their weight of infusing solution. This is aqueous, containing about 10 per cent sodium chloride, 0.5 per cent of an antimycotic and sufficient glycerol (between 33–40 per cent) to achieve a water activity of about 0.82–0.6. The mixture is heated to 70 °C for 15 min in cans (Pavey, 1972). After a subsequent period of about 15 h at room temperature, the meat pieces are surface-dried and stored in impermeable Cryovac bags. Such meat will remain acceptable for several months at 38 °C, although it undergoes textural and colour changes. At first it becomes more tender, there being a concomitant breakdown of collagen: later there is increasing toughness which is associated with Maillard-type cross-linking reactions (Obanu *et al.*, 1975).

Chapter 9

The storage and preservation of meat: III Direct microbial inhibition

In Chapters 7 and 8 methods of preserving meat were considered which depended essentially on discouraging microbial growth through the creation of unfavourable environments in the meat. As should be apparent from Chapter 6, other modes of preservation are possible. These involve action more directly inhibitory or lethal to moulds and bacteria. The latter may be destroyed by ionizing radiation or poisoned, either by specific microbial poisons (antibiotics) or by substances of general toxicity for biological tissues which are virtually harmless to consumers at the levels effective against micro-organisms (chemical preservatives).

9.1 Ionizing radiation

The concept of employing ionizing radiation to preserve food has developed since around 1940. In the period 1954–64, the US Quartermaster-Generals Department carried out long-term studies on various meat products including ground beef and pork and bacon. It was concluded that foods which had been irradiated with up to 5.6 Mrad by γ-rays from Co^{60} or by electrons of energy up to 10 MeV were wholesome (US Surgeon General, 1965). Subsequently, in 1968, the Food and Drug Administration withheld approval for irradiated ham; but an international project on the future of irradiation was initiated in 1970, and in 1980 the participating bodies (including the FAO and WHO) proposed that irradiation with a dose less than 1 Mrad should be accepted as a process for preserving all major categories of food (Anon., 1981).

Of the many types of ionizing radiation which are known, only high energy cathode rays or soft X-rays from generators and γ-rays from radioactive sources (e.g. Co^{60}) are useful in practice for treating foodstuffs (Hannan, 1955). Whatever the type of radiation used, it is important that the energy level of the rays should not exceed about 10 MeV since otherwise an induced radioactivity may arise in certain elements in the meat (McElhinney et al., 1949), although this effect would not be large at energy levels below about 15 MeV (Hannan, 1955). In fact, energy levels above a few MeV are not practicable.

The relative characteristics of cathode rays, soft X-rays and γ-rays will not be discussed here; the former are more useful for the treatment of surfaces, the latter where treatment in depth is necessary. Within broad limits the important factor is the *total* dose received by the product, but there is evidence that if it is delivered at a high rate there may be a greater biological effect and less chemical change than at a low rate (Hannan, 1955).

The advantages of ionizing radiation for food preservation include their highly efficient inactivation of bacteria, the low *total* chemical change they cause and the appreciable thickness of material which can be treated after packing in containers – even those made of metal. Some of the disadvantages will be considered below. Meats preserved by irradiation are superior in quality to thermally processed meats (e.g. canned products), the only other shelf-stable meat products available for immediate consumption (Wierbicki, 1980).

9.1.1 Chemical and biochemical aspects

As the adjective signifies, ionizing radiations produce ions and other chemically excited molecules in the exposed medium; but this is only the first of a series of chemical effects many of which are not beneficial and have to be offset against their antimicrobial action. The activated molecules react further and in unusual ways, forming free radicals, polymers and, in the presence of oxygen, peroxides. In meat, and in other foods, where there is a substantial aqueous phase, destruction of organic molecules also takes place indirectly – largely through their reaction with the H atoms and OH radicals of irradiated water molecules, whereby they are reduced or oxidized respectively (Dainton, 1948; Hannan, 1955).

Irradiation damage to molecules is approximately proportional to their molecular weight. Thus, the very large DNA molecules, which are essential to microbial survival, are particularly vulnerable. Since the nutrients derived on digestion from proteins, carbohydrates and fats have molecular weights of *ca.* 150–200 Da, the damage they sustain is about a million-fold less than sustained by DNA. A sterilizing dose which reduces the numbers of *Cl. botulinum* by 10^{12} changes only about 0.2 per cent of the proteins, 0.3 per cent of the carbohydrates and 0.4 per cent of the lipids in the product concerned (Brynjolfsson, 1980). Irradiation in the frozen state reduces all changes by 75 per cent.

Proteins are the principal organic constituents of meat. The changes produced in them by ionizing radiations are determined both by the intrinsic nature of the proteins and by the dose. Such large protein molecules as titin and nebulin are more susceptible to irradiation than myosin or actin (Horowits *et al.*, 1986). Random coil structures in proteins are less resistant than helical arrangements (Elias, 1985); collagen is thus more labile than globular proteins (Bailey and Rhodes, 1964). In general, the amount of change detected by the consumer arises from alterations in only a small proportion of the total molecules exposed. The effects in meat are less than they would be in some other foods, since much of the water it contains is bound – thus limiting secondary reactions. Moreover, many substances are present which can act as free radical acceptors. There is little destruction of amino acids as combined in the proteins, but soluble amino acids are deaminated (Strenström and Lohmann, 1928) forming keto acids and aldehydes, and, in the case of S-containing amino acids, H_2S (Dale and Davis, 1951). Nevertheless, structural modifications occur in the proteins, even with low doses, which do not cause any apparent alteration (Fricke, 1938). Observed chemical changes include loss of solubility at the iso-electric point (or 'denaturation'), polymerization and degradation to aggregates of lower molecular weight (Svedberg and Brohult, 1939). As with proteins denatured by other means, there is a rise of pH (Batzer et al., 1959a).

With a dose of 5 Mrad (50 kGy) (approximately that required for microbial sterility) meat proteins show a noticeable loss of water-holding capacity (Schweigert, 1959). The changes are paralleled by the response of isolated myofibrils to added ATP. At low ionic strength myofibrils which have been subjected to 5 Mrad synaerese less on addition of ATP than do non-irradiated controls: at high ionic strength irradiated myofibrils swell less on the addition of ATP (Lawrie et al., 1961). The latter effects indicate, amongst other features, a measure of enzymic inactivation. Such effects have been intensively studied, and in some cases they are due to the oxidation of the SH-groups of the enzyme proteins (Barron and Dickman, 1949). It has been pointed out that irradiation of meat with 1–10 kGy could be useful in retaining quality since proteolysis by endogenous enzymes would thereby be diminished (Lakritz and Maerker, 1988). At the higher dose, overall proteolytic activity was found to be reduced by 40 per cent, although β-glucuronidase was not affected. Nevertheless, most enzymes require much more than 5 Mrad for inactivation, and this can be a serious problem in the storage of irradiated foods, since, although there may be sufficient freedom from bacteria to permit storage at high temperatures, the latter may accelerate adverse enzymic change. In particular, proteolytic enzymes may survive up to 70 Mrad (Schweigert, 1959), although the proteolytic activity of beef muscle is reduced 50 per cent by 1.6 Mrad (16 kGy).

Notwithstanding the relative resistance to irradiation of amino acids in proteins, a soluble protein has been prepared from heated meat which gives a so-called 'wet-dog' odour when it is irradiated, and there may be concomitant destruction of about 13 per cent of the amino acids (Headin et al., 1961). Rhodes (1966) found little evidence for amino acid destruction with doses up to 20 Mrad. The effect is more noticeable in beef than in pork and appears to be associated with a gelatin-like protein derived from collagen (Headin et al., 1961).

Collagen shrinks when irradiated in a dry state and becomes soluble in water if irradiated wet (Perron and Wright, 1950), and, indeed, irradiation caused softness and tenderness of texture as an immediate effect (Coleby et al., 1961). The hydrothermal shrink temperature of collagen decreases with increasing dosage. After 5 Mrad it falls from 61 °C to 47 °C, and after 40 Mrad to 27 °C, i.e. it shrinks at room temperature (Bailey et al., 1962). The effect is probably due to the destruction of some of the hydrogen bonds which hold together the triple helix (cf. Fig. 3.5).

Changes in the pigment proteins of meat on irradiation are sometimes beneficial. Thus, under some conditions, myoglobin may yield a bright red compound having an absorption spectrum similar to that of oxymyoglobin but rather more stable (Ginger et al., 1955). It is formed more particularly in pork: the myoglobin of beef tends to oxidize to brown metmyoglobin on irradiation (Coleby et al., 1961). On the other hand, irradiation of cooked meat reconverts the brown colour to red (cf. § 10.1). In the presence of air, some of the meat pigments may be converted into green sulphmyoglobin (Fox et al., 1958), especially where the ultimate pH is about 5.3, by H_2S produced from smaller molecules such as amino acids.

In a detailed study of colour changes in wrapped meat packs irradiated with 5 kGy, and stored at 4 °C, Miller et al. (2000) found evidence for the development of carbon monoxide myoglobin. This was more evident in pork than in beef or lamb. Carbon monoxide is produced in meat when subjected to ionizing radiation (Dauphin and Saint-Lèbe, 1977).

In a comprehensive review of the effects of ionizing radiation on meat colour, Brewer (2004) concluded that the susceptibility of the iron atom in myoglobin to energy input was largely responsible for colour changes. Its electrons can exist in various spin states and its orbital electrons exist at more than one energy level, making the environment of the iron atom especially susceptible to energy-donating compounds. The bright red colour of fresh meat can be enhanced, during irradiation, by the addition of anti-oxidants, the pH and oxygen status of the environment, by gas packaging and temperature control.

It has already been mentioned that the soluble amino acids are much more liable to attack than those in proteins, yielding ammonia, H_2S, etc. The principal components of the volatile off-flavour produced on irradiation are methyl mercaptan and H_2S (Batzer and Doty, 1955). Studies with S^{35} have shown that most of the methyl mercaptan is derived from methionine. Although H_2S is produced from both methionine and glutathione, it is mainly derived from other amino acids (Martin et al., 1962). The quantities of methyl mercaptan, ethyl mercaptan, diethylsulphide, isobutyl mercaptan produced all increase as the dose rate rises from 1 to 6 Mrad (Merritt et al., 1959). Meat of high ultimate pH appears to resist the changes that produce S-containing volatiles on irradiation. Thus, in a study of vacuum-packed pork, stored after irradiation with 4.5 Mrad, Ahn et al. (2001) found that DFD pork was more resistant to subsequent undesirable changes than PSE pork or that of normal ultimate pH – a factor of importance, since the antimicrobial action of irradiation would be more marked in meat of high pH (cf. § 6.3.3).

Ionizing radiation brings about changes in meat lipids which resemble those of oxidative rancidity (Coleby, 1959). In the absence of oxygen, fatty acids are decarboxylated (Whitehead et al., 1951) and, if unsaturated, they are polymerized (Burton, 1949); but in the presence of oxygen, hydroperoxides and carbonyls are formed (Dugan and Landis, 1956). The quantity of carbonyls produced increases with increasing dose (Table 9.1; Batzer et al., 1959b). Since, however, it does not increase with increasing fat content, it is clear that most of the carbonyls produced on irradiation are not derived from the oxidation of neutral fat. Ionizing radiation may break C–C links, forming various aliphatic hydrocarbons.

There is a regular increase in acetaldehyde, acetone and methylethylketone with increasing dose (Merritt et al., 1959). Nevertheless, Ahn et al. (2000), in a

Table 9.1   Effect of irradiation dose and fat content of beef on carbonyl production ($\times 10^{-5}M$ carbonyl/g meat)

Irradiation dose (Mrad)	% Fat in meat			
	6	8	13	23.3
0	1.27	3.22	0.98	0.92
2	4.31	4.93	2.13	2.30
4	6.60	8.74	3.40	3.57
6	8.63	11.50	4.65	5.06
8	11.20	12.88	6.46	7.25
10	11.50	10.35	7.42	4.95

comparison of non-irradiated porcine *l. dorsi* with that subjected to 5–10 kGy (0.5–1 Mrad), and subsequently stored at 4 °C, found no increase in volatiles derived from lipid oxidation, but higher levels of those arising from the breakdown of S-containing amino acids. The slightly stronger odour of the irradiated samples was not disliked. Although cholesterol oxidation products are formed in fresh, raw meat, their level is enhanced by ionizing radiation (Nam *et al.*, 2001).

The carbohydrates of meat tend to be oxidized in the 6 position, yielding gluconic acids and aldehydes (Phillips, 1954).

Although irradiation affects only a relatively small proportion of the molecules, this fact becomes rather more important when the molecules concerned are present in small amounts – as in the case of vitamins, although, of course, this applies also to other methods of preservation. Vitamin C and thiamin are particularly affected, destruction of the latter representing, perhaps, the greatest nutrient loss with irradiated meat (Groninger *et al.*, 1956), although vitamin B_{12} is the most radiosensitive vitamin (Markakis *et al.*, 1951).

Irradiation of cured meats converts nitrate to nitrite, but the amount of nitrite produced is insufficient to constitute a hazard, especially as it also is destroyed by irradiation (Hougham and Watts, 1958).

The potential drawback of ionizing radiation in meat processing is not so much the destruction of the proteins (which is negligible) or of vitamins (which is not negligible but of no nutritional importance) or even the production of off-flavours and off-odours (which can be serious organoleptically). It is the possibility of producing minute quantities of biologically potent and toxic chemicals, e.g. carcinogens from sterols (Weiss, 1953). As experiments involving a long-term ingestion of irradiation foods continue, however, this danger appears to be receding.

9.1.2   Organoleptic aspects

9.1.2.1   Immediate effects

It will have been apparent from § 9.1.1 that, depending on the dose, various organoleptic changes will arise on irradiating meat. Odour and flavour can be adversely affected by the production of H_2S and mercaptans, carbonyls and aldehydes – this being worse in beef than in pork or lamb (Huber *et al.*, 1953); colour by the production of metmyoglobin and sulphmyoglobin; and texture and water-binding capacity by denaturation changes in the structural proteins. The tenderizing effect caused by changes in the collagen molecules would only be apparent at doses so high that the meat would be inedible. To achieve sterility in the meat – and the possibility of indefinite storage without refrigeration – upwards of 5 Mrad would be required (§ 6.4.4) and this would cause marked deterioration in the attributes of eating quality. Attempts have consequently been made to minimize such adverse effects. Although the presence of oxygen exacerbates irradiation-induced chemical change, in general, its removal has not given appreciable benefits (Hannan, 1955). Radiochemical changes can be decreased considerably, however, by irradiating in the frozen state and the meat quality is much improved if the commodity is frozen and held at a very low temperature before irradiation (Table 9.2). In these circumstances removal of oxygen is advantageous (Huber *et al.*, 1953).

The beneficial effects of low temperatures may well be due to the virtual removal of the aqueous phase, thus preventing secondary chemical changes (Hannan, 1955). A somewhat larger dose may be required to achieve sterility in meat at such low

Table 9.2   Effect of prestorage at various temperatures on odour, flavour and colour of raw beef given 1.5 Mrad (Huber *et al.*, 1953)

Prestorage treatment		Organoleptic rating[a]		
Time (h)	Temp. (°C)	Odour	Flavour	Colour
24	3	2.5	2.8	1.5
24	−15	3.0	3.5	1.5
72	3	3.0	2.8	2.8
72	−15	4.8	4.5	4.5
72	−35	5.0	4.0	4.5
72	−180	4.5	4.5	4.5
96	3	3.8	3.0	3.5
96	−15	4.8	4.3	4.5

[a] Average taste panel of five members: controls rated as 5.

temperatures, but the order of this requirement is much less than the decrease achieved in irradiation damage to the meat (Hannan, 1955). Irradiated frozen meat could subsequently be stored at a relatively high temperature but, of course, it would be subject to the disadvantages of drip in these circumstances. Off-flavour development is minimal if irradiation is carried out below −20 °C (Merritt *et al.*, 1978).

The possibility of minimizing irradiation damage by the incorporation of protective compounds has been considered on the assumption, that they would react with the activated molecules and free radicals produced and thus prevent them attacking the organic molecules of the meat. The potential additives and their irradiation products must be non-toxic. Ascorbic acid, nitrite, sulphite and benzoate have been used in this context (Huber *et al.*, 1953; Pratt and Eklund, 1954). It is interesting to note that irradiation odour was markedly decreased by adding ascorbate after irradiation. This suggests that it may act by directly reducing irradiation products rather than by reacting with free radicals during irradiation. The major difficulty with the use of protective compounds in practice is that of ensuring their effective distribution throughout the meat.

Another approach is the use of in-package odour scavengers; activated charcoal has been employed with some success (Tausig and Drake, 1959). Pre-irradiation storage above the freezing point does not lessen organoleptic damage (Batzer *et al.*, 1959a).

9.1.2.2   Storage changes

Since the purpose of any method of meat preservation is to permit its storage in a form as near as possible to that of the fresh commodity, nonmicrobial changes during storage are as important as the immediate effects of the preservative process employed. The prolonged storage period at relatively high temperatures, which is possible with radiation sterilized meats, permits various chemical and biochemical changes which would be precluded during the more limiting conditions with other methods of preservation. Among the purely chemical changes is non-enzymic browning (§ 8.1.3). Both because of the increased storage times and temperatures, and because irradiation produces carbonyls, Maillard-type browning is greater in irradiated meat (Lea, 1959).

The immediate effects of irradiation in causing fat oxidation have been mentioned: during storage of meat irradiated at low temperature there is a possibility of further oxidation due to an after-effect, first noted by Hannan (1955).

Coleby *et al.* (1961) found that the initial irradiation odours and flavours of beef and pork exposed to 5 Mrad gradually changed, during storage at 37 °C, to stale and bitter flavours; and the not unpleasant pink colour of irradiated pork tended to turn brown.

Some protein denaturation occurs on irradiation. This increases during storage, especially at high temperature, and the resultant loss in water-holding capacity causes considerable exudation (Cain *et al.*, 1958; Schweigert, 1959).

The most important detrimental changes during the storage of irradiation sterilized meat, however, are those due to surviving activity of proteolytic enzymes (§ 5.4.1).

The storage for 1 year at 37 °C of beef and pork steaks, which had been irradiated in depth and had received 5 Mrad (50 kGy), caused a marked increase in total soluble nitrogen. Protein equivalent to about 25 per cent of the original total nitrogen was broken down to peptides and amino acids and insoluble aggregates of crystalline tyrosine formed on the meat surface (Lawrie *et al.*, 1961). This proteolysis principally involved sarcoplasmic proteins, for there was no increase in soluble hydroxyproline (indicative of connective tissue breakdown) and, microscopically, the myofibrils *appeared* to be unchanged. Although it was evident that the actomyosin complex had not been proteolysed, its altered nature was reflected in a diminution of ATP-ase activity and accounted for some of the loss in water-holding capacity. Apart from exudation, the stored, irradiated meat had a coagulated, crumbly texture similar to that of lightly cooked meat. Zender *et al.* (1958) and Radouco-Thomas *et al.* (1959) found that the texture of sterile rabbit muscle was almost completely broken down during storage at 25 °C or 37 °C over some months. To some extent the greater breakdown can be attributed to the fact that the rabbit was sterilized by irradiation only on the surface, and that within it proteolysis was unimpaired. Doty and Wachter (1955) indicated that the proteolytic enzymes of meat are diminished somewhat by irradiation. There appeared to be species differences in the degree of post-mortem proteolysis since, under comparable conditions, beef is not so extensively broken down as rabbit (Sharp, 1963).

It is evident that there are some proteolytic enzymes in meat which substantially survive even a sterilizing dose of ionizing radiation. Consequently, degradative changes and organoleptic deterioration will occur if the meat is held at the storage temperatures which sterilizing doses of irradiation make possible from the microbiological standpoint. Attempts have been made to overcome this defect in the product. One approach is to heat the meat, before irradiation, to inactivate the proteolytic enzymes. Cain *et al.* (1958) subjected fresh and heated beef and pork (71 and 77 °C, respectively) to 2–3 Mrad (20–30 kGy) and stored the meat for about 8 months at 22 °C. Although there was extensive fluid loss, protein breakdown and formation of tyrosine crystals in the fresh meats, the corresponding preheated meats were acceptable in texture and flavour. These workers also noted that bacon stored satisfactorily for 8 months at 22 °C after irradiation – possibly because its proteolytic enzymes had been denatured by the curing salts. Enzyme inactivation by preheating may be achieved by exposure to a relatively high temperature for a short time (e.g. 163 °C for 2 h) or to a lower temperature for a longer time (e.g. 50–80 °C for 20 h): the latter causes undesirable texture changes due to partial breakdown of connective tissue (Whitehair *et al.*, 1964). More recently, the attainment of a temperature of 70–75 °C in the centre of products has been regarded as sufficient to inactivate enzymes, and, following irradiation in a hermetically sealed container with a sterilizing dose of either γ-rays, X-rays or electrons, to ensure a long storage life at ambient temperature without quality deterioration (Wierbicki, 1980). After such pre-heating, of course, meat can no longer be considered fresh and thus one of the intended advantages of irradiation – the preservation of meat in a state close to that of the fresh commodity – is automatically lost. If taste panel testing of cooked meat were the only criterion of acceptability, this would be less serious than the aforementioned organoleptic changes in sterilized fresh meat stored at high temperature.

9.1.3 Radiation pasteurization

There is growing realization that, whereas the prolonged storage of fresh meat at high temperatures after sterilizing doses of ionizing radiation would be offset by proteolytic and other deterioration, a substantial extension of the storage life of meat preserved by refrigeration would be possible if this were combined with a low dose. Although such 'pasteurizing' doses would increase the useful life of the meat, they would be too low to cause organoleptic changes (Ingram, 1959). There would be little advantage with frozen meat, since this is not directly affected by microorganisms (although pasteurizing doses of irradiation might increase its storage life after thawing), but spoilage of fresh or chilled meat, stored at 0–5 °C, can be usefully discouraged. This is especially so since the microorganisms which tolerate such temperatures are particularly sensitive to irradiation (Ingram, 1959). It is apparent that meat stored at 0–5 °C, after irradiation with 50,000–1,000,000 rad, can be held from 5 to 10 times as long as unirradiated meat before microbial spoilage (Morgan, 1957; Shea, 1958; Niven, 1963).

Despite the general susceptibility of the cold-tolerant micro-organisms to pasteurizing doses of irradiation, certain more resistant species are bound to survive even in small numbers, e.g. *B. thermosphacta* (McLean and Sulzbacher, 1953). Such micro-organisms spoil meat by producing a sour odour rather than the stale, musty odour from the pseudomonads which grow in stored, unirradiated chilled meat.

Although irradiation with pasteurizing doses can prolong the refrigerated storage life of meat microbiologically, other concomitant changes of non-microbial origin may become apparent. Doses as low as 50,000 rad (0.5 kGy) can cause flavour changes detectable by trained individuals: doses above 200,000 rad intensify those off-flavours and make them noticeable to a higher percentage of consumers (Niven, 1963). Symptoms of accelerated oxidation may be found in beef irradiated with 25,000–100,000 rad (Lea *et al.*, 1960) indicated by the development of a tallowy odour and flavour; and the yellow carotenoid may be perceptibly bleached. Lea *et al.* concluded that, where the surface of the meat was exposed to the atmosphere, the margin between desirable and undesirable effects of pasteurizing doses of irradiation (as in the carriage of chilled beef between Australia and the United Kingdom) scarcely justified its adoption. On the other hand, where the surface of the meat is covered with an oxygen impermeable wrap, oxidative changes are largely eliminated and recontamination of the irradiated surfaces is avoided. In these circumstances the process may be beneficial. An extensive experiment with pasteurizing doses, in combination with chill temperatures (1 °C), was carried out on lamb and beef shipped from the United Kingdom to Australia and New Zealand (Rhodes and Shepherd, 1966). 400,000 rad was the maximum dose of surface irradiation which could be tolerated without causing detectable colour or odour changes. Such a dose delayed microbial growth on the surfaces of lamb carcasses (in packs free of air), or in beef joints, for more than 8 weeks. Although, since frozen lamb is an accepted standard commodity, the procedure would not be commercially attractive with lamb, it could have real advantages with beef – a commodity which commands a markedly higher price when chilled than when frozen.

Such observations have been fully substantiated by many other investigations. Thus, Lefebvre *et al.* (1992) demonstrated that irradiation of ground beef with 1–5 kGy extended the storage life at 4 °C by 4–15 days, microbial growth being progressively suppressed, although the flora tended to change from Gram-negative bacilli to Gram-positive cocci as the dose of irradiation increased. Irradiation was associated with increased peroxide values during storage and taste panels detected some loss of flavour, but the latter was insignificant in ground beef irradiated at 1 kGy (100,000 rad) (Lefebvre *et al.*, 1994).

The antimicrobial benefits of ionizing radiation clearly must be evaluated against the concomitant effects in causing discoloration and rancidity due to the oxidation of unsaturated fatty acids. Formanek *et al.* (2003) demonstrated that such undesirable effects can be minimized when α-tocopherol is fed to the animal, or added to meat after mincing.

A bacteria-proofed container, applied before treatment, is necessary to prevent recontamination; auto-oxidation is simultaneously eliminated by making the packaging material relatively impermeable to oxygen. Thus the need for carbon dioxide to prevent the growth of moulds would be dispensed with, eliminating the considerable expense of gas storage. However, the mechanical problem of avoiding rupture of the pack in handling arises. Similarly, pasteurizing doses of irradiation, combined with light refrigeration, may be of benefit with prepackaged cuts. At the moment, the practical usefulness of irradiation preservation of meat would seem to be within such contexts.

9.1.4 Policy and detection

On the basis of joint meetings of an expert committee of FAO/IAEA/WHO (1981) it was declared that no hazard occurs in food irradiated with doses up to 1 Mrad (10 kGy) and, in the UK, a comprehensive report was published by the Advisory Committee on Irradiated and Novel Foods in 1986 which sanctioned the use of irradiation in food treatment, but with certain qualifications. These were that the maximum average dose should not exceed 1 Mrad; that the energy of the rays employed should not exceed 5 MeV (if em rays were used) or 10 MeV (if an electron beam was used); that the patterns of consumption by the public be monitored in relation to nutrient and toxicological contents; and that the irradiated food should be so labelled. The presence of *E. coli* O157:H7 in ground beef led the Federal Drug Administration in the USA to approve the irradiation of red meat in 1997, and the use of ionizing radiation is now widespread in that country to control pathogens in fresh and frozen meat (Satin, 2002); but there is still some resistance to its use in Europe.

Under the UK Food (Control of Irradiation) Regulations of 1991, certain classes of food may be irradiated up to a maximum dosage (the value being 7 kGy for poultry) and under the Food Labelling (Amendment) (Irradiated Food) Regulations of 1990, all irradiated foods are required to have a display indicating that they have received such treatment.

For labelling to be effective, it is clearly important that there should be a procedure available which is capable of detecting whether or not a food has been irradiated. Since a major feature of ionizing radiation is the negligible chemical change produced in the target, detection is most difficult; but several exceedingly sensitive methods have been suggested. As indicated earlier, ionization involves the production of free radicals which are usually short-lived. These persist, however, when produced in hard material such as bone, and can be detected by electron spin resonance (Phillips, 1988). Radiolytic products from DNA (e.g. *cis*-thymidine glycol) can be detected immunologically at a level as low as 10^{-15} moles. Dihydrothymidine is produced under anoxic conditions by the interaction of water-derived radicals and thymidine (Deeble *et al.*, 1990). It is a highly specific index of radiation treatment. Again, the $OH^\bullet$ radicals produced by ionizing radiation from phenylalanine in proteins yield *o*-, *m*- and *p*-tyrosine. Since only the last of these occurs naturally, the titre of the *o*- and *m*-tyrosine can be used as a measure of the irradiation received by the product (Elias, 1985). Radiolysis of the saturated triglycerides of foods produces 2-alkylcyclobutanones and methods have been developed for the specific detection of 2-dodecyl- and 2-tetradecylbutanone (Stevenson *et al.*, 1993). Irradiation of triglycerides also produces much larger quantities of carbon monoxide than would be found in control material (Furuta *et al.*, 1992). A relationship can be shown between the concentration of water-derived radicals from connective tissue and the relative viscosity of hyaluronic acid solutions.

It should be pointed out that, apart from the usefulness of ionizing radiation in preserving food for human consumers, the procedure can offer protection indirectly. Thus, about 250 tonnes of frozen horsemeat is imported annually for the manufacture of pet foods. Sixty per cent of it is contaminated with *Salmonella* spp, but the safety of the animal consumers, and of their owners, can be ensured by treating the horsemeat with 0.65 Mrad (e.g. from a Co^{60} source).

9.2   Antibiotics

Since antibiotics are chemicals, it is arguable whether they should be categorized separately from chemical preservatives (§ 9.3). Insofar as they are selective in their action between microbial species, however, and may foster the development of resistant strains of micro-organisms, they differ from such general antimicrobial chemicals as organic acids and alcohols.

The events leading to the use of certain so-called broad-spectrum antibiotics in food preservation have been outlined in § 6.4.3. The choice of a suitable antibiotic depends on the type of spoilage to be controlled, on the stability and solubility of the antibiotic at the pH of the food, on its stability to heat and on its lack of toxicity. Two aspects of the use of antibiotics are particularly important (Deatherage, 1955) and should be reiterated. Firstly, since they are mainly bacteriostatic rather than bactericidal, they are most effective where the total bacterial population is low – discouraging their indiscriminate use with highly contaminated or partially spoiled meat. Secondly, since they do not sterilize foods, they delay rather than prevent spoilage; and they may alter the bacterial flora since the latter will differ in sensitivity. One disadvantage of antibiotics, therefore, is the possible development of resistant micro-organisms in the meat after susceptible bacteria, which normally compete with them for the available nutrients, are eliminated. As already indicated (§ 6.4.3), this difficulty can be overcome to some extent by employing, for preservative purposes, antibiotics which are not applied in human medicine; but, in commercial practice, elimination of normal spoilage organisms, and thereby of the unpleasant superficial symptoms of contamination, could conceivably permit the growth of dangerous bacteria whose toxic by-products would not be easily detectable before consumption of the meat. This danger can be circumvented by combining the use of antibiotics with refrigeration (Barnes, 1956), since most pathogens will not grow rapidly at chill temperatures, and it is in this context that antibiotics are most useful. Nevertheless, under experimental conditions, antibiotics have also been used successfully against potential deep spoilage organisms where meat is to be exposed to a high temperature for a short time, rather than subjected to prolonged storage. Notwithstanding their disadvantages, they can give the benefits of enhanced storage life without themselves causing any chemical or biochemical change in the meat. Few are effective against yeasts and moulds.

With antibiotics there is a remote possibility of toxicity from residues which are not destroyed during cooking (§ 6.4.3). This danger can be overcome to some extent by legislating for permissible limits for residues; and generally the level of antibiotics initially present in meat declines during storage and before the meat is consumed (Weiser et al., 1954).

Virtually no work has been done on the relative efficacy of antibiotics in different muscles, but there is some suggestion that they may be destroyed more rapidly in muscles of high ultimate pH – or where bacterial growth causes the pH to rise (E.M. Barnes, personal communication).

In general, the storage of fresh meat at chill temperatures is curtailed by the development of micro-organisms on the exposed surfaces belonging to the genera *Pseudomonas* and *Achromobacter* (§§ 6.2 and 6.3.1), although deep spoilage organisms may be involved in some cases where cooking has been delayed or there is bone taint. Since, under such conditions, the microflora is mixed, the use of a 'broad-spectrum' antibiotic (e.g. oxytetracycline or chlortetracycline) can appreciably delay spoilage by both types of organisms. The meat so preserved will of course be thereby more liable to the non-microbial changes already indicated in the case of fresh and chilled meat (§ 7.1.1).

The antibiotics may be injected (intravenously or intraperitoneally) into the animals pre-slaughter (McMahan et al., 1955–6; Ginsberg et al., 1958), perfused after death (Deatherage, 1955, 1957), sprayed on to the carcasses or cut surfaces and, if packaging operations are involved, incorporated into the film (Firman et al., 1959). Alternatively, the meat may be dipped in an antibiotic-containing solution (Tarr et al., 1952).

Many examples of the efficacy of such treatment could be given. Thus Goldberg et al. (1953) showed that the storage life of ground beef could be extended to 9 days at 10 °C by the incorporation of 0.5–2 ppm of chloramphenicol, chlortetracycline or oxytetracycline. Controls, and samples treated with penicillin, bacitracin or streptomyocin, spoiled in 5 days. In particular, the storage of fresh comminuted meat, such as pork sausages, can be doubled or trebled by the use of antibiotics (Wrenshall, 1959).

Table 9.3 Microbial counts on steaks treated with
chlortetracycline (CTC) in various ways and stored at 5°C
(after Firman *et al.*, 1959)

Treatment			Micro-organisms ($\times 10^6$/g meat) (days at 5°C)		
Preslaughter injection	Surface	Type of film	0	5	10
–	–	Plain	0.013	8.56	897
–	–	CTC	0.012	0.73	276
–	CTC	Plain	0.002	0.14	225
–	CTC	CTC	0.001	0.19	88
CTC	–	Plain	0.005	0.001	14
CTC	CTC	Plain	0.006	0.012	6
CTC	–	CTC	0.0001	0.002	1
CTC	CTC	CTC	0.005	0.005	0.25

The relative efficacy of antibiotics as administered by preslaughter injection, by spraying and by the use of impregnated wraps is shown in Table 9.3.

It is evident that each procedure has antimicrobial action but that preslaughter injection is the most effective: distribution of the antibiotic is clearly more complete when it can be carried by the animal's bloodstream into the most minute interstices of the muscle. Moreover, the inhibitory antibiotic is present *in situ* before contaminating organisms can start to multiply.

Although the tetracyclines extend the refrigerated storage life of meat, moulds and yeasts tended to grow after about 14 days at 2°C (Barnes, 1957). On the other hand, Niven and Chesebro (1956) showed that sorbic acid could be used to suppress yeasts in minced beef. Should the temperature of such meat products rise to 15°C, antibiotic-resistant pathogens, unhindered by the presence of normal spoilage organisms, may be able to multiply, e.g. *Salmonella Typhimurium* (Hobbs, 1960). Animals are more likely to be carrying antibiotic-resistant organisms if they have been fed with antibiotic supplements (Barnes, 1958; and cf. § 2.5.2.2).

The tetracycline antibiotics and irradiation appear to be complementary in extending the refrigerated life of fresh meat (Niven and Chesebro, 1956). Low-level irradiation of meat tends to change the spoilage flora from a Gram-negative one to a Gram-positive one (McLean and Sulzbacher, 1953) – and the tetracyclines are particularly effective in retarding the growth of the latter. Table 9.4 illustrates an experiment where small pieces of beef were irradiated with 10,000 rad or treated with 10 ppm oxytetracycline and then irradiated. The complementary effect of irradiation and antibiotics against the bacteria is evident.

The problems for control by antibiotics in canned meats are very different from those in fresh meat. An antibiotic which is relatively heat stable (e.g. nisin, subtilin or tylosin) is needed and it must be especially effective against spore-forming bacteria (Barnes, 1957; Hawley, 1962). During prolonged storage the retention of nisin activity is enhanced by acid conditions and by short-time, high-temperature heat treatments. For low-acid foods, such as meat, the use of nisin and of similar antibiotics can only be considered in conjunction with sufficient heat to kill *Cl. botulinum*.

Table 9.4   Effect of gamma irradiation (10,000 rad) alone or in combination with oxytetracycline (10 ppm) on bacterial flora of beef muscle stored at 2 °C (Niven and Chesebro, 1956)

Treatment	Micro-organisms ($\times 10^3$/g meat) (days)			
	0	8	14	20
Control	90	700,000	4,000,000	–
Irradiated	0.5	100	60,000	800,000
Irradiated + antibiotic	0.05	1	60	2,000

Antibiotics might be usefully employed to control spoilage bacteria which are very heat resistant and which cannot otherwise be killed unless the product is given a heat treatment so excessive that its texture is damaged (Hansen and Riemann, 1958). In semi-preserved products, such as canned hams, mild heat treatment (or pasteurizing) will suffice to destroy any surviving faecal streptococci which would resist the antibiotics and would otherwise necessitate refrigerated storage for the product. The possibility of diminishing heat treatment, or curing salts, through the use of antibiotics could lead to some danger from *Cl. botulinum*, however, and antibiotics are, therefore, not permitted with semi-preserved meats (Hawley, 1962), although certain antibiotics are known to lower the heat resistance of spores of *Cl. botulinum* and *Cl. sporogenes* (Leblanc *et al.*, 1953).

Antibiotics have also been employed to permit the conditioning of meat (§§ 5.4.2 and 10.3.3.2) in a relatively short period at high temperature – both alone and in combination with pasteurizing doses (45,000 rad) of ionizing radiation (Wilson *et al.*, 1960).

Insofar as competition between micro-organisms may involve the production of chemicals by one type which are inimical to the survival of another (as well as competing in the same pool of nutrients), bacterial cultures have been deliberately employed to extend the shelf life of various meat products (Luecke, 1990). Such protective cultures can suppress the growth of food poisoning organisms. However, although the use of antibiotics can effectively preserve meat, their widespread use is currently unlikely because of concerns about the development of resistance strains of micro-organisms. Genetic manipulation of starter cultures (such as lactic acid bacteria) in fermented meat products could increase their protective and probiotic capacity as represented by catalase and bacteriocin (Hammes and Hertel, 1998).

9.3   Chemical preservatives

When hygienic principles were little practised, and less understood, chemical preservatives were not infrequently used in foods to offset what is now recognized as microbiologically dangerous action. Nevertheless, this remedy has drawbacks because chemical preservatives may be non-specific protoplasmic poisons and as undesirable for the consumer as for the micro-organisms against which they are directed; moreover, their effect may be cumulative rather than immediate. The term was defined by the United Kingdom Preservatives Regulations of 1962 as follows:

> 'Preservative' means any substance which is capable of inhibiting, retarding or arresting the process of fermentation, acidification or other deterioration of food or of masking any of the evidence of putrefaction; but does not include common salt (sodium chloride), lecithins, sugars or tocopherols; nicotinic acid or its amide; vinegar or acetic acid, lactic acid, ascorbic acid, citric acid, malic acid, phosphoric acid, pyrophosphoric acid or tartaric acid or the calcium, potassium or sodium salts of any of the acids specified in this subparagraph; glycerol, alcohol or potable spirits, isopropyl alcohol, propylene glycol, monoacetin, diacetin or triacetin; herbs or hop extract; spices or essential oils when used for flavouring purposes.

More recently, however, the addition of nicotinamide and ascorbic acid to meat, as colour preservatives, was prohibited. Formaldehyde, as such, is also prohibited, since it is demonstrably toxic (Wiley, 1908), but is permitted up to 5 ppm if extracted from wet strength wrapping materials.

Fresh meat, when in the intact carcass, is not usually severely contaminated except on the surface (cf. § 6.1). On the other hand, in preparing products containing comminuted or minced meats, there is every opportunity for massive bacterial contamination from the hands of operatives and from equipment, and this certainly occurred frequently in the past. It was in such a context that chemical preservatives were particularly employed in relation to meat.

Very few chemicals are now permitted as preservatives – and these only in minute specified quantities. Apart from nitrate, nitrite, sorbic acid and tetracyclines (which have already been considered in relation to curing and antibiotics), the United Kingdom Preservatives in Food Regulations 1962 lists only seven, namely sulphur dioxide, propionic acid, benzoic acid, methyl-p-hydroxy-benzoate, ethyl-p-hydroxy-benzoate, diphenyl, o-phenylphenol and copper carbonate. Of these seven only sulphur dioxide is permitted in meat preservation, up to 450 ppm being in sausage and sausage meat (Anon., 1972a). Its effect is antimicrobial, and at the permitted level it has no beneficial effect on meat colour, so the public would not recieve an erroneous impression of the meat's freshness (Kidney, 1967). Although the use of sulphur dioxide as a preservative at permitted levels is generally regarded as safe, it has been associated, nevertheless, with an increase in the severity of asthmatic and other respiratory conditions in some individuals (Simon, 1998) and means of further limiting dependence on sulphites have been sought. Chitosan, a polysaccharide from the shells of crabs and shrimp, and carnocin, an antimicrobial agent produced by *Carnobacterium piscicola*, have been shown to be effective antimicrobials in combination with low levels of sulphite in preserving chilled sausage meat (Roller *et al.*, 2002). Boric acid was also employed until relatively recently both in sausage meat and in curing, but its use in the United Kingdom was discontinued by the aforementioned Act of 1928. (This prohibition was temporarily suspended during the Second World War.)

Spices and essential oils are excluded from the UK Preservative Regulations 'when used for flavouring purposes' (above). Various essential oils have preservative properties, however, and have been used to extend the storage life of meat products (McNeil *et al.*, 1973). These include eugenol in cloves and allyl isothiocyanate in mustard seed. Shelef, Naglik and Bogen (1980) demonstrated that 0.3 per cent of sage or rosemary was inhibitory and 0.5 per cent bactericidal. With the increasing resistance of consumers to the use of synthetic preservatives in food there is renewed interest in the identification and use of naturally occurring chemicals that have antimicrobial properties. Thus, Roller (2002) has reviewed the antifungal and antimicrobial properties of the polysaccharide chitosan. Its efficacy, especially in combination with other antimicrobial agents, warrants further development.

Marinades (cf. § 10.2.1.1), defined as pickles containing vinegar and wine with herbs and spices which have been traditionally used to steep fish and meat, would be excluded from the above definition of 'preservative'; but, of course, these are a mode of chemical preservation depending primarily on the antimicrobial action of acetic acid. The scientific basis of the process has been studied by Gault (1991) since there has been a recent increase in the range of meat products offered in supermarkets and these include a range of marinades, either in final form or which can be prepared from ingredients supplied. Instructions for the latter include marinading at chill temperature for 1–2 hours and subsequent grilling. Although not all marinades lower the pH to the extent required for microbial stability or optimal eating quality, there is now sufficient scientific understanding of the process to permit successful marketing of acid-marinaded meat products having specific attributes.

As already indicated (cf. §§ 6.4.3 and 9.2), various micro-organisms produce organic acids and alcohols by anaerobic fermentation of food substrates and these, by inhibiting other organisms that are concomitantly present and which could spoil the food or make it toxic, can act in its preservation. The particular antimicrobial chemicals produced during fermentation depend on the nature of the food (and other aspects of the environment) and on the micro-organism. In recent years the traditionally recognized benefits of lactic acid bacteria in promoting health (probiosis) have been more positively exploited (Salminen et al., 1998), and certain strains appear to have anticarcinogenic action (Zabala et al., 2001). Lactic acid is frequently the effective inhibitory agent. Its efficacy, and that of other such agents may be enhanced by curing salts and dehydration. Fermentation of fresh and salted meats is an ancient process (Brothwell and Brothwell, 1969; Zeuthen, 1995), although as traditionally operated, was uncontrolled and slow. Faster methods have been developed which use starter cultures rather than relying upon indigenous micro-organisms (Bacus, 1984). Moreover, meat fermentation by bacteria and fungi is now being applied to hitherto non-fermentated products to accelerate flavour development, extend storage life and control pathogens (Smith and Palumbo, 1981) (Lücke, 2000). Staphylococcus spp. limit fat oxidation and aldehyde production, and contribute to the production of desirable flavour components, such as esters (Mantel et al., 1998). A review of the nature and variety of fermented meats has been published (Campbell-Platt and Cook, 1995).

Perhaps, under the general heading of chemical preservatives, carbon dioxide should be included (Callow, 1955a); also ozone (Kefford, 1948; Kaess, 1956), both of which have been used to discourage the growth of surface micro-organisms on beef carcasses during prolonged storage at chill temperatures. Although ozone leaves no toxic residues in the meat, its use in the store can be dangerous for personnel. Moreover, it accelerates the oxidation of fat and is more effective against air-borne micro-organisms than against those on the meat. The latter destroys the gas (Kaess, 1956).

Chapter 10

The eating quality of meat

Considering the diversity and duration of the events which determine the nature of meat, it seems curious that the consumer's palate can only react to the commodity over a few minutes during mastication. No conscious sensation is derived from the process of digestion over the subsequent 10 hr or so when the amino acids, fatty acids, vitamins, minerals and other constituents are being liberated and absorbed into the body. Nevertheless, however fleeting, the organoleptic sensations may enhance or impair the efficacy of digestion by their reflex action on the production of gastric and intestinal juices – and thus the nutritive value of the food, as the work of Claude Bernard and Pavlov suggested in the nineteenth century and as most textbooks on physiology imply or testify. The attributes of eating quality will be considered in this chapter.

Of the attributes of eating quality, colour, water-holding capacity and some of the odour of the meat are detected both before and after cooking and provide the consumer with a more prolonged sensation than do juiciness, texture, tenderness, taste and most of the odour which are detected on mastication.

The increased sale of meat as relatively small prepackaged cuts for the individual consumer, has emphasized the inherent differences in eating quality between muscles (as well as revealing variability from unidentified causes). The concomitant development of 'seaming out' muscles (or groups of muscles) from the (frequently still warm) carcass which is now undertaken widely instead of the more traditional carcass dressing (cf. § 5.2.2), makes quality control more feasible. By introducing standards for the quality of retail meat, and guidelines for their attainment (e.g. a minimum postslaughter ageing time and optimal processing operations) general improvements have been achieved in New Zealand (Bickerstaffe *et al.*, 2001).

In a detailed reassessment of some 30 anatomically defined bovine muscles, in respect of their biochemical composition and its relation to eating quality, Jeremiah *et al.* (2003a, b, c) emphasized that they differed significantly in tenderness, juiciness, flavour and overall palatability, and that these attributes were related to such conditions as losses in cooking and after freezing and thawing. Jeremiah *et al.* concluded that the mode of cooking would be required to be tailored to the intrinsic nature of individual muscles, rather than to type of joint insofar as the latter comprised various muscles of mixed nature, if consumers were to enjoy maximum satisfaction. (This view has been supported by Kolle *et al.*, 2004.)

Clearly, whatever the scientific basis of the attributes of eating quality in meat, their significance will be determined by regional preferences and by the views of the individual consumer. Some prefer markedly tough meat: others prefer excessive tenderness. Nevertheless, between the member countries of the European Union less extreme opinions prevail and attempts to identify common standards of meat quality have been made in the interests of international trade. A major study of the eating quality of grilled *l. dorsi* and casseroled *semimembranosus* muscles from Galloway and Charollais steers was undertaken by Dransfield *et al.* (1984). This involved the assessment of beef from the same animals by the meat research laboratories of eight member countries, using both local scales and those of the other countries. It was evident Irish and English panels tended to value flavour more highly than tenderness and juiciness, whereas the latter attributes were of predominant importance to Italian panellists. French and Belgian taste panels showed a preference for the flavour of aged beef from older animals. It was concluded that it was not feasible to recommend a standard scale or cooking procedure for the eight participating countries, but the assessment of texture was consistent and comparable.

In a comparison of the attributes of eating quality of 15 species used for meat in Norway, Rødbotten *et al.* (2004) found that colour was the most significant after texture, flavour being the least important attribute. Liver flavour and 'gamey' flavour were the best descriptions of flavour between species.

Increasingly detailed information on the structure and function of genes and on the pattern of their expression via mRNA transcription, protein synthesis and the metabolic pool, which the emerging disciplines of genomics and bioinformatics (cf. § 3.3.1) have provided, is potentiating the deliberate selection of the characteristics of muscular tissue which determine the nature of meat. Thereby the differences in the attributes of eating quality and nutritive value required to satisfy the exacting and individual requirements of modern consumers of meat could be met (Eggen and Hocquette, 2003).

In recent years there has been a considerable development of so-called 'organic' rearing of animals, this being reported to produce meat of superior eating quality. So far, scientific studies have failed to demonstrate that the meat has any organoleptic advantage over that produced conventionally (cf. Millet *et al.*, 2005).

10.1   Colour

Since 1932, when Theorell crystallized the principal pigment of muscle and it was shown that myoglobin was not identical with the haemoglobin of the blood, it has been accepted that the colour of meat is not substantially due to haemoglobin unless bleeding has been faulty (§ 5.2.2). The appearance of the meat surface to the consumer depends, however, not only on the quantity of myoglobin present but also on the type of myoglobin molecule, on its chemical state and on the chemical and physical condition of other components in the meat. Each of these, in turn, is determined by a variety of factors.

10.1.1   The quantity and chemical nature of myoglobin

Factors determining the quantity of myoglobin were incidentally indicated in § 4.2. As one generalization, it is clear that a high level of muscular activity evokes the elaboration of more myoglobin – reflecting, in this respect, differences due to species, breed, sex, age, type of muscle and training. Thus, muscles of the hare have more myoglobin than those of the rabbit; those of racing thoroughbreds have more than those of draught horses; those of bulls have more than those of cows; and those of steers have more than those of calves. A very pale red colour is a predominant consumer prerequisite for the raw flesh of the young bovine (veal) (cf. § 4.3.4). Much of the pigment present in veal is derived from residual haemoglobin of the blood (Klont *et al.*, 1999).

The constantly operating muscle of the diaphragm has more myoglobin than the occasionally and less intensively used *l. dorsi*; and free range animals have more muscle pigment than their stall-fed counterparts. Another kind of factor is the plane and nature of nutrition – a high plane, and a diet low in iron, both leading to low myoglobin concentrations (although by different mechanisms). Japanese consumers prefer beef that is less red than that which prevails in Europe. It has been found that the myoglobin content of the muscles can be reduced by feeding green tea to the cattle (Zembayashi *et al.*, 1999). The quantity-determining factor which is most difficult to understand is that causing the variability, occasionally encountered within a given muscle, when the myoglobin concentration may be several hundred-fold different over distances of 1 cm.

Species differences in the myoglobin molecule have been observed. The hues of red oxymyoglobin and of brown metmyoglobin from beef and pork are not identical; but this subject has been little investigated. When freshly cut, the surface of pork *l. dorsi* forms oxymyoglobin at a faster rate than that of beef (Haas and Bratzler, 1965).

Most of the striking differences in the colour of meat surfaces arise from the chemical state of the myoglobin molecules. It is not appropriate to consider here details of the chemistry of the muscle pigment, which are available elsewhere (e.g. Lemberg and Legge, 1949), but brief comments are desirable. Some of the chemical states in which myoglobin may be encountered in meat are shown in Table 10.1. The myoglobin molecule consists of a haematin nucleus attached to a protein component of the globulin type: the molecular weight is about 17,000. The haematin portion comprises a ring of four pyrrole nuclei co-ordinated with a central iron atom. The iron may exist in both reduced and oxidized forms. In the ferrous form it can combine with gases such as oxygen and nitric oxide. The ability to combine with oxygen is lost when the globin portion of the molecule is denatured and the tendency for the iron to oxidize to the ferric form is then greatly increased; but union of the iron with nitric oxide is strengthened (Lemberg and Legge, 1949). Whilst, therefore, the oxidation of purplish-red myoglobin or of bright red oxymyoglobin to brown metmyoglobin is accelerated by any factors which cause denaturation of the globin (Brooks, 1929b, 1938; Watts, 1954) by the absence of reducing mechanisms and by low oxygen tension (cf. Fig. 7.3), these same circumstances enhance the stability of the red colour of cured meat, converting nitric oxide myoglobin into nitric oxide haemochromogen (Watts, 1954; Killday *et al.*, 1988). In these pigments the iron is in the ferrous form, but nitrite will also react with metmyoglobin to form a red compound (Barnard, 1937) – metmyoglobin nitrite.

Table 10.1 Pigments found in fresh, cured or cooked meat

Pigment	Mode of formation	State of iron	State of haematin nucleus	State of globin	Colour
1. Myoglobin	Reduction of metmyoglobin; de-oxygenation of oxymyo-globin	Fe++	Intact	Native	Purplish-red
2. Oxymyoglobin	*Oxygenation* of myoglobin	Fe++ (or Fe+++?)	Intact	Native	Bright red
3. Metmyoglobin	*Oxidation* of myoglobin, oxymyoglobin	Fe+++	Intact	Native	Brown
4. Nitric oxide myoglobin (nitrosomyoglobin)	Combination of myoglobin with nitric oxide	Fe++	Intact	Native	Bright red (pink)
5. Nitric oxide metmyoglobin (nitrosomet-myoglobin)	Combination of metmyoglobin with nitric oxide	Fe+++	Intact	Native	Crimson
6. Metmyoglobin nitrite	Combination of metmyoglobin with excess *nitrite*	Fe+++	Intact	Native	Reddish -brown
7. Globin myohaemo-chromogen	Effect of *heat*, denaturing agents on myoglobin, oxymyoglobin; irradiation of globin haemichromogen	Fe++	Intact (usually bound to denatured protein other than globin)	Denatured (usually detached)	Dull red

8. Globin myohaemichromogen	Effect of *heat*, denaturing agents on myoglobin, oxymyoglobin, metmyoglobin, haemochromogen	Fe^{+++}	Intact (usually bound to denatured protein other than globin)	Denatured (usually detached)	Brown (sometimes greyish)
9. Nitric oxide myohaemochromogen	Effect of *heat*, denaturing agents on nitric oxide myoglobin	Fe^{++}	Intact	Denatured (absent)	Bright red (pink)
10. Sulphmyoglobin	Effect of H_2S and oxygen on myoglobin	Fe^{++}	Intact but one double bond saturated	Native	Green
11. Metsulphmyoglobin	Oxidation of sulphmyoglobin	Fe^{+++}	Intact but one double bond saturated	Native	Red
12. Choleglobin	Effect of hydrogen peroxide on myoglobin or oxymyoglobin; effect of ascorbic or other reducing agent on oxymyoglobin	Fe^{++} or Fe^{+++}	Intact but one double bond saturated	Native	Green
13. Nitrihaemin	Effect of large excess *nitrite* and *heat* on 5	Fe^{+++}	Intact but reduced	Absent	Green
14. Verdohaem	Effect of reagents as in 7–9 in excess	Fe^{+++}	Porphyrin ring opened	Absent	Green
15. Bile pigments	Effect of reagents as in 7–9 in large excess	Fe absent	Porphyrin ring destroyed Chain of pyrroles	Absent	Yellow or colourless

In the myoglobin, oxymyoglobin and metmyoglobin of fresh meat, or the nitric oxide myoglobin and metmyoglobin nitrite of cured meat, the haematin nucleus is intact and the protein is in a native form, but the colour and valency of the iron vary. On heating, as in cooking, the globin is denatured, but the haematin nucleus still remains intact as in the red globin haemochromogen, or more commonly brown globin haemichromogen and in pink nitric oxide haemochromogen. It has been suggested that the colour complexes in cooked meat are denatured haemo-proteins wherein the protein may be one or other of the several denatured proteins present, and not only globin (Ledward, 1971b). Irradiation of brown globin haemichromogen converts it to red globin haemochromogen (Tappel, 1957a). Reduction of the haematin nucleus occurs when myoglobin is exposed simultane-ously to hydrogen sulphide and oxygen (forming green sulphymyoglobin) on the one hand, or to hydrogen peroxide and ascorbic acid (or other reducing agents) on the other (forming green choleglobin) – as caused by the growth of certain micro-organisms, or in tissue injury *in vivo*. Sulphmyoglobin formation is more likely to be observed in meat having an ultimate pH above 6 since, at lower pH values, bacteria capable of producing H_2S are unable to do so (Nicol *et al.*, 1970; Shorthose *et al.*, 1972). If these conditions are intensified the porphyrin ring may be opened, although the iron remains, forming green verdohaem; and finally, on further or more intense exposure, the iron will be lost from the porphyrin, which will split from the protein moiety and open out, forming the chain of pyrroles characterizing yellow or colourless bile pigments. Excess of nitrite in cured meats can produce the crimson-red colour of metmyoglobin nitrite, and if the latter is treated with an even greater excess of nitrite and heated in acid conditions, green nitrihaemin is formed.

In fresh meat, before cooking, the most important chemical form is oxymyoglo-bin. Although it occurs on the surface only, this pigment is of major importance, since it represents the bright red colour desired by purchasers. In uncooked meat the cytochrome enzymes (§§ 4.2.1 and 5.3) are generally still capable of utilizing oxygen for a considerable period post-mortem. Although there is no oxygen in the depths of the meat, the gas can diffuse inwards for some distance from meat sur-faces exposed to the air, and a point of balance is established between the rates of diffusion and of uptake by the cytochrome enzymes – and by myoglobin, in forming oxymyoglobin. The depth of penetration d is given by $d \times \sqrt{2c_0 D/A_0}$, where c_0 is the pressure of oxygen on the surface and D and A_0, respectively, the coefficients of dif-fusion and consumption (Brooks, 1938). The bright red colour of oxymyoglobin will predominate and be apparent to the observer – from the outside in to the point where the ratio oxymyoglobin : myoglobin is about $1:1$, i.e. about 84 per cent of the total depth of oxygen penetration (Brooks, 1929b). Since different muscles have dif-ferent inherent surviving respiratory activity, d will vary under a given set of con-ditions. Thus, after exposure of cut surfaces to the air for 1 h at 0 °C, the depth of the oxymyoglobin layer was found to be 0.94 mm in horse *psoas* muscle in which respiratory activity is relatively high and 2.48 mm in horse *l. dorsi* in which respira-tory activity is relatively low (Lawrie, 1953b). In bacon, where the respiratory enzymes are largely inactivated by the high salt concentration, the depth of pene-tration is about 4 mm (Brooks, 1938). Again, since the coefficient of diffusion decreases less than the respiratory activity for a given fall in temperature, the depth of the bright red layer of oxymyoglobin will be greater at 0 °C than, for example, at 20 °C (Brooks, 1929b, 1935; Urbin and Wilson, 1958) – hence the tendency for the colour of meat surfaces to become somewhat brighter when stored at lower temperatures.

As will be apparent from the above considerations, the principal pigment of cooked meats is brown globin haemichromogen and, in the case of cooked bacon, red nitric oxide haemochromogen. Tappel (1957b), however, believes that a globin nicotinamide haemichromogen also contributes to the colour of cooked meat. The bright red (pink) colour of Parma ham – achieved without the addition of nitrate or nitrite – is due to Zn-protoporphyrin IX (Wakamatsu et al., 2004). Brown pigmentation in cooked meats, unlike that in fresh, is normally a desirable attribute of meat quality. The temperature of cooking naturally affects the degree of conversion of the pigments. Thus, beef cooked to an internal temperature of 60 °C has a bright red interior; that cooked to an internal temperature of 60–70 °C has a pink interior; and that cooked to an internal temperature of 70–80 °C or higher is greyish brown (Jensen, 1949). Denaturation of myoglobin in meat is considerable at temperatures which cause negligible denaturation of the pigment in solution (Fig. 10.1; Bernofsky et al., 1959). Below 65 °C, myoglobin denaturation, as measured by the percentage extractability of the pigment, may arise from enzymic action or co-precipitation rather than from the temperature. Myoglobin is one of the more heat-stable of the sarcoplasmic proteins. It is almost completely denatured, however, between 80 and 85 °C, and this has been made the basis of a test for determining whether or not meat has been heated to 90 °C (Roberts, 1971). Meat so heated is unlikely to be a source of viable foot-and-mouth virus. As a precaution against poisoning from meat infected with pathogens such as E. coli O157:H7, it was recommended by UK health authorities that minced meat products should be heated for at least 2 minutes at 70 °C, until no red colour remained, it being the assumption that the organism would be destroyed before myoglobin was denatured and converted to myohaemochromogen. An investigation of lamb products by Lytras et al. (1999) demonstrated, however, that denaturation of myoglobin occurred considerably before the temperature required to inactivate E. coli O157:H7 was attained. (Moreover there was some difference between muscles in the formation of the brown pigment.) From its colour the meat could thus appear 'cooked' before

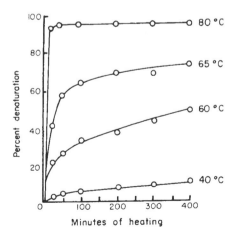

Fig. 10.1   The effect of temperature on percentage denaturation of myoglobin from beef, as measured by decreased water extractability (Bernofsky et al., 1959). (Courtesy Dr J. B. Fox.)

potential danger from the organism was removed. It has been suggested that determination of the relative proportion of deamidated actin components may also provide a measure of the severity of heat treatment sustained by cooked meat (King, 1978). Parsons and Patterson (1986a) demonstrated that the content of soluble nitrogenous components remaining in saline extracts of heated bovine *l. dorsi* and *semimembranosus* muscles decreased systematically as the heating temperature was raised from 40 to 90 °C. In a subsequent study (Parsons and Patterson, 1986b) using differential scanning calorimetry, they showed that the three peaks in the thermograms of raw beef gradually changed on heating as the proteins denatured. There was a correlation between the maximum temperature applied and the onset of denaturation and between the area of the thermograms and the duration of heating. These features were believed to be diagnostic of the heat treatment received by the meat.

Other factors contribute to the brown colour of cooked meat, including the caramelization of carbohydrates and Maillard-type reactions between reducing sugars and amino groups. The latter is particularly marked in pork, where considerable production of reducing sugars occurs through amylolytic action post-mortem (Sharp, 1957, 1958) and largely determines the degree of browning in this meat, which has relatively little myoglobin (Pearson *et al.*, 1962). During cooking, there is a significant release of non-haem iron from haem pigments and these enhance lipid oxidation (Igene *et al.*, 1979).

10.1.2 Discoloration

If the ultimate pH of the meat is high, the surviving activity of the cytochrome enzymes will be greater (Lawrie, 1952b). Moreover, because the muscle proteins will be considerably above their iso-electric point, much of the water in the muscle will still be associated with them and the fibres will be tightly packed together, presenting a barrier to diffusion. As a result of these two factors, the layer of bright red oxymyoglobin becomes vanishingly small and the unpleasant, purplish-red colour of myoglobin itself will predominate to such an extent that the meat will appear dark (dark-cutting beef, 'glazy' bacon). Furthermore, the high ultimate pH alters the absorption characteristics of the myoglobin, the meat surfaces becoming a darker red (Winkler, 1939). Such meat will also appear dark because its surface will not scatter light to the same extent as will the more 'open' surface of meat of lower ultimate pH.

On the other hand, in the so-called PSE condition in pigs (§§ 3.4.3 and 5.4.1) the meat is very pale. One reason is the relative absence of myoglobin; another is chemical change in the pigment. The latter occurs either because the rate of pH fall has been very fast (and the sarcoplasmic proteins, including myoglobin, exposed to low pH whilst post-mortem temperatures were still high) or the ultimate pH is very low (Brooks, 1930). In both cases the myoglobin is exposed to conditions causing its oxidation to metmyoglobin which has a low colour intensity. In addition, the structure is 'open' and scatters light. The colour stability of muscles from pigs of NN genotype is significantly greater than that from nn muscles: the chemical status of the latter promotes faster metmyoglobin formation (Tam *et al.*, 1998). Refraction may contribute to the inverse relationship between the pH of meat and its paleness (Swatland, 2002, 2003).

Metmyoglobin is the most commonly occurring undesirable pigment on meat surfaces: its brown colour is noticeable when about 60 per cent of the myoglobin exists in this form (Brooks, 1938). As already indicated, the production of metmyoglobin from myoglobin or oxymyoglobin is accelerated by all conditions which cause denaturation of the globin moiety (Brooks, 1929b; Watts, 1954). These include (as well as low pH) heat, salts and ultraviolet light. Prolonged storage, as is possible at chill temperatures, or shorter holding at higher temperatures, cause surface desiccation (§§ 7.1.1.1 and 9.1.2.2) thus increasing salt concentration and promoting the formation of metmyoglobin. Low temperatures delay metmyoglobin formation both directly and indirectly by suppressing the residual activity of the oxygen-utilizing enzymes. Further details of the molecular changes involved in metmyoglobin formation may well be revealed by the use of monoclonal antibodies (Levieux and Levieux, 1996).

The formation of metmyoglobin is maximal at about 4 m oxygen pressure (cf. Fig. 7.3, p. 206: Brooks, 1935). The layer of brown metmyoglobin is thus formed a little below the meat surface (Brooks, 1935). On the other hand, oxidation of the pigment of cured meat, nitric oxide haemochromogen, is directly proportional to increasing oxygen tension (cf. §§ 8.3.2.2 and 8.3.3). It is thus not surprising that the rate of discoloration of cured meats increases directly with the pressure of oxygen under which they are packed, whereas that of fresh meat is inhibited by increasing oxygen pressure in the pack (Rickert et al., 1957). Lamb has a higher residual oxygen demand than either beef or pork, and this may be related to its greater tendency to discolour when stored as fresh meat (Atkinson and Follet, 1971).

Metmyoglobin, once formed, can be reduced both anaerobically (Lawrie, 1952b) and aerobically by surviving enzymes of the cytochrome system and with nicotinamide adenine dinucleotide (NADH) as coenzyme (Watts et al., 1966). Oxygen uptake, and the concentration of this coenzyme, are highly correlated in beef and lamb (Atkinson and Follet, 1973). The metmyoglobin reductase, dependent on nicotinamide-adenine as coenzyme, which is present in muscle, is inactivated at ca. 50 °C; but there is evidence for a more robust form of this enzyme (or of some other heat-tolerant system) which can operate at up to 70 °C (Osborn et al., 2003). Its action may explain the occasional occurrence of an undesirable pink or red discoloration in the interior of cooked, uncured meat.

Metmyoglobin reduction by NADH-cytochrome b_5 reductase involves cytochrome b_5 at the mitochondrial membrane and cytochrome b at the sarcoplasmic reticulum (Arihara et al., 1995). While there is a high negative correlation between *aerobic* reducing activity and metmyoglobin formation, however, there is little between the latter and metmyoglobin-reducing activity *anaerobically* (Ledward, 1972). O'Keefe and Hood (1982) found relatively little correlation between the tendency of the pigment of different muscles to oxidize and their power to reduce metmyoglobin once it had formed.

The mechanism is probably similar to that in red blood cells whereby any tendency for methaemoglobin to form is normally opposed by effective reducing systems involving flavoprotein enzymes (Gibson, 1948; Watts et al., 1966). Because the bright red of oxymyoglobin is desirable, most prepackaged fresh meat is placed in an oxygen-permeable wrap; but after a few days, even at chill temperatures, some of the surface pigment begins to oxidize to metmyoglobin or to myohemichromogen, through incipient denaturation of the globin moiety. This originally discouraged central packaging of fresh meat in oxygen-permeable wraps. If the meat is vacuum-packed, however, no oxygen can get in and the surviving activity of the cytochrome enzymes reduces the small amount of metmyoglobin which forms in these circumstances, replacing it by the purplish red of myoglobin (Dean and Ball,

1960). Vacuum-packed meats (e.g. in oxygen-*impermeable* shrinkable film) can be stored under chill conditions for some weeks. They can be allowed to reoxygenate before sale (thus restoring the bright-red colour of oxymyoglobin) when the film is removed. Centralized prepackaging of fresh meats has been established on such a basis.

Controlled within-package atmospheres of CO_2 or nitrogen, now frequently employed to prolong the storage life of chilled meat, will often contain (initially) traces of oxygen, whereby the meat surface will tend to form metmyoglobin. The latter, however, will be converted back to myoglobin by the above reducing enzyme systems of the muscles. Such discoloration, albeit transient, could be a problem if the meat were opened to the atmosphere after too short a period of equilibration (Gill and McGinnis, 1995). The latter have pointed out that the relationship between the degree and duration of such discoloration, on the one hand, and the volumes and oxygen concentration of within-package atmospheres on the other, will require better definition if transient metmyoglobin formation is to be avoided.

The cherry-red colour of the carbon monoxide derivative, which is more stable than oxymyoglobin, has been suggested as an alternative for prepacked meats. The colour stability of refrigerated meat can be retained for 15 days at 3 °C by this method (Flain, 1964). Although the cherry-red colour of carboxymyoglobin will present an attractive appearance for longer than oxymyoglobin under packaging conditions, deterioration in the microbiological status may be masked thereby (Kropf, 1980). The amount of carbon monoxide which is effective in colour retention (*ca.* 0.4%) in an atmosphere of 60% CO_2, 40% N_2, does not constitute a toxic hazard (Sørheim *et al.*, 1999) but its incorporation has not been generally accepted. Some success in retaining the bright-red colour of oxymyoglobin has attended the use of mixtures of carbon dioxide and oxygen as in-pack atmospheres. The carbon dioxide inhibits microbial growth (Taylor, 1971). Not surprisingly, the intrinsic biochemical differences between individual muscles (cf. § 4.3.5) affect the relative tendency of their pigment to oxidize to metmyoglobin under post-mortem conditions. Thus, the surface of both bovine and porcine *psoas* and *gluteus medius* muscles tend to form metmyoglobin faster than *l. dorsi* when exposed to air, even at comparable pH levels (Hood, 1971; Owen and Lawrie, 1975; Gill and McGinnis, 1995). This reflects the greater residual oxygen uptake of the former which, in turn, reflects the great intrinsic cytochrome oxidase and succinic dehydrogenase activities and also their usually higher ultimate pH (Lawrie, 1952a, b). Under oxygen, however, bovine *psoas*, *gluteus medius* and *l. dorsi* muscles show a comparable resistance to discoloration at 4 °C (MacDougall and Taylor, 1975).

It is now known that nitric oxide naturally occurs in the body, being produced in many cells by nitric oxide synthetase (Kou and Schroeder, 1995); and participating in a number of vital biochemical reactions. It can react with superoxide to form peroxynitrite (Koppenol, 1999) and this reaction probably contributes to the formation of metmyoglobin when fresh meat discolours (Connolly and Decker, 2004).

Discoloration of both fresh and cured meats has become a serious problem with modern methods of prepackaging and display. Light of visible wavelength, which does not affect the pigment of fresh meat over 3 days, dissociates the nitric oxide from the cured meat pigment, and cured meats may be discoloured after only 1 h exposure to visible light (Watts, 1954). Partially cooked cured products are especially susceptible. Incandescent, tungsten-filament and fluorescent lighting all cause the same degree of fading for a given time of exposure and light intensity. While ultraviolet light does not appear to have any greater effect than visible light on the fading of cured meats, it will cause brown discoloration in fresh meat – possibly through denaturation of the globin (Haurowitz, 1950). Freezing affords no protection against discoloration by light.

Attempts to avoid browning due to metmyoglobin formation, by incorporating ascorbic acid in meat products to reduce the oxidized pigment as soon as it forms (Bauernfeind, 1953) – and before extensive denaturation of the globin – or by the use of niacin (Coleman and Steffen, 1949), which is said to form a stable red pigment of myoglobin, have been made. The efficacy of ascorbic acid in this context has been utilized by pre-slaughter injections (D. E. Hood, personal communication). Whilst nicotinamide slows the rate of metmyoglobin formation, amounts as great as 60 mg/per cent are required, and nicotinic acid actually accelerates metmyoglobin formation (Kendrick and Watts, 1969). Both these expedients have been forbidden by legislation in the United Kingdom. That the brown pigment can be reduced again does not automatically mean that it will thereafter take up oxygen to form oxymyoglobin. The circumstances causing the pigment to oxidize may also have denatured the globin, and brown globin haemichromogen, although it can be reduced to reddish globin haemochromogen, cannot then form a co-ordination complex with oxygen. Meat of pH above 6 is considered unsuitable for holding in evacuated, gas-impermeable packs since bacterial production of H_2S leads to the formation of green sulphmyoglobin (Nicol et al., 1970; Shorthose et al., 1972). The relative absence of muscle glycogen in the immediate pre-mortem period, to which a high ultimate pH is usually due, is also responsible for failure to produce an appreciable quantity of glucose post-mortem. As a result, micro-organisms are obliged to utilize amino acids for energy instead, and at a lower population than in meat of normal ultimate pH. Thereby, off-odours become detectable when microbial counts are 10-fold less than those required to produce off-odours in the latter (Newton and Gill, 1978).

Among other undesirable colours in meat reference may again be made to those discolorations caused by microbial growth (§ 6.2) and to the excessive degree of unpleasant browning, accompanied by bitterness, occurring in dehydrated meats (especially pork) during storage (§ 8.1.3). Again, although the freezing of meat after the onset of rigor mortis is not detrimental to colour, there is considerable darkening of the lean and whitening of the fat observable in the fresh and cooked product when the meat has been blast frozen whilst post-mortem glycolysis is proceeding (Howard and Lawrie, 1956, 1957b). The nature of the colour change is unknown. Apart from browning of exposed surfaces through desiccation, increased time and temperature of holding post-mortem also tend to decrease the ability of freshly exposed surface to form oxymyoglobin (Bouton et al., 1957).

Pink or green discolorations are occasionally encountered in the fat of cured meat. These are probably due to the metabolic products of halophilic bacteria (§ 6.2; Jensen, 1945). The fat of fresh meat from old dairy cows is sometimes distinctly yellow, due to the accumulation of carotenoid pigments in the tissue and much of the present popular demand for young meat animals arises from the paleness of their intramuscular fat. If fed on grain instead of pasture for about two months prior to slaughter cattle lay down a whiter fat. In yellow fat, as the carotenoid content increases, the ratio of cis-monounsaturated fatty acids also increases (Zhou et al., 1993). Yellow or brownish discoloration of back fat in bacon has long been a problem. Three lipofuscin-like pigments have been isolated from discoloured areas, but conditions responsible for their accelerated production have not yet been elucidated (Juhász, *et al.*, 1976). The yellow colour of pork backfat increases with increase in the content of linoleic and α-linoleic acids (Maw *et al.*, 2003). For meat products, such as sausages, the addition of porphyrins prepared from blood would give an attractive red colour at 100 ppm, and, being of natural origin, would not constitute an artificial contaminant (C. L. Walters, personal communication).

Occasionally a rainbow-like iridescence appears on meat products which consumers assume (wrongly) represents some aspect of spoilage. Swatland (1988) considered the phenomenon was related to hydration of the tissue and Wang (1991) showed that the iridescence increases as the water-holding capacity decreases. Surface iridescence varies between different types of muscle, being particularly prevalent in bovine *semitendinosus* and associated with younger animals and relative low values of ultimate pH (Kukowski *et al.*, 2004). Absorbance of light by mitochondria in porcine muscles may well contribute to the red colour in addition to the myoglobin content and the pH (Swatland, 2004).

10.2   Water-holding capacity and juiciness

In that it affects the appearance of the meat before cooking, its behaviour during cooking and juiciness on mastication, the water-holding capacity of meat is an attribute of obvious importance. This is particularly so in comminuted meats such as sausages, where the structure of the tissue has been destroyed and is no longer able to prevent the egress of fluid released from the proteins. Diminution of the *in vivo* water-holding capacity is manifested by exudation of fluid known as 'weep' in uncooked meat which has not been frozen, as 'drip' in thawed uncooked meat, and as 'shrink' in cooked meats, where it is derived from both aqueous and fatty sources.

Lawson (2004) showed it was possible that water forced from between myofibrils *during* the contraction of rigor mortis, could enter channels formed between the fibre and the cell membrane because of the destruction of the integrin complex by calpain (cf. § 3.2.2.). From such channels water could flow to the exterior as weep or drip. If, however, the calpain-mediated degradation of integrin occurred post-rigor, it was hypothesized that water is expelled into the surrounding connective tissue which, since it has an affinity for water, could lead to diminished loss of fluid. Longitudinal muscular contraction, together with myofibrillar shrinkage, are the major factors driving water out of the myofibrillar system (Bertram *et al.*, 2004a).

Most of the water in muscle is present in the myofibrils, in the spaces between the thick filaments of myosin and the thin filaments of actin/tropomyosin. The interfilament space has been observed to vary between 320 Å and 570 Å in relation to pH, sarcomere length, ionic strength, osmotic pressure and whether the muscle is pre- or post-rigor (Offer and Trinick, 1983). This corresponds to a threefold change in volume, and it signifies a far greater change in the interfilament water content than can be accounted for by the binding of water to muscle proteins. Indeed Hamm (1960) pointed out that not more than 5 per cent of the total water in muscle could be directly bound to hydrophilic groups on the proteins. This amount is little altered by changes in the structure and charges of the latter although it is important in processing since its presence accelerates renaturation of proteins during dehydration and freezing (Greaves, 1960). The stepwise release of water from meat by the application of different temperatures has indicated that water is bound by the proteins in several layers (Hamm, 1960; Wierbicki *et al*, 1963).

The complexity of the system has also been shown by NMR studies. Water exists in at least two environments in muscle, and in each of these a proportion is 'bound' or 'free' (Pearson *et al.*, 1974). During the onset of rigor mortis, there is little change in the 'bound' water, but the proportion of so-called 'free' water in the extracellular region increases at the expense of the 'free' intracellular water. Careful histological studies have shown that at least two extracellular environments develop in muscle during rigor mortis (Offer *et al.*, 1983–85a).

In detailed studies of myofibrils, Offer and Trinick (1983) presented evidence in support of their view that most of the water in muscle is held by capillary forces between the thick and thin filaments. The relative unimportance of the binding of water to the surface of proteins was demonstrated by the fact that maximum water holding by myofibrils occurred under conditions when a considerable amount of the A-band proteins had been extracted. Interfilament spacing determines the major water-holding capacity of myofibrils, and that spacing is mainly determined by long-range electrostatic forces. Constraints opposing swelling when the negative charges on the protein filaments are increased at elevated pH include the attachments of the actin filaments to the Z-line, those of the myosin filaments to the M-line proteins and the cross-links between the actin and myosin filaments themselves (Millman and Nickel, 1980; Offer and Trinick, 1983). In explaining the swelling of myofibrils due to the uptake of water when they are bathed in salt solutions (as in curing), Offer and Trinick believe that the increased interfilament spacing is due not only to increasing negative charges on the filaments but also to an effect of the salt on the restraining links. As long as the cross-links between the thin (actin) and thick (myosin) filaments remain attached, the lattice cannot swell appreciably. Conversely, if the lattice does swell, the cross-links cannot remain attached. When the cross-links dissociate they must all do so at the same time to allow swelling. Swelling must therefore be a highly co-operative phenomenon (Offer and Trinick, 1983).

In pre-rigor muscle, wherein cross-linking between the actin and myosin filaments is prevented by the high ATP concentration (cf. § 4.2.3), expansion of the filament lattice in the presence of salt will be greater than in post-rigor muscle and will then be restrained only by the holding of the thick and thin filaments by the M- and Z-line proteins respectively. With sufficient salt concentration, the A-band will be extracted (cf. § 10.2.1.2).

Various procedures have been employed to determine the water-holding capacity of meat. The filter press method of Grau and Hamm (1953) has been used for many years and the gravimetric bag method of Honikel (1998) more recently. Because these methods involve some pressure damage to the structure, Bertram *et al.* (2001) suggested that they measured free water rather than water-holding capacity; and they have shown that low field NMR transverse relaxation (termed T_2) is an effective, non-invasive alternative for its determination. The technique revealed a component at 30–45 ms (referred to as T_{21}) and one at 100–180 ms (T_{22}). During rigor mortis the T_{21} water population is affected and contributes to myofibrillar water characteristics. T_{22} represents extracellular water *in vivo*.

10.2.1   Uncooked meat

10.2.1.1   *Factors determining exudation*

The exudation of 'weep' or 'drip' will depend on the quantity of fluid released from its association with the muscle proteins on shrinkage of the lattice of thin and thick filaments, whereby the water held by capillarity will be diminished, and on the extent to which, if released, fluid is permitted access to the exterior. Some considerations apply generally to all muscles. Thus, since post-mortem glycolysis in a typical muscle will normally proceed to an ultimate pH of about 5.5 – and this is the iso-electric point of the principal proteins in muscle – some loss in water-holding capacity is an inevitable consequence of the death of the animal (§§ 5.3 and 5.4.1). The *extent* of post-mortem pH fall will, therefore, affect the water-holding capacity, and the higher the ultimate pH the less will be the diminution in water-holding capacity (Cook *et al.*, 1926; Empey, 1933).

An unusually low ultimate pH is found in the muscles of certain breeds of pig (Lawrie *et al.*, 1958; Monin and Sellier, 1985) and this causes the development of very watery flesh in the meat of such animals post-mortem. Thus, both limited and excessive extents of post-mortem glycolysis affect the water-holding capacity of meat. The former is reflected by dry, firm and dark-cutting (DFD: cf. §§ 5.1.2 and 10.1.2) and the latter by pale soft exudative (PSE: cf. §§ 2.2 and 3.4.3) conditions.

The sarcoplasmic proteins to which some of the water-holding capacity is due (Hamm, 1960, 1966) are especially affected by post-mortem pH fall (Scopes, 1964). Moreover, the loss of ATP and the consequent formation of actomyosin as muscles go into rigor mortis will cause loss of water-holding capacity at any pH (cf. § 4.2.3). This arises both because the water holding capacity of actomyosin is less than that of the myosin and actin from which it forms (Millman, 1981) and also because the lower ATP level initiates denaturation in those proteins whose integrity *in vivo* is particularly dependent on the provision of energy (§ 5.3). Additionally, the extent of sarcomere shortening pre-rigor contributes to loss of water-holding capacity during post-rigor storage (Honikel *et al.*, 1986). Fluid from the myofilaments is released, dilutes the sarcoplasm (Pearson *et al.*, 1974; Penny, 1977), lowers the intra-cellular osmotic pressure and, thereby, increases the extracellular space (Offer and Trinick, 1983).

The *rate* of post-mortem pH fall is also an important determinant of water-holding capacity (Lawrie, 1960b; Penny, 1977). Denaturation of the sarcoplasmic proteins is worsened the faster the rate of pH fall (Scopes, 1964; Bendall and Wismer-Pedersen, 1962). A fast rate of pH fall (i.e. of ATP breakdown) will increase the tendency of the actomyosin to contract as it forms (Bendall, 1960) and thus express to the exterior fluid which has become dissociated from the proteins. When a fast rate of pH fall post-mortem is due to elevated temperatures (§ 4.2.2) the enhanced loss of water-holding capacity observed is partly due to increased denaturation of the muscle proteins, and partly to enhanced movement of water into extracellular spaces (Penny, 1977).

An abnormally rapid rate of post-mortem glycolysis also occurs in the muscles of certain pigs and produces pale soft exudative meat (muscle proteins being exposed to low, but not abnormal, ultimate pH values whilst still at near *in vivo* temperature). This aspect of PSE was considered in some detail in §§ 2.2 and 3.4.3. The condition arises in muscles when control of the release of Ca^{++} ions from the sarcotubular system is impaired because of faulty stereochemical configuration in junctional foot protein (cf. § 2.2).

Conditioning the meat (§ 5.4.1) increases its water-holding capacity (Cook *et al.*, 1926) and this at various environmental pH values (Fig. 10.2; Hamm, 1959). Although the pH of the meat itself may rise in these circumstances, this does not account for the phenomenon (Hamm, 1960). The increase in the water-holding

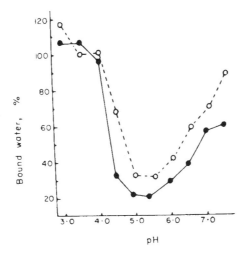

Fig. 10.2 The effect of conditioning on pH hydration curve of beef muscle (Hamm, 1960).
● = 1 day post-mortem. ○ = 7 days post-mortem. (Courtesy Prof. R. Hamm.)

capacity is more likely to be caused by changes in the ion–protein relationships, there being a net increase in charge through absorption of K^+ ions and release of Ca^{++} ions (Arnold *et al.*, 1956). According to the hypothesis of Kristensen and Purslow (2001), who observed slow proteolysis of vinculin and desmin, and rapid proteolysis of talin, during conditioning, such destruction of the cytoskeletal members removes the linkage between shrinkage of the muscle fibres *per se* and that of the myofibrils, whereby the force expelling water from within the cell is eliminated and re-entry of the water-potentiated, water-holding capacity is increased.

Apart from these general effects, the water-holding capacity of meat is affected by several of the factors which cause differentiation in muscles, such as species, age and the muscular function (§ 4.3). It is found, for instance, that the water-holding capacity of pork is higher than that of beef (Körmendy, 1955; Schön and Stosiek, 1958b; Hamm, 1975). Again, although the age of the animal does not appear to be an influence with pork, it is with beef, calves having a greater water-holding capacity (Schön and Stosiek 1958a). To some extent, these differences are a further reflection of differences in the rate and extent of pH fall (Fig. 4.4), the ultimate pH in pork and veal tending to be higher than that in beef (Lawrie, 1961; Lawrie *et al.*, 1963b). Some of the differences between and within muscles in water-holding capacity (Taylor and Dant, 1971) can be similarly explained – but not all. Thus, the water *content* of different muscles in beef and pork varies (Table 4.21(i) and 4.21(ii)), but this could be due to pH differences (Lawrie *et al.*, 1963b). On the other hand, both in beef and pork the *l. dorsi* has a lower water-holding capacity than the *psoas* (Hamm, 1960). This is so even when the rate and the extent of pH fall are identical (Howard *et al.*, 1960a), suggesting that there are different types of protein present (Lockett *et al.*, 1962; cf. Table 4.28).

Muscles having a high content of intramuscular fat tend to have a high water-holding capacity (Saffle and Bratzler, 1959). The reasons for this effect – which is real enough – are unknown: possibly the intramuscular fat loosens up the microstructure, thus allowing more water to be entrained (Hamm, 1960). Within a

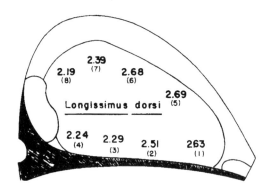

Fig. 10.3 Relative water-holding capacity with pork *l. dorsi* muscle (Urbin *et al.*, 1962). (Courtesy Dr M. C. Urbin.)

given muscle, the water-holding capacity may vary appreciably, even when the ultimate pH is virtually constant (Fig. 10.3; Urbin *et al.*, 1962).

All the factors affecting the water-holding capacity of muscle apply equally well to frozen and unfrozen meat. With frozen meat, however, removal of water from within the muscle cells during the process of freezing, as normally carried out commercially, provides an additional potential reservoir of fluid which appears as 'drip' on thawing (§ 7.1.2.2) although it can be ameliorated by the same means as 'weep' and can be largely avoided with very fast rates of freezing. These are not feasible commercially.

It is to be expected (and made clear in Fig. 10.2) that just as pH values above the iso-electric point of the muscle proteins enhance water-binding capacity, so will those below it. The latter, of course, would not arise naturally in meat since the enzymes affecting post-mortem glycolysis tend to be inactivated as the pH falls to 5.4–5.5 which is the iso-electric point of the muscle proteins. Only very rarely does the pH fall below 5.0. The preservation of foods by organic acids (cf. § 9.3) as in the traditional preparation of marinaded meat by vinegar and spices, however, involves conditions which enhance the water-holding capacity of muscle proteins on the acidic side of their iso-electric point. Rao *et al.* (1989) made a detailed study of such systems using various beef muscles in acetic acid solutions of 0.01–0.25 M. The water-holding capacity increased over the range 5.1 down to 4.0 in the six muscles investigated, that in *l. dorsi* having a significantly higher swelling ratio than that of the other muscles in the range between pH 4.3 and 4.0. Between pH 5.1 and 4.4 swelling increased in all muscles, both along and across the fibre axis. As the pH of the marinading solution approached 4.0, however, fibre swelling occurred predominantly in the 'white' type muscles, whereas fibre shrinkage occurred in predominantly 'red' type muscles. Interactions between the swelling of the muscle fibres and that of the connective tissue determined the total swelling of the muscles between pH 4.5 and 4.0.

The relative responsiveness of muscles on marinading to swell also reflects their total content of protein and the proportion of the latter which is connective tissue. Thus the swelling of muscle fibres dominates over that of collagen in muscles with a relatively high content of total protein such as *l. dorsi* at pH levels below 4.3. In this same pH range, however, the swelling of *supraspinatus*, in which the total protein content is relatively low and that of connective tissue high, is dominated by that of collagen (Gault, 1991). Both the perimysial collagen and that of the reticular fibres of the endomysium swell in marinading, but the effect is more marked in muscles in which the endomysium is thin (e.g. *l. dorsi*). Generally, however, the effect of marinading in swelling the myofibrillar proteins appears to be more important than that of the connective tissue proteins.

Gault (1991) demonstrated that mildly acid conditions during marinading are associated with an increased toughness in the cooked meat when the pH was *ca.* 5.0. He attributed the effect to the increased denaturation of myofibrillar proteins and the increased shrinkage of collagen when heated in this pH range. The effect was noted irrespective of the age of the animal providing the meat (although peak shear force was greater in that of older animals) or of the extent of sarcomere shortening (although the peak shear force was less in meat with longer sarcomeres).

Marinading in 75 millomol $CaCl_2$ was reported to enhance beef tenderness (possibly by stimulating calpains) but to create a bitter taste (Lourdes-Perez *et al.*, 1998). Since marinading requires a considerable period of time, the direct injection of acids has been considered. Eilers *et al.* (1994) showed that the injection of lactic acid improved the tenderness of tough meat and, more recently, Berge *et al.* (2001) demonstrated that the injection of 10 per cent (w/v) 0.5 M lactic acid, either pre- or post-rigor, tenderized beef by accelerating the release of lysosomal enzymes, increasing degradation of myosin heavy chains and decreasing the heat stability of perimyseal collagen, although they noted that discoloration could limit acceptability of the product.

10.2.1.2   Measures minimizing exudation
In relation to sausage (and other comminuted meats), small retail cuts and, more recently, prepackaged cuts, there has, naturally, been much interest in means of diminishing 'weep' or 'drip'. From the considerations above, it is clear that the use of meat having a high ultimate pH, or in which post-mortem glycolysis (ATP breakdown) is slow, and rapid cooling of the carcass before the onset of rigor mortis, will enhance water-holding capacity. Of course, if the temperature of the carcass is reduced *too* quickly during post-mortem glycolysis, cold-shortening with toughening and loss of water-holding capacity will ensue (cf. § 5.4.1). Clearly the bulk of the portion of meat which is being considered, and the capacity of the refrigeration facility, will determine whether the effect on water-holding will be beneficial or deleterious. Again, the actual exudation will be less if the area of cut surface of meat is minimized, if it is cut along instead of across the grain, and, in the case of prepacked meat, if the wrapping film used is not shrunk too tightly on to the meat. With frozen meat it is desirable to use the fastest possible freezing rate *after* the onset of rigor mortis, but pre-rigor freezing is liable to cause the excessive 'drip' of thaw rigor (§ 7.1.2.2). Preslaughter injections, designed to raise the ultimate pH (Howard and Lawrie, 1956; Penny *et al.*, 1963, 1964; Hatton *et al.*, 1972), can lead to commercially significant decreases in exudation from fresh and preserved meat.

Sausage and comminuted meats are, on the one hand, more liable to exude fluid (even if the water-holding capacity of the proteins is intrinsically high) because the structure of the meat is destroyed in their preparation, thus removing its contribution to the physical retention of fluid. On the other hand, however, the nature of these products permits direct manipulation of the meat to enhance its water-holding capacity artificially. Before considering special effects, the behaviour of water itself as an additive should be noted. The ratio of water to meat affects the overall

Table 10.2 Influence of water to meat ratio on water retention of pork muscle (after Sherman, 1961)

Water:meat ratio	Per cent water retention at 0 °C
No water added	−4.0
1:2	−1.0
1:1	−1.5
3:2	0.5
2:1	9.5
3:1	5.0
4:1	−5.0
5:1	−5.0
6:1	−6.0

Table 10.3 The relative effects of various sodium salts and chlorides of different metals in enhancing the water-holding capacity of muscle homogenates: ionic strength 0.4 (Hamm, 1960)

pH	Order of efficacy
6.4 and 5.5	(a) *Sodium salts* $F^- < Cl^- < Br^- < CNS^- < I^-$,
6.4	(b) *Chlorides* $Ca^{++} < Ba^{++} < Mg^{++} < K^+ < Na^+ < Li^+$
5.5	$K^+ < Na^+ < Mg^{++} < Ca^{++} < Li^+ < Ba^{++}$

waterholding capacity of the mix. The latter, as measured by a centrifugal method (Sherman, 1961), is maximal when the ratio is about 2:1 (Table 10.2).

The incorporation of the salts of strong acids, such as sodium chloride, in the comminuted meat mix is important in enhancing water-holding capacity (Gerrard, 1935). The efficacy of different sodium salts and of different chlorides is shown in Table 10.3, for an ionic strength less than 0.4. While the order for the sodium salts is the same at both high and low pH, that for the various cations is markedly different, the divalent species being less effective than the monovalent at relatively high pH, the converse tending to be true at relatively low pH.

The more strongly ions are bound by the protein the stronger will be the hydrating effect (Hamm, 1957). The effect of anions in shifting the iso-electric point to more acid values and in enhancing the water-holding capacity above the original isoelectric point was shown by Hamm in 1960. It has been presumed that the water-holding capacity of connective tissue proteins is similarly enhanced by ions.

Offer and Trinick (1983) suggested that the water up-take by myofibrils in strong salt solutions is caused both by expansion of the lattice of thick and thin filaments as the increasingly negatively charged components repelled one another and also by disruption of the forces which determine the regular arrangement of the filaments at the Z- and M-lines and between the heads of the myosin molecules and the adjacent actin filaments. Both sodium chloride and pyrophosphate exert these two effects (Bendall, 1954; Greene, 1981).

The propensity of muscle fibres to swell in hypertonic solutions is initially opposed by the restraint of the endomysium, but with increasing time post-mortem the degree of swelling increases. Wilding et al. (1986) attributed this to a weakening of the endomysium per se during conditioning, but Knight et al. (1989) demonstrated that the proportion of fibres without endomysial sheaths increases at this time. Such stripped fibres swell more in hypertonic media than those in which the endomysium is intact. Earlier evidence of a weakening in the link between muscle fibres and their endomysium during conditioning was given by Stanley (1983) who showed that endomysial tubes empty of myofibrillar material could only be prepared after ageing. It would thus appear that myosin extraction by salt would be more readily effected from fibres which have lost their endomysia. Muscles may contain fibres of several types (cf. § 4.3.5) and it is thus not surprising that the myofibrils they contain react differently to hypertonic treatment (Offer et al., 1983–85a), a feature which could be important in explaining and controlling the variability of meat in curing.

At high ionic strength, salt has a dehydrating effect: hydration is at a maximum when the ionic strength is about 0.8–1.0. This corresponds to 5 and 8 per cent of sodium chloride for meat without, and with, 60 per cent added water, respectively (Callow, 1931; Hamm, 1957). Knight and Parsons (1988), however, in observing the swelling of myofibrils in concentrated salt solutions, attributed the effect to entropic swelling pressure caused by a steric resistance to the rotational movement of the tails of myosin molecules imposed by the actin filaments to which they were attached: the swelling was greatest in ca. 6 per cent (M) sodium chloride. They attributed the apparent dehydrating effect of higher salt concentrations to the precipitation of myosin, a feature which would reverse its depolymerization in M sodium chloride and cause shrinkage.

Certain salts of weak acids, in particular phosphates and polyphosphates, are also added to comminuted meats to enhance water-holding capacity, especially in continental-type sausages. Hamm and Grau (1958) found the following order of increasing efficacy of the sodium salts – monophosphate, cyclotriphosphate, diphosphate, tetraphosphate and triphosphate. The efficacy of triphosphate addition appears to depend upon its being enzymically broken down to diphosphate (Hamm, 1975). In agreement with the finding of Arnold et al. (1956) that increased hydratability in conditioned meat parallels loss of Ca^{++}, they suggested that the effect of such phosphates depended on their sequestering action on divalent cations, although this has been refuted (Sherman, 1961). Bendall (1954) concluded that the effects of most of these phosphates was largely one of ionic strength and pH, whereas that of pyrophosphate (in the presence of 1 per cent sodium chloride), which was much greater, was specific and due to the splitting of actomyosin into actin and myosin and to the formation by the latter of a sol, the effect reflecting that of ATP itself. This view was confirmed by Hellendoorn (1962).

In this context Offer and Trinick (1983) have shown that pyrophosphate substantially reduces the salt concentration required to produce maximum swelling when myofibrils are placed in solutions of sodium chloride. In the absence of pyrophosphate the protein in the middle of the A bands is extracted, but, in its presence, all of the A band protein dissolves. In a subsequent study by Knight and Parsons (1988), when myofibrils were supported between grid bars rather than resting on a coverslip, it was observed that some material in the centre of the sarcomeres resisted extraction by M sodium chloride: it appears to be titin or nebulin (i.e. the material of the 'gap filaments': cf. § 3.2.2). The extraction of both A band

and Z proteins, which occurs in M sodium chloride, is progressively inhibited at higher salt concentrations, presumably because myosin is salted out.

The loss of water-holding capacity arising from the formation of actomyosin when the *in vivo* ATP level falls and rigor mortis ensues (§ 4.2.3), and the efficacy of pyrophosphate in improving it in minces, suggests that there would be some benefit in keeping the ATP high post-mortem. Unfortunately, no means of preventing rigor mortis by preslaughter treatment have been found, and the injection of pyrophosphate is fatal because of hypocalcaemia (Howard and Lawrie, 1957a). But comminuted meats permit manipulation to retain the high pre-rigor water-holding capacity of the myofibrillar proteins (Körmendy, 1955; Savic and Karan-Djurdjic, 1958). Successful procedures depend on there being a sufficiency of ATP remaining in the muscle. The addition of 2 per cent sodium chloride to pre-rigor meat appears to prevent the onset of rigor mortis and, thereby, the slight shortening of sarcomeres and the loss of waterholding capacity which always accompany actomyosin formation, provided the salt penetrates the tissue sufficiently swiftly before the ATP level has fallen to the point of rigor onset. Strangely enough, it does not prevent the subsequent breakdown of ATP (Hamm, 1957). Indeed it is accelerated by the salt. Since sodium acetate has no such effect, Hamm believes that the binding of chloride ions is responsible. In practice, advantage can be taken of this effect by freezing comminuted pieces pre-rigor storing the frozen meat below –18°C, to limit ATP breakdown, and subsequently thawing them in the presence of salt. In the absence of salt, the thawing of meat frozen pre-rigor causes excessive exudation (§ 7.1.2.2). The retention of water-holding capacity is even greater – and lasts much longer – if the pre-rigor meat is comminuted with salt *then* frozen (Hamm, 1966). This is because in the region of the freezing point (*ca.* –1°C) ATP breaks down slower in salted meat than unsalted. It is better therefore to salt pre-rigor meat before freezing than to add the salt during the preparation of the sausage emulsion (Honikel and Hamm, 1978). A similar enhancement of the water-holding capacity of sausage can be obtained if the salted meat is freeze-dried in the pre-rigor state (Honikel and Hamm, 1978). The phenomenon has been reviewed in detail by Hamm (1981). Whereas the onset of rigor mortis has little effect on the water-holding capacity of unsalted meat homogenates, as measured either by cooking loss or by the exudate from unheated tissue, it causes severe loss of water-holding capacity in homogenates which have been salted post-rigor. This reflects the action of salt in preventing the union of actin and myosin in pre-rigor meat and its predominant denaturing action post-rigor. It is of interest to note that the occurrence of cold-shortening does not affect the water-holding capacity of salt comminuted meat, provided salting has taken place before the onset of rigor mortis. Since the salt added to pre-rigor meat has an inhibitory effect on the enzymes involved in post-mortem glycolysis, the pH attained is about 0.3–0.4 units higher than that in corresponding comminuted meat which has been salted post-rigor; and this means that the added salt is less detectable to the palate.

Much of the success of comminuted meat production depends on the ability of the muscle proteins to hold fat as well as water. The factors determining the stability of sausage meat emulsions are thus important. A major function of salt in these products is to loosen the myofibrillar proteins and to increase their ability to emulsify fat, especially at pH values near their iso-electric point (Swift and Sulzbacher, 1963). O'Neill *et al.* (1989a, b) made a detailed study of muscle proteins as emulsifiers in meat systems. Myosin was more surface-active than actin or actomyosin at water–air interfaces. The fragments derived from myosin by proteolytic enzymes were less surface-active than the parent myosin molecule, although fraction S1 (cf. § 3.2.2) was more active than fragments from the myosin tail.

Myofibrillar proteins derived from 'fast' (white) muscles (e.g. *cutaneous trunci*) produce gels which are different in nature in emulsified meat products than those from 'slow' (red) muscles (e.g. *masseter*) (Young *et al.*, 1992). Products made from *cutaneous trunci* gelled at temperatures 10 °C lower than those made from *masseter*. The reduced tendency for myofibrils from *masseter* muscles to aggregate at high (cooking) temperatures in comparison with those from *cutaneous trunci* is associated with a greater water-holding capacity (Egelansdal *et al.*, 1995).

There are limitations to the amount of fat which the protein water gel can hold if the sausage is to maintain its structure during handling and processing. These limits are determined by a number of factors. Thus although mild warming of the emulsion in a chopper aids in releasing soluble proteins, temperatures above 22 °C may cause the emulsion to break down (Wilson, 1960). Over-chopping increases the surface area of the fat particles to the extent that the water–protein phase is unable to hold them in the emulsified state. Over-mixing of the emulsion, particularly at temperatures from 18 to 22 °C, may cause moisture and fat to separate. The addition of cooked rind, which contains partly denatured collagen and elastin, enhances the water-holding ability of sausage meats to some extent (Heidtmann, 1959). Mixing microbial transglutaminase with isolates of myofibrillar protein and soya enhances the adhesion of the gels whereby a greater quantity of the meat protein can be replaced by that of soya (Ramirez-Suarez and Xiong, 2003). During the heating of sausage mix, the coagulating network of proteins or filaments surrounds the melting fat particles which cannot therefore coalesce. The larger the meshes of the coagulated network, the less coalescence can occur and the greater the water-holding capacity (Hamm, 1975). Little is known about the role of the phase transitions of fats on the stability of emulsions. Differential thermal analysis has been used to examine such phenomena with a view to achieving better emulsion control. It has been found that there are two primary ranges of melting in both beef and pork fats (Townsend *et al.*, 1968). The ranges are 3–14 °C and 18–30 °C for beef fats and 8–14 °C and 18–30 °C for pork fats. The stability of emulsions above 18.5 °C coincides with the onset of melting of the higher melting portion of the fats. The state of the proteins used is among the factors of importance. Thus, for example, the emulsifying capacity of beef *semitendinosus* muscle decreases from 30 min to 4 days postmortem: thereafter, on storage at –4 °C, it increases (Graner *et al.*, 1969).

Sugars, which are sometimes added to continental-type sausage meat, have no effect in enhancing water-holding, although they have such an effect in intact muscular tissue (Hamm, 1960).

10.2.2 Cooked meat

10.2.2.1 Shrink on cooking

The factors affecting loss of 'weep' or 'drip' from uncooked meat also apply to the water-holding capacity of cooked meat: relative differences are retained on heating (Bendall, 1946; Hamm and Deatherage, 1960a, b). The losses due to the shrinkage on cooking, however, will be greater – to an extent determined by such extraneous circumstances as method, time and temperature of cooking – since the high temperatures involved will cause protein denaturation and a considerable lowering in

Table 10.4   Percentage cooking losses from good- and poor-quality beef joints (Bouton *et al.*, 1958)

Category of loss	Good grade		Poor grade	
	Sirloin roast	Topside roast	Sirloin roast	Topside roast
Total	28.9	41.3	41.7	48.3
Fat	17.8	1.4	4.8	-1.4[a]
Water	9.4	36.9	34.0	47.7
Residue	1.7	3.0	2.9	3.0

[a] Fat increment arises from that added to pan on cooking.

waterholding capacity (Baker, 1942; Wierbicki *et al.*, 1954; Paul and Bratzler, 1955). Moreover, some of the shrink or juice on cooking will represent nonaqueous fluid, since the high temperatures will melt fat and tend to destroy the structures retaining it.

As an example of how the factors in § 10.2.1 are reflected in shrinkage and loss of juice on cooking, it may be mentioned that the induction of a high ultimate pH in muscle will diminish that particular portion of the cooking loss which is due to the exudation of moisture (Bouton *et al.*, 1957),* and the benefits of adding pyrophosphate to comminuted meats are retained on cooking (Bendall, 1954). On the other hand, a fast rate of pH fall will increase moisture loss in cooking. Thus, Sayre *et al.* (1964) found that where the pH of pork muscle at 40 min post-mortem (pH_1) was lower than 5.9, cooking losses were about 40–50 per cent, whereas when the pH_1 was above 6.0, losses were only 20 per cent. Again, the effect of conditioning meat in enhancing water-holding capacity is to some extent reflected in diminished cooking losses, but this effect is not apparent with all joints (Bouton *et al.*, 1958). Fjelkner-Modig and Tornberg (1986), using pulsed NMR, found that the juiciness and tenderness of pork when fried at 80 °C was largely determined by the distribution of water between the intra- and extracellular phases. The greater juiciness of fried pork from Hampshire pigs in comparison with those of the Swedish Yorkshire breed was associated with a greater percentage of intracellular water in the former.

Losses from good quality meat tend to be less *overall* than those from poor-quality meat (Table 10.4). Although the former lose more fat (which can be expected in view of their greater fat content) they lose less moisture, possibly because the structural changes caused by the presence of the fat enhance water-holding capacity (Saffle and Bratzler, 1959).

Table 10.4 also indicates the effect of the type of joint in determining cooking loss. In the sirloin, where there is considerable intramuscular fat, the shrink is largely due to fat.

Chemical changes in meat proteins on subjection to dry heat up to 80 °C were indicated in §§ 8.1.1 and 8.2.2. There is a loss of free acidic groups and of water-

* The greater loss of moisture from meat of normal pH tends to cause more evaporative cooling; and under comparable conditions meat of high pH tends therefore to reach a rather greater internal temperature on cooking (Lewis *et al.*. 1967).

Table 10.5   Effect of internal meat temperature on cooking
loss in beef ·

	Internal temperature of meat (°C)		
	60	70	80
Total cooking loss (% wet weight)	10.5	28.8	40.5
Moisture loss (% wet weight)	5.6	9.6	14.0

holding capacity, and a rise of pH, as the temperature is increased from 0 to 80 °C
(Fig. 8.1).

There is also an increase in the buffering power of the meat in the region 5.0–7.0.
Bendall (1946, 1947) attributed these effects to denaturation changes, particularly
in the sarcoplasmic proteins. They were consistent with the fission of protein chains
at labile linkages involving imidazole, –SH and –OH groups, followed by hydrogen
bonding between carboxyl and amino groups.

However common the empirical observation that increasing temperature
increases cooking loss, few controlled experiments have been carried out on this
topic. Sanderson and Vail (1963) cooked beef muscles to constant internal temperature of 60, 70 and 80 °C and observed that, in these circumstances, the total cooking
loss increased, only some of the increment, however, being due to loss of moisture
(Table 10.5).

Meat cooked quickly to a given internal temperature has a lower cooking loss
and is more juicy than that cooked slowly to the same temperature (Bramblett and
Vail, 1964).

Although the conversion of collagen to gelatin at 100 °C will tend to increase
water-holding capacity (Hamm, 1960), this is offset by severe changes in the sarcoplasmic and myofibrillar proteins, for the overall water-holding capacity drops
markedly as the temperature is raised between 80 and 100 °C (Table 8.4).

These changes are sufficiently severe to make it relatively immaterial how long
they are applied; on the other hand, with cooking temperatures below 70–80 °C
shrinkage increases with increasing time of cooking (Savic and Suvakov, 1963).
Time–temperature curves of meat during cooking reveal a plateau at about 70 °C
suggesting that some chemical change is occurring at this point. It does not appear
to be due to connective tissue breakdown (Lawrie and Portrey, 1967).

The quantity of juice obtained on heating increases further between 107 and
155 °C (Tischer et al., 1953). This probably reflects some of the protein breakdown,
with destruction of amino acids, which occurs in such ranges of temperature (Beuk
et al., 1948). During roasting of the meat, coagulation of the proteins on the surface
inhibits loss of fluid, and the more rapid the heating, the faster the formation of this
layer and the lower the shrink (Andross, 1949). A similar explanation accounts for
the lower shrink in meat cooked after immersion in boiling water rather than after
slow raising of the temperature of initially cold water. Grilling and dielectric heating
lower the loss of juice still further (Causey et al., 1950). In recent years there have
been increasing complaints from consumers about the amount of whitish fluid that
exudes on cooking bacon, especially as the exudate darkens on further heating. The

phenomenon possibly reflects the higher moisture, and lower salt, content of modern bacon. The exudate contains sarcoplasmic proteins, with only traces of those of the myofibrils (Sheard *et al.*, 2001).

Because the majority of investigations on the effects of cooking on the loss of fluid from meat have tended to be empirical, Bendall and Restall (1983) made a controlled study of the behaviour of myofibres at different levels of aggregation. Loss of water reflected the configurational changes which characterized the type of aggregate concerned. Whereas thin muscle strips and small bundles of myofibres tended to shorten above 64 °C, individual myofibres did not do so, but only decreased in diameter, when heated in aqueous media at temperatures up to 90 °C.

Expulsion of water from individual myofibres was slow between 40–53 °C, but was rapid at 60 °C, apparently under the influence of a new force. This suggested that some protein was denaturing and shrinking at the latter temperature; and Bendall and Restall (1983) identified it with type IV/V collagen in the basement membrane (cf. § 3.2.1). Since about 60 per cent of the water of the myofibre had been expressed at this stage, the subsequent shrinkage of the endomysial collagen above 64 °C failed to express further moisture.*

When small bundles of myofibres were heated below 60 °C they behaved similarly, decreasing in diameter only. Above 64 °C, however, they shortened, attaining 30 per cent of their initial length at 90 °C with a loss of 70 per cent of their initial water content. Thin muscle strips shortened even more markedly at above 64 °C, the severity of the shortening being determined by the amount of perimysial collagen in the muscle from which they had been prepared.

Based on their observations with such model systems, Bendall and Restall (1983) concluded that the behaviour of a large piece of meat when stewed (i.e. heated in an aqueous medium) could be explained in four stages. Firstly, a slow loss of fluid from the constituent myofibres into extracellular spaces as sarcoplasmic and myofibrillar proteins denature between 40 and 53 °C, there being no concomitant shortening. Secondly, rapid fluid loss from myofibres as the temperature rose to 60 °C as the collagen of the basement membrane heat shrinks. Thirdly, heat shrinkage of the endomysial, perimysial and epimysial collagens between 64 and 90 °C, there being much shortening, decrease in myofibre diameter and increased cooking loss. Finally, during prolonged heating, conversion of epimysial, then of endomysial and perimysial, collagens, to gelatin – and concomitant tenderizing – occurs.

10.2.2.2   Juiciness

The degree of shrinkage on cooking is directly correlated with loss of juiciness to the palate (Siemers and Hanning, 1953). Juiciness in cooked meat has two organoleptic components. The first is the impression of wetness during the first few chews and is produced by the rapid release of meat fluid; the second is one of sustained juiciness, largely due to the stimulatory effect of fat on salivation (Weir, 1960). This function of the latter explains why, for example, the meat of young animals

---

* Offer and Trinick (1983) further elucidated the mechanism for loss of fluid on cooking by a microscopic examination of individual myofibrils. They showed that these actively shrink in diameter on being heated to 45–60 °C. Above 65 °C they actively shorten: concomitantly, shrinkage of the collagen network contributes to expression of sarcoplasmic fluid to the exterior from the anular spaces between fibres and their endomysial sheaths.

gives an initial impression of juiciness but, due to the relative absence of fat, ultimately a dry sensation (Gaddis *et al.*, 1950). Good-quality meat is more juicy than that of poor quality, the difference being at least partly attributable to the higher content of intramuscular fat in the former (Gaddis *et al.*, 1950; Howard and Lawrie, 1956). An association between juiciness and intramuscular fat has been noted in comparing rib roasts from several groups of steers and bulls which had the same sire (Bryce-Jones *et al.*, 1963): juiciness varied significantly between the groups. Several studies have shown the benefits, in overall eating quality as well as in juiciness, of an increased content of intramuscular fat (Bejerholm and Barton Gade, 1986; Warriss *et al.*, 1996).

There is some suggestion that juiciness reaches a minimum when the pH level of the meat is about 6 (Howard and Lawrie, 1956); this possibly reflects the greater ability of the muscle proteins to bind water in this pH region; but, if this were the entire explanation, juiciness would be expected to decrease still further with even higher pH levels.

The process of freezing does not itself affect juiciness (Law and Vere-Jones, 1955), there being no difference in this respect between meat which has been chilled or frozen and held for the same length of time. On the other hand, there is an effect of storage. Thus, the beef held at $-10\,°C$ for 20 weeks was much less juicy than corresponding beef held for a few days at $0\,°C$ (Howard and Lawrie, 1956). This effect is also apparent during conditioning, roasts and grills of beef being most juicy some 24 h after slaughter and thereafter decreasing in juiciness in the following order: 3 days at $0\,°C$, 2 days at $20\,°C$, 14 days at $0\,°C$ (Bouton *et al.*, 1958).

Freeze drying, even when operated under optimum conditions, causes some loss in juiciness (Hamm and Deatherage, 1960a, b); this can be offset to some extent by the induction of a high ultimate pH in the meat (Table 10.6).

The ranking order shows that juiciness was greatest in the fresh (frozen) meat of high ultimate pH, and somewhat less in corresponding dehydrated material; but both were considerably more juicy than the meat of low ultimate pH both before and after freeze drying.

Table 10.6 The sum of the juiciness rankings given by eight tasters for control (low ultimate pH) and adrenaline-treated (high ultimate pH) beef before and after freeze dehydration (after Penny *et al.*, 1963)

Muscle	A. Control (low ultimate pH)		B. Adrenaline treated (high ultimate pH)	
	Fresh (frozen)	Dehydrated	Fresh (frozen)	Dehydrated
Semimembranosus	23	23	13	19
Biceps femoris	26	22	15	12
Psoas	23	25	12	16
L. dorsi (lumbar)	19	24	11	14
L. dorsi (thoracic)	26	25	10	20
Deep pectoral	25	24	10	19
Total	142	143	71	100

10.3 Texture and tenderness

10.3.1 Definition and measurement

Of all the attributes of eating quality, texture and tenderness are presently rated most important by the average consumer and appear to be sought at the expense of flavour or colour. This notwithstanding it is most difficult to define what is meant by either term.

According to Hammond (1932a), texture, as seen by the eye, is a function of the size of the bundles of fibres into which the perimysial septa of connective tissue divide the muscle longitudinally (§ 3.2.1). Coarse-grained muscles – in general those that have the greatest rate of post-natal growth – such as *semimembranosus*, have large bundles, fine-grained muscles (e.g. *semitendinosus*) have small bundles. The size of the bundles is determined not only by the numbers of fibres but also by the size of the latter. Coarseness of texture increases with age but in muscles where the fibres are small it does not become quite so apparent as in those where they are large. In general, coarseness of texture is greater in the muscles of male animals, and in those of large frame; breed also has an effect (Hammond, 1932a). The thickness of the perimysium in beef differs between muscles and contributes to variation in tenderness (Brooks and Savell, 2004).

The size of the fibre bundles is not the only factor determining coarseness, however. The *amount* of the perimysium round each bundle is important, the perimysial layer being thick in coarse muscles (Ramsbottom and Strandine, 1948). Since the elements defining texture are aspects of connective tissue, it might have been expected that there would have been a direct correlation between coarseness of grain and toughness after cooking. This is not so, however (Ramsbottom and Strandine, 1948); yet there is an indirect correlation between muscle fibre diameter and tenderness (Hiner *et al.*, 1953). Such observations emphasize the complexity of texture and tenderness as attributes of eating quality.

The overall impression of tenderness to the palate includes texture and involves three aspects: firstly, the initial ease of penetration of the meat by the teeth; secondly, the ease with which the meat breaks into fragments; and thirdly, the amount of residue remaining after chewing (Weir, 1960).

There have been many attempts to devise objective physical and chemical methods of assessing tenderness which would compare with subjective assessments by taste panels. The difficulty of doing so is considerable. Thus physical methods have included the basis for measuring the force in shearing (Warner, 1928; Kramer, 1957; Winkler, 1939), penetrating (Tressler *et al.*, 1932; Lowe, 1934), 'biting' (Lehmann, 1907; Volodkevitch, 1938), mincing (Miyada and Tappel, 1956), compressing (Sperring *et al.*, 1959) and stretching the meat (Wang *et al.*, 1956). Chemical methods have involved determination of connective tissue (Lowry *et al.*, 1941; Neumann and Logan, 1950) and enzymic digestion (Smorodintzev, 1934) amongst other criteria. The compression of meat through a small orifice (Sperring *et al.*, 1959) gives, on raw meat, an objective assessment closest to the tenderness ratings obtained by taste panel. The concepts of fracture mechanics have been used to relate quantitative measurements of raw meat texture to qualitative observations in cooked bovine *m. semitendinosus* (Purslow, 1985). It is evident that the muscle fibre bundle is an important structural feature in fracture behaviour and that the strength of the perimysial connective tissue containing the bundle has a major influence on the toughness of the cooked meat, confirming, in objective terms, the subjective impressions of 50 years ago.

The degree of tenderness can be related to three categories of protein in muscle – those of connective tissue (collagen, elastin, reticulin, mucopolysaccharides of the matrix), of the myofibril (actin, myosin, tropomyosin) and of the sarcoplasm (sarcoplasmic proteins, sarcoplasmic reticulum). The importance of their relative contribution depends on circumstances such as the degree of contraction of the myofibrils, the type of muscle and the cooking temperature. Measurements of the shear, compression and tensile force reflect changes in the myofibrillar structure. After the initial yield, applied forces reflect the state of the connective tissue. The latter may be determined by measuring adhesion values (Bouton *et al.*, 1975).

Since the sarcoplasmic proteins are water-soluble, it might seem that they could not contribute to meat texture. However, *in situ*, their concentration is *ca.* 25 per cent in the sarcoplasm (Scopes, 1970), they coagulate on heating, a proportion is bound to structural elements in the muscle cell (Clarke *et al.*, 1980) and the viscosity of F-actin is modified thereby (Morton *et al.*, 1988). It must be presumed, therefore, that their contribution cannot be entirely dismissed. It is worth noting that the viscosity of muscle press juice, which is substantially composed of sarcoplasm, is twice as great as that of blood plasma (Jalango *et al.*, 1987), the functionality of which as a binding agent in heated foods is well known.

10.3.2 Preslaughter factors

Species is the most general factor affecting tenderness. To some extent this is a reflection of texture (§ 10.3.1). Thus the large size of cattle, in relation to sheep or pigs, is, generally, associated with a greater coarseness of their musculature (Hammond, 1932a). Although it has been the impression that pork contains little connective tissue compared with beef (Mitchell *et al.*, 1927), Tables 4.21(i) and 4.21(ii) show, respectively, that the hydroxyproline content of corresponding muscles of beef and pork vary from 350 to 1430 µg/g and from 420 to 2470 µg/g. Since hydroxyproline can be equated to connective tissue, this fact, and the discrepancy between the greater tenderness of veal and its high content of connective tissue in relation to beef, indicates that the type of connective tissue, as well as its quantity, is important.

Texture may also be implicated in breed differences in tenderness. The relatively greater tenderness of the meat from Aberdeen Angus cattle can be partly explained by their small size – this being reflected in fine grain (Hammond, 1963a). Nevertheless, other factors are involved, since dwarf beef was judged less tender than beef from normal-sized animals (Jacobson *et al.*, 1962) on the one hand, while the hypertrophic muscles of 'doppelender' beef were at least as tender as those from normal animals (I. F. Penny, personal communication). The latter feature is also evident in cross-bred progeny. Thus, the toughness of the *semitendinosus* muscles of the offspring of 'doppelender' Aberdeen Angus bulls and Jersey cows was found to be less than in those where the sires were normal (Bouton *et al.*, 1976).

Carpenter *et al.* (1955) showed that the introduction of the Brahman breed decreased beef tenderness. The decrease in tenderness, associated with an increased contribution from *bos indicus* in the cross-breed, is recognized as an important factor in the management of tenderness in Australia (Thompson, 2002). More detailed data on the effect of breed are given in Table 10.7 (Palmer, 1963). Although no marked differences between breeds has been found in the gross content of connective tissues, other factors such as the chemical nature of collagen

Table 10.7   The relationship of breed to tenderness (taste panel)

Breed	No. of cattle	Per cent cattle giving score of[a]				
		5–6	4–5	3–4	2–3	1–2
Angus	84	59	27	11	2	–
Brahman	196	9	27	40	21	3
Brangus	18	10	48	38	5	–
Devon	12	58	42	–	–	–
Hereford	48	40	48	11	2	–
Shorthorn	122	20	37	38	4	2

[a] Range 1–6: 1 being inedible, 6 excellent.

Table 10.8   Influence of sire on mean tenderness (shear force) of beef roasts from groups[a] of cattle[b]

Sire group	7–8th rib	9–10th rib
1	5.8	6.1
2	6.1	6.2
3	7.4	6.8
4	7.6	7.6
5	9.4	7.3

[a] Increased force signifies less tenderness.
[b] Six cattle in each group.

(cf. §§ 3.2.1 and 4.3.5) could be implicated in creating differences in the tenderness of their meat.

Even within a breed, however, tenderness is heritable to an extent of over 60 per cent (Cartwright *et al.*, 1958), again indicating that texture is by no means its sole determinant. Different sires are associated with different degrees of tenderness (Bryce-Jones *et al.*, 1963); data in Table 10.8 indicate that such differences are co-variant in different portions of the carcass.

The connective tissue content of a given muscle may vary indeed between the individual pigs within a litter (Lawrie and Gatherum, 1964).

As indicated before (§ 3.2.1), variations in tenderness of a given muscle between pigs of the same age and slaughter weight may not correlate well either with their total collagen content or with the amounts of immature or mature collagen cross-links present. This discrepancy may partially be explained by differences in the factors controlling the rate of post-mortem glycolysis and of autolysis, but other, as yet identified, physiological factors may be responsible (Avery *et al.*, 1996). Thus, Maltin *et al.* (1997) found a positive correlation between the diameter of fast twitch glycolytic fibres from *l. dorsi* of the pig and toughness, as determined instrumentally.

In general, increasing age connotates decreasing tenderness, although, as already mentioned, also a decrease in connective tissue content (Bate-Smith, 1948; Hiner and Hankins, 1950; and cf. Tables 4.15 and 4.19). This apparent contradiction may

Table 10.9   Tenderness (taste panel) ratings[a] of various muscles from beef steers of two ages (after Simone et al., 1959)

Muscle	18 months	30 months
Adductor	4.67	3.85
Semimembranosus	3.91	3.35
L. dorsi (level 6–8th rib)	6.21	5.95
L. dorsi (level 9–11th rib)	6.16	5.57

[a] High signifies greater tenderness.

Table 10.10   Release of soluble proteins, hydroxyproline and ninhydrin-positive material from the connective tissue of biceps femoris of beef by collagenase after 12h incubation

Animal age, months	Sol. protein	Hydroxyproline	Ninhydrin-positive
	(μg/ml incubation medium)		
$1\frac{1}{2}$	230.7 ± 9.2	28.8 ± 0.9	456 ± 10
13–16	122.7 ± 8.5	11.3 ± 0.2	148 ± 7

probably be explained by the fact that the connective tissue in young animals has less cross-bonding (Boucek et al., 1961; Goll et al., 1963; Light and Bailey, 1989). Typical data for beef animals at 18 and 30 months of age are given in Table 10.9.

The results of other workers are similar. Decrease of tenderness appears to be less marked with beef from animals older than 18 months, differences between 40 and 90-month-old animals being relatively small (Tuma et al., 1962). With increasing animal age, the proportion of salt- and acid-soluble collagens decrease in bovine muscle; such differences have been demonstrated by starch gel electrophoresis. The extent of intra- and intermolecular cross-linking between the polypeptide chains of collagen concomitantly increases (Carmichael and Lawrie, 1967a, b; Bailey, 1968). Further reflections of the changing character of collagen with increasing animal age include a decreasing solubility on heating (19–24 per cent of total collagen is soluble in calves, 7–8 per cent in 2-year-old steers, and 2–3 per cent in old cows: Sharp, 1963, 1964) and decreasing susceptibility to attack by enzymes (Goll et al., 1963); cf. Table 10.10. Progressive increases in toughness have also been shown with the cooked meat from sheep aged from 2 months to 8 years (Bouton et al., 1978b).

Young and Braggins (1993) endeavoured to clarify the relative importance of the concentration and the solubility of collagen in determining the tenderness of sheep meat. They concluded that the former was the predominant determinant of eating quality, whereas solubility was more closely associated with the objective determination of shear force.

In situ examination of meat by ultraviolet optical probes has revealed an increase in the incidence of fluorescence peaks as the perimysium increases in beef animals between 12–17 months of age. A subsequent decrease in fluorescence of the meat of animals between 17–24 months reflects the separation of perimyseal layers by muscular tissue but the fluorescent peaks broaden as the perimyseal layers thicken (Swatland, 1994).

As Bouton *et al.* (1978a) have indicated, age/tenderness relationships reflect not only direct chronological changes in muscular and connective tissues, but also associated effects due to the increasing bulk and fatness of carcasses with age. These influence the differential effect of cooling conditions on the extent of 'cold-shortening' in specific muscles.

In a comparison of the eating quality of bull and steer beef, using a series of entire and castrated twins, Bryce-Jones *et al.* (1963, 1964) found that steer meat was more tender, the difference being especially marked in *l. dorsi* and *semitendinosus* muscles, and at the level of the 7–8th vertebrae.

Table 10.9 also indicates that there are distinct differences in tenderness between muscles. Many years ago intermuscular variability in tenderness was comprehensively studied in 50 beef muscles by Ramsbottom and Strandine (1948). For the raw muscles, the shear (i.e. the force required in lb to shear a sample of $\frac{1}{2}$ inch (12 mm) diameter) ranged from 3.8 in the *l. dorsi* to 20.0 in the cutaneous members, and in the cooked muscles from 7.1 in *psoas major* to 15.6 in *sternocephalicus*. In agreement with the latter data, the taste panel rated *psoas major* as most tender, whereas *sternocephalicus* was one of the toughest. The data for both beef and pork (Table 4.21) indicate that *psoas major* had least hydroxyproline of the muscles studied by Lawrie *et al.* (1963a, 1964) although stroma nitrogen, representing insoluble protein of various origin, was higher in *psoas* than in *l. dorsi* (Table 4.28). Loyd and Hiner (1959) showed that muscles differed in their contents of (apparent) collagen and elastin and that there was a significant inverse correlation between the hydroxyproline of these alkali-insoluble protein fractions and tenderness (Table 10.11).

Insofar as different muscles have differing proportions of epimysial, perimysial and endomysial connective tissue, and these are characterized by differing types of collagen (Bailey and Sims, 1977) – the polypeptide chains of which would be more or less firmly cross-linked (cf. § 3.2.1), and affected by heat to different extents (Bendall and Restall, 1983) – it is not surprising that they differ considerably in tenderness. Apart from the effect of their connective tissue on this parameter, of course, muscles' relative susceptibility to shorten before or during rigor mortis (cf. § 10.3.3) will accentuate tenderness differences between them. Again, increased understanding of the intrinsic biochemical differences between individual muscles is being directed to improving the eating quality of such cuts as the steakmeat (chuck) and round (cf. Fig. 3.1). In a more recent survey of the tenderness of 40 bovine muscles, Belew *et al.* (2003) found that the Warner-Bratzler shear force ranged from 2.03 and 2.73 kg in diaphragm and *psoas major*, respectively, to 5.12 and 7.74 kg in *m. pectoralis profundis* and *m. flexor digitorum superficialis*, respec-

Table 10.11   Hydroxyproline content of collagen and elastin fractions from beef muscles (mg/100 g muscle)

Muscle	Collagen hydroxyproline	Elastin hydroxyproline
Semitendinosus	84	3
L. dorsi	11	1
Psoas major	9	1

tively. Belew *et al.* also found significant differences between 20 locations along the length of bovine *m. gluteobiceps*.

Tenderness within a given muscle may also vary significantly (Bryce-Jones *et al.*, 1964). For example, there is a systematic decrease in tenderness in proceeding from the proximal to the distal end of beef *semimembranosus* (Paul and Bratzler, 1955; Ginger and Weir, 1958). The tenderness of beef *biceps femoris* increases from insertion to origin (Rogers, 1969), and lateral portions of pork *l. dorsi* are more tender than medial portions (Urbin *et al.*, 1962). An investigation of the *biceps femoris*, *semitendinosus*, *semimembranosus* and *adductor* muscles of beef showed that definable *intra*muscular differences in tenderness were significant, especially within *biceps femoris* (Reuter *et al.*, 2001).

It has been suggested that systematic and significant differences in tenderness between locations within muscles could be exploited, known regions of toughness being removed for manufactured products (Denoyelle and Lebihan, 2003).

The other principal protein found in connective tissue is elastin. Its properties have been fully described by Partridge (1962). The molecule has a central core containing two unusual amino acids, desmosine and isodesmosine, which are derived from lysine. Being most resistant to heat and thus to degradation on cooking, elastin would, at first sight, appear to constitute an important factor contributing to meat toughness. Fortunately the amount of elastin in most muscles, other than that associated with blood vessels, is very small. Nevertheless, notwithstanding its minor concentration, the intractible nature of elastin cannot be entirely ignored in assessing meat texture, especially in those muscles where it appears to be more prevalent (e.g. *semitendinosus*). The mucoprotein of the ground substance, in which the fibres of collagen and elastin are embedded, is a minor constituent; its distribution parallels that of elastin (McIntosh, 1965).

It is probably justifiable to include the sarcoplasmic reticulum in the category of connective tissue. It surrounds individual myofibrils, and enhanced cohesion of myofibrils parallels 'woodiness' in freeze-dried meat (Penny *et al.*, 1963), toughness in meat held aseptically at 37°C for 30 days (Sharp, 1963), and toughness in fish stored in ice (Love and Elerian, 1963).

Intramuscular fat (marbling) tends to dilute the connective tissue of elements in muscle in which it is deposited, and this may help explain the greater tenderness reported for beef from well-fed good-quality animals (Beard, 1924).*

There is tentative evidence that both the quantity and quality of collagen may be modified by the nutrient regime of growing animals. Rapid growth in young animals would be expected to foster a higher proportion of less cross-linked collagen and thus increased tenderness (Bailey and Light, 1989). Although the increased growth rate effected by the administration of anabolic hormones does not appear to affect tenderness consistently, that of β-agonists has been associated with toughness (cf. § 2.5.2.1).

Since the connective tissue content of the muscles is not increased by the administration of β-agonists (Fiems *et al.*, 1989; Warriss *et al.*, 1991b), it must be presumed that the collagen is more cross-linked than normal (Dawson *et al.*, 1990). Similar

---

* Another factor, however, may be a decreased susceptibility of heavier fatter carcasses to cooling fast enough to encounter the toughening of 'cold-shortening' (cf. § 7.1.1.2(c)).

toughening has been found in veal calves treated with clenbuterol (Berge, Culioli and Ouali, 1993).

Increasing consumer awareness of differences in tenderness (and in other attributes of eating quality) between specific muscles is promoting developments in their marketing, especially in the USA ('muscle profiling': Jones *et al.*, 2001) and in selecting the most palatable muscles from joints that, overall, have hitherto been regarded as being of lower quality. Such muscles as *infraspinatus* and *teres major* show most promise for development into palatable steaks; whereas *semimembranosus* and *vastus lateralis*, for example, are less suitable for upgrading.

10.3.3   Post-slaughter factors

In general, the preslaughter factors which affect tenderness do so by determining the amounts and distribution (texture) and the type of connective tissue. As we have seen, there is a common but not invariable indirect correlation between connective tissue and tenderness. Within a given muscle, however, where amounts and type of connective tissue are constant, there can be considerable differences in tenderness caused by post-slaughter circumstances. The most immediate of these is post-mortem glycolysis.

10.3.3.1   Post-mortem glycolysis

Howard and Lawrie (1956) found that the *rate* of pH fall post-mortem was inversely related to tenderness of the meat on subsequent cooking and Marsh (1962) indicated that there was a direct relationship between the time elapsing before rigor mortis and tenderness. Increased tenderness is observed when the pH falls slowly, either because of naturally occurring variability (Marsh *et al.*, 1981) or because of deliberate manipulation (Howard and Lawrie, 1956), and a relatively high pH is thus maintained for some time in combination with near *in vivo* temperatures, and may well induce early conditioning changes (Marsh *et al.*, 1981) by such enzymes as CASF (cf. § 5.4.2).

Such observations were made in circumstances when the shortening of high temperature rigor could not develop. Locker (1960a) showed that for muscle not attached to the skeleton or otherwise held taut, the loss of tenderness during the onset of rigor mortis was directly related to the degree of shortening *at that time* (i.e. to the degree of interdigitation of actin and myosin filaments: Fig. 4.2). The degree of shortening, or of tension development, during the onset of rigor mortis in muscle which is free to shorten, is a direct function of temperature (Bendall, 1951; Marsh, 1954, 1962) down to about 15 °C (Locker and Hagyard, 1963). If such isolated muscle is exposed to temperatures lower than about 14 °C at this time, there is again an increasing tendency to shorten – it being as great at 2 °C as at 40 °C – and this is associated with decreased tenderness on cooking (cf. § 7.1.1.1).

Although the marked rigor shortening of muscle at high temperatures had long been known, the cold-shortening phenomenon is a more recent factor. The cold-shortening/toughness relationship is by no means linear. On exposure of excised, pre-rigor muscle to temperatures which cause cold-shortening, the degree of toughness in the cooked meat increases as the degree of pre-rigor shortening increases from 20 to 40 per cent of the initial length; thereafter, as the degree of shortening increases to 60 per cent, toughness once more decreases (Marsh and Leet, 1966). Shortening up to 40 per cent of initial length signifies a greater degree of interdigitation of actin and myosin, a greater measure of crossbonding during the onset of rigor mortis, and is reflected by a greater degree of toughness in the meat. Electron micrographs show that, in such muscles, the ends of the myosin filaments buckle against (or pierce) the Z-lines. It is conceivable that the myosin filaments then link through the Z-line, leading to toughness (Dickinson, 1969; Voyle, 1969). It is suggested that, in shortened muscles, the myosin filaments fuse to form a continuum (Locker, 1976). Thus, whereas in cooking non-shortened muscle the myosin filaments would coagulate in discrete masses and myofibrils would break at the I band (from which myosin is absent), this weak point would not occur in shortened sarcomeres in general.

Voyle (1969) and Dransfield (1994) cautioned against accepting too readily the interdigitation of myosin and actin filaments as the sole explanation for toughness. Indeed Bouton *et al.* (1973c), whilst confirming that shear values are highly dependent on the degree of myofibrillar contraction in muscles of normal ultimate pH, also noted that adhesion values (which reflect the state of the intrafibrillar connective tissue) are significantly increased in contracted fibres. As an indication that collagen may make a more positive contribution to the degree of toughness in shortened muscle, Rowe (1974) found evidence for alterations in the perimysial connective tissue which paralleled the degree of shortening of the sarcomeres. As the muscle contracted, the loose configuration of the collagen changed to a well-defined lattice. In studying *raw* beef muscles. Rhodes and Dransfield (1974) found that the resistance to shear *increased* with extension above rest length. They explained this phenomenon by postulating that the cross-sectional area of stretched sarcomeres contained a relatively increased proportion of connective tissue. As Davey and Winger (1980) pointed out, although there is much evidence to associate shortened sarcomeres with toughness in the cooked meat, it has been observed that, when ox *sternomandibularis* muscle is rigor-shortened at 37°C or is cold-shortened at 2°C and then allowed to enter rigor mortis at 37°C, there is no increased toughness over corresponding non-shortened muscle (Locker and Daines, 1975, 1976). In such instances early conditioning changes may be involved in tenderizing the meat, as suggested above. Savell *et al.* (1977a) demonstrated that markedly greater tenderness ratings could be established in meat having sarcomeres of a given length, when greater proteolytic action had taken place in them.

It is important to note that, if muscle is cooked for four hours at 100°C (when connective tissue is destroyed), and whilst stretched to the point where there is no overlap of actin and myosin filaments, tensile strength remains at about two-thirds of corresponding unstretched muscle (Locker, 1976). This suggested that the gap filaments might be a major factor determining the tensile strength of cooked muscle, although broken down during ageing or conditioning (§ 5.4.2). Yet King (1984) demonstrated that connectin breaks down during heating of muscle at 60–80°C – and to a greater extent than during ageing at 2°C for three weeks. He thus concluded that the coagulation of myosin, either alone or in association with other myofibrillar proteins, was more important for the surviving structure of cooked meats than a continuum of gap filaments, as suggested by Locker and Wild (1982).* Subsequently the latter workers reassessed the problem. They postulated that

---

* The breakdown of connectin is more intense at 60°C than at 80°C, carboxyl proteases being more active at the former temperature (King, 1984). It also occurs in 'cold-shortened' muscle wherein the tenderizing changes of conditioning are inhibited.

nebulin, rather than being the protein of the *N*-lines (§ 3.2.2), is present in the gap filaments, together with titin, and that it is the breakdown of nebulin which accounts for the lability of the gap filaments in conditioning (Locker and Wild, 1984).

It is of interest that the curves representing the inverse relationship between sarcomere length and toughness, whilst similar in shape, differ in the range of toughness values according to the connective tissue content of the muscle (J. R. Bendall, personal communication). The decrease in toughness observed in cooking muscles which have shortened beyond 40 per cent of their initial length may signify disruption or tearing of the structure (Marsh and Leet, 1966). Indeed electron micrographs have indicated that the phenomenon is probably due to the fracturing of certain sarcomeres by those which have severely shortened (Dickson *et al.*, 1970). Later studies (Marsh *et al.*, 1974) by electron microscope revealed that, with shortening greater than 50 per cent, a series of nodes developed in the fibre. These were regions of supercontraction: between them there was fracturing of the fibre which appeared sufficient to account for the decreased toughness. Similar considerations may apply to the tenderizing effect of high pressure (~100 MPa) on muscle, despite severe concomitant shortening (McFarlane, 1973).

If shortening of the muscle is *prevented* during the onset of rigor mortis as, for example, when the muscles are not excised but held rigid on the carcass, the temperature effect is not directly reflected by the degree of toughness in cooking. Nevertheless, even when held on the carcass, as in normal commercial practice, certain muscles still have a limited freedom to contract, and certain parts of muscles may shorten whilst other parts lengthen, thus causing local toughening even if the overall length is fixed (Marsh and Leet, 1966; Marsh, 1964; Marsh *et al.*, 1968). (Clearly, it is not possible to cool the entirety of a large carcass to 15 °C within a few moments of death when the *in vivo* temperature is about 37 °C.) An interesting aspect of the limited shortening to which some muscles would be liable, even when held on the carcass, has been demonstrated (Herring *et al.*, 1965). Thus, when beef carcasses were suspended horizontally during post-mortem glycolysis, the sarcomere lengths in such muscles as *psoas major* and *rectus femoris* were greater – and toughness was less – than when carcasses were suspended in a vertical position. Similar findings have been reported by Hostetler (1970). Suspension by the *obturator foramen* proved the most effective. Davey *et al.* (1971) found that the tenderness of *biceps femoris*, *semimembranosus* and *l. dorsi* (lumbar) was more than doubled by processing lamb carcasses in a standing rather than a vertical position. Decreased shortening of *l. dorsi* and increased tenderness on cooking has also been reported with pork carcasses when these are suspended by the pelvis rather than by the Achilles tendon. Fisher *et al.* (2000) demonstrated that pelvic suspension of pig carcasses markedly increased tenderness, and enhanced the brine uptake and yield, of hams prepared from the *semimembranosus* and *gluteobiceps* muscles, without adversely affecting the quality of those from *biceps femoris*. The stretching which various muscles undergo in horizontally suspended carcasses reduces that element of toughness due to contracted myofibrils and also to a minor degree the adhesion of intrafibre connective tissue (Bouton *et al.*, 1973a).

The efficacy of electrical stimulation immediately after death in swiftly lowering the pH of muscles to the point at which they no longer react to cold by shortening and toughening was considered in detail in § 7.1.1.2. As indicated, Marsh *et al.* (1987), in experiments where the rate of post-mortem glycolysis was altered by various combinations of modes of electrical stimulation and of temperature, found that a rate whereby a pH of 5.9–6.0 was attained at *ca.* 3 h post-mortem (pH$_3$), was associated with optimal tenderness. The use of electrical stimulation to maximize pH fall early post-mortem, although it prevents that element of toughness due to 'cold-shortening', may not induce a pH of 5.9–6.0, and tenderness may thus be less than optimal.

It has been pointed out that cold-shortening is not likely to affect muscles deep in the carcass, as in the rump area of beef (L. Buchter, personal communication). Because of the initial high temperature in this region and its insulation, post-mortem glycolysis will be swift and will have proceeded to completion long before refrigeration can lower the temperature below 15 °C. Indeed the speed of glycolysis in this region may lead to an exudative condition in beef, somewhat comparable to PSE in the pig. This can be avoided by removal of the meat from the hot carcass and swift cooling in air at 15 °C.* Moreover, at this temperature and under these conditions there will be no 'cold-shortening' and, possibly because the shortening which tends to accompany the onset of rigor mortis when it occurs at body temperature will also be avoided, such meat may be somewhat more tender than that left on the carcass (Schmidt *et al.*, 1970), and will have an enhanced water-holding capacity (Taylor and Dant, 1971). Thus, the bulk of the meat, in relation to the refrigeration capacity applied, will determine whether refrigeration during post-mortem glycolysis will enhance or detract from various aspects of meat quality (cf. § 10.2.1.2).

At temperatures below the freezing point, damage to the sarcotubular system would be relatively severe in both 'red' and 'white' muscles, and its ability to recapture calcium ions adversely affected. If freezing had been sufficiently rapid after death to fix ATP concentrations at pre-rigor levels, this would be manifested as the phenomenon of 'thaw-rigor' (unless thawing were extremely fast: cf. § 7.1.2.2). It is found, in fact, that both 'red' and 'white' muscles are susceptible to the massive shortening of 'thaw-rigor' (above). A comparison of the behaviour of 'red' (*semitendinosus*) and 'white' (*psoas*) muscles of the rabbit substantiates the view that 'cold-shortening' and 'thaw-rigor' both reflect abnormal post-mortem stimulation of the contractile actomyosin ATP-ase (Lawrie, 1968; Newbold, 1980), whatever may be the precise mechanisms.

It should be noted that whilst muscle which has been frozen pre-rigor and then thawed rapidly, thereby undergoing 'thaw-rigor' and becoming very tough when cooked, it is tender if it is cooked directly in the pre-rigor state without thawing. This has been attributed to the retention of much of the muscle's water because the high temperature fixes the pH at near *in vivo* values (Miles and Lawrie, 1970).

Shortening and toughening of muscles can be arrested, even if they are frozen pre-rigor, provided either that they are thawed on the carcass or, if they are excised, that they are slowly thawed at about –2 °C, when the rigidity of the remaining ice will prevent shortening (Marsh, 1964; Marsh and Thompson, 1958). Dransfield (1996) attributed the tenderness of pre-rigor frozen muscle held at –3 °C (when the rate of ATP breakdown is relatively rapid) to the proteolytic action of the calpains. If pre-rigor frozen muscle is held for a month at –12 °C, there is an even slower rate of ATP breakdown; but over this period it is sufficient to remove the prerequisite for 'thaw-rigor' (Davey and Gilbert, 1976b). The breakdown appears to be due to

---

* It should be noted that the *rate* of heat removal is probably as important in causing 'cold-shortening' as the cold temperature itself. Thus the high rate of cooling of excised muscles in a water bath could cause shortening and toughening even at 15 °C (unpublished).

direct enzymic action; post-mortem glycolysis does not occur at this temperature (rapid thawing of such meat leads, of course, to the marked shortening of thaw-rigor). Empirical observation also illustrates the effect of a fast rate of post-mortem glycolysis. Thus Gray (1966) refers to mid-Western Canada where temperatures in winter may fall to –40 °C and the cooling power of the air is greatly increased by strong winds. Steers would be killed in the field and the meat cut up and allowed to freeze. No culinary procedure could produce anything but a tough roast from such beef which, as is now evident, froze pre-rigor and underwent thaw-rigor on cooking.

Again, in the country areas of Nigeria, beef is frequently tough. To avoid the rapid proliferation of bacteria in the prevalent hot, humid environment, carcasses are dismembered for sale in local markets. Rigor mortis thus occurs at an ambient temperature of 30–37 °C and the unrestrained muscles undergo shortening (Prof. D. A. Ledward, personal communication). It is evident that high-temperature rigor is an important factor contributing to toughness in these circumstances.

The *rate* of post-mortem glycolysis has another effect on tenderness in addition to that on shortening during the onset of rigor mortis. Where the rate of pH fall is inordinately fast, as with the PSE condition in pigs (Bendall and Wismer-Pedersen, 1962), sarcoplasmic proteins are denatured and precipitate on to those of the myofibrils; and the latter may also be denatured to some extent, since they become less soluble in these circumstances. It has been shown that the ratio of insoluble myofibrillar protein to total protein in muscle is directly correlated with toughness (Hegarty *et al.*, 1963).

The *extent* of post-mortem glycolysis, apart from its rate, also has an effect on the tenderness of beef, pork and lamb (Howard and Lawrie, 1956; Lewis *et al.*, 1962; Bouton *et al.*, 1973b; Watanabe *et al.*, 1996). As the ultimate pH increases from 5.5 to 6.0, tenderness appears to decrease; at ultimate pH levels above 6, however, tenderness increases once again (Fig. 10.4). It has been amply confirmed in beef

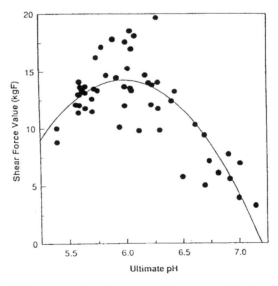

Fig. 10.4   Relationship between shear force and ultimate pH of bovine *l. dorsi et lumborum* muscles. (After Watanabe *et al.*, 1996; reproduced by kind permission of Dr C. E. Devine and Elsevier Science Ltd.)

(Jeremiah *et al.*, 1991; Purchas and Aungsupakorn, 1993) and lamb (Devine *et al.*, 1993) that tenderness is minimal at ultimate pH values between 5.8 and 6.2. Watanabe and Devine (1996) subsequently showed that the breakdown of titin and nebulin is minimal at an ultimate pH in this region and concluded that one effect of ultimate pH on tenderness is exerted through its action on proteolytic enzymes.* Both shear force and adhesion, as respective measures of the contributions of myofibrils and connective tissue to toughness, decrease as the ultimate pH rises (Bouton *et al.*, 1973c). In the region of pH 6.8, tenderness becomes excessive and is associated with a jelly-like consistency in the meat which lowers its overall acceptability (Bouton *et al.*, 1957). The exact tenderness/pH relationship varies between different muscles. Thus, in mutton, the pH of minimum tenderness was 5.64, 5.90 and 6.05 for *biceps femoris*, *semitendinosus* and *l. dorsi* respectively (Bouton and Shorthose, 1969). If pre-rigor meat is heated quickly enough, so that the enzymes effecting post-mortem glycolysis are inactivated faster than the heat can accelerate their activity, a high pH will result. If it is of the order of 7 this would be expected to enhance tenderness (cf. Fig. 10.4) to a much greater extent than the shortening of the excised muscle would diminish it (even though such shortening during cooking is especially severe with pre-rigor meat; Marsh, 1964). It has been shown, in fact, that the relative tenderness of pre-rigor cooked meat is directly related to the level of pH which has been attained at the moment of cooking (Miles and Lawrie, 1970). The enhanced tenderness may be a reflection of the greater water content and water-holding capacity of the muscle proteins (§ 5.4.1) and of the consequent swollen nature of the muscle fibres at high pH. Some of the tenderness of pre-rigor meat can be similarly explained. Changes in the distribution of water between the intracellular environment and extracellular spaces may also be a factor, additional to those of myofibrillar contraction and the nature and orientation of connective tissue, in determining tenderness (Currie and Wolfe, 1980). As indicated in § 10.2.1.1, the water-holding capacity of meat increases for at least 2 pH units on each side of the iso-electric point of the muscle proteins. It is evident that a high pH, at least within the physiological range, is associated with enhanced tenderness: it has now been shown that pH values on the acidic side of the iso-electric point are also associated with increased tenderness (Gault, 1985). Such pH values would not be encountered naturally, of course, but during the manufacture of pickled products.

10.3.3.2   Conditioning (see also §§ 5.4 and 7.1.1.2)
That the tenderness of meat increases when it is conditioned (e.g. stored at chill temperatures for 10–14 days) has long been recognized (Lehmann, 1907); and, of course, such meat as venison is regularly aged for this purpose. The decrease in tenderness which is associated with the onset of rigor mortis (§ 10.3.3.1) is gradually reversed as the time of post-rigor conditioning increases. To reiterate the views in § 5.4.2, this is not due to the dissociation of the actomyosin formed during the onset

---

* The tenderness of the flesh from animals which had been chased to exhaustion, and in which, presumably, glycogen reserves had been significantly lowered, was commented on by Hunter in 1794. The effect of excitability in lowering *in vivo* glycogen reserves and raising the ultimate pH of the musculature is implicity recognized in Thomas Hardy's *Mayor of Casterbridge* (1886). He refers to a town square in which there was '. . . a stone post . . . to which the oxen had formerly been tied for baiting with dogs to make them tender before they were killed . . .'.

of rigor mortis (Marsh, 1954), and the absence of any increase in end groups (Locker, 1960b) shows that the myofibrillar proteins are not appreciably proteolysed in these circumstances. Moreover, the absence of soluble hydroxyproline-containing substances in meat, even after one year at 37 °C, indicates there is no extensive proteolysis in connective tissue proteins (Sharp, 1959). Although there is little proteolysis of connective tissue proteins, certain cross-links in the telopeptide region of collagen molecules are apparently broken, possibly due to the action of lysosomal enzymes (Etherington, 1971, 1972). In respect of the myofibrillar proteins, although no massive proteolysis occurs, it has already been indicated (§ 5.4.1) that a number of subtle alterations occur. Thus, the calcium-activated sarcoplasmic factor (calpains) attacks troponin T (above pH 6), the Z-lines (desmin), the M-line proteins, tropomyosin and the so-called gap filaments (connectin) (Locker et al., 1977; Penny and Dransfield, 1979; Penny, 1980; Young et al., 1980); and lysosomal enzymes attack troponin T (below pH 6) as well as the cross-links of the non-helical telopeptides of collagen and the ground substance. There is extensive proteolysis of the soluble sarcoplasmic proteins (Hoagland et al., 1917), and the cytoskeletal proteins (Kristensen and Purslow, 2001). These changes, together with a loss of calcium ions, and uptake of potassium ions, by the muscle proteins (Arnold et al., 1956), cause their water-holding capacity to increase during conditioning.

Proteomic techniques have revealed, in great detail, that changes in the patterns of proteins occur during post-mortem glycolysis and conditioning. It is envisaged that these changes in pattern will be related subsequently to a series of specific tenderness levels in meat (Lametsch et al., 2003).

Whatever the nature of the particular protein changes during conditioning which are significant in relation to the increased tenderness, it has long been clear that muscles contain proteolytic enzymes (§ 5.4.2) which operate much more readily at 37 °C than at 5 °C (Sharp, 1963), and that, in general, higher temperatures of conditioning produce a given degree of tenderizing in a considerably shorter time than do lower temperatures. This effect was studied by Bouton et al. (1958) and by Wilson et al. (1960). The former workers found that conditioning for 2 days at 20 °C gave the same degree of tenderizing as 14 days at 0 °C, and that the benefits of conditioning were more marked with beef of poor quality (cf. also Moran and Smith, 1929), which was initially tougher although the final degree of tenderness achieved was similar in beef of good and poor quality.

Wilson et al. (1960) employed antibiotics to control bacterial spoilage and were thus able to study temperatures as high as 49 °C. Semimembranosus muscles from the rounds of beef carcasses which had been infused with oxytetracycline (to a concentration of 30–50 ppm) were employed. The muscles were prepared as ¾ inch steaks and vacuum sealed in plastic film. After appropriate conditioning periods at 2 °C, 38 °C, 43 °C and 49 °C, the meat was cooked and assessed for tenderness by a taste panel. Some of the results are given in Table 10.12.

As Table 10.12 shows, the tenderness score was increased by all conditioning procedures over that of controls. Moreover, the meat held for 2 days at 38 °C, or for 1 day at 43 °C or 49 °C, was more tender than that kept for 14 days at 2 °C. The tenderness increment was particularly high in the meat held at 49 °C, but the latter had a somewhat undesirable flavour. Conditioning at 38 °C was difficult to control, even with the dose of ionizing radiation given to steaks at this temperature (in addition to the antibiotics) because of the greater risk of bacterial growth. The optimum time and temperature required to have the same degree of tenderizing as that arising

Table 10.12 Mean tenderness values for beef steaks conditioned in various ways (after Wilson *et al.*, 1960)

Time and temperature of conditioning	Tenderness	
	Initial	**Residual**
Non-conditioned controls	5.2	5.2
14 days at 2 °C	5.9	5.8
Non-conditioned controls[a]	5.3	5.5
2 days at 38 °C	6.0	6.1
Non-conditioned controls	5.1	5.2
1 day at 43 °C	6.3	6.2
Non-conditioned controls	5.2	5.4
1 day at 49 °C	7.4	7.2

[a] Given 45,000 rad ionizing radiation.

during 14 days at 0 °C was 1 day at 43 °C. Nevertheless, the *rate* of conditioning decreases gradually as holding temperatures rise from 40 °C to 60 °C. It then decreases sharply, ceasing altogether at 75 °C (Davey and Gilbert, 1976b). The observations of Penny and Dransfield (1979) are relevant in this context. Although proteolysis of troponin T correlated with increasing tenderness in beef muscles when conditioning took place at temperatures between 3 and 15 °C – and the rates of proteolysis increased with increasing temperature – the concomitant increase of tenderness was *proportionately* less at higher temperatures of ageing. This may reflect protein denaturation as an additional factor in the latter circumstances and recalls the observations of Sharp (1963) (§ 5.4.1), who found that muscles which had been stored at 37 °C homogenized less readily than those stored at 0 °C. However, it is known that CASF (calpains) (Dayton *et al.*, 1976) and cathepsin B (Swanson *et al.*, 1974) lose activity above 40 °C and 50 °C, respectively. At higher temperatures (*ca.* 60 °C) carboxyl proteases are active in proteolytic breakdown of muscle proteins (King and Harris, 1982); but, nevertheless, their capacity to break down connectin is less at 80 °C than at 60 °C.

Another interesting aspect of the large temperature coefficient of conditioning changes is the use of electrical stimulation to avoid cold-shortening in meat which is refrigerated swiftly post-mortem. As indicated in § 7.1.1.2, electrical stimulation produces a low pH rapidly in the musculature at a time when the temperature is still at *in vivo* levels. The combination of low pH and relatively high temperature activates lysosomal proteases, and, before the pH falls below 6, the temperature possibly activates calpains. Electrical stimulation promotes significant conditioning changes to occur during the short period when the meat is still at *in vivo* temperature (Devine and Graafhuis, 1995). If the meat is cooled too quickly after electrical stimulation, however, this advantage is lost and the tenderizing action does not operate, although the toughness due to cold-shortening can be avoided thereby.

Very fast chilling has been defined in the European Union as chilling to a temperature of –1 °C by 5 h post-mortem. As Joseph (1996) has pointed out, the very low temperatures required to achieve this in the refrigerating environment would cause much variation in the biochemical and biophysical status of muscles, especially in those nearest the source of the refrigeration. On one hand calcium ions, released by the cold shock, would tend to cause the toughening of cold-shortening; on the other hand they would tend to enhance tenderness either directly or by stimulating the action of proteolytic enzymes. In remote areas of developing countries there are considerable advantages in using solid CO_2 as an alternative to mechanical chilling for hot, deboned meat (Gigiel, 1985). The procedure, however, promotes cold-shortening and toughness in the meat (Swain et al., 1999).

If high temperature conditioning is applied to meat immediately after slaughter, this can induce marked shortening of the muscles as they go into rigor mortis and subsequent toughness, an adverse effect long known in the laboratory and observed in practice in lamb carcasses (Davey and Curson, 1971). When, however, muscles are restrained from shortening, they are more tender if they undergo rigor mortis at 37 °C than at 15 °C (Locker and Daines, 1975). This may possibly be due to enhanced activity, at this temperature, of calpains (which operate optimally at near in vivo pH), since there is evidence that, in muscles held at 37 °C (but which were not free to shorten), tenderness was greatest in those wherein the pre-rigor pH, for-tuitously, was slow to fall (Marsh et al., 1981; Marsh, 1983). Harris and McFarlane (1971) found that beef l. dorsi muscle tenderized more rapidly than semimembra-nosus when aged at 0–1 °C for up to 6 weeks (cf. § 5.4.1). This was true whether or not the muscles were stretched by hanging the carcasses by the obturator foramen. Stretching was found to give a tenderizing effect equivalent to that obtained by ageing for 2 weeks at 0–1 °C when using the conventional method of suspension. Ouali and Talmant (1990) and Monin and Ouali (1991) have extensively reviewed the reasons for differentiation between muscles in the rates and extents of ageing which they undergo post-mortem (cf. § 5.4.2).

Further evidence for differences between muscles in their behaviour during con-ditioning has been provided by Bailey and Light (1989) based on the extractability of perimysial collagen (Table 10.13) and Stanton and Light (1990) have shown that although the extractability of endomysial collagen is much greater before condi-tioning in bovine psoas major and gastrocnemius than in extensor carpi radialis and supraspinatus, the subsequent increase in solubility of such collagen during condi-tioning is markedly more extensive in the latter two muscles. Simões et al. (2005) found that the tenderness of biceps femoris was the most accurate predictor for the overall carcass tenderness after ageing for 7 days at ca. 0 °C.

In general, endomysial collagen is much more labile during conditioning than that of the perimysium; and, within the endomysium, type III collagen is preferen-tially attacked in comparison with type I (Stanton and Light, 1990).

Table 10.13 Extraction of perimysial collagen from various bovine muscles before and after conditioning

Muscle	Unconditioned	Conditioned
Psoas major	0.8	10.5
Supraspinatus	0.4	4.6
Gluteus medius	0.6	2.3
Gastrocnemius	0.8	1.7
Pectoralis profundus	0.4	1.3

10.3.3.3 Cooking
Whether cooking will cause an increase or a decrease in tenderness depends on a variety of factors, including the temperature to which the meat is raised, the time of the heating and the particular muscle being considered.

Whilst, in general, cooking makes connective tissue more tender by converting collagen to gelatin, it coagulates and tends to toughen the proteins of the myofibril. Both these effects depend on time and temperature, the former being more important for the softening of collagen and the latter more critical for myofibrillar toughening. Prolonged cooking times and relatively low temperatures are thus justified for meat which has much connective tissue and conversely (Weir, 1960). The tenderizing effects of prolonged cooking is additional to that of ageing (conditioning) (Davey *et al.*, 1976).

The degree of solubility of collagen increases with temperature. At about 60–65 °C collagen shortens and is converted into a more soluble form (Bendall, 1946; Bear, 1952; Machlik and Draudt, 1963). The shrinkage temperature of collagen is fairly characteristic. Meat juices, and the sarcoplasmic proteins therein, appear to play some part, however, in the effect (Giffee *et al.*, 1963) since the shrinkage temperature is 65 °C, when collagen is heated in water (Winegarden *et al.*, 1952). The helical structure can be seen to unwind on heating for 10 minutes at 64 °C (Anon., 1974b; Snowden and Weidemann, 1978). The percentage of collagen (beef) solubilized by heat increases gradually from about 60 °C to 98 °C. At the latter temperature conversion to gelatin is marked (cf. Table 10.14; Paul, 1975). Gelatin formation is swift with pressure cooking at 115–125 °C (Bendall, 1946); but with retorting, there is a marked loss of collagen solubility above 100 °C (Palka, 1999).

Partial reversion of the collagen to gelatin transformation explains the finding that, whereas meat cooked at 80 °C is tougher when measured at 20 °C than at 70 °C, that cooked at 55 °C shows no difference with the temperature of assessment (Ledward and Lawrie, 1975). Some of the discrepancies between objective assessment of meat tenderness at room temperature and subjective assessment by taste panel, using warm meat, may well be explained thereby.

In experiments with beef *sternomandibularis* muscle (Locker *et al.*, 1977), in which the time of heating at 70 °C was varied, it was reported that when muscles were cooked beyond 40 min, myosin and actin denatured, but not collagen or the gap filaments (cf. § 3.2.2). The latter appeared to withstand heating for 4 h at 100 °C (Although conditioned muscles, they are attacked by the calpains and then disintegrate on cooking: cf. § 5.4.1). Evidence for the heat stability of the gap filaments was based on electrom microscopy (Locker and Wild, 1982); but subsequent work by King (1984), using sodium dodecylsulphate gel electrophoresis), revealed a degree of breakdown of connectin (i.e. gap filament) which was difficult to reconcile with its being responsible for the integrity of the muscle structure after heating.

Scanning calorimetric studies (Martens *et al.*, 1982; Ma and Ledward, 2004) show that actin is relatively stable to heat, not being denatured until the temperature is above 75 °C. The relatively high denaturation of titin (Pospiech *et al.*, 2002) may contribute to the increase in toughness on cooking.

There is increased tenderness with increased solubilization of collagen in braising, but relatively little softening, despite increased collagen solubility, on roasting (Table 10.14).

The attainment of a given temperature by microwave energy is associated with less denaturation of the myofibrillar and sarcoplasmic proteins than when attained

Table 10.14   Effect of temperature cooking on shear force and collagen solubilization in various beef muscles (after Paul, 1975)

Muscle		Mean internal temperature (°C)			
		(a) Roasting			
		58	67	75	82
L. dorsi and triceps brachii	Shear force (kg)	4.5	3.3	3.5	3.5
	Collagen solubilized (% total)	2.7	4.8	6.4	7.7
Semitendinosus	Shear force (kg)	3.5	3.1	3.3	3.3
	Collagen solubilized (% total)	4.3	6.0	8.0	11.0
Biceps femoris	Shear force (kg)	4.3	3.9	3.6	3.6
	Collagen solubilized (% total)	6.3	8.7	8.5	13.6
		(b) Braising			
		70	98	98 (held 30 min)	98 (held 90 min)
L. dorsi and triceps brachii	Shear force (kg)	3.9	3.4	2.9	2.2
	Collagen solubilized (% total)	3.7	11.3	21.5	44.9
Semitendinosus and Biceps femoris	Shear force (kg)	4.9	3.8	3.0	2.2
	Collagen solubilized (% total)	4.9	10.2	23.3	52.0

by conventional heating; but this can be attributed to the further progression of the same type of effects due to the increased time of cooking necessary with the latter process (Roberts and Lawrie, 1974). On the other hand, there is evidence that microwave heating preferentially increases the solubilization of collagen (McCrae and Paul, 1974).

These conflicting influences help to explain why different muscles react differently to cooking (cf. Table 10.14). Thus, beef *l. dorsi* is tender and *biceps femoris* (which has about twice as much collagen) is tough when boiled to 61 °C; but the converse holds when they are braised at 100 °C (Cover and Hostetler, 1960). While *l. dorsi* cooked at 60, 70 or 80°C showed no difference in shear force, there was a considerable decrease in the force required to shear *semitendinosus* and *semimembranosus* when cooked at 70 and 80°C compared with 60°C (Sanderson and Vail, 1963), the latter two muscles having more connective tissue than *l. dorsi*. In studying the effect of cooking porcine steaks (*l. dorsi*) and leg roasts (*gluteobiceps*) to final internal temperatures of 65, 72.5 and 80°C, Wood *et al.* (1995) detected some loss of tenderness at the higher temperatures.

In their study of bovine *sternomandibularis* muscles, Davey and Gilbert (1974) found that increasing cooking temperature produced two separate phases of toughening. The first occurred between 40 and 50°C, and was apparently due to the denaturation and insolubilization of the contractile proteins. The second occurred

between 65 and 75 °C, and was apparently due to fibre shrinkage as collagen denatured. Above 75 °C – and with increasing time of cooking – toughness diminished
as collagen breakdown occurred. If the predominant collagen in a muscle has
thermally-labile cross-links, then heating will cause greater solubility and decreased
shear values; whereas, if the collagen has heat-stable crosslinks, increased tension
and toughness will result on heating (Bailey and Sims, 1977). The relative contributions made by the collagens of the basement membrane, epimysium, perimysium
and endomysium – and by the degree of heat-stable cross-linking in each – will
influence the toughness of a given muscle when heated at constant sarcomere
length.

Following shrinkage on heating, the muscle fibres are held together by denatured
perimysial collagen, the strength of the latter depending on the proportion of its
heat-stable cross-links. At the higher temperatures attained during prolonged
cooking, shear values decrease, probably due to cleavage of peptide bonds and of
mature cross-links, especially the former (Bailey, 1989).

In using the scanning electron microscope, Rowe (1989) found that the collagen
fibres of the endomysium appeared to become beaded after heating at 50 °C for an
hour, possibly because of the concomitant denaturation of the closely associated
myofibrillar proteins. The sarcolemma denatured after one hour at 60 °C and disin-
tegrated at 70 °C (when the endomysial collagen had shrunk). On the other hand,
the basement membrane appeared to survive at 100 °C for an hour.

Davey and Gilbert (1975b) showed that, for a given degree of shortening during
the onset of rigor mortis, and a given cooking temperature, the meat of young beef
animals was more tender than that of older ones. During the cooking of veal the
collagen readily dissolves to set as a gel on cooling. On the other hand, using the
same muscle and temperature, the collagen from older animals is insoluble and
the meat tough. The highly cross-linked intramuscular collagen of the older animals
binds the myofibrils together even when it is denatured, and generates greater
tension during heat contraction (Bailey, 1974).

The effect of shortening during the onset of rigor mortis on subsequent toughness in cooked meat has been considered in detail above (§ 10.3.3.1). The shortening observed in these circumstances reflects the extent to which individual
sarcomeres *contract*, i.e. the degree of interdigitation of actin and myosin filaments
in each sarcomere. Cooking produces further, additional shortening by causing the
sarcomeres to *shrink* overall. Such shrinkage of sarcomeres does not occur at temperatures up to 60°C but it does at 79 °C (Giles, 1969). At the latter temperature
the *M*-lines and *I*-bands become disrupted and changes are observed in the collagen fibres (which, at 70 °C, are above their shrink temperature). Whereas with
unshortened muscle, cooking shrinkage occurred along the fibres and precedes fluid
discharge (transverse swelling accommodates entrapped fluid), such discharge
occurs during the cooking of highly shortened muscle, and across the fibres. Cooking
shortening at 80 °C, of muscle which has not shortened during the onset of rigor
mortis, contributes to toughening to the same degree as cold-shortening (Davey and
Gilbert, 1975b, c) (cf. also § 10.2.2.1).

The high concentration of the sarcoplasmic proteins in the aqueous phase of
muscle (§ 3.2.2), and their lability post-mortem (§ 5.4.1), have already been mentioned. There is evidence that their precipitation on cooking above *ca.* 40 °C causes
the muscle fibres to adhere to one another and leads to gelling (Tornberg and
Persson, 1988). They make an important contribution to the adhesion of cooked

sausage batters (Farouk *et al.*, 2002). Thus, although the sarcoplasmic proteins are extractable by water, their potential contribution to texture/tenderness *in situ* should not be dismissed.

However desirable from the point of view of increased tenderness, pressure cooking may be associated with detrimental changes in the biological value of the meat proteins. Thus, Beuk *et al.* (1948) showed that when pork was autoclaved at 112 °C for 24 h, 45 per cent of the cystine was destroyed. It appeared that the essential amino acids were unaffected, but this was as determined by acid hydrolysis of the meat. Enzymic digestion showed a lowered availability of several essential amino acids – tryptophan being especially affected. This would be reflected in their value during digestion. It should be emphasized, however, that no lowering of the nutritive value of meat would occur at temperatures below 100 °C (Rice and Beuk, 1953), and the latter is substantially above the normal temperatures attained by meat during cooking. In Australia, critical control points (cf. §§ 6.1.2.2 and 6.4.1), based on the various pre- and post-slaughter factors that affect tenderness, have been identified and incorporated into a grading scheme for the management of meat tenderness (referred to as Meat Standards Australia: Thompson, 2002).

10.3.3.4 Processing

Subsequent processing may alter meat tenderness. The effects of pre-rigor freezing have already been considered in relation to post-mortem glycolysis. Alterations of tenderness through freezing post-rigor meat are also known. Although the rates of freezing normally used in commerce have no effect in this respect, meat blast frozen at a rate sufficient to freeze the deepest portion of the carcass in 24 h (§ 7.1.2.2) tends to be tougher on cooking than corresponding meat which has been chilled for the normal 2–3 days before freezing (Howard and Lawrie, 1956) – possibly because the latter has had what amounts to a short conditioning period (§§ 5.4 and 10.3.3.2). If, however, the rate of blast freezing is increased, so that the deepest part of the carcass freezes in only 18 h, the meat is found to be as tender as corresponding meat frozen after 2–3 days chilling (Howard and Lawrie, 1957b). Since the absence of a chilling period in the former would again operate against tenderness, it must be presumed that this is more than offset by microstructural changes effected by the greater rate of freezing. Their nature is unknown for, although freezing makes post-rigor meat more tender when the rate is fast enough to cause intrafibrillar ice formation (Hiner *et al.*, 1945; Hiner, 1951), such rates are virtually impossible with beef quarters. On the other hand, the necessary rates *could* occur in freezing steaks for prepackaged sale. An alternative or additional explanation of the toughening observed in the blast freezing of hot beef carcasses could be the induction of 'cold-shortening' (cf. § 10.3.3.1 above). The observed increase in tenderness with even faster blast freezing could signify immobilization of the surface musculature by freezing before cold-shortening could occur.

In the accelerated freeze drying of meat, even under optimum operating conditions, the rehydrated product is somewhat less tender and more 'woody' than fresh meat. This appears to be partly attributable to the effect of the plate temperature on the sarcoplasmic and myofibrillar proteins (§ 8.2.2) and it is largely obviated by the induction of a high ultimate pH. The latter, whilst producing excessive tenderness in fresh meat, restores a desirable degree of tenderness to the meat after accelerated freeze drying (Tables 8.9 and 8.10).

Ionizing radiation at sterilizing doses (about 5 Mrad; 50 kGy) or above, causes changes in the meat proteins which increase tenderness (Coleby *et al.*, 1961). This is probably due to changes in the collagen molecule, for the shrink temperature of isolated collagen decreases from 61 °C to 47 °C with 5 Mrad and to 27 °C with 40 Mrad (Bailey *et al.*, 1962).

Experiments in Australia (Anon, 1971) showed that it was possible to tenderize both beef and mutton by subjecting the muscles from freshly slaughtered animals to very high pressures for short periods. Pressure of 100 MN m^{-2} applied for 2–4 min reduced the shear value (Warner–Bratzler) for various cuts by three to four-fold.

Microscopic examination showed that such high pressures cause severe contraction and disorganization of the muscles and Macfarlane (1973) demonstrated that the application of prerigor pressure (100 MPa) at 30 °C produced shortening of the muscles of the same order as that obtained in 'cold-shortening', without the accompanying increase in toughness. Subsequently, it was found that combined pressure heat treatments (150 MPa at 60 °C for 30 min) effects a substantial decrease in shear force even in cold-shortened meat (Bouton *et al.*, 1977a), and it was shown that it is the myofibrillar proteins which are primarily affected by the treatment (Bouton *et al.*, 1977b). Adhesion values, which are believed to derive from connective tissue, are not affected. The imidazole groups of histidine appear to be implicated in the pressure solubilization effect (Macfarlane and Mackenzie, 1976) (cf. § 12.1.2) Ma and Ledward (2004) studied the phenomena further in post-rigor beef and confirmed the marked tenderization seen at high (60–70 °C) temperature and pressures around 200 MPa, although they also observed some loss of adhesion under these conditions. The causes could include increased enzymic action on the drastically modified myofibrillar proteins or on the slightly modified collagen. Jiminez-Colmenero, *et al.* (2001) thought the effect could be due to enzymic action on the myosin heavy chain or connectin. It would appear, however, that changes in the collagen are also important (Ma and Ledward, 2004). If the pressure and temperature are applied sequentially, and not in combination, no such tenderization is seen.

10.3.4 Artificial tenderizing

Attempts to make meat tender artificially are by no means new. They have included beating the meat, cutting it into small portions so that the strands of connective tissue were severed, marinading it with vinegar, wine or salt and enzymic tenderizing – inadvertently employed at least 500 years ago by the Mexican Indians when they wrapped meat in pawpaw leaves during cooking. In recent years such attempts have become more systematic. The tradition of wrapping food in fern leaves for protection during transit is still found in certain parts of Colombia. A typical odour and texture concomitantly develop which are considered attractive. Proteolytic enzymes in the fern increase degradation of the perimyseal connective tissue, enhancing tenderness (Sotelo *et al.*, 2004).

Recognition that certain plants, fungi and bacteria produced nontoxic proteolytic enzymes (Balls, 1941; Hwang and Ivy, 1951) was followed by their incorporation into commercial meat tenderizers. These were first used as dips. As such, they were somewhat unsatisfactory since they overtenderized the surface, producing a mushy texture (and sometimes unusual flavour), and, since they were unable to penetrate within the meat, left the interior unaffected. One method of overcoming this difficulty was to introduce the enzyme solution into the pieces of meat before cooking through fork holes (Hay *et al.*, 1953). Another was to pump the major blood vessels of the meat cuts post-mortem with enzyme-containing solution. A third was to rehydrate freeze-dried steaks in a solution containing proteolytic enzymes (Wang and Maynard, 1955). This ensured a much better distribution of the enzyme than did dipping or perfusion, but was still not ideal. Preslaughter injection of the live animal was proved to be the most effective method of introducing proteolytic enzymes into meat so that they penetrate uniformly into the furthest interstices of the tissue. This method was patented by Beuk *et al.* in 1959. A concentration of about 5 to 10 per cent of tenderizing enzyme is advocated, and the total quantity injected approximates to 0.5 mg/lb (0.25 mg/kg) live weight, although it varies according to the enzyme employed. There is some suggestion that the more active muscles, which have more connective tissue, get more enzyme because they have a greater vascularity. At enzyme levels suitable for tenderizing the muscles, the tongue, and organs such as liver may accumulate excessive quantities of enzyme and disintegrate on cooking. Animals are slaughtered 1–30 min after injection. In general, the injected enzymes do not harm the live animal. This is because the pH of the blood is considerably above their optimum pH, because they depend on – SH groups for activity and these are inoperative at *in vivo* oxygen tensions, and because they do not reach their optimum temperature of activity until, during cooking, the range 70–85 °C is attained (Gottschall and Kies, 1942). Nevertheless, the oxygen supply is limited within the cartilage matrix, allowing *in vivo* enzymic activity. As a result the injection of papain (temporarily protected by oxidation against inactivation during its passage in the blood) is manifested in rabbits by a dropping of the ears (McLuskey and Thomas, 1958). Again, structural and histochemical changes have been detected in animal livers after injection of papain and ficin at doses of 200 mg/kg body weight (Nestorov *et al.*, 1970). The toughness of lamb, induced by 'cold-shortening', can be offset by preslaughter injections of papain at commercial dose levels (Rhodes and Dransfield, 1973).

There is significant breed effect. The introduction of Brahman blood into cattle leads to a greater resistance to the tenderizing effects of papain (Huffman *et al.*, 1967).

Some of the enzymes which have been used in the tenderizing of meat are listed in Table 10.15.

Table 10.15   Relative potency of preparations of proteolytic enzymes on muscular tissue (after Wang *et al.*, 1957)

Enzyme preparation	Activity against		
	Actomyosin	Collagen	Elastin
Bacterial and fungal			
Protease 15	+++	–	–
Rhozyme	++	–	–
Fungal amylase	+++	Trace	–
Hydralase D	+++	Trace	–
Plant			
Ficin (fig)	+++	+++	+++
Papain (pawpaw)	++	+	++
Bromelin (pineapple)	Trace	+++	+

Table 10.16 Increasing tenderness, and decreasing residue, through enzymic treatment of beef (after Wang *et al.*, 1958)

Enzyme	Concentration (%)	Mean taste panel score	
		Tenderness	Residue
Fungal amylase	0	6.5	5.5
	0.045	7.4	5.4
Bromelin	0	6.5	5.5
	0.0003	7.6	4.5
Ficin A	0	6.2	6.1
	0.0003	7.9	5.9
Ficin B	0	5.8	5.8
	0.0003	6.5	5.2
Papain	0	5.5	5.6
	0.0003	7.0	4.6

It will be seen from Table 10.15 that the bacterial and fungal proteolytic enzymes act only on the proteins of the muscle fibre. They first digest the sarcolemma, causing disappearance of nuclei, then degrade the muscle fibre, eventually causing loss of cross-striations. The action of the proteolytic enzymes of plant origin is preferentially against connective tissue fibres. They first break up the mucopolysaccharide of the ground substance matrix, then progressively reduce the connective tissue fibres to an amorphous mass. It should be emphasized that these enzymes do not attack native collagen: they act upon the collagen as it is denatured by heat during cooking (Partridge, 1959). Elastin is not altered during conditioning or cooking and the activity of the proteolytic enzymes against elastin fibres suggests the presence of an elastase (Wang *et al.*, 1958). Unlike the tenderizing changes during conditioning (§§ 5.4 and 10.3.3.2), the enzymes used in artificial tenderizing break down connective tissue proteins to soluble, hydroxy-proline-containing molecules.

Some idea of the relative efficacy of these enzymes in tenderizing meat is given in Table 10.16 It will be noted that, in agreement with its lack of effect on connective tissue, the fungal enzyme has no effect on the residue as assessed by the taste panel after mastication.

As an alternative to the addition of proteolytic enzymes, meat might be artificially tenderized by stimulation of the muscles own proteolytic (catheptic) activity. Induced vitamin E deficiency would enhance the activity of the lysosomal enzymes (Tappel *et al.*, 1962); their liberation from the containing cell particles may be increased by excess vitamin A.

As another alternative, the formation of the mucopolysaccharide of the ground substance matrix and of collagen could be suppressed by the administration of cortisone and through vitamin C deficiency respectively (Whitehouse and Lash, 1961; Stone and Meister, 1962).

Sodium chloride itself, and other salts, have a tenderizing action on meat which is not inconsiderable (Wang *et al.*, 1958; Kamstra and Saffle, 1959) and post-mortem perfusion of joints with salt solutions has been of some success in this context (Bouton and Howard, 1960). Some of these effects are due to an enhanced water-holding capacity – either direct, or, as in the case of phosphate, through a concomitant raising of the pH (Bendall, 1954). Even the injection of water enhances tenderness (Williams, 1964a).

The reported effects produced by calcium chloride on meat tenderness, however, have been conflicting. Thus, when injected preslaughter, calcium salts accelerate the onset of rigor mortis and induce toughening (Howard and Lawrie, 1956). On the other hand, when meat is infused with calcium chloride solutions post-mortem, enhanced tenderness has resulted and this has been attributed to stimulation of calpains during subsequent conditioning (Whipple *et al.*, 1994; Polidori *et al.*, 2000; Gonzalez *et al.*, 2001; Geesink *et al.*, 2001) (cf. § 5.4.2), even when such infusion does not take place until 24 h post-mortem (Boleman *et al.*, 1995). Rehydration of dehydrated meat with a calcium chloride solution, however, appears to increase tenderness by increasing myofibrillar fragmentation (Gerelt *et al.*, 2002). Other workers have found that the infusion of calcium chloride increases exudation and toughness (Farouk and Price, 1994), especially if injected into the meat pre-rigor (Rousset-Akrim *et al.*, 1996). Increased tenderness, after calcium infusion at 0.5 to 6 hours post-mortem, has been attributed not to enhanced proteolysis by calpains, but to the reduction in cold-shortening through acceleration of post-mortem glycolysis by the Ca^{++} ions (Rees *et al.*, 2002b). It is evident that our understanding of the action of salts on meat tenderness is still incomplete.

10.4   Odour and taste

10.4.1   Definition and nature

Flavour is a complex sensation. It involves odour, taste, texture, temperature and pH. Of these, odour is the most important. Without it one or other of the four primary taste sensations – bitter, sweet, sour or saline – predominates.* Odour and taste are most difficult to define objectively. It is true that gas chromatography has permitted precise measurement of the volatiles from foodstuffs, but this has not infrequently confused the issue. The compounds isolated have not always corresponded with recognized subjective odour responses.

It is becoming appreciated that the nature of flavour involves simultaneous multisensory perception – an integration of responses from touch and auditory receptors with those for the chemical senses, odour and taste, as well as sight and various psychological factors, such as experience and expectation. (The agreeable odour of roast beef is nauseating when it emanates from a flower – *Iris foetidissima*, Moncrieff, 1951.) Multimodal neurones, which respond to stimulation from more than one sensory receptor, have been identified (Cook, 2003).

Considerable progress in clarifying the nature of consumers' perception of flavour has been made by 'nosespace' analysis, a development in which the air expelled from the food during mastication is collected and analysed by gas chromatography-mass spectroscopy (Linforth and Taylor, 1993). The use of electronic noses and tongues in flavour analysis has increased rapidly. Although there are still

---

* The Japanese recognize a fifth classification of taste, 'umami', which describes the savoury quality of monosodium glutamate and certain peptides. Such substances act as flavour enhancers: their receptors are sterically different from those responsible for the four common taste sensations (Kuninaka, 1981).

problems with the technique, it is anticipated that these will be resolved in due course (Deisinge *et al.*, 2004).

The oral stimulus from the viscosity of food is related to the perception of its flavour: a combination of texture measured by electromyography and 'nosespace' analysis is further elucidating its nature (Cook, 2003). A recent approach involves controlling the rate and quantity of selected flavours when individual solutions of each are mixed and fed via a multichannel device into the mouth (Cook, 2003); and the analysis of saliva, collected by cotton buds at intervals during chewing (Davidson *et al.*, 2000), has increased understanding of the sensations perceived by the consumer.

The evaluation of odour and taste still depends mainly on the taste panel. Variability between individuals in intensity and quality of response to a given stimulus, and in a given individual due to extraneous factors, makes the choosing of taste panel members, and the conditions of operation of the panel, matters of importance. The question has received considerable attention (Ehrenberg and Shewan, 1953; Peryam, 1958; Dawson *et al.*, 1963; Williams and Atkins, 1983). The operation of meat taste panels in particular has been investigated by Harries *et al.* (1963). It is not difficult to appreciate that there may well be genuine disagreement over the more subtle aspects of odour and taste.

The mechanisms by which human beings normally detect odour and taste will not be detailed here since full descriptions are available in physiological textbooks, but a few comments should be made. Response to odour occurs in the olfactory cells of the nasal surfaces and is conveyed from these to the brain for interpretation by the olfactory nerves. It is now known that the olfactory sensory neurones link the external surface of the mucosal membrane, where they are associated with an extensive system of cilia and, via axons, to the olfactory bulb of the brain. The cilia contain proteins, consisting of seven helical segments, which span the thickness of the membrane, linking specific receptor sites for odoriferous molecules on the exterior with proteins on the cytoplasmic side which are united with guanosine triphosphate (G-proteins) and, in a series of reactions involving cyclic AMP, transmit signals along the axons to the brain (Goodenough, 1998). There appears to be a 1:1 relationship between the frequency of molecular vibration of the stimulating odoriferous compounds and the properties of the corresponding responses of the olfactory bulb (Wright *et al.*, 1967). In adults, response to taste occurs in specialized cells on the tongue, the soft palate and the top of the gullet. As in the case of odour, it probably involves chemical reactions between the molecules concerned and the nerve endings in the taste cells – interpretation of the sensation again being made in the brain. Proteins have been isolated from the taste buds of cows and pigs, which form complexes with bitter and sweet substances directly in proportion to the actual bitter or sweet tastes of the latter (Dastoli *et al.*, 1968). While it has not been unequivocally shown how the dozen or so classes of responses (Bate-Smith, 1961) of the different olfactory cells are reflected morphologically, taste cells can be roughly localized, different areas of the tongue responding to the four primary sensations (bitter, sweet, sour and salt). There are also secondary taste reactions which can be described as metallic or alkaline (Moncrieff, 1951).

In considering the objective determination of taste, it is desirable to remember that, even with the primary sensation of bitterness, one person in three considers phenylthiocarbamide tasteless, although it is intensely bitter for two-thirds of the population (Blakeslee, 1932).

In ideal circumstances response to odour is about 10,000 times more sensitive than that to taste. Thus, while ethyl mercaptan can be detected in air at a concentration of 3×10^{-9} per cent, the sensation of bitterness, which is the most acute taste, is detectable from strychnine at a concentration in water of 4×10^{-5} per cent. Odour and taste in foodstuffs are important both aesthetically and physiologically for, if pleasant, they stimulate the secretion of digestive juices.

10.4.2   General considerations

It has long been a matter of common observation that the organoleptically desirable taste and odour of meat develops on cooking. The taste of raw muscle is bland, being only slightly sweet, salt, sour or bitter, according to its biochemical state and origin.

Fifty years ago it was established that water-soluble dialysates of muscle, which contained inosinic acid and a glycoprotein, gave a meaty odour on heating (Crocker, 1948; Batzer et al., 1960), and that odours resembling those of cooked meat could also be produced on heating the various amino acids of the glycoprotein with glucose and inosine (Batzer et al., 1962). Gel permeation chromatography of extracts of raw beef yields a dozen fractions; half of these produce a recognizable boiled beef aroma on heating (Mabrouk et al., 1969). The two fractions giving the strongest odour represented nearly 80 per cent of the diffusate and contained methionine, cysteic acid and 2-deoxyribose. By heating an S-containing amino acid (cysteine or methionine) with ribose, Morton et al. (1960) demonstrated that a pork-like aroma could be produced; whereas if other amino acids, quantitatively equivalent to 1–3 times the weight of the S-containing amino acid, were included, an aroma closer to that of beef resulted. In further patents, inorganic S-containing components and carbohydrate derivatives were specified (May and Morton, 1961; May, 1961). Thus sulphides yielded a bacon-like odour when heated with 2- or 3-methylbutyraldehyde (Wiener, 1972); and when thiamine, diacetyl and hexanal were heated with S-containing polypeptides, a poultry-like aroma developed (Giacino, 1968). (Meaty odours could also be obtained on heating nitrite with a mixture of amino acids: Hoersch, 1967.)

These relatively empirical observations have been supported and extended by more specific chemical studies of heat-induced changes in amino acids, carbohydrates and fats, both in isolation and in mixtures. Upwards of 750 compounds have now been identified in the volatiles from heated beef (Ho, 1980; MacLeod and Seyyedain-Ardebili, 1981; MacLeod, 1986) (Table 10.17).

Many types of heat-induced reactions lead to the production of meat flavours. Those believed to be particularly important have been summarized (Van den Ouweland et al., 1978) and include the pyrolysis of peptides and amino acids, the degradation of sugars, the oxidation, dehydration and decarboxylation of lipids, the degradation of thiamin and ribonucleotides and interactions involving sugars, amino acids, fats, H_2S and NH_3.

Volatile products are derived from amino acids on pyrolysis via Strecker degradation which involves deamination and decarboxylation of amino acids into aldehydes containing one carbon atom less, and by Maillard reactions, which are initiated by interactions between amino and carbonyl groups, leading to a complex series of degradation products (including N-substituted, 1-amino, 1-deoxy-2-ketones). Thus, for example, acetaldehyde forms on pyrolysis of phenylalanine, β-

Table 10.17   Volatile components of cooked beef aroma
(after MacLeod, 1986)

Type	No. Identified
Aliphatic hydrocarbons	73
Alicyclic hydrocarbons	4
Terpenoids	8
Aliphatic alcohols	46
Aliphatic aldehydes	55
Aliphatic ketones	44
Alicyclic ketones	8
Aliphatic carboxylic acids	20
Lactones	32
Aliphatic esters	27
Aliphatic ethers	5
Aliphatic amines	20
Chlorinated compounds	10
Benzenoid compounds	86
S-compounds (non-heterocyclic)	68
Furans and derivatives	43
Thiophenes and derivatives	40
Pyrroles and derivatives	20
Pyridines and derivatives	17
Pyrazines and derivatives	54
Oxazoles and oxazolines	13
Thiazoles and thiazolines	29
Other S-heterocycles	13
Miscellaneous	12

alanine, cysteine and methionine, methylpropanal from valine, 3-methylbutanal from leucine, indole from tryptophan, toluene and ethylbenzene from phenylalanine, and carbon disulphide and sulphur dioxide from cysteine and methionine (Fujimaki *et al.*, 1969; Arroyo and Lillard, 1970; Kato *et al.*, 1971; Watanabe and Sato, 1971; Patterson, 1975). The aromas derived from heating mixtures of amino acids do not correspond exactly to those produced on heating meat extracts containing these amino acids (Merritt and Robertson, 1967), and this may be because the *sequence* of amino acids is important.

The carbohydrates of meat are also important in producing flavour on heating. They lose the elements of water in two stages (at 180 °C and 220 °C), forming furfural from pentoses and hydroxymethylfurfural from hexoses. At about 300 °C there is caramelization with the formation of a large number of odoriferous compounds, including furans, alcohols and aromatic hydrocarbons (Fagerson, 1969).

Although amino acids and carbohydrates can each yield odoriferous compounds when heated in isolation, relatively little heating is required when these are mixed together (Wasserman and Spinelli, 1972).

Notwithstanding the evident capacity of water-soluble precursors to produce flavours resembling those of different meat species on heating, fats or fat-soluble precursors were also shown to be implicated in accounting for species differences and in contributing generally to meat flavour. There are, of course, considerable differences between species in intramuscular fat (cf. Table 4.6 and Dahl, 1958a). Thus, the volatile carbonyls from heated pork fat include octanal, undecanal, hepta-2:4-dienal and nona-2:4-dienal. Few of these are derived from heated lamb fat and none from heated beef fat (Hornstein *et al.*, 1960, 1963). Deca-2:4-dienal is present in appreciable quantities in the volatiles from beef and pork fat, but absent from lamb fat. 4-methyloctanoic acid appears to contribute to the odour of mutton and goat flesh (Wong *et al.*, 1975a); and Young *et al.* (1997) demonstrated that the odour and flavour of sheepmeat were specifically related to branched chain fatty acids (e.g. 4-methyloctanoic and 4-methylnonaoic). Their intensity is increased by pasture-derived 3-methylindole and alkyl phenols. Long-chain aldehydes, such as 2-undecanal, reflect a pasture diet: a grain diet is characterized by the presence of 2,3-octanediol. Grass contains significant quantities of linoleic acid and this partially explains the difference in eating quality between grass-fed and concentrate-fed cattle and sheep. (Wood *et al.*, 2003).

Traditionally it has been believed that the full flavour of meat cannot be developed without its associated fat; but the possible health hazards of the ingestion of fat have led producers to rear leaner animals. Fortin *et al.* (2005), in a study of pork, concluded that unless there is a minimum of 1.5 per cent intramuscular fat, the meat lacks flavour (and tenderness).

Pork and beef can be clearly distinguished, using canonical analysis, by quantitative differences in the pattern of lipid-derived volatiles in the head-space above the cooked meats (Mottram *et al.*, 1982).

The importance of phospholipids rather than triglycerides in explaining the contribution of lipids to cooked meat flavour has been shown by Mottram and Edwards (1983). Removal of the intramuscular triglycerides had relatively little effect on the pattern of volatiles but subsequent removal of the phospholipids caused a loss of aliphatic aldehydes. There was also a marked decrease in pyrazines, which suggests that, in cooked meat, lipids are normally involved in Maillard reactions and inhibit the production of pyrazines.

Using model systems, Campo *et al.* (2003) assessed the importance of oleic, linoleic and α-linoleic acids, with or without the presence of cysteine and ribose, in relation to the development of odour in cooked meat. 'Fishy' notes were experienced only with mixtures including linoleic acid – an effect exacerbated by the presence of ferrous iron.

The off-flavour referred to as 'warmed-over flavour' in cooked meat was ascribed to the oxidation of lipids by Tims and Watts (1958) and, more specifically, to that of phospholipids catalysed by both haem and non-haem iron (Love, 1983). The rancidity develops much faster than that in uncooked meat during refrigerated storage. Any process which damages muscle membranes, such as chopping or emulsification, exacerbates the condition, but it can be retarded by the use of antioxidants, such as nitrite, phosphate and naturally occurring herbs and spices (e.g. rosemary, which contains various factors which inhibit rancidity). The characteristics, and the means of control, of 'warmed-over flavour' were reviewed by Gray and Pearson (1987).

Preslaughter stress was found to reduce the sensory perception of 'warmed-over flavour' in pork, an observation attributed to the high ultimate pH developed (Byrne *et al.*, 2001). Rapid identification of the sensory and chemical components responsible can be made using low field nuclear magnetic resonance (Brøndum *et al.*, 2000).

Species-specific flavours may also be related to subtle differences in the amino acid and carbohydrate content of adipose tissue (Wasserman and Spinelli, 1972).

The gas chromatographic profile for the volatiles from lean whale meat has a component which is not present in other mammalian species (Hornstein *et al.*, 1963). This has proved to be trimethylamine, however, which is exogenous to the whale, being derived from bacterial degradation of the trimethylamine oxide of ingested krill (cf. Table 4.3 and Sharp and Marsh, 1953).

Apart from amino acids, carbohydrates and fats, thiamin appears to be an important precursor of meat aroma (MacLeod and Seyyedain-Ardebili, 1981). The thermal degradation of thiamin produces at least eight volatile compounds which have been identified in cooked meat aroma (Galt and MacLeod, 1984) – including H_2S, formic acid and heterocyclic furanoid compounds. The odour threshold of bis-2,3-furyl disulphide is only 2 parts per 10^{14} of water vapour (MacLeod, 1984).

Clearly, whether derived from water-soluble or fat-soluble precursors, the pattern of volatiles produced on heating meat must be important for flavour. Ammonia, acetaldehyde, acetone and diacetyl, volatile fatty acids (formic, acetic, propionic, butyric and isobutyric), dimethylsulphide and H_2S were early detected (Yueh and Strong, 1960), but as identification techniques have progressed, the full complexity of the pattern has become more apparent. The concentration of H_2S, and of other reactive sulphur compounds such as methanethiol, increases with increase in heating time (Persson and Von Sydow, 1974). Cooked pork contains 1.2 to 1.3 times as much H_2S as cooked beef, possibly because of its higher level of free sulphydryl groups (Martehenko and Kozenyasheva, 1974). These are a source of H_2S during the cooking of meat (Hofmann and Hamm, 1969). Aliphatic sulphur-containing components account for more than 50 per cent of the total volatiles from beef treated for one hour, but with extension of heating time to 4h, a considerable increase in heterocyclic sulphur compounds is observed (Galt, 1981).

The incorporation of sulphur into heterocyclic ring systems (e.g. thiazoles) is a general feature of increased time or temperature of heating (Schwimmer and Friedman, 1972; Kato *et al.*, 1973). A 3,5-dimethyl-1,2,4-trithiolane has been identified in the volatiles from boiled beef (Chang *et al.*, 1978), 2-alkylthiophane in those from roast beef (Min *et al.*, 1979) and 4,6-dimethyl-2,3,5,7-tetrathiaoctane in those from roast pork (Ho, 1980).

Two unusual S-containing compounds have been found only in the aroma volatiles of cooked mutton, viz. bis-mercaptomethylsulphide and 1,2,3,5,6-pentathiepane(lenthionine). They possibly arise from the interaction of formaldehyde with H_2S (MacLeod, 1984). It is likely that 4-methyl- and 4-ethyloctanoic acids, which are found at relatively high concentrations in lamb tissues, contribute significantly to the flavour of this species (Brennand and Lindsay, 1992).

Pyrazines (paradiazines) are a second group of heterocyclic compounds which are particularly important for meat aroma. They also increase with time of heating (Koehler and Odell, 1970). They arise on heating aliphatic aminohydroxy compounds and they figure prominently in more recent attempts to identify compounds significant for meat flavour. Both pyrrolopyrazine (Flamant *et al.*, 1977) and 4-acetyl-2-methylpyrimidine (Ohloff and Flamant, 1978) are associated with roast beef odour. The bicyclic pyrazines have roasted or grilled animal notes (Flamant *et al.*, 1976).

The more severe heating involved in roasting appears to be associated with increased production of both thiazoles and pyrazines. Nevertheless, in a detailed study of the effect of mode of cooking on beef volatiles, MacLeod and Coppock (1976) found that pyrazines tended to be associated with underdone, boiled or

Table 10.18 Effect of mode of cooking on beef volatiles
(after MacLeod and Coppock, 1976)

Well-cooked, boiled	Underdone, boiled or microwave

High MW hydrocarbons
  tetra-, penta-, hexa- and
  hepta-decanes
	Low MW hydrocarbons
	  heptane, octane, decane,
	  undecane, hept-l-ene,
	  undec-4-ene

Benzenoids
  Benzene, n-propylbenzene,
  toluene, o- and p-xylenes,
  ethylbenzaldehyde
	Pyrazines
	  dimethyl-, ethyl- and
	  dimethylethyl-pyrazines

Furans
  2-ethyl- and 5-n-pentyl-
  furans	

Misc.
  3-methylbutanol, pyridine,
  2-methylthiophen
	Misc.
	  acetone, methylbutanol

microwave cooking, whereas benzenoid compounds were associated with well-cooked or boiled beef (Table 10.18).

Pyridines are produced from the reaction of aldehydes with ammonia, from the pyrolysis of amino acids (such as β-alanine and cystine), and from Maillard reactions (Galt, 1981). Under roasting conditions proline and glucose react to produce pyridines (Tressl *et al.*, 1979). Pyridine formation tends to be related to species: the relatively high concentration of alkylpyridines in the volatiles from lamb may well be a significant factor in determining the acceptibility of its flavour (Buttery *et al.*, 1977).

Beef volatiles produced during the more intensive cooking procedures also contain furans and high MW hydrocarbons, whereas low MW hydrocarbons were more characteristic of underdone beef (Table 10.18). A not dissimilar pattern is found in the volatiles from cooked pork (Heidemann and Wismer-Pedersen, 1976). Fat is assumed to be the precursor of furan and methylfuran (Persson and von Sydow, 1974). Several other sources have been implicated: viz. the Amadori rearrangement of carbohydrates and amino acids in the Maillard reaction, the thermal decomposition of thiamin and the breakdown of ribonucleotides (Baines and Mlotkiewicz, 1984). The meat-like odour of furanoid compounds varies with the degree of substitution of the furan ring (Ho, 1980). 4-Hydroxy-5-methyl-3(2H)-furanone, which may be isolated from beef broth, can react with H_2S to produce a complex mixture having the odour of roast beef (Van den Ouweland *et al.*, 1978).

During cooking, various alcohols and esters appear to be generated. Thus *iso*-hexanol, *iso*-heptanol and *iso*-decanol, and a range of acetic acid esters, arise on cooking pork (Kevel and Kogma, 1976) (cf. Table 10.19).

Heating also causes changes in the pattern of nucleotides in meat (Wismer-Pedersen, 1966; Macy *et al.*, 1970; Gorbatov and Lyaskovskaya, 1980). It is of interest to recall that Japanese workers concluded that mononucleotides are substantially responsible for meat flavour. It appears to be essential that the purine moiety should be substituted with a hydroxyl group at position 6 and that the ribose

Table 10.19   Esters identified in raw and cooked pork
(Kevel and Kogma, 1976)

Ester	Raw meat	Cooked meat
	peak area (cm²)	
Methyl acetate	0.36	–
Propyl acetate	0.44	4.29
Ethyl butyrate*	0.30	1.08
Butyl acetate	–	3.48
Iso-amyl acetate*	119.60	3.58
Iso-amyl butyrate*	–	3.68
Heptyl acetate*	–	Traces
Octyl acetate	–	0.42
Ethyl pelargonate	–	17.80
Iso-amyl caprilate	–	3.60
Iso-amyl pelargonate	–	0.84

* Presumed identification.

be substituted with a phosphate group at position 5 (Kodama, 1913; Yoshida and Kageyama, 1956). As well as glutamic acid, inosine and hypoxanthine have been available as flavouring condiments in Japan. Nevertheless, although from time to time this group of substances has been supposed to contribute to flavour – either because of a decrease or an increase in their concentration – Patterson (1975) concluded that they played a minor role.

It will be clear that both the duration and temperature of cooking must influence the nature and intensity of odour and taste in meat. Since, except with pressure cooking, the interior of the pieces of meat which are being cooked cannot rise above 100 °C until all the water has been driven off (Crocker, 1945), it will have relatively little flavour in comparison with the exterior, where the high temperature and the relative absence of moisture produce various substances having appreciable odour and taste. It may be that the high temperature attained in the depth of the meat with pressure cooking explains why it is said to have a somewhat better flavour (Fenton *et al.*, 1956). Leg roasts of lamb cooked to an interior temperature of 65 °C have an odour and taste more characteristic of lamb than if cooked to an internal temperature of 75 °C (Weir, 1960).

In a reassessment of the effects of cooking temperature on the flavour of the leaner pork meat which consumers have demanded, Wood *et al.* (1995) confirmed, with both loin steaks (*l. dorsi*) and leg roasts (*gluteobiceps*), that an increase in the final internal temperature from 65 to 80 °C enhanced pork flavour and diminished abnormal flavour.

10.4.3   Variability in odour and taste

The flavour of meat is subject to variability due to both intrinsic and extrinsic factors. Of the former, differences due to species have already been referred to. There is some evidence for differences due to breed (Jacobson *et al.*, 1962; Warriss *et al.*, 1996). Elmore *et al.* (2000) found that levels of pyrazines and sulphur-containing compounds (volatiles derived from Maillard reactions) were much higher in the meat from Soay than in that of Suffolk lambs. Even within a breed it is sometimes possible to detect inherited aspects of meat flavour (Bryce-Jones *et al.*, 1963).

Increasing animal age is associated with increased intensity of flavour, as the bland flavour of veal and the characteristic flavour of beef testify. Strength of flavour in beef tends to increase up to 18 months of age and thereafter to reach a plateau (Tuma *et al.*, 1962). This fact must be presumed to be determined by age-related changes in the precursors (cf. § 10.4.2) but detailed information on the substances responsible is awaited. The distinctive flavours of sheep and goat meat are much less noticeable when derived from the young animals of these species (Schönfeldt *et al.*, 1993).

Since the meat of older animals tends to contain more fat – and that of a more saturated character – no doubt fat-soluble precursors are amongst those involved. As the percentage of intramuscular fat increases in porcine *l. dorsi* there is a concomitant increase in the percentage of mono-unsaturated fatty acids and a decrease in that of polyunsaturated fatty acids, and this is associated with improved flavour (Cameron and Enser, 1991). Nevertheless, although fat is clearly essential for flavour, there is an optimum and no evidence that excess fat causes progressive increments (Patterson, 1975).

Since there are systematic biochemical differences between muscles, it is not surprising that these should have different flavours when cooked, as the relative blandness of that from *psoas* muscle (tenderloin) and the strength of that from diaphragm (skirt) indicates. Again, bovine *l. dorsi* has a stronger flavour than *semitendinosus* (Howard and Lawrie, 1956; Doty and Pierce, 1961). Carmack *et al.* (1995) ranked 12 muscles from steers of a single maturity level and grade in order of flavour intensity after cooking and found that the *biceps femoris* had the highest score and the *supraspinatus* the lowest score. Both Mabrouk *et al.* (1969) and Dannert and Pearson (1967) demonstrated that there were differences between different beef muscles in flavour-precursor molecules; but, there has been relatively little investigation of this topic so far.

There is also one major biochemical variable which affects muscle flavour – the ultimate pH. In general, the higher the ultimate pH, the lower is the flavour intensity as determined by taste panel – possibly because the consequently swollen structure interferes with access to the palate of the substances involved (cf. Fig. 10.5; Bouton *et al.*, 1957). A similar effect has been noted in cured meats. Thus bacon of relatively high ultimate pH appears less salty to the palate than that of low pH, even when the salt content is the same (Ingram, 1949a). Apart from this effect of high pH, however, substantial differences have been detected in the steam volatiles from normal lamb and beef (ultimate pH 5.5–5.8) and that having an ultimate pH above 6 (Park and Murray, 1975). In the latter, as well as quantitative differences, specific increments in trimethylamine, ammonia and collidine have been shown (Ford and Park, 1980). In stored, vacuum-packed pork, a high ultimate pH (*ca.* 6.3–6.6) is associated with the production of relatively high concentrations of hydrogen sulphide, methanethiol, dimethylsulphide and various other sulphur-containing volatiles in comparison with corresponding products of normal ultimate pH (*ca.* 5.5–5.6) (Edwards and Dainty, 1987) and this reflects differences in the predominant microbial flora.

A second important biochemical variable arises from the changes which occur when meat is held for some time after the ultimate pH has been reached to 'age' or 'condition' it. During this period it becomes more tender, but the flavour also tends

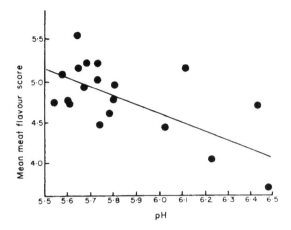

Fig. 10.5 Relationship between mean flavour (as determined by taste panel) and ultimate pH in beef muscle. (After Bouton *et al.* 1957.)

to increase or alter (Howe and Barbella, 1937; Harrison, 1948; Bouton *et al.*, 1958). Objective study has shown that there is, concomitantly, a marked increase in high MW hydrocarbons, benzenoid components and pyrazines (Coppock and MacLeod, 1977). Changes in the free fatty acids during ageing no doubt contribute to the flavour changes observed. Thus the level of oleic acid in the intramuscular fat of *l. dorsi* has been observed to increase during 21 days storage of beef at 2 °C (Hood and Allen, 1971). Moreover, progressive nucleotide breakdown (§ 4.2.3), whereby ADP and AMP are ultimately split to ribose, hypoxanthine, phosphate and ammonia, may be a contributing factor.

The intrinsic factors considered above tend to cause expectable, and generally desirable, variability in meat flavour – although what is desirable is partly determined by custom. Thus, the caprylic acid odour of goat flesh is disliked in Western Europe.

Of extrinsic factors causing variability in flavour, diet can be significant, although this often refers to relatively undesirable features derived from specific components rather than to the level or intensity of feeding. According to Rhodes (1969) there was little evidence for flavour differences between the cooked meat from grass-fed animals or those fed intensively on concentrates (other than those that can be expected from animal age), unless the pasture includes plants with specific flavour-inducing compounds (Hedrick *et al.*, 1980). An international investigation by Sanudo *et al.* (2000) showed, however, that Spanish consumers preferred the flavour of lamb containing a relatively high content of n-6 polyunsaturated fatty acids derived from concentrates, whereas British consumers preferred the stronger flavour of lamb associated with n-3 polyunsaturated fatty acids derived from grass; but consumer preferences were clearly dominated by previous experience and custom.

Most processing operations, albeit designed to retain organoleptic quality as well as to prevent microbial spoilage, tend to cause flavour deterioration (and will thus be considered in § 10.4.4). On the other hand, attempts at deliberate enhancement of meat odour and taste have mainly been confined to cured and comminuted meats and sausages, which frequently contain added spices, condiments (including sodium glutamate), sugars, etc. (Gerrard, 1935; Wilson, 1960).

The flavour of cured meats is acknowledged as being different from that of the uncured commodity – and sought for this reason. Since pork is the meat most frequently preserved by curing, it is not surprising that most of the research on the effects of curing on flavour has been on bacon and ham.

A comparison of the volatiles from cooked ham and cooked pork shows that curing strongly inhibits hexanal formation in comparison with other carbonyls (Cross and Ziegler, 1965). Nitrite is one of the components in the curing salts which has been implicated; it also appears to inhibit the production of various other higher MW carbonyls (Bailey and Swan, 1973). On the other hand, a higher percentage of carbonyls of chain length C_2–C_5 is found in the headspace of cooked cured meat (Heidemann and Wismer-Pedersen, 1976). There thus appears to be a direct relationship between the aroma of cured pork and a low concentration of the secondary products of lipid oxidation (Cross and Ziegler, 1965), but volatile sulphur-containing compounds are also important. There is considerable evidence that a high ratio (> 5:1; Heidemann and Wismer-Pedersen, 1976) of H_2S/mercaptans develops on cooking cured pork and is associated with its typical flavour. Increasingly sophisticated methodology has elucidated the role of nitrite in altering the pattern of flavour volatiles in bacon from that characteristic of uncured pork. Thus, alkyl nitrates and nitriles have been detected in the former (Mottram et al., 1984), and these may result from modification by nitrite of the pattern of thermal oxidation of lipids.

The cooking of raw bacon and ham causes an increase in ammonia and a decrease in methyl- and dimethyl-amines (Gorbatov and Lyaskovaskaya, 1980).

Curing also causes an increase in free amino acids which is further enhanced on cooking (Balabukh and Lyaskovskaya, 1978), and their conversion to various volatiles no doubt contributes to flavour development. Indeed, the addition of amino acids (such as methionine) to the curing brine has been suggested as a means of improving the flavour of cured pork products (Heidemann and Wismer-Pedersen, 1977).

Attempts at extraneous enhancement of meat odour and taste have mainly been confined to cured meats and sausages, which frequently have added spices, condiments (including sodium glutamate), sugars, etc. (Gerrard, 1935; Wilson, 1960). Much of the flavour of bacon and of continental-type sausages depends on the metabolic activities of micro-organisms whose growth is fostered by traditional manufacturing procedures (Tanner, 1944). To some degree this flavour is derived from the hydrolysis of fats and the breakdown of free fatty acids. The order in which the latter are attacked by the micro-organisms depends on the microbial species which are present (Wahlroos and Niinivaara, 1969). Systematic attempts have been made to introduce micro-organisms deliberately in order to produce particular desired flavours. For instance, Niven et al. (1958) proposed the use of Pediococcus cerevisiae as a sausage starter culture, and McLean and Sulzbacher (1959) demonstrated that the addition of a species of Pseudomonas to a meat curing brine significantly altered its flavour. In a comprehensive study, using various combinations of lactic acid bacteria strains and Staphylococcus spp., Berdagué et al. (1993) demonstrated the predominant importance of starter cultures in determining the production of specific flavour volatiles in dry fermented sausages.

It seems likely that very considerable advances will be made in the enhancement and control of meat odour and taste. This might be done by the addition to the comminuted product of controllable micro-organisms capable of fostering flavour, or even by the preslaughter administration of desired, flavour-producing chemicals or micro-organisms (e.g. *Thamnidium elegans*; Williams, 1964b). Although preslaughter administration may not yet be realistic, the addition of exogenous enzymes, in the manufacture of dry fermented sausages, is already undertaken (Ansorena *et al.*, 2002a).

For many years the water-soluble flavour constituents of meat have been processed in the preparation of concentrated meat extracts. The non-volatile components in these are very similar to those present in the fresh meat (Wood and Bender, 1957; Bender *et al.*, 1958). In each, there are appreciable quantities of salts, lactic acid, carnosine, creatine and hypoxanthine. The concentrates, which are subjected to prolonged heating, contain more creatine and are considerably darker – amino acids and sugars having reacted to form Maillard-type compounds. Additional volatile components in the concentrates include H_2S and isovaleraldehyde (Bender and Ballance, 1961).

10.4.4   Undesirable odour and taste

In a study at the Food Refrigeration Process Engineering Research Centre of Bristol University, beef which had been stored frozen for over 10 years was found to have a flavour which was as acceptable as that of fresh meat. Nevertheless there is a gradual loss in flavour during storage and this may occur even in the frozen condition (Howard and Lawrie, 1956) – possibly due to the slow loss of highly volatile substances. In this connection it is interesting to note that a faint odour of diacetyl is not infrequently observed in frozen meat stores at $-10\,°C$. Such losses are unavoidable, but undesirable odour and taste may arise during the storage of meat because of microbial growth, chemical deterioration on the surface or tainting by extraneous agents.

Odours produced by micro-organisms growing on meat surfaces are not so objectionable as those due to the metabolic products of anaerobes (§ 6.2): the former tend to be sour rather than putrid. The lipases of such micro-organisms will attack fat, splitting off fatty acids with more or less unpleasant consequences according to their nature (§ 7.1.1.3). The exact nature of the off-odours will, of course, depend on the types of micro-organisms growing, and these in turn will be determined by such factors as the temperature of storage and the nature of the product (fresh, cured, comminuted), as described in Chapter 6. Relatively high temperatures and the absence of oxygen will produce putrid off-odours through the breakdown of proteins, as in prepacked bacon stored at $20\,°C$ (Cavett, 1962), or in bone taint (§ 6.1.2.2) – in those deep-seated portions of the carcass which have not been cooled sufficiently quickly after death and where there is a reservoir of suitable micro-organisms in the lymph nodes (Nottingham, 1960).

Jackson *et al.* (1992) reported that, in an atmosphere of 20 per cent carbon dioxide/80 per cent oxygen, packaged beef loins developed stale off-odours when stored at $3\,°C$ for 28 days: 1-hexene, 1-heptene, ethyl acetate and benzene contributed to this effect. The predominant micro-organism was *Ps. putida*. Comparable beef stored under vacuum, under 100 per cent carbon dioxide or in 40 per cent carbon dioxide/60 per cent nitrogen developed much less off-odour, the predominant micro-organisms being *Lactobacillus plantarum* in the first two packs and *Leuconostoc mesenteroides* in the third. The volatiles associated with these two micro-organisms appeared to be a mixture of acetone, toluene, acetic acid, ethyl acetate and chloroform. A high ultimate pH in the muscles causes the predominant microflora of vacuum-packed pork to be Gram-negative organisms which produce a series of volatile, sulphur-containing compounds with concomitant off-odour development. Such does not arise from the lactic acid bacteria which predominate in this product when it has a normal ultimate pH (Edwards and Dainty, 1987). A species of *Proteus inconstans* has been isolated which produces 'cabbage odour' – due to methane diol – in sliced, vacuum-packed bacon (Gardner and Patterson, 1975). Microbiologically derived taints in meat occur in the order: buttery, cheesy, sweet, fruity and putrid.

Free fatty acids, produced by microbial action or otherwise, will accelerate the development of oxidative rancidity. The latter will occur even at $-10\,°C$ during long storage. The conditions predisposing towards oxidative rancidity in intramuscular fat have been thoroughly investigated (Lea, 1939; Watts, 1954). Relatively little is known, however, concerning the compounds responsible for off-flavours which are produced by comparatively minor reactions, although they are principally carbonyls. Since there are over 200 different species of carbonyls present in such circumstances (Evans, 1961), it is not surprising that positive identification of the most trouble-some members has proved difficult. Species differences in the development of off-odours and tastes arise from the different spectra of fatty acids produced by lipolysis, and of carbonyls produced during oxidative rancidity The phospholipids of meat fats are the most unstable constituents and they may well play a major role in accelerating flavour deterioration (Younathan and Watts, 1960). The separated cephalin fractions of both beef and pork intramuscular fats produce fishy odours, but, during the oxidation of the corresponding unfractionated fats, their effect is less noticeable with pork (Hornstein et al., 1961). The deliberate induction of a high ulti-mate pH has been shown to greatly retard the oxidation of fat in pork even at rel-atively elevated frozen storage temperature (cf. § 7.1.1.3). On the other hand, when formaldehyde-treated feed containing a high percentage of unsaturated fatty acids is fed to pigs, their fat is made even more unsaturated, and this is associated with an oily aroma (Ford et al., 1975). Such feeding to ruminants also leads to an enhanced tendency to undergo oxidative rancidity (although this can be controlled by permitted antioxidants). Moreover, the meat has a sweet aroma and flavour which some consumers find objectionable (Ford et al., 1974). The sweet aroma and flavour has been attributed to increased amounts of *cis*-6-γ-dodecenolactone (Park and Murray, 1975) and the oily aroma to increased levels of *trans, trans*-2,4-decadienal.

Elmore et al. (2000) varied the content of n-3 polyunsaturated fatty acids in the muscles of lambs by feeding supplements of linseed oil (which increased the level of α-linoleic acid) and of fish oil (which increased the levels of eicosapentenoic and docosahexaenoic acids). These also increased the amount of aromatic volatiles derived from auto-oxidation of the polyunsaturated fatty acids on cooking.

Despite the hydrogenating action of rumen micro-organisms, the degree of unsat-uration of ruminant fat can be enhanced by feeding linseed and fish oils (without exacerbating fat oxidation), if ribose and cysteine are also fed. Maillard reactions between them and the fatty acids strengthen meat flavour (Campo et al., 2003). Such feeding studies have not, however, increased the ratio of polyunsaturated fatty acids to saturated ones. Feeding a mixture of two parts soya oil to one of linseed oil raises the polyunsaturated : saturated fatty acid ratio to 3 without significantly raising the ratio of n-6 : n-3 acids (Enser et al., 2001).

Even without protection against the hydrogenating action of rumen micro-organisms, the pattern of fat laid down can be affected by the nature of the diet. Thus significant differences in the percentages of hexanoic acid, and of its branched chain isomers, and in those of the unsaturated octadecenoic acids (oleic, linoleic and linolenic), have been found in the subcutaneous fat of lambs fed on pastures of white clover or perennial ryegrass (Purchas *et al.*, 1986).

Insofar as so-called 'organic' food is desirable, forage-fed beef is said to have advantages over grain-fed; but its high content of polyunsaturated fatty acids, and 'grassy' flavour, make it less palatable, especially after frozen storage, since there is more lipid deterioration. If propyl gallate and a beef flavouring agent are incorporated when restructuring steaks from forage-fed beef, however, its palatability is enhanced (Deveaux *et al.*, 2003).

Cattle fattened immediately before slaughter, on crops which have been treated with dieldrin, may acquire a taint from this chemical, whilst animals grazing on pastures containing certain weeds (including peppercress and ragweed) are unable to excrete indole and skatole derived from the metabolism of tryptophan. This causes skatole taint (Empey and Montgomery, 1939) in the flesh. Skatole taint can also be a problem with the meat of boars. It seems that the large quantity of oestrogens produced in boar testes decreases voluntary feed intake, increases gut transit time and thus fosters the activity of those micro-organisms which produce skatole from tryptophan (Claus, *et al.*, 1994). A number of off-flavours are detected in the meat of sheep when they graze certain pastures for some weeks before slaughter (Table 10.20). The effects are more noticeable at particular times of the year, at certain stages of growth of the plants and within specific soil conditions. A period of about two weeks on neutral feed will generally overcome any such problems in lambs (Ford and Park, 1980).

When the flesh of certain pigs is heated, an unpleasant odour arises which is commonly referred to as boar odour – although it has been reported in the flesh of both sexes (Self, 1957). The agent responsible is fat soluble, but unsaponifiable (Craig *et al.*, 1962). Deatherage (1965) found that boar odour could be eliminated by implanting the animals with hexoestrol sometime before death. Using a combination of gas liquid chromatography and mass spectroscopy Patterson (1968a) has identified the substance responsible as 5α-androst–16-ene–3-one. It is present in the flesh of most boars over 200 lb (100 kg) live weight, but not in that of gilts or hogs. He suggested it is related to the corresponding alcohol, which is present in boar submaxillary gland (Patterson, 1968b). Physiological aspects of the formation of androstenone in the boar were comprehensively reviewed by Claus *et al.* (1994). It is of interest to note that 44 per cent of men are unable to detect 'boar odour' but

Table 10.20 Effect of fodder plants on flavour of meat from sheep (Anon., 1973b)

Plant	Meat flavour
Lucerne, white clover, sweet glycine	Sharp odour; objectionable when strong
Perennial ryegrass, panic grass, kikuya grass	Strong meat odour/flavour (acceptable to some consumers)
Green oats	Strong meat flavour, pungent odour (acceptable to lamb consumers)
Rape	Sickly odour and flavour, undesirable to most

only about 8 per cent of women cannot do so (Griffiths and Patterson, 1970). It has been shown that there is no effect of taint on eating quality until the level of androstenone rises above 1 µg/g (Patterson and Stenson, 1971). This signifies that about 50 per cent of boars would yield satisfactory meat.

An international assessment of the significance of skatole and androstenone in causing boar taint was undertaken by a group of seven collaborating, European countries. The study was based on the analysis of data from 4000 entire male pigs and 400 gilts (Bonneau et al., 2000a). Overall, most consumers were dissatisfied with the odour and flavour of the flesh from entire males. High levels of skatole were the predominant cause of boar odour, whereas skatole and androstenone contributed equally to unpleasant flavour. There were national differences. Thus, whereas British consumers were generally satisfied with the odour and flavour of meat from both entire males and gilts, Danish and Dutch consumers strongly objected to the odour of that from males. French, German, Spanish and Swedish consumers found both the odour and flavour of the meat from entire males unacceptable (Bonneau et al., 2000b). It was concluded that, in the short term, a reduction in the levels of skatole would achieve a limited improvement in consumer satisfaction, but that, in the longer term, a reduction in the levels of both skatole and androstenone would be necessary to overcome the problem of boar taint (Bonneau et al., 2000b).

The investigation of Amor-Frempong et al. (1997) indicated that instrumentation for on-line classification (and sorting) of pigs should be based on response criteria to androstenone, skatole and indole, rather on the concentration of these compounds. Subsequently Amor-Frempong et al. (1998) compared the responses to boar odour by a sensory panel with those by an electronic 'nose'. Their findings indicated that the latter was comparably effective.

Meishan pigs from China have distinct advantages, maturing at an early age, producing large litters and providing meat that is more tender and highly flavoured than that of pure Western breeds (Touraille et al., 1989); but they have higher levels of androstenone and skatole in the backfat, which are conducive to boar taint (Prunier et al., 1987). Castration causes a fall in the androstenone level and an increase in the titre of the enzyme cytochrome P4502EI, which enhances the breakdown of skatole in the liver (Whittington et al., 2004). Since neither animal age nor weight is correlated with taint in the average boar of potential commercial value, a simple test to make an early detection of taint involves the heating of a sample of fat to about 375 °C using an electrically operated soldering iron.

Under certain conditions of heating, H_2S liberated from the meat proteins can react with mesityl oxide (derivable from the acetone in tin lacquers) to give various compounds, including 4-methyl-4-mercaptopenta-2-one, which produces a most offensive 'catty' odour (Aylward et al., 1967). From time to time phenolic substances, used in dipping sheep, have been detected in the meat at time of consumption.

Meat, especially meat fat, which has been in refrigerated storage for a considerable time, as in shipment from the Southern Dominions to the United Kingdom, may become unmarketable due to taints absorbed from extraneous sources such as diesel oil and fruits. Activated charcoal, placed in the cold store, will frequently reabsorb the taint from the meat (Macara, 1947).

The off-odours developed in meat stored above or below the freezing point are not a direct consequence of refrigeration, but other commercial processes may cause flavour changes, e.g. dehydration, freeze dehydration and irradiation. Dehydrated and freeze-dried meat is not only particularly susceptible to oxidative rancidity in

Table 10.21 Concentrations of volatiles from beef canned
to $F_0 = 10$ at different temperatures (ppb)

	115 °C	121 °C	131 °C
H_2S	6900	6400	4400
Methylmercaptan	1400	1200	780
Dimethylsulphide	810	770	530
2-methylpropanal	83	54	9
2-methylbutanal	79	43	30
2-ethylfuran	180	120	89

the presence of oxygen but also, especially at high temperatures, to the development of mealy and paint-like odours. In the absence of oxygen, bitter tastes develop
because of Maillard-type reactions. Prophylactic procedures have been discussed in
§§ 8.1.3 and 8.2.3. Irradiation causes both immediate and storage changes in the
odour and taste of meat (§§ 9.1.2.1 and 9.1.2.2). The production of H_2S, mercaptans,
carbonyls and aldehydes, especially in beef (Huber *et al.*, 1953), is largely responsible. In this context, within-package odour scavengers (Tausig and Drake, 1959), the
addition of protective compounds such as ascorbic acid (Huber *et al.*, 1953) and irradiation at temperatures far below the freezing point (Hannan, 1955) have been used
with some success. On storage, meat which has been sterilized by irradiation develops stale and bitter flavours. Some of these are due to surviving activity of the meat's
proteolytic enzymes which produce free tyrosine from proteins. Prolonged cooking,
by causing marked breakdown of the meat proteins and the production of H_2S, is
undesirable – though it may be necessary for tough meat (Weir, 1960). The odour
and taste of beef roasts, cooked to an internal temperature of 82 °C over 2 h in an
oven at 177 °C, generally received a lower taste panel rating than grills from the
same animal cooked to the same internal temperature over 2 h in an oven at 288 °C
(Howard, 1956; Howard and Lawrie, 1956; Bouton *et al.*, 1957, 1958). In canning,
meat is subjected to high temperature for considerable periods determined by the
product and intention, and there are concomitant changes in the concentrations of
aroma components. Perrson and Von Sydow (1974) compared the effects of canning
beef when processed to $F_0 = 10$ by three temperatures. The shorter heating time
required at higher temperature decreased the concentrations of those compounds
which could be related to the off-flavour in canned beef, i.e. aldehydes and sulphur-
containing compounds (Table 10.21). The addition of lysine or arginine to the beef
before canning decreased the concentration of aldehydes and that of fumarate or
malonate decreased the concentration of S-compounds. On storage of the canned
beef over 12 months at 20 °C the intensity of typical meat aroma decreased, but
reheating at 121 °C restored some of the original flavour attributes.

Chapter 11

Meat and human nutrition

11.1 Essential nutrients

Regarded nutritionally, meat is a very good source of essential amino acids, and, to a lesser extent, of certain minerals. Although vitamins and essential fatty acids are also present, meat is not usually relied upon for these components in a well-balanced diet. On the other hand, an organ meat, such as liver, is a valuable source of vitamins A, B_1 and nicotinic acid. Even in respect of its accepted nutrient role, however, little is yet known about possible differences in the value of meat from different species, breeds and muscles. Although the role of muscular tissue is the same wherever it occurs, and consists predominantly of contractile proteins, the amino acid composition of which is said not to vary grossly between species (Crawford, 1968), the accessories of the contractile process are certainly not identical even between the muscles of a given species (cf. § 4.3). There are differences in the contents of ancillary proteins, of free amino acids, of fatty acids and of various other substances, and in their character. These can be presumed not to be without nutritional significance, albeit subtle.

It is well known of course that a muscle containing much connective tissue will provide meat which is relatively resistant to digestion and absorption, and that this will be worsened by faulty cooking; but how important this may be in relation to the absorption of the nutrients of the meat has been little studied. Moreover, since connective tissue proteins have a lower content of essential amino acids than those of contractile tissue, meat having a high percentage of collagen or elastin will also have *relatively* lower intrinsic nutritive value. There is less of the essential sulphur-containing amino acids in connective tissue (Bender and Zia, 1976), and an inverse relationship in meat samples between hydroxyproline (as an index of connective tissue) and tryptophan (Dahl, 1965). Nevertheless, Kofranyi and Jekat (1969) demonstrated that, for human consumers, connective tissue may not be nutritionally disadvantageous until the ratio of connective tissue nitrogen to muscular tissue nitrogen is greater than 1. Indeed a mixture of 84 per cent muscle nitrogen and 16 per cent gelatin nitrogen was shown to have a biological value of 99, whereas that of 100 per cent beef muscle was 92 (in relation to whole egg protein as 100). Incidentally, although the replacement of meat by beef tendon in emulsion-style sausages lowers the overall acceptability of the product, preheating the tendon at as low as 60 °C permits a higher level of incorporation before adverse effects are noted (Sadler and Young, 1993).

The increasing development of the mechanical deboning of meat, and the inclusion of mechanically recovered meat in food products, has made an assessment of its nutritional value important. When whole animal carcasses are mechanically recovered, the overall composition is close to that of hand-deboned meat (Field, 1974), but the content of calcium, ash and iron is higher in the former (Newman, 1980–81). It is likely to contain bone collagen. Clearly, the nutritional value of mechanically recovered meat will vary with the bone source and the levels of connective tissue and calcium present.

Insofar as connective tissue may be added to meat products, however, in amounts greater than those naturally associated with the muscular tissue there is clearly a need, in the interests of consumers, for methodology to identify added collagen. In this regard, artificially prepared collagen (rind) can be differentiated histochemically from natural collagen by its birefringent colour under polarized light (Flint and Firth, 1983).

11.1.1   Amino acids

The amino acid composition of the proteins of the principal types of meat is shown in Table 11.1. In respect of the essential amino acids, beef would appear to have a somewhat higher content of leucine, lysine and valine than pork or lamb, and a lower content of threonine. Despite the minor nature of these species differences, however, it should be pointed out that the meat represented in Table 11.1 is of

Table 11.1   Amino acid composition in fresh meats (Schweigert and Payne, 1956) (as % crude protein)

Amino acid	Category	Beef	Pork	Lamb
Isoleucine	Essential	5.1	4.9	4.8
Leucine	Essential	8.4	7.5	7.4
Lysine	Essential	8.4	7.8	7.6
Methionine	Essential	2.3	2.5	2.3
Cystine	Essential	1.4	1.3	1.3
Phenylalanine	Essential	4.0	4.1	3.9
Threonine	Essential	4.0	5.1	4.9
Tryptophan	Essential	1.1	1.4	1.3
Valine	Essential	5.7	5.0	5.0
Arginine	Essential for infants	6.6	6.4	6.9
Histidine	Essential for infants	2.9	3.2	2.7
Alanine	Non-essential	6.4	6.3	6.3
Aspartic acid	Non-essential	8.8	8.9	8.5
Glutamic acid	Non-essential	14.4	14.5	14.4
Glycine	Non-essential	7.1	6.1	6.7
Proline	Non-essential	5.4	4.6	4.8
Serine	Non-essential	3.8	4.0	3.9
Tyrosine	Non-essential	3.2	3.0	3.2

Table 11.2   Tryptophan and lysine in pork muscles (Hibbert and Lawrie, unpublished data)

Muscle	Tryptophan (mg/g)	Lysine (mg/g)
L. dorsi (4–6 lumbar)	0.015	0.089
L. dorsi (13–15 thoracic)	0.012	0.078
L. dorsi (8–12 thoracic)	0.018	0.061
Psoas	0.017	0.072
Semimembranosus	0.015	0.083
Rectus femoris	0.013	0.059
Supraspinatus	0.021	0.071

random origin. It is certainly feasible that more significant differences may exist between specific muscle locations, or that breed, and animal age, have important effects. It has been reported, for example, that the contents of arginine, valine, methionine, isoleucine and phenylalanine increase (relative to the concentrations of other amino acids) with increasing animal age (Gruhn, 1965).

Further, there is evidence that the content of certain essential amino acids may differ at different parts of the carcass. Some data on tryptophan and lysine in certain pork muscles are given in Table 11.2.

The amino acid content may be affected by processing (e.g. heat, ionizing radiation: §§ 7.2.2 and 9.1.1 above); but, unless processing conditions are both severe and prolonged, such destruction is minimal. Rather more important is the possibility that certain amino acids may become unavailable (Bender, 1966). Thus Dvorak and Vognarova (1965) found that after heating beef for 3 h at a series of temperatures, 90 per cent of the available lysine was retained at 70°C and only 50 per cent at 160°C. There is a linear relationship between loss of available lysine in canned beef and the severity of the process (Ziemba and Mälkki, 1969). A 20 per cent fall in available tryptophan and methionine was observed in canned pork after only 40 min at temperatures above 70°C (Hibbert, 1973). Bender and Husaini (1976) found no loss in available methionine when beef was autoclaved for 1 h at 115°C, but when it was processed in the presence of other food constituents, such as wheat flour and glucose, there was a loss in net protein utilization which could be related to a fall in available methionine. Amino acids can also become unavailable during the prolonged storage made possible by canning. Bender (1966) found that veal, canned in 1823, had a biological value of only 27 when examined in 1959, although analysis after acid hydrolysis indicated that there had been no destruction of amino acids, i.e. prolonged storage led to grossly reduced availability. On the other hand, over practical storage periods, he detected no diminution in biological value (after the initial loss caused by the process itself). Smoking and salting may also slightly diminish amino acid availability. Storage of freeze-dried meat for 1 year at 20°C, in air and with about 5 per cent available water, has been found to cause a loss of 50 per cent in available lysine, although this is unusual (Bender, 1966).

11.1.2 Minerals
Mineral components in several meats are shown in Table 11.3 (after McCance and Widdowson, 1960). Of these, potassium is quantitatively the most important,

followed by phosphorus, except in cured meat where sodium from the added salt predominates.

Another general feature is the increase which occurs on cooking (which is mainly due to moisture loss). In respect of species differences the high content of iron in beef no doubt reflects the greater concentration of myoglobin in this species than in mutton or pork. In this respect the lean of sperm whale and seal will contain about 30 mg Fe/100 g wet weight because of the high myoglobin concentration in these species (cf. § 4.3.1). The proportion of iron present as soluble haem was found to decrease from 65 per cent (for uncooked meat) to 22 per cent when cooked at 60 °C – and decreased further with increasing cooking temperature. The proportion of insoluble iron increased rapidly and significantly (Purchas et al., 2004b). Nevertheless these changes were not regarded as sufficient to impair nutritive value.

Corresponding data for the mineral content of various offal tissues is shown in Table 11.4 (after Paul and Southgate, 1978; Kiernat et al., 1964).

It is evident that the iron, copper and zinc contents of kidney and liver are much higher than those in muscular tissue. The titre of iron in the liver of the pig is noticeably greater than those in ox and sheep but there are no other marked interspecies differences.

Table 11.3   Mineral content of meat and meat products (after McCance and Widdowson, 1960)

Meat	Mineral (mg/100 g)							
	Na	**K**	**Ca**	**Mg**	**Fe**	**P**	**Cu**	**Zn**
Beef, steak (raw)	69	334	5.4	24.5	2.3	276	0.1	4.3
Beef, steak (grilled)	67	368	9.2	25.2	3.9	303	0.2	5.9
Mutton, chop (raw)	75	246	12.6	18.7	1.0	173	0.1	2.1
Mutton, chop (grilled)	102	305	17.8	22.8	2.4	206	0.2	4.1
Pork (raw)	45	400	4.3	26.1	1.4	223	0.1	2.4
Pork, chop (grilled)	59	258	8.3	14.9	2.4	178	0.2	3.5
Bacon (raw)	975	268	13.5	12.3	0.9	94	0.1	2.5
Bacon, back (fried)	2790	517	11.5	25.7	2.8	229	0.1	3.6

Table 11.4   Mineral content of offal tissues (after Kiernat et al., 1964; Paul and Southgate, 1978)

Source	Mineral (mg/100 g)							
	Na	**K**	**Ca**	**Mg**	**Fe**	**P**	**Cu**	**Zn**
Brain	140	270	12	15	1.6	340	0.3	1.2
Kidney, Sheep	220	270	10	17	7.4	240	0.4	2.4
Ox	180	230	10	15	5.7	230	0.4	1.9
Pig	190	290	8	19	5.0	270	0.8	2.6
Liver, Sheep	76	290	7	19	9.4	370	8.7	3.9
Ox	81	320	6	19	7.0	360	2.5	4.0
Pig	87	320	6	21	21.0	370	2.7	6.9

The importance of the zinc content of meat has been recently emphasized. It has been reported that infants on an entirely vegetarian diet may suffer from retarded cognition activity because of zinc deficiency.

In mammalian tissue selenium is found in selenoproteins, including the enzyme glutathione peroxidase, which is involved in protecting cells against oxidation by catalysing the reduction of hydrogen and lipid peroxidases (Burk, 1997). In a comparison of the glutathione peroxidase activity of different organs, Daun and Akesson (2004) found that the concentration of the enzyme was highest in porcine liver and kidney, but much lower in muscle. In cattle values were relatively low in organ meat and high in muscle.

11.1.3   Vitamins

The content of vitamins in various meats is shown in Table 11.5.

It is clear that the content of vitamin B_1 in pork (and even in bacon) is considerably higher than that in other meats, and that there is a relatively high concentration of folic acid in beef.

Typical values for the vitamin content of various offal tissues are given in Table 11.6.

It is evident that organ meats, in general, possess markedly higher contents of vitamins than muscular tissue. This is especially so in respect of the concentrations of vitamins A and B_{12}. Interspecies differences are relatively small. In this context it is of interest to note that the concentration of thiamin in pig offal is no greater than that in the offal of ox or sheep, despite its ten-fold higher concentration in muscular tissue of the species. Meat can be regarded as an important dietary source of vitamins B_1 and B_2 and in the United Kingdom meats provide about 40 per cent of the average nicotinic acid intake. The lability in processing of vitamin B_1, in particular, has been studied extensively. Some data, comparing conventional heating procedures with microwave heating, are given in Table 11.7 (Hallmark and van Duyne, 1961).

Despite shorter cooking times, losses were rather greater by the microwave procedure, but overall losses of vitamin B_1 were similar by both methods. The relative

Table 11.5   Vitamin content of various raw meats (after McCance and Widdowson, 1960)

Vitamin units/ 100 g raw flesh	Beef	Veal	Pork	Bacon	Mutton
A (I.U.)	trace	trace	trace	trace	trace
B_1(thiamin) (mg)	0.07	0.10	1.0	0.40	0.15
B_2(riboflavin) (mg)	0.20	0.25	0.20	0.15	0.25
Nicotinic acid (mg)	5	7	5	1.5	5
Pantothenic acid (mg)	0.4	0.6	0.6	0.3	0.5
Biotin (µg)	3	5	4	7	3
Folic acid (µg)	10	5	3	0	3
B_6(mg)	0.3	0.3	0.5	0.3	0.4
B_{12} (µg)	2	0	2	0	2
C (ascorbic acid) (mg)	0	0	0	0	0
D (I.U.)	trace	trace	trace	trace	trace

Table 11.6   Vitamin content of various offal tissues (after Kiernat *et al.*, 1964; Paul and Southgate, 1978)

Source		Vitamin (units/100 g raw tissue)									
		A (I.U.)	B₁ (mg)	B₂ (mg)	Nicotinic acid (mg)	Biotin (µg)	Folic acid (µg)	B₆ (µg)	B₁₂ (µg)	C (mg)	D (µg)
Brain		Tr.	0.07	0.02	3.0	2	6	0.10	9	23	Tr.
Kidney,	Sheep	100	0.49	1.8	8.3	37	31	0.30	55	7	–
	Ox	150	0.37	2.1	6.0	24	77	0.32	31	10	–
	Pig	110	0.32	1.9	7.5	32	42	0.25	14	14	–
Liver,	Sheep	20,000	0.27	3.3	14.2	41	220	0.42	84	10	0.50
	Ox	17,000	0.23	3.1	13.4	33	330	0.83	110	23	1.13
	Pig	10,000	0.31	3.0	14.8	39	110	0.68	25	13	1.13
Lung,	Sheep	–	0.11	0.5	4.7	–	–	–	5	31	–
	Ox	–	0.11	0.4	4.0	6	–	–	3	39	–
	Pig	–	0.09	0.3	3.4	–	–	–	–	13	–

Table 11.7   Comparison of cooking losses and vitamin B₁ retention in conventional and microwave cooking

Sample	Cooking method	Internal temp. (°C)	Cooking losses water and fat (% initial wt)	Vitamin B₁ retention in meat and dripping (% initial)
Beef	Conventional	62.5	18–20	81–86
	Microwave	71	29–39	70–80
Pork	Conventional	85	34	80
	Microwave	85	37	91
Beef loaves	Conventional	85	24	76
	Microwave	85	27	80
Ham loaves	Conventional	85	18	91
	Microwave	85	28	87

retention of vitamins B₁ and B₂ in several kinds of meat when cooked conventionally is shown in Table 11.8 (after Noble, 1965).

Vitamin B₁ is mainly lost from meat by leaching. Losses average about 15–40 per cent on boiling, 40–50 per cent on frying, 30–60 per cent on roasting and 50–70 per cent on canning (Harris and von Loesecke, 1960). Vitamins B₆, B₁₂ and pantothenic acid have a similar order of lability to vitamin B₁. As distinct from B vitamins, 90–100 per cent of vitamin A is retained after heating to internal temperatures as high as 80 °C.

Since the biological activity of 25-hydroxy vitamin D is several-fold greater than that of vitamin D itself, its specific concentration in meat and its associated tissues (cf. Tables 11.5 and 11.6). for which accurate analytical methods are available, is probably more significant than that of the latter.

11.1.4   Fatty acids

The unsaturated fatty acids, linoleic (C 18:2), linolenic (C 18:3) and arachidonic (C 20:4) appear to be essential. They are necessary constituents of cell walls,

Table 11.8   Retention of vitamins B_1 and B_2 on cooking (cuts of beef and veal cooked at 149°C; pork cooked at 175°C)

Type of meat	Time of cooking (min/lb)	% retention B_1	% retention B_2
Beef			
Short rib	30	25	58
Chuck	35	23	74
Round (roast)	27	40	73
Round (steak)	18	40	65
Veal			
Chops	–	38	73
Round (steak)	–	48	76
Pork	(Total time, min)		
Chops	50	44	64
Spare rib	120	26	72
Tenderloin	40	57	83

mitochondria and other intensely active metabolic sites. Whilst the body can produce oleic acid from saturated precursors, it cannot readily produce any of the above, unless one of them is available in the diet. Oleic, linoleic and linolenic acids each belong to a different family of compounds in which unsaturation occurs at the n–9, the n–6 and n–9, and the n–3, n–6 and n–9 carbon atoms, respectively, in the hydrocarbon chain numbering from the methyl carbon (n). They are thus referred to as the n–9, n–6 and n–3 series. (or the ω-3, ω-6 and ω-9 series). Linoleic acid is abundant in vegetable oils (such as soya and corn oils) and at about 20 times the concentration found in meat; and linolenic acid is present in leafy plant tissues. Eicosapentaenoic acid (C20:5, n–3) and docosahexaenoic acid (C22:6, n–3) are normally present at low concentration in meat tissues, but there are high concentrations in fish and fish oils. This has been associated with the rarity of coronary thrombosis among the Greenland Inuit despite their very high intake of fat and cholesterol. Differences in the component fatty acids are reflected in the iodine number of the fats. Those of plant origin have iodine values averaging about 120, whereas those of meat animals average about 60, that of pork being somewhat higher, and that of lamb somewhat lower, than the value for beef (cf. Tables 4.3 and 4.5). That there are differences between muscles within a given species in the concentrations of unsaturated fatty acids, and between different fractions within a single muscle, has already been indicated (§ 4.3.5).

Comparative data on the content of polyunsaturated fatty acids and of cholesterol in the muscular tissue and offal of the common meat species are given in Table 11.9. It is clear that the titre of linoleic acid is markedly greater in the lean meat of pigs than in that of either the ox or sheep. Such species differences are also reflected in the composition of kidney and liver. The latter tissue in all three species is a particularly rich source of polyunsaturated fatty acids. Brain, as Crawford (1975) has emphasized, has a uniquely high content of C 22 polyunsaturated acids. It is also noteworthy in Table 11.9 that the cholesterol concentration in offal (and particularly brain) is very much greater than that in muscular tissue.

Table 11.9 Polyunsaturated fatty acids and cholesterol in lean meat and offal (after Paul and Southgate, 1978) (as % total fatty acids)

Source	C18:2	C18:3	C20:3	C20:4	C22:5	C22:6	Cholesterol (mg/100 g)
Beef	2.0	1.3	Tr.	1.0	Tr.	–	59
Mutton	2.5	2.5	–	–	Tr.	–	79
Pork	7.4	0.9	–	Tr.	Tr.	1.0	69
Brain	0.4	–	1.5	4.2	3.4	0.5	2200
Kidney, Sheep	8.1	4.0	0.5	7.1	Tr.	–	400
Ox	4.8	0.5	Tr.	2.6	–	–	400
Pig	11.7	0.5	0.6	6.7	Tr.	–	410
Liver, Sheep	5.0	3.8	0.6	5.1	3.0	2.4	430
Ox	7.4	2.5	4.6	6.4	5.6	1.2	270
Pig	14.7	0.5	1.3	14.3	2.3	3.8	260

To avoid possible health dangers from the consumption of the flesh of ruminants, a greater degree of unsaturation could be introduced into their fats. Normally the feeding of highly unsaturated vegetable fats to sheep and cattle with this intention would be invalidated because ingested fats would be reduced by the rumen bacteria. If, however, they are first treated with formaldehyde, they resist reduction and can raise the degree of unsaturation in the ruminant fat stores very considerably (Cook et al., 1970). The reported benefits to human consumers of increasing the proportion of mono-unsaturated acids to saturated acids in the diet in comparison with that of increasing the ratio of polyunsaturated acids has fostered attempts to modify the fat of the pig by feeding high-oleic sunflower oil (Ziprin et al., 1990) (cf. § 11.3). Again, selection for 'doppelender' development in cattle, it has been suggested, would not only provide meat with greater efficiency but would also ensure that the fat was more highly unsaturated (Ashmore and Robinson, 1969).

There is also evidence that polyunsaturated components of animal fats are essential for brain development, especially in the foetus (Crawford, 1975). When linoleic and linolenic acids are ingested, they are metabolized by animal liver to produce two families of long chain polyunsaturated fatty acids which are specific to animals, respectively, the n-6 and n-3 series. It is significant that brain cells contain these acids and not the parent linoleic and linolenic. Moreover, the chain-elongation and desaturation of linoleic acid also gives rise to the prostaglandins which are important in controlling blood pressure and for other essential purposes. These include prostacyclin, produced in the arterial lining, which has vasodilatory and antiaggregating properties in the blood, and thromboxane, which is thrombotic, promotes aggregation and is produced in the platelet membranes. The lipids of free-living mammals such as the eland are predominantly polyunsaturated and phospholipid in nature, whereas those of intensively reared animals are mainly saturated triglycerides. These facts suggest not that vegetables should be eaten and meat avoided, but that meat from wild or unimproved species should be preferentially sought (Crawford, 1975) (cf. § 1.3). In this context it is of interest to note that the concentration of C_{20} and C_{22} polyunsaturated fatty acids in whale muscle is about seven-fold greater than in that of domestic animals (Tveraaen, 1935) a reflection of the lipid composition of the krill upon which this species subsists.

Notwithstanding the beneficial attributes of polyunsaturated fatty acids, it should be noted that lipid oxidation products are believed to adversely affect the health of cells. Fortunately muscular tissue contains several enzymes that protect cells against such change, the most important of which is glutathione peroxidase (Halliwell *et al.*, 1995). Moreover, there is evidence that conjugated linoleic acid has an anti-mutagenic effect (Kritchevsky, 2000).

Where as in recent years consumers have been advised to limit their intake of saturated fats (and to achieve a ratio of polyunsaturated:saturated fatty acids greater than 4), the type of polyunsaturated fatty acid is now being emphasized and a higher ratio of n-3:n-6 fatty acids is advocated (Wood *et al.*, 2003). There is also now concern about the consumption of *trans*-unsaturated fatty acids in which the double bonds are in the *trans*-stereometric position. These acids tend to form during high-temperature hydrogenation of oils for use in food products.

11.2   Toxins and residues

As already described (§§ 8.3.1 and 8.3.3), the adventitious presence of nitrate in the salts used to cure meats in early times, and its microbiological reduction to nitrite by halophilic organisms, was responsible for the desirable pink colour of the products, for their flavour and for their safety from *Cl. botulinum*. Because nitrite can destroy blood pigments and vitamin A (Roberts and Sell, 1963), the residual level in cured meats was restricted to 500 ppm about 35 years ago. The effect of nitrite is very serious for infants since foetal haemoglobin is particularly susceptible to oxidation until they are three months old and the enzyme systems capable of reducing metmyoglobin back to myoglobin are often deficient in the very young (National Research Council, 1981).

Awareness that nitrite can react with secondary and tertiary amines to produce carcinogenic nitrosamines, such as N-nitrosodimethylamine (Lijinsky and Epstein, 1970), led to a further reduction in the permitted level of residual nitrite to 200 ppm and to attempt to cure meat without nitrite, although this latter possibility cannot be contemplated unless some other means of eliminating *Cl. botulinum* can be found. The most prevalent secondary amines in raw pork appear to be piperidine, diethylamine, pyrrolidine and dimethylamine (Bellatti and Parolari, 1982). During the maturation of cured pork products, the level of dimethylamine rises from *ca.* 0.1 ppm to 3 ppm.

A comparative assessment of the amine content of sausages from Northern and Southern Europe showed that tyramine and phenylethylamine were present at higher concentrations in the latter, possibly due to the decarboxylase activity of certain strains of the starter organisms *Kocuria varians* and *Staphylococcus carnosus* (Ansorena *et al.*, 2002a).

It is clearly important, however, to maintain perspective in such problems. Thus about 65 per cent of the nitrite ingested by humans is that present in human *saliva* (Greenberg, 1975), and it would be virtually impossible to eliminate its precursor, nitrate, from any diet. Nitric oxide is produced in skeletal muscles (and in other tissues) by nitric oxide synthatase (Brannan and Decker, 2002); and it has been suggested that nitrates, and the nitrites derived from them, enhance the body's defences against gastroenteritis by suppressing pathogens (Dykhuisen *et al.*, 1996). Nevertheless, the WHO (1977) recommends that the level of nitrate in drinking water should not exceed 11 ppm. The quantity of nitrate in vegetables is ten-fold greater than in cured meats, although their relatively high content of ascorbic acid would tend to inhibit nitrosation (Walters, 1983). Moreover, at the low concentrations of residual nitrite in cured products, amines form nitrosamines only with difficulty (Walters, 1973). Random surveys of hams have revealed less than 1 ppb of *N*-nitrosodimethylamine (Fiddler *et al.*, 1971), although heating of fat, especially at high temperature, increases its concentration (Patterson and Mottram, 1974).

It should be mentioned that *N*-nitrosodimethylamine has been found in the blood of 97 per cent of healthy individuals (Laknitz *et al.*, 1980). In those with hypochlorhydria, however, the relatively alkaline conditions in the intestinal tract permit growth of nitrate-reducing bacteria whereby concentrations of nitrite are enhanced considerably and may become carcinogenic *per se* (Newberrie, 1979). On the other hand, it has been suggested that it is the natural production of nitrite in the human intestinal tract, a capacity developed in early infancy, which usually affords protection against environmental spores of *Cl. botulinum*, which are ubiquitous (Tannenbaum *et al.*, 1978), and certain unexplained cot deaths have been attributed to the absence of this capacity.

When meat products are smoked, polycyclic aromatic hydrocarbons (including carcinogenic substances such as 3,4-benzpyrene) may precipitate on to surfaces. Smoke produced when the temperature of the smouldering wood exceeds 500 °C is more likely to produce them (Potthast, 1975), but it is only when fat falls on to hot cinders during charcoal grilling of meats that the levels become significant.

Various hormones are now administered to enhance the growth of animals (cf. § 2.5.2.1) and some of these, such as hexoestrol, are believed to be carcinogenic (Gass *et al.*, 1964). Synthetic oestrogens have thus been prohibited in a number of countries. It seems unlikely, however, that residues of these would be present at significant levels in meats. Early studies indicated that there were no detectable residues of the hormones in the flesh of treated cattle (Perry *et al.*, 1955) or pigs (Braude, 1950), provided they were used in accordance with instructions. Radioimmune assays have enabled very precise assessments to be made.

The technique has placed the problem into a more rational perspective since it has been shown, for example, that the levels of testosterone in treated animals may be substantially less than those in untreated animals of different age or sex (Hoffman and Kung, 1976). Moreover, it has been calculated that it would be necessary to consume 200 tonnes of beef liver or 200 tonnes of lean meat from cattle implanted with diethylstilboestrol to obtain the amount of oestrogen administered in a single birth control pill (Jukes, 1976). Legislative and public health aspects of the use of anabolic agents have been reviewed (Coulston and Wills, 1976).

Attempts have been made to standardize the mode of monitoring meat for residues within the member states of the European Union. In 1985 the European Parliament proposed to prohibit the use of certain substances having a hormonal action in increasing the growth of animals, whilst permitting the use of oestradiol-17β, progesterone, testosterone, trenbolone and zeranol until there was a unanimous decision of the European Court on their safety. A Select Committee of the House of Lords (1985–86) noted that a total ban on the use of hormones for growth promotion would reduce the quantity of lean meat available to consumers.

It is feasible that meat could be the vehicle for various mycotoxins produced by moulds. These could be acquired when animals ate contaminated feeds. They could also arise in such products as mould-fermented sausages. Although selected moulds are encouraged to grow on the latter during the maturation period, undesirable species capable of producing toxins could also thrive. Thus the presence of ochratoxins, produced by *Aspergillus ochraceus* (and by various *Penicillium* spp.) and ingested from mouldy feed, causes swine disease and carcass condemnations (Krogh, 1977). Aflatoxins are also produced by *Aspergillus* spp. These are believed to be carcinogenic to humans. The significance of ochratoxins for human consumers is still unknown.

Certain lactic acid bacteria have been shown to produce undesirably high levels of histamine and tyramine in fermented sausages (Maijala and Eerola, 1993). In such sausages it appears that the concentrations of putrescine, cadaverine and tyramine depend not only on the nature of the raw materials used but also on the interaction between processing temperature and the particular starter organisms employed (Maijala *et al.*, 1995). There is increasing evidence that fermented meat products, although hitherto considered safe, may cause outbreaks of gastroenteritis involving such micro-organisms as *Salmonella* and verocytotoxigenic *E. coli* (Moore, 2004).

Since bone tends to concentrate heavy metals, such as lead, barium and strontium, it might seem that an increased use of mechanically recovered meat could be a hazard. However, the overwhelming balance of evidence indicates that these elements are not present in the product at sufficient concentrations to be significant for health (Newman, 1980–81).

The use of pesticides in agriculture, especially those which are persistent, such as the organochlorine group, could lead to their deposition in the tissues of animals grazing treated pastures or feeds, and, accordingly, various surveys of pesticide residues in meat have been made. Thus, Madarena *et al.* (1980) assessed meat for residues of 14 organochlorine pesticides. These included BHC isomers, the DDT group and cyclodienes. Total organochlorine residues in beef, pork, rabbit and horse were, respectively, 10, 80, 110 and 160 ppb.

11.3   Meat-eating and health

It is evident (cf. § 11.1) that meat provides the majority of the nutrients required for health by human consumers. Nevertheless, there continues to be vigorous controversy about the effects on health of its long-term consumption, since it has been alleged to be associated with the development of carcinoma, cardiovascular disease and hypertension.

Since diet modifies the enteromicrobial ecology of the alimentary tract, it would be supposed that carnivorism would promote the predominance of different microflora having a different capacity to affect its histology than herbivorism (Paterson, 1975). There is some evidence that increased consumption of meat, with concomitant alterations in other dietary constituents, and increased gut transit time, may play a rôle in the development of carcinoma of the large intestine. In areas of high incidence of the disease the intestinal flora has an increased proportion of anaerobic organisms, such as *Bacteroides* which produce 7-α-dehydroxylase. This enzyme converts cholic acid to the suspected carcinogen, deoxycholate. Insofar as meat consumption tends to reflect affluence, the impression that cancer of the bowel is more prevalent in richer societies may appear logical: conversely, cancer of the stomach appears to be more prevalent among the less afffluent.

The balance of present opinion, however, indicates that meat consumption *per se* is not a factor in carcinogenesis (Pearson, 1981; Roberts, 1999). When adjusted for all malignant neoplasms (other than lung cancer, which is positively associated with cigarette smoking), mortality rates in the USA were found not to have changed in the period 1940–1975 although the consumption of meat and poultry had greatly increased in this period (Leveille, 1980).

On the other hand, in that meat could be the repository of extraneous carcinogens (e.g. residues of benzpyrenes from smoking operations, anabolic hormones) it could constitute a hazard, as could any other food so contaminated (cf. § 11.2).

Since the intake of the saturated fatty acids, lauric, myristic and palmitic, raises plasma levels of cholesterol (Grande, 1975), and these are found in relatively high concentrations in animal fats, an association between this phenomenon and meat consumption might be anticipated. It has been suggested, therefore, that a high ratio of unsaturated/saturated fatty acids in the diet would be desirable since this might lower the individual's susceptibility to cardiovascular diseases, in general, and to coronary heart disease and cerebral vascular disease, in particular (Keys *et al.*, 1960). Partly in response to the views of the Committee on Medical Aspects of Food Policy (1984), altered breeding, feeding and butchery methods have led to a marked fall in the fat content of beef, pork and lamb (Higgs, 2000), from *ca.* 20–26 per cent to *ca.* 4–8 per cent in the 1990s.

Cholesterol is transported in the plasma by three types of lipoprotein – very low density, low density and high density. High levels of high-density lipoproteins and low levels of low-density lipoproteins have been associated with a low incidence of coronary diseases (Jackson *et al.*, 1975). Whereas, however, it has been found that increasing the content of polyunsaturated fatty acids in the diet lowers the level of low-density lipoprotein cholesterol in the blood, it also lowers that of the protective high-density lipoproteins (Vega *et al.*, 1982). This latter adverse effect appears to be due to the n-6 acids (e.g. linoleic and arachidonic): acids of the n-3 series (e.g. linolenic, eicosapentaenoic and docosahexenoic) are now believed to be responsible for the beneficial effects of polyunsaturated acids in protecting against coronary heart disease (Leaf and Weber, 1988) even in those with a high intake of fat and cholesterol (Dyerberg and Bang, 1979). Since there is also evidence that monounsaturated (n-9) fatty acids have an effect in lowering serum cholesterol levels which is similar to that of the n-3 acids (Mattson and Grundy, 1985), it may be desirable to consider their concentration in any attempt to specify desirable ratios of saturated to unsaturated fatty acids in meat or other foods. Although high levels of serum cholesterol have been positively correlated with death from cardiovascular diseases in women and younger men, mortality is negatively related to this parameter in men over 45 (Kennel and Gordon, 1970).

Insofar as the polyunsaturated n-3 fatty acids have a marked tendency to oxidize, the titre of lipid oxidation products could be expected to be relatively high in food sources containing them and in the tissues of human consumers thereof. Lipid peroxides have been associated with various aspects of coronary heart disease (Addis, 1986). There are indications that it is the oxides of cholesterol which are toxic rather than cholesterol (Taylor *et al.*, 1979; Paniangvait *et al.*, 1995). These can be found in fresh meat, but ionizing radiation increases their level. Whereas cholesterol oxidation can occur without concomitant lipid oxidation in beef, the oxidation of fatty acids accelerates that of cholesterol (Nam *et al.*, 2001). A survey of processed meats

revealed that cholesterol oxides were absent from meat samples, but certain unidentified components were found (Higley *et al.*, 1986).

   If fish oils were to be incorporated into feeds to modify the lipids in the tissues
of non-ruminants in the believed interests of the human consumer (cf. § 11.1.4), it
would be necessary to control lipid oxidation. Processing meat pre-rigor has been
suggested for this purpose (Judge and Aberle, 1986).

   It is worth noting that the feeding of polyunsaturated fatty acids has been associated with an increased incidence of carcinoma and immuno-suppression (Dayton
et al., 1969; Bennet *et al.*, 1987).

   It is clear that the relationship between cardiovascular disease, the consumption
of fat and serum cholesterol levels is complex, and, in the present state of our knowledge, correlations which can be demonstrated within individuals should not be
presumed to apply to the population as a whole and vice versa. In the 1970s and
1980s various organizations cautioned against the consumption of, especially, red
meat insofar as the fat it contains is more highly saturated than that in pork or
poultry and this was believed to be a factor predisposing to cardiovascular disease.
The fact that the redness signifies a high content of readily assimilable iron and that
whereas anaemia can potentially affect all consumers, only some individuals are at
risk from cardiovascular problems, were ignored. Australian research has provided
evidence that substantial intake of lean red meat lowers several factors associated
with cardiovascular disease, such as hyperlipidaemia and elevated blood pressure,
despite increasing the proportion of arachidonic acid in plasma phospholipids
(O'Dea and Sinclair, 1985). Moreover, a diet rich in lean beef has been shown to
reduce the level of serum lipoprotein cholesterol. Again, although lean red meat
contributes to an increase in the prothrombotic arachidonic acid, it also increases
the levels of dihomogammalinolenic and eicosapentaenoic acids, which antagonize
the effects of arachidonic acid in promoting the aggregation of platelets (Sinclair
et al., 1994).

   Following detailed analysis of the muscles of cattle, sheep and pigs available to
consumers in the UK, Enser *et al.* (1996) concluded that they constitute a valuable
source of polyunsaturated fatty acids (especially of the C_{20} and C_{22} n-3 acids) in the
diet.

   That there is a connection between a high intake of salt and hypertension has
long been recognized and low-salt diets have been used to control the latter. Clearly
a substantial amount of salt could be derived from the consumption of cured meats,
but fresh meat *per se* is a minor source of dietary salt (Pearson, 1981).

   The exact way in which the individual human consumer will digest, absorb and
utilize the amino acids of meat (or those of any other protein-containing food)
cannot be predicted precisely: the existence of considerable variation, even amongst
the vast majority who are usually regarded as 'normal', is proven (Williams, 1956).
Some individuals, however, differ in their metabolic response to the proteins of
foods in a manner which is sufficiently marked as to be 'abnormal', and, insofar as
meat is an important source of essential amino acids, it is desirable to refer briefly
to some disturbances of protein metabolism which constitute pathological conditions (Carson, 1970). These include disorders of the digestive enzymes (e.g. cystic
fibrosis of the pancreas) and faulty mechanisms of intestinal amino acid transport
(e.g. cystinuria, in which there is faulty absorption of cystine and dibasic amino acids;
and Hartnup disease, where transport of neutral amino acids and malabsorption of
tryptophan are features). In addition, there is a large number of genetically

determined conditions in which the intermediary metabolism of one or several amino acids is defective.

Allergic reactions to the proteins of meat have been found, hitherto, in individuals having a general intolerance to food proteins (Kekomaki *et al.*, 1967). Work by Han *et al.* (2000) has suggested that bovine serum albumen and bovine gamma globulin are potentially allergenic to certain individuals.

Although diets rich in protein (including meat) elevate serum levels of uric acid – and thus restriction in their consumption may be advised in certain circumstances – their reported association with gout is not clear (Zöllner, 1975) as the condition is now thought to be hereditary.

Amongst the more positive attributes of meat is its ability not only to supply iron – and that from meat is absorbed 3–5 times more readily than iron from plant foods (Rogowski, 1980) – but also to enhance the absorption of iron from non-meat sources which are concomitantly consumed (Cook and Monsen, 1976). Iron appears to be absorbed as non-haematin compounds having a molecular weight of less than 10,000 dalton (Hazell *et al.*, 1981). (It is clearly important to be able to distinguish between haematin and non-haematin iron. This can be done using a combination of atomic absorption spectroscopy and inductively coupled plasma optical emission spectroscopy: Lumley, 2001.) In meat there are also small quantities of such minerals as manganese, which protects against degenerative diseases of the bone (Underwood, 1977); of zinc, which promotes growth, sexual maturity and wound healing; and of cobalt, which is essential for the synthesis of vitamin B_{12}. There is some evidence that high levels of creatine in the diet (*ca.* 30 g/day, over a week), may enhance muscular performance in athletes (Harris *et al.*, 1993).

Quite apart from its effect on health through the provision of nutrients, meat is being increasingly recognized as a 'functional' food, i.e. one which can beneficially affect physiological processes in the consumer and, thereby, potentially mitigate or prevent disease (Jiminez-Colmenero *et al.*, 2001). Its nature, and that of its derived products, can be tailored to fulfil currently perceived compositional and organoleptic desiderata, both by conventional means and by more novel procedures such as the planned use of probiotic organisms (cf. §9.3) and genetic modification.

In respect of organ meats, there have been occasional reports of toxicity. Thus, the livers of polar bears and of husky dogs contain very high levels of vitamin A, and excessive consumption of these organs by Arctic explorers has been associated with symptoms of hypervitamosis. The feeding of vitamin A to domestic meat animals to increase growth efficiency, and its concentration thereby in their liver, has been blamed for facial malformations in babies whose mothers ate liver during pregnancy; but, on present evidence, it would be premature to believe that the undoubtedly excellent nutrients in the liver of domestic meat animals, not least the content of vitamin B_{12}, should be eschewed by the general consumer – or even by the vast majority of pregnant women – because of its vitamin A content.

Possible dangers from the use as feed for animals of offal from the brain and spinal cord have been recognized recently in relation to a disease which affects the nervous system of cattle and which is referred to as bovine spongioform encephalitis (BSE). The symptoms of the condition are similar to those observed in scrapie, a disease long known in sheep, which can be transmitted to other species. Indeed the feeding to cattle of inadequately treated brain and spinal cord from scrapie-infected sheep was suggested as the cause of bovine spongiform encephalitis. According to a detailed report, commissioned by the UK government, BSE may have arisen, *de novo*, from a spontaneous mutation in a single cow, the condition being spread by recycling tissue from this animal in the feed given to others. (If a mutation were involved, however, it would seem more likely that it would be caused in a number of animals by some common predisposing factor which has not been identified so far.)

In the brains of mice which have been injected with sheep scrapie, ubiquitin–protein complexes have been detected in intracellular filamentous structures. These are similar to those found in the brains of human patients suffering from Alzheimer's disease, suggesting that there may be a process of neurological degeneration common to both conditions (Lowe *et al.*, 1990). The formation of ubiquitin–protein complexes in nervous tissue reflects the cryoprotective activity of ubiquitin (cf. §5.1.2). The insoluble plaques which are responsible for the death of nerves in transmissible spongioform encephalopathies appear to consist of fibres in which β-pleated sheets of protein are aggregated right-handed helices. The amyloid fibres are formed from intermediate structures which arise during transitions in the folding topography of the protein.

The agents responsible for such conditions are simpler than viruses, produce no immune reaction and are more stable to heat and other denaturing factors than normal proteins. They are resistant to most proteases; but a protease has been designed to degrade them (Anon., 2001). The infective molecules are distorted forms of the normal proteins of the brain cell membrane (prions). They appear to multiply, without the involvement of nucleic acid, by causing the molecular configuration of the normal prions to adopt their distorted format and, in turn, these impress their altered configuration onto other normal brain membrane proteins. Prions seem to have some function in the metabolism of copper, binding it at the cell surface (Jones, 2002).

It appears unlikely that the consumption of such abnormal prions from the brains or spinal cord of cattle or sheep would cause spongioform encephalitis in the human brain since their amino acid sequence differs significantly from that of the human prion (Prusiner, 1995). There is no current proof that BSE can be transmitted to human consumers of bovine offal from the brain or spinal cord. A species barrier has apparently prevented scrapie from infecting humans for hundreds of years. Nevertheless, it should be noted that a mental condition develops in certain natives of New Guinea who habitually include brains in their diet; and that whereas Alzheimer's disease, and a similar condition, Creutzfeldt–Jacob disease (CJD), are found mainly in the elderly, a so-called 'new variant' (nvCJD) affects the young. It differs from the other transmissible spongioform encephalopathies in that the disease-associated form of the prion protein can be readily detected in the nervous system (Ironside, 1999). As a precautionary measure, the use of potentially infected offal in meat products was banned by the authorities in Britain. Several methods are now available to accurately assess the level of glial fibrillary acidic protein – as an indicator of spinal cord – in meat products (Schmidt *et al.*, 2002).

Neither muscular tissue itself, nor offal, are high in calories, and since they provide satiety of long duration (Rogowski, 1980) their consumption *per se* will not lead to overeating or obesity.

There can be no question that meat is an excellent source of the nutrients required for health. That it may cause carcinoma, cardiovascular or other diseases in the otherwise healthy individual is far from being proven, the evidence being contradictory, and the concept biologically unlikely. Thus, the incidence of such diseases is low among the Masai, for example, whose consumption of meat and related products is particularly high (Yudkin, 1967); elephants are vegetarian, yet suffer severely from atherosclerotic conditions. Meat frequently features in the diet of centenarians.

It is evident that we know too little, as yet, about the biochemical requirements of the individual human consumer to accept generalizations against meat-eating.

Chapter 12

Prefabricated meat

12.1 Manipulation of conventional meat

Although hitherto unexploited sources such as non-meat proteins (§ 12.2) and abattoir offal (§ 12.3) are currently being utilized or investigated for the fabrication of 'meat', there have been concomitant developments in the modes of usage of conventional meat which merit consideration in the present text.

12.1.1 Mechanically recovered meat

Carcass dressing operations leave varying proportions of meat adhering to bones: the world total has been estimated to be 2 million metric tons (Field, 1976). Such meat is of identical nature to conventional carcass meat. In the past, since bones were not accorded the degree of hygienic handling essential to minimize microbial spoilage and other undesirable organoleptic changes, the associated meat quality deteriorated. The mechanical devices now available to separate meat from bone, together with adequate refrigeration and limited storage, make it possible to produce meat suitable for human consumption. The devices used either separate the meat by pressing a finely ground mixture against perforations in a stainless steel drum or by applying high pressure (10–25 MPa) to roughly broken bones, when the meat becomes plastic and flows from the bones (Newman, 1980–81). Mechanically recovered meat, obtained by either method, is free from organoleptically detectable bone. Heat generation is inevitable, but the incorporation of refrigeration sufficient to maintain the temperature below 10 °C limits microbial growth, oxidative rancidity and browning due to the oxidation of haem pigment (Ostovar et al., 1971). During mechanical deboning, the structure of the myofibril is considerably altered at both Z- and M-lines (Schnell et al., 1974), and the connective tissue content is reduced (Ranken and Evans, 1979).

When bones of high marrow content are processed, the lipid and haem concentration of the product is increased. These circumstances worsen the tendency of the mechanically recovered meat to undergo oxidative rancidity (Froning and Johnson, 1973; Demos and Mandigo, 1996), and this is exacerbated further if pork bones are

Table 12.1 Typical composition of meat recovered mechanically (ME) or manually (MA) from bone (after Newman, 1980–81) (Components as per cent wet weight)

Bone source	Total lipid		Crude protein		Ash		Calcium	
	ME	MA	ME	MA	ME	MA	ME	MA
Pork ham	39.0	37.9	10.2	15.7	4.0	0.5	1.4	0.03
loin	29.5	23.6	14.0	16.7	1.8	0.7	0.4	0.04
Beef rump	41.9	11.8	10.0	17.6	4.3	0.8	1.5	0.08
loin	33.4	22.5	11.6	16.4	4.3	1.0	1.5	0.01
Mutton breast	36.5	38.1	15.0	15.5	1.2	1.0	0.1	0.02

used, since the lipid therein is relatively unsaturated (Meiburg *et al.*, 1976). The dark colour of mechanically recovered meat, generally, limits its incorporation into certain products such as UK sausage. On the other hand, its relatively high pH (especially when marrow bones are involved) increases the extractibility of the muscle proteins and the emulsifying power of the meat (Field, 1976). The high pH, together with the finely divided nature of mechanically recovered meat, makes it very liable to microbial spoilage during storage, but this can be readily controlled by maintaining the temperature below 5 °C (for short time usage) or below the freezing point (for longer storage) (Meiburg *et al.*, 1976).

Typical data on the composition of meat recovered either mechanically or manually from bones are given in Table 12.1. It will be apparent that mechanically recovered meat contains more lipid, less protein and very much more ash and calcium than manually deboned meat. Nevertheless, there appears to be no single chemical index for the accurate identification of mechanically deboned meat: the total haem pigment is probably the most useful (Meech and Kirk, 1986). This view was confirmed in an extensive investigation of the methods available for the detection of mechanically recovered meat. The chemical composition was the least reliable criterion of those studied (Crossland *et al.*, 1995), whereas gel electrophoresis (Savage *et al.*, 1995), microscopy (Pickering *et al.*, 1995a) and immunological techniques (Pickering *et al.*, 1995b) proved reliable indices. Recently multivariate analysis of isoelectrically focused protein profiles has shown promise as a detection procedure (Skarpeid *et al.*, 2001).

12.1.2   High-pressure modification

The well-known susceptibility of enzymic reactions to high pressure, and the effects of the latter on the secondary, tertiary and quaternary structure of proteins (Jøsephs and Harrington, 1968), led Australian workers to study the response of muscle under such conditions (Macfarlane, 1984) (cf. § 10.3.4). Many variables were found to affect the latter. These included the severity of pressure, its duration, its rate of change, the concomitant temperature, the extent of development of rigor mortis at time of pressurization and the type of muscle.

It had been shown by Ikkai and Ooi (1969) that, in the presence of ATP, pressure dissociated actomyosin into its constituent actin and myosin, and Macfarlane (1973) demonstrated that a pressure of *ca.* 100 MPa (*ca.* 1000 atm), when applied to pre-rigor muscle for 2–4 min at room temperature had a marked tenderizing effect on the cooked meat. Although such tenderizing could not be achieved in post-rigor muscle at 0–20 °C, pressures of 100–150 MPa were effective on subsequent holding at 45–60 °C (Bouton *et al.*, 1977a). Cold-shortened muscle could also be tenderized by a combination of pressure heat, although this cannot be achieved by ageing (Davey *et al.*, 1976: cf. § 5.4.1). Moreover, the tenderizing action was achieved, in pre-rigor muscle, despite shortening of *ca.* 35 per cent. (cf. § 10.3.4)

Various explanations have been advanced for the tenderizing action of high pressure. Respecting the general effects of high pressure and temperature on proteins, whereas electrostatic bonds and hydrophobic interactions are susceptible to the former, they are less affected by high temperature. However, covalent bonds and SH groups are labile at high temperature but more resistant to high pressure (Ledward and Mackey, 2002). Kennick and Elgasim (1981) regarded the F- to G-actin transformation and the depolymerization of myosin to be important in this respect, since heating denatures these in the dissociated state instead of as acto-myosin aggregates. In pre-rigor muscle, high pressure (150 MPa, 5–10 min, 35 °C) affects the sarcoplasmic reticulum, causing proteolysis of the 100,000 dalton ATP-ase protein and of calsequestrin (Horgan, 1980–81), whereby the Ca^{++} concentration rises and contraction and post-mortem glycolysis are greatly stimulated, the latter being enhanced by activation of phosphorylase (Horgan and Kuypers, 1983). Accelerated conditioning no doubt arises due to the early release of lysosomal enzymes (Elgasim and Kennick, 1980) through damage to the lysosomal membranes in the presence of excess Ca^{++} ions (Macfarlane and Morton, 1978) from the disorganized sarcoplasmic reticulum.

Homma *et al.* (1994) demonstrated that an increase in the activity of such enzymes as the cathepsins B, D and L, with increasing pressure up to 400 MPa, was due to their increased release from lysosomes; and even at 100 MPa the release of cathepsins is more marked during the normal course of conditioning (Kubo *et al.*, 2002). Nevertheless cathepsin H is inactivated above 200 MPa and cathepsin D above 500 MPa (Montero and Gómez-Guillén, 2002). Subsequently Homma *et al.* (1995) showed that the calpain system was also affected by high pressure. Above 200 MPa the activity of both μ- and m-calpains decreased rapidly, but since that of the calpastatin was even more labile, the activity of the calpains was dominant and this may contribute to the increase of tenderness caused by the high pressure.

Pressurization of muscles having a high ultimate pH has little effect on the proteins of the sarcoplasmic reticulum. It is only in muscles which undergo appreciable post-mortem glycolysis that proteolysis of these proteins occurs under pressure and thus implicates enzymes of lysosomes in their breakdown (Horgan, 1987). Electron microscopy indicates that the sarcolemmal and endomysial sheaths are separated from the myofibrils (Macfarlane and Morton, 1978) and this feature must contribute to the increased tenderness; but connective tissue is not affected (Ratcliff *et al.*, 1977). According to Locker and Wild (1984) there is a concomitant change in the 'gap filaments'. Certainly pressure/heat treatment produces a greater degradation of connectin ('gap filaments') than heat alone (Macfarlane *et al.*, 1986) and Kim *et al.* (1992) showed that α-connectin was converted to β-connectin by pressure alone. Subsequently, using immunoelectron microscopy, Suzuki *et al.* (2001) showed that connectin in the M-line was substantially disoriented by even short exposure (*ca.* 5 min) to high pressure. The tenderizing effect of high pressure is not affected, however, by pretreatent with tenderizing enzymes. Such would have been expected if connectin were substantially responsible for myofibrillar toughness. The apparent connectin (titin) content of beef *l. dorsi* muscle remains unaltered despite wide variations in the Warner–Bratzler shear values of steaks prepared from aged and unaged meat (Fritz *et al.*, 1993). Suzuki *et al.* (1993) demonstrated that high pressure has no effect on the structure of collagen as judged by electrophoretic and thermographic data.

Pressures of the order of 400 MPa, nevertheless, cause proteoglycans to break down into smaller molecules (Ueno *et al.*, 1999).

In a study comparing the effect of various combinations of high pressure and heating, Ma and Ledward (2004) concluded that when meat is heated between 60 and 70 °C under pressure (200 MPa), the enhanced tenderness observed is mainly due to proteolysis: structural changes in the collagen and myofibrillar proteins were of less importance.

Using NMR imaging, Bertram *et al.* (2004b) demonstrated that the structural changes responsible for tenderizing (and changes in the location of water) during high-pressure treatments differ from those that occur during ageing.

Apart from its tenderizing effect, the high-pressure treatment of meat, if it could be applied in industry, would greatly accelerate operations in abattoirs by making possible the production of vacuum-packed, tenderized meat from the hot-deboned carcass. It would be less easy, however, to apply the procedure to post-rigor meat since the need for heat would yield a cooked appearance. Through its effect in accelerating or inhibiting post-mortem glycolysis, pressure treatment could be employed to produce meat of any desired pH in the normal range (Macfarlane, 1984). The biological value of the pressurized meat is not altered, but digestibility is increased (Elgasim and Kennick, 1980). High-pressure treatment appears to produce the same pattern of flavour precursor molecules as does conditioning (Suzuki *et al.*, 1994).

Of more immediate practical importance is the solubilizing action of high pressure on myofibrillar proteins (Macfarlane *et al.*, 1984). The imidazole groups of histidine appear to be implicated (Macfarlane and McKenzie, 1976). Pressures of *ca.* 150 MPa at 0–3 °C, enhanced the adhesion between meat particles, in a manner similar to that of trisodium polyphosphate in the presence of salt, but the effect operated at lower pH values. Even in the absence of salt, when increased solubility of the myofibrillar proteins was not a factor, a binding effect was produced. Studies by Suzuki and Macfarlane (1984) indicated that the increased heat-setting properties of myofibrillar proteins achieved under high pressure is due to depolymerization of the myosin monomers whereby they reaggregate in a different complex on release of pressure. This suggests that the use of pressure could lower the quantity of salt or polyphosphate required to solubilize myofibrillar proteins, and thus operate as a binding agent, in the production of reformed meats. Thereby the amounts of salt needed in foods could be reduced.

The application of high pressure changes the colour of pork and beef (Shigehisa *et al.*, 1991). The discoloration appears to be due to globin denaturation (and possible loss of haem), at *ca.* 200–360 MPa, and to subsequent oxidation to metmyoglobin (Carlez *et al.*, 1995). On the other hand the nitrosomyohaemoglobin of cooked ham is resistant to such pressure-induced change (Goutefongea *et al.*, 1995). The application of moderate pressure (80–100 MPa) to meat for a short time, before exposing it to the air, prolongs the retention of the bright-red colour of oxymyoglobin at the surface – provided such exposure is within a few days of slaughter of the animal: its application after *ca.* 3 weeks is not associated with this benefit (Cheah and Ledward, 1997). It appears that the muscle enzyme system that accelerates met-myoglobin formation is initially inhibited more than that of the enzyme system that reduces metmyoglobin back to myoglobin.

Under high pressure, the proteins of muscle produce gels, which are different from those caused by high temperature; and, in the future, these differences may well be exploited in altering the texture of meat products to develop novel attributes of desirability for the consumer (Ledward and Mackey, 2002).

Although it has been reported that pure fats and oils are stabilized against oxidation by exposure to high pressure (Cheah and Ledward, 1995) exposure of pork to pressures of *ca.* 800 MPa cause the lipids of pork to oxidize more rapidly than those of controls (Cheah and Ledward, 1996). Pressures above *ca.* 400 MPa cause the formation of denatured ferric haemoprotein (as in cooking) and it is thus feasible that the latter acts as a pre-oxidant in meat which has been subjected to high pressure and, if so, this effect might limit the application of the technology concerned. The effect of high pressure on micro-organisms is considerable and has been outlined in § 6.3.6.

The above considerations refer to the application of *hydrostatic* pressure. More recently the effects of applying *hydrodynamic* pressure have been studied. Hydrodynamic pressure can be created by a controlled explosion whereby a pressure/shock wave passes through water and any object in the water with similar mechanical impedance to that of water (Kokosky, 1998). Exposure of meat to much hydrodynamic pressure increases tenderness to a level similar to that obtained by conditioning the meat for 3–5 weeks (Solomon *et al.*, 1997). Structurally, the hydrodynamic pressure detaches sarcomeres at the Z-line and the A/I junction (Zuckerman and Solomon, 1998). It decreases the content of myofibrillar protein and concomitantly increases that of soluble protein (Spanier and Romanowski, 2000), as is the case when high hydrostatic pressure is applied (Macfarlane *et al.*, 1984).

12.1.3   Reformed meat

Various developments since the Second World War have caused many meat consumers to favour lean steaks rather than meat from joints, with its associated fat. Because of the high cost of steaks from the relatively limited locations in the carcass which yield meat of the desired characteristics, attempts to produce, from the less expensive cuts, portions of meat having the desirable organoleptic attributes of traditional steaks have been marked since 1970. The high content of connective tissue in such meat has necessitated its comminution or subdivision by mechanical means, and its subsequent reforming or restructuring into steak-like portions. The process permits the control of product colour, texture and fat distribution. The size of the comminuted pieces has determined which term is appropriate, but in the present context 'reforming' will be employed. Restructured meat products have been extensively reviewed in a volume edited by Pearson and Dutson (1987).

Reformed meat can be prepared from fresh and cured meat by the procedure of tumbling or massaging (as now used for hams; cf. § 8.3.1) and by compressing together thin slices (or flakes) after partial freezing. The mode of comminution affects the properties of the meat: flaking is the most frequent method. When carried out at −2 °C flaking yields ribbon-like pieces, whereas an emulsified mass is obtained at 5 °C (Huffman and Cordray, 1983). There is a significant decrease on subsequent cooking loss as the particle size in comminution decreases and an increase in tenderness and in cohesiveness. On the other hand, over-comminution also causes loss of cohesiveness, the products reformed from ground meat being inferior to those from flaked meat (Chesney *et al.*, 1978). Wafer-thin slices permit increased extraction of myofibrillar proteins which foster binding and better cohesion on cooking above 45 °C, possibly due to the formation of hydrogen and electrostatic bonds. More recent studies have indicated, however, that the increased binding seen on heating is more likely to be due to disulphide bonds and hydrophobic interactions, although hydrogen bonds formed on cooking may also increase the strength of binding (Lee and Lanier, 1995; Ledward and Varley, 1999; N. Howell, personal communication).

The relatively small muscles of the less expensive locations in the carcass, such as the distal parts of the limbs and the neck, naturally vary in their biochemical and chemical constitution (§ 4.3). Thus, they have differing proportions of connective tissue, their yield of extractable protein – which is so important for reformed meat – may vary from 3 to 45 per cent of the total protein content (Saffle and Galbreath, 1964) and their ultimate pH is characteristically different.

Although the incorporation of salt (as would be anticipated from §§ 8.3.1, 8.3.2 and 10.2) enhances protein extraction, and thereby the texture of reformed products, it also increases the tendency for oxidative rancidity and discoloration through metmyoglobin formation (Huffman, 1981). From Table 12.2 it is evident that an intermediate level of salt (*ca.* 1 per cent) gives the benefits of lower cooking loss and greater tenderness without excessive rancidity (as determined by thiobarbituric acid). The extractibility of myofibrillar proteins (and thus the binding capacity of the system during comminution) is greatest when the meat is refrigerated to just above the freezing point, but temperatures below −4 °C enhance metmyoglobin formation (Mandigo, 1983).

The term surimi originally referred to a Japanese product made from mechanically-deboned white fish which was subsequently refined by removal of water-soluble proteins (including enzymes and pigments) and salts. Surimi thus has an enhanced content of myofibrillar protein in comparison with the original fish and is important as a functional ingredient in the food industry. Moreover, because enzymes and haem pigments have been removed, it has an enhanced stability. After a similar leaching process, minced pork, beef and mutton have been shown to yield useful functional products (Lee *et al.*, 1987; Torley *et al.*, 1988) which have lower fat and cholesterol and better rheological properties than other manufacturing grade meats, and a bland taste, although beef and mutton surimi have a rather dark colour. A general account of the nature of surimi prepared from meats, with particular reference to the chemical composition and microscopic nature of beef surimi, was

Table 12.2 Mean values for organoleptic characteristics of flaked pork steaks (after Mandigo, 1983)

Characteristic	Per cent salt			
	0	0.75	1.50	2.25
Rancidity (TBA)	0.11	0.50	0.84	0.94
Cooking loss (%)	37.5	21.0	14.6	13.4
Tenderness (sheer: kg)	1.04	0.84	0.69	0.71

given by Knight (1992), who suggested procedures for altering its texture in desired modes.

Reformed meat requires the incorporation of relatively high levels of lipid to develop the same degree of juiciness as intact meat (Cross and Stansfield, 1976). The mechanical disruption involved alters the mode of flavour release and the flavour of reformed meat tends to differ from the meat before treatment.

Although the high water-holding capacity of pre-rigor meat would seem to be an advantage in the manufacture of reformed meat, the comminution process would cause the swift onset of rigor mortis, unless the meat was comminuted in the presence of salt immediately post-mortem (cf. § 10.2.1.2), and this could prove difficult in industrial production. Raising the pH of post-rigor meat by the addition of alkali confers some but not all of the beneficial quality attributes of pre-rigor meat (Anon., 1983–85).

The comminution process permits the incorporation of non-meat (§ 12.2) or offal (§ 12.3) proteins and those of mechanically recovered meat (§ 12.1.1). These can contribute substantially to the cohesiveness of the reformed meat and lower its cost. Of proteinaceous materials, only bovine plasma, wheat gluten and soya protein are able to bind the portions of meat together in the absence of salt (Siegel *et al.*, 1979), but polysaccharides, such as alginate, can do so (Mearus and Schmidt, 1986).

The incorporation of 0.4 per cent of the platelet protein F XIIIa which is a transaminase, with 0.2 per cent phosphate and 1 per cent sodium chloride, has been shown to yield marked textural advantages in binding meat pieces in restructured products. The enzyme can be produced on a large scale, as a recombinant protein, by *Saccharomyces cerevisiae* fermentation (Nielsen *et al.*, 1995).

The pressure required in reforming the comminuted meat will be determined by the initial characteristics of the meat, its temperature and the machinery used. It ranges from 2–7 MPa.

12.2   Non-meat sources

The severe shortage of protein of high biological value in developing countries, and the high cost of meat in those which are not fortunate economically, has fostered great interest in the possibility of fabricating protein-rich foods from plant sources in a palatable form. Artificial meat-like products, which have controlled texture, flavour, colour and nutritive value (Sjoströn, 1963), have been marketed which can substitute directly for meat (meat analogues) or can economically extend the bulk, and help the texture, of conventional meat products and non-meat foods.

Texturization of recovered proteins has been achieved by three principal methods – fibre spinning, thermoplastic extrusion and heat gelation (Kinsella, 1978; Lawrie and Ledward, 1983). The first of these processes has been instinctively exploited by spiders and silkworms for millions of years, but it was not developed by humans until the end of the nineteenth century. The spinning of meat-like fibres from non-meat proteins was patented about 30 years ago (Boyer, 1954). Although their thermoplastic extrusion to form meat-like pieces is even more recent (Atkinson, 1970), a significant fraction of fabricated foods now depends on the latter process (Harper, 1979).

Among the vegetable proteins which have been exploited in this way are the glutens of wheat and the globulins of groundnut, cottonseed, peanut, sesame, yeast and soya bean. Protein isolated from the latter is of comparable biological value to that of meat, it is valuable in the manufacture of emulsion-type products (because of its power to emulsify, stabilize, texturize and hydrate comminuted meats) and it can be used as an ingredient of spun fibres. In most extraction procedures, defatted vegetable protein (e.g. soya bean) is heated with a slightly alkaline aqueous solution. The protein which dissolves is mechanically separated from residues and the major protein constituents precipitated by acid, forming a curd (which is subsequently neutralized by food-grade alkali). If an alkaline suspension of the soya protein curd is kept for some time at high pH, its capacity to aggregate as fibres is enhanced. The suspension is spun into an acid bath, when long filaments form. The native proteins of soya and of many other vegetables can also form a liquid homogenous suspension or mesophase (Tombs, 1972) with water and salt. The mesophase proteins can be extruded into water to form filaments. This procedure has the advantage that strong alkaline pH levels are not required.

In thermoplastic processes the proteinaceous material (often defatted soya flour, not further purified) is fed into the hollow barrel of an extruder where a tapered screw forces it, under high pressure and temperature, towards a narrow orifice into the exterior. Under these conditions, starch components gelatinize, proteins partially denature and the tractile mass is restructured and aligned. These prefabricated products combined with other food components, colouring and flavour can be made up as simulated meats (Meyer, 1967), which are marketed as 'meat steaks'. Synthetic 'ham', 'beef', 'pork' and 'bacon' have been available in the USA since before 1960. Such are also made available as dehydrated meat 'bits' for use in soups, stews, sausages and other comminuted foods for the general market, including institutional feeding (Coleman and Creswick, 1966). They may have a protein content as high as 30 per cent and have only 1 per cent of fat (none of it being of animal origin). In frankfurter sausages the replacement of 4 per cent of meat protein with that from soya has been found to reduce the cost of the final product by 33 per cent; and the yield (based on the meat used) was increased by 40 per cent (Cook et al., 1969).

Because of the intrinsic cheapness of proteins from vegetables and microbial origin and (depending on the source) the absence of nutritional disadvantages, there is bound to be a great increase in their use as substitutes for expensive proteins of animal origin. Clearly there is also an increasing possibility that the recommended levels of substitution could be exceeded, and a concomitant need to establish means of quantitatively determining the origin of proteins in food products (Food Standards Committee, 1975). Although serological methods can be employed to distinguish and quantify native proteins, processing (especially heating) renders them ineffective. It is still possible to use gel electrophoresis, in conjunction with disaggregating agents such as β-mercaptoethanol and urea, to assess the proteins origin (Mattey et al., 1970). Beyond a certain intensity of processing, however, such means of identification become ineffective. On the other hand, if there were some component of meat which was robust to processing, which was characteristically present in myofibrillar proteins and which was absent from non-meat proteins, an accurate assessment of a products lean meat content could be made. 3-methylhistidine forms a standard component of myofibrillar proteins (Hardy et al., 1970). Being an amino acid it survives even severe processing and it appears to be absent from proteins of microbial or plant origin (Rangeley and Lawrie, 1976). The titre of 3-methylhistidine has been successfully employed, for example, to determine the content of meat protein in canned mixtures of beef and soya (Hibbert and Lawrie, 1972) and soup powders (Jones *et al.*, 1982).

Although the titre of protein-bound, connective tissue-free, 3-methylhistidine appears to be very similar between corresponding skeletal muscles of species as different as whales, rabbits and domestic meat animals, and between the muscles of the prime cuts of the carcass, it is low in smooth muscle (Rangeley and Lawrie, 1977). Moreover, Jones *et al.* (1985) demonstrated that it was also low in the *masseter/malaris* (cheek) muscles of cattle. The latter phenomenon reflects the fact that the myosin of such muscles contains only *ca.* 17 per cent of the 3-methylhistidine titre which is characteristic of the skeletal muscles of the bovine (White and Lawrie, 1985). The 3-methylhistidine content of the actins of bovine *masseter/malaris*, however, is similar to that in other skeletal muscles in this species and in other meat animals (Johnson *et al.*, 1986; Johnson and Lawrie, 1988). While this indicates that the 3-methylhistidine content of actin would be even more precise as an index of lean meat in processed foods than that of the contractile proteins overall (actomyosin), the latter index is useful as a predictor of the meat from prime cuts. It has also been suggested that actin could be unequivocally quantified by its electrospray mass spectrometry pattern (Taylor *et al.*, 1993b).

The authenticity of meat products, in terms of their meat content, and in relation to non-meat components and permitted or non-permitted offal, the species from which they were alleged to have been derived and the processing treatment received by the meat (e.g. freezing, ionizing irradiation, mechanical recovery, ageing), is still a matter of ongoing concern in national and international trade. The parameters analysed, and the modes of analysis used, in determining such authenticity, were reviewed by Hitchcock and Crimes (1985), Hargin (1996) and Lockley and Bardsley (2000) (cf. §4.3.1). They described the techniques now available (and varying in complexity and cost) which are based on the ubiquity of DNA in the nuclei and the mitochondria of most body cells, on the uniqueness of its sequence of bases in the individual and on its much superior robustness to processing in comparison to proteins. Moreover, it is possible to simplify the analysis of DNA synthesized during amplification by polymerase chain reactions, by incorporating a fluorescent group into the primer oligonucleotides (this can eliminate the need for an electrophoretic separation step in the analysis). These DNA-based techniques will continue to be developed and, in all but exceptional situations, make protein-based methods obsolete.

Recently, monoclonal antibodies against the thermostable muscle protein, troponin I, have been successfully employed to differentiate species and to distinguish muscular tissue from gelatin and the proteins of blood and milk (Chen *et al.*, 2002).

For some consumers (such as vegetarians) an organoleptically attractive, nutritious product, entirely free from meat, would be a desideratum. One commercially successful commodity of this nature is manufactured from the RNA-reduced cells of a *Fusarium* species (Schwabe) in a continuous fermentation process. The harvested mycoprotein has a fibrous texture and a moisture content similar to that of meat.

12.3   Upgrading abattoir waste

In view of the frequent drawbacks of low acceptability, absence of organoleptic quality and high cost in meat-like products prefabricated from vegetable or bacte-

Table 12.3 Typical protein concentrations in under-utilized tissues of meat animal (after Young and Lawrie, 1974)

Tissue	% Protein
Lung (bovine and ovine)	16–17
Lung (porcine)	14–15
Stomach (ovine)	12–14
Stomach (porcine)	14–15
Rumen (bovine)	10–13
Reticulum and omasum (bovine)	9–10
Abomasum (bovine)	7–9
Blood plasma	7–8

rial proteins, it would be highly desirable to reassess the potential for making edible and attractive foods from the substantial amounts of slaughterhouse protein which are currently wasted. Much of it is of high biological value and it would seem both economic and more rational nutritionally to investigate this possibility before relying too heavily on the unconventional. The subject has been comprehensively reviewed by Young (1980).

Young (1980) calculated that, in a typical carcass of an Aberdeen Angus steer, about 5 kg of protein is present in currently non-utilized offal – stomachs, lungs and blood. When this is multiplied by the 3 million cattle units consumed in the UK annually, and to it is added corresponding data from lambs, sheep and pigs, and the 15 per cent of total bone collagen which could be used for human food (Jobling and Hughes, 1977), a total of 43 million kg of protein per year is obtained. This is equivalent to 215 million kg of lean meat, i.e. about one-eighth of the annual consumption. Typical values for the protein content of offal are given in Table 12.3.

A general prerequisite for the upgrading of offal proteins is their recovery from the source. When proteins are in suspension or solution they can be recovered by flocculation, ion exchange chromatography or ultra-filtration (Denmead et al., 1973–74). With blood itself, in which protein concentration is relatively high, heat coagulation can be applied, but the products are unaesthetic and there are considerable losses of nutritive value and of functionality. In 1976, Dill devised a procedure for obtaining both serum and globin proteins from blood as white powders, colour having been removed by acetone extraction of the haem component, but the functionality of the products was relatively low.

The functionality of blood plasma proteins can be largely retained by partial freeze-drying (Young and Lawrie, 1974) or by ion exchange chromatography (Howell, 1981). The latter process yields three major protein fractions which possess different attributes when used as egg albumen substitutes in cake mixes, according to the time and temperature of the baking process. Moreover, there are species differences, porcine plasma proteins being more useful at temperatures between 80 and 90 °C and bovine plasma proteins above 95 °C (Howell and Lawrie, 1981).

In the past, when attempts were made to recover solid slaughter wastes for feeds or fertilizers, or to recover fat, neither the nutritive value nor the functional properties of the proteins were considered. Azeotropic distillation, controlled enzymic hydrolysis and extraction by alkali or anionic detergents can now be employed to recover protein from such sources with minimum loss.

Proteins can be recovered in good yield from bovine, ovine and porcine lungs and stomachs (Young and Lawrie, 1974), using alkaline extraction over 8 hours at 0 °C and reprecipitation with acid. Comparable recoveries can be obtained by extracting at 20 °C over 2 h (Swingler and Lawrie, 1979). The yield can be increased considerably if the temperature of alkaline extraction is raised to 60 °C, this being partly due to enhanced solubilization of collagen; but an increased amount of lysino-alanine is then detected. The latter problem – the nutritional significance of which is yet unproven – can be avoided by extracting proteins from offal by sodium dodecyl sulphate (SDS) (Gault, 1978). The extracted SDS–protein complex is precipitated by ferric chloride, and the SDS removed with acetone or methanol and potassium chloride (Lundgren, 1945; Ellison et al., 1980). The proteins thus recovered are bland and of improved functionality.

Insofar as there is a growing market for protein hydrolysates (as flavour enhancers, functional ingredients or merely as nutritional additives to foods of low protein quality), it is of interest that these can be readily produced by controlled enzymic treatment (Webster et al., 1982).

As in the case of proteins from non-meat sources (cf. § 12.2), offal proteins can be texturized to resemble the fibres of lean meat by fibre spinning or to simulate meat pieces by high-pressure, high-temperature extrusion.

Fibres containing 17–18 per cent protein can be spun from such sources. Those spun from stomachs and lungs tend to have less mechanical strength than those from blood plasma, but since they have higher contents of isoleucine and methionine (which are deficient in the latter), both textural and nutritional benefits can be achieved by spinning fibres from mixed sources (Young and Lawrie, 1975; Swingler et al., 1978).

There are advantages in spinning fibres from mixtures of offal protein and polysaccharides such as alginate (Imeson et al., 1979) since these can confer improved textural properties when used as a partial substituent for lean meat in sausages (Rusig, 1979). Moreover, the alginates can confer binding without the use of salt.

Not surprisingly, spinning is associated with a significant reduction in the number of micro-organisms originally present in the offal from which the fibres are prepared. Spun protein fibres were found to be microbiological sterile for at least 3 years at 0 °C (Swingler et al., 1979).

Although cereal, soya and other plant proteins can be thermoplastically extruded, this is not possible with proteins recovered from offal when these are used alone (Mittal, 1981). They can be extruded successfully, however, when incorporated at up to 35 per cent in mixtures with soya protein, and, in such products, the offal protein does not require to be purified as much as when fibres are to be spun from them. Proteins extracted from offal by anionic detergents are much less suitable for thermoplastic extrusion than those extracted by alkali (Mittal and Lawrie, 1984).

Lipid–protein interactions are important determinants of the structure of offal protein. Thus, their suitability for thermoplastic extrusion (in association with a carrier such as soya protein) is greatly altered by the amount of residual lipid in the extracted protein. The controlled removal of lipid from the latter by the use of fat solvents of different polarity causes systematic and extensive changes in the texture of the products subsequently prepared from the proteins by thermoplastic extrusion (Areâs, 1983; Areâs and Lawrie, 1984).

Fresh bone, virtually free from adhering meat, has become commercially available, under hygienic conditions, through the operation of new mechanical deboning processes (cf. § 12.1), and recent recovery methods can fractionate such bone into protein, lipid and calcium phosphate of a quality suitable for use as food ingredients (e.g. edible bone collagen) (Jobling and Jobling, 1983). Typically, pork and beef bones yield *ca.* 21 per cent protein, 12–15 per cent lipid and 21–29 per cent ash.

Since the proteins which can be readily extracted from stomachs and lungs are derived mainly from smooth muscle in these tissues, and since the latter closely resembles striated muscle in composition (§ 4.1.1), fibres from such sources would approximate closely to lean meat, notwithstanding their derivation from prohibited offal. They could thus pose both analytical and legislative problems.

Whilst the use of prefabricated meats from animal or vegetable sources in convenience foods, for special diets and as complementary feeding in underdeveloped areas, may be expected to increase steadily, most authorities agree that they will not displace the demand for carcass meat in the foreseeable future. This is not only a question of eating quality. It reflects recognition that, notwithstanding their lower capacity for producing protein than vegetables or bacteria, meat animals represent the only economically feasible means of utilizing the protein in plant sources growing on poor ground.

And it must be recognized that a great proportion of the world's land surface is only fit for grazing and not for cultivation. As Blaxter (1968) has emphasized, ruminants in particular can convert the fibrous portions of plants, which cannot be used directly for human food, into high-quality protein. The self-contained fermentation system of ruminants can produce desirable protein almost as efficiently as industrial fermentation plants can produce protein of a type which has yet to be established as organoleptically acceptable to the general consumer.

Bibliography

ABRAHAM E. P., CHAIN, E. B., FLETCHER, C. M., FLOREY, H. W., GARDNER, A. D., HEATLEY, N. G. and JENNINGS, M. A. (1941) *Lancet*, **ii**, 177.
ABRAHAMSON, H. (1949) *J. Biol. Chem.* **178**, 179.
ACEVADO, I. and ROMAT, A. (1929) *Bol. Min. Agric., Buenos Aires* **28**, 221.
ADACHI, R. R., SHEFFER, L. and SPECTOR, H. (1958) *Food Res.* **23**, 401.
ADAMS, J. M., JEPPESEN, P. G. N., SANGER, F. and BARRELL, B. G. (1969) *Nature, Lond.* **223**, 1009.
ADAMS, R. D., DENNY-BROWN, D. and PEARSON, C. M. (1962) *Diseases of Muscle: A Study in Pathology*, 2nd ed, Henry Kimpton, London.
ADDIS, P. B. (1969) *Proc. 22nd Ann. Recip. Meat Conf.*, Pomona, Calif., p. 151.
ADDIS, P. B. (1986) *Fd. Chem. Toxic.* **24**, 1021.
AHN, D. U., JO, C. and OLSON, D. G. (2000) *Meat Sci.* **54**, 209.
AHN, D. U., NAM, K. C., DU, M. and JO, C. (2001) *Meat Sci.* **57**, 419.
AITKEN, A., CASEY, J. C., PENNY, I. F. and VOYLE, C. A. (1962) *J. Sci. Fd. Agric.* **13**, 439.
AKERS, J. M. (1969) *Food Manuf.* **44**, No. 1, 22.
ALASNIER, C., REMIGNO, H and GANDEMER, G. (1996) *Meat Sci.* **43**, 213.
ALISON, M. R., POULSOM, R., FORBES, S. and WRIGHT, N. A. (2002) *J. Path.* **197**, 419.
ALLEN, D. M. (1974) *Proc. 27th Ann. Recip. Meat Conf.*, Nat. Livestock and Meat Bd., Chicago, p. 56.
ALLEN, E. (1968) *Proc. 21st Ann. Recip. Meat Conf.*, Athens, Georgia, p. 306.
ALLEN, E., BRAY, R. W. and CASSENS, R. G. (1967a) *J. Food Sci.* **32**, 26.
ALLEN, E., CASSENS, R. G. and BRAY, R. W. (1967b) *J. Anim. Sci.* **26**, 36.
ALLEN, L. M. and PATTERSON, D. S. R. (1971) In *2nd Symposium Condition and Meat Quality in Pigs* (Animal Res. Inst. Zeist), p. 90.
ALLEN, W. M., HERBERT, C. N. and SMITH, L. P. (1974) *Vet. Rec.* **94**, 212.
ALONSO, M. D., LOMAKO, T., LOMAKO, W. and WHELAN, W. J. (1995) *FASEB J.* **9**, 1126.
AMOR-FREMPONG, I. E., NUTE, G. R., WHITTINGTON, F. W., and WOOD, J. D. (1997) *Meat Sci.* **47**, 63.
AMOR-FREMPONG, I. E., NUTE, G. R., WOOD, J. D., WHITTINGTON, F. W. and WEST, A. (1998) *Meat Sci.* **50**, 139.
ANDERSON, A. W. (1955) *Munic. Engng.* **40**, 1119.
ANDERSON, A. W., NORDEN, H. C., CAIN, R. F., PARRISH, G. and DUGGAN, D. (1956) *Food Tech.* **10**, 575.
ANDO, N. and NAGATA, Y. (1970) *Proc. 16th Meeting European Meat Res. Workers, Varna*, p. 859.
ANDREWS, F. N., BEESON, W. M. and JOHNSON, F. D. (1954) *J. Anim. Sci.* **13**, 99.
ANDROSS, M. (1949) *Brit. J. Nutr.* **3**, 396.
ANIL, M. H., MCKINSTRY, J. L., WOTTON, S. B. and GREGORY, N. G. (1995) *Meat Sci.* **41**, 101.
ANIL, M. H., WHITTINGTON, P. E. and MCKINSTRY, J. L. (2000) *Meat Sci.* **55**, 313.
ANON. (1816) 'The Experienced Butcher', Cited by Rixson (2000) *loc. cit.*
ANON. (1938) *C.S.I.R.O., Aust., Sect. Food Pres. Circ.*, No. 2, p. 72.
ANON. (1940) *Analyst*, **65**, 257.
ANON. (1944) *U.S. Dept. Agric. Circ.*, No. 706.
ANON. (1950) *The Australian Environment*, 2nd ed., C.S.I.R.O., Melbourne.
ANON. (1952) *Analyst*, **77**, 543.

ANON. (1955) *Wld. Hlth. Org. Tech. Rept. Ser.*, No. 99.

ANON. (1957a) *Beretn. Akad. tek. Videns.*, *Kbh.*, No. 27.

ANON. (1957b) Meat Hygiene, *FAO Agric. Studies*, No. 34.

ANON. (1961) *Analyst* **86**, 557.

ANON. (1962) *A.R.C. and M.R.C. Report of the Joint Committee on Antibiotics in Animal Feeding*, H.M.S.O., London.

ANON. (1963) *Analyst* **88**, 422.

ANON. (1967) *Joint Survey of Pesticide Residues in Foodstuffs Sold in England and Wales* (Assoc. of Public Analysts), p. 22.

ANON. (1968) *Ann. Rept. Australian Meat Bd.*, p. 117.

ANON. (1971) *2nd Symposium Condition and Meat Quality in Pigs*. Animal Res. Inst. Zeist.

ANON. (1972a) Food Additives and Contaminants Committee, *Report on the Review of the Preservatives in Food Regulations*, 1962, H.M.S.O., London.

ANON. (1972b) *Ann. Rept. C.S.I.R.O.*, Meat Res Lab., pp. 18, 34.

ANON. (1973a) *Production Year Book*, F.A.O., Rome, **27**.

ANON. (1973b) *Meat Res. in C.S.I.R.O.*, p. 2.

ANON. (1974a) *Meat Res. Inst. Ann. Rept.* 1972–73, p. 11.

ANON. (1974b) *Meat Res. in C.S.I.R.O.*, p. 9.

ANON. (1977–79) Bienn. Rept. ARC Meat Research Inst., p. 58.

ANON. (1981–83a) Bienn. Rept. ARC Meat Res. Inst., p. 100. (H.M.S.O.: Lond.)

ANON. (1981–83b) Bienn. Rept. ARC Meat Res. Inst., p. 93. (H.M.S.O.: Lond.)

ANON. (1981) Wld. Hlth. Org., Techn. Rept. Ser. No. 659.

ANON. (1983–85) Bienn. Rept. A.F.R.C. Res. Inst., p. 77.

ANON. (1983) *Daily Telegraph*, *Lond.* 29th Febr.

ANON. (1986) *Analyst* **111**, 969.

ANON. (1987–88) Ann. Rept. Meat Ind. Res. Inst. NZ Inc., p. 8.

ANON. (1991) *Analyst*, **116**, 761.

ANON. (1993a) *FAO Production Yearbook*, 1993. Vol. 43.

ANON. (1993b) *Analyst* **118**, 1217.

ANON. (1993c) *Ann. Rept. Inst. Food Res.*, Reading, p. 16.

ANON. (1994–95) *Project Guide*, p. 127 (Meat Res. Corp., Sydney, Australia).

ANON. (1995) *Analyst* **120**, 1823.

ANON. (1996) *Analyst* **121**, 889.

ANON. (2001) *Chem. in Britain* **37** (9), 201.

ANON. (2003) *FAO Production Yearbook*, vol. 57.

ANON, M. C. and CALVELO, A. (1980) *Meat Sci.* **4**, 1.

ANSON, M. L. and MIRSKY, A. E. (1932–33) *J. Gen. Physiol.* **16**, 59.

ANSORENA, D., MENTEL, M. C., ROKKA, M., TALON, R., EEROLA, S., RIZZO, A., RAEMAEKERS, M. and DEMEYER, D. (2002a) *Meat Sci.* **61**, 141.

ANSORENA, D., ASTIASARAN, I. and BELLO, J. (2002b) In *Research Advances in the Quality of Meat and Meat Products*, Research Signpost, Ed. F. Toldra, p. 157.

ANTHONY, W. B. (1969) Animal Management Conf., Syracuse, N.Y.

APPERT, N. (1810) *The Art of Preserving Animal and Vegetable Substances for Many Years*, Patris & Cie, Paris.

ARAKAWA, N., GOLL, D. E. and TEMPLE, J. (1970) *J. Fd. Sci.* **35**, 703.

ARÊAS, J. A. G. (1983) Ph.D. Dissertation, Univ. Nottingham.

ARÊAS, J. A. G. and LAWRIE, R. A. (1984) *Meat Sci.* **11**, 275.

ARIHARA, K., CASSENS, R. G., GREASER, M. L., LUCHENSKY, J. J. and MOZDZIAK, P. E. (1995) *Meat Sci.* **39**, 205.

ARNOLD, N., WIERBICKI, E. and DEATHERAGE, F. E. (1956) *Food Tech.* **10**, 245.

ARNOLD, R. N., SCHELLER, K. K., ARP, S. C., WILLIAMS, S. N., BUEGE, D. R. and SCHAEFER, D. M. (1992) *J. Anim. Sci.* **70**, 3055.

ARROYO, P. T. and LILLARD, D. A. (1970) *J. Fd. Sci.* **35**, 769.

ASHMORE, C. R. and ROBINSON, D. W. (1969) *Proc. Soc. Exptl. Biol. Med.* **132**, 548.

ASSMAN, H., BIELENSTEIN, H., HOBBS, H. and JEDELOH, B. (1933) *Deutsch. med. Wchn.* **59**, 122.

ASTOLFI, A., DE GIOVANNI, C., LANDUZZI, L., NICOLETTI, G., RICCI, C., CROCI, S., SCOPECE, L., NANNI, P. and LOLLINI, P. L. (2001) *Gene* **274**, 139.

ATKINSON, J. L. and FOLLET, M. J. (1971) *Proc. 17th Meeting European Meat Res. Workers. Bristol.* p. 685.

ATKINSON, J. L. and FOLLET, M. J. (1973) *J. Fd. Technol.* **8**, 51.

ATKINSON, W. T. (1970) U.S. Pat. No. 3,488,770.

AURADE, F., PINSET, C., CHAFEY, A., GROS, F. and MONTARRAS, D. (1994) *Differentiation* **55**, 185.

AVERY, N. C., SIMS, T. J., WARKUP, C. and BAILEY, A. J. (1996) *Meat Sci.* **42**, 355.

AWAD, A., POWRIE, W. D. and FENNEMA, O. (1968) *J. Food Sci.* **33**, 227.

AYLWARD, F., COLEMAN, G. and HAISMAN, D. R. (1967) *Chem. Ind.* 1563.

BABIKER, S. A. (1984) *J. Arid Environment* **7**, 377.

BABIKER, S. A. and LAWRIE, R. A. (1983) *Meat Sci.* **8**, 1.

BACH, L. M. N. (1948) *Proc. Soc. Exp. Biol. Med.* **67**, 268.

BACHMANN, P. (1980) *Cell Tissue Res.* **206**, 431.

BACON, R. T., BELK, K. E., SOFOS, J. N., CLAYTON, R. P., REAGAN, J. O. and SMITH, G. C. (2000) *J. Fd. Prot.* **63**, 1080.

BACUS, J. (1984) *Food Technol.* **38**(6), 59.

BADAWAY, A. M., CAMPBELL, R. M., CUTHBERTSON, D. P. and FELL, B. L. (1957) *Nature, Lond.* **180**, 756.

BAGGER, S. V. (1926) *J. Path. Bact.* **29**, 225.

BAILEY, A. J. (1968) *Nature, Lond.* **160**, 447.

BAILEY, A. J. (1974) *Path. Biol.* **22**, 675.

BAILEY, A. J. (1988) *Proc. 34th Intl. Congr. Meat Sci. Technol.*, Brisbane, p. 152.

BAILEY, A. J. (1989) *Proc. 42nd Ann. Recip. Meat Conf.*, p. 127. (Natnl. Livestock & Meat Bd.: Chicago)

BAILEY, A. J., BENDALL, J. R. and RHODES, D. N. (1962) *Intl. J. Aappl. Radn. Isotopes* **13**, 131.

BAILEY, A. J. and LIGHT, N. D. (1989). *Connective Tissue in Meat and Meat Products* (Elsevier Appl. Sci.: Lond.)

BAILEY, A. J., PEACH, C. M. and FOWLER, L. J. (1970) *Biochem. J.* **117**, 819.

BAILEY, A. J., RESTALL, D. J., SIMS, T. J. and DUANCE, V. C. (1979) *J. Sci. Fd. Agric.* **30**, 203.

BAILEY, A. J. and RHODES, D. N. (1964) *J. Sci. Fd. Agric.* **15**, 504.

BAILEY, A. J. and ROBINS, S. P. (1973a) In *Biology of the Fibroblast* (Eds. E. KULONEN and J. PIKKARAINEN), Acad. Press, New York, p. 385.

BAILEY, A. J. and ROBINS, S. P. (1973b) *Matrix Biol.* **1**, 130.

BAILEY, A. J. and ROBINS, S. P. (1976) *Sci. Prog., Oxf.* **63**, 421.

BAILEY, A. J. and SIMS, T. J. (1977) *J. Sci. Fd. Agric.* **28**, 565.

BAILEY, C. (1972) ARC Meat Res. Inst., Memo. No. 19.

BAILEY, C. and COX, R. P. (1976) *The Chilling of Beef Carcases*, Inst. Refrig. London.

BAILEY, C., JAMES, S. T., KITCHELL, A. G. and HUDSON, W. R. (1974) *J. Sc. Fd. Agric.* **25**, 81.

BAILEY, K. (1946) *Nature, Lond.* **157**, 368.

BAILEY, K. (1954) *The Proteins*, vol. II Part B (Eds. H. NEURATH and K. BAILEY), p. 951, Academic Press, New York.

BAILEY, M. E. and KIM, M. K (1974) *Proc. 20th Meeting Europ. Meat Res. Workers*, Dublin, p. 35.

BAILEY, M. E. and SWAN, J. W. (1973) *Proc. Meat Ind. Res. Conf.*, Amer. Meat Inst., Chicago, p. 29.

BAIN, N., HODGINS, W. and SHEWAN, J. M. (1958) *2nd Intl. Symp. Fd. Microbiol. Camb.* (Ed. B. P. EDDY), p. 1, London, H.M.S.O.

BAINES, D. A. and MLOTKIEWICZ (1984) In *Recent Advances in the Chemistry of Meat* (Ed. A. J. BAILEY), p. 119 (Roy. Soc. Chem.: Lond.)

BAIRD, D. M., NALBANDOV, A. V. and NORTON, H. W. (1952) *J. Anim. Sci.* **11**, 292.

BAIRD-PARKER, A. C. (1987) In *Elimination of Pathogenic Organisms from Meat and Poultry* (Ed. F. M. J. SMULDERS), p. 149 (Elsevier Appl. Sci., Amsterdam)

BAKER, L. C. (1942) *Chem. Ind.* **41**, 458.

BAKER, M. L., BLUNN, C. T. and PLUM, M. (1951) *J. Hered.* **42**, 141.

BALABUKH, A. H. and LYASKOVSKAYA, Y. N. (1978) *Myasn. Ind. S.S.R.* (3), 33.

BALDWIN, E. (1967) *Dynamic Aspects of Biochemistry*, 5th ed., Cambridge Univ. Press.

BALL, C. O. (1938) *Food Res.* **3**, 13.

BALL, C. O. (1959) *Food Tech.* **13**, 193.

BALLS, A. K. (1941) *U.S. Dept. Agric. Circ.*, No. 631.

BANFIELD, F. H. and CALLOW, E. H. (1934) *Ann. Rept. Fd. Invest. Bd., Lond.*, p. 72.

BANFIELD, F. H. and CALLOW, E. H. (1935) *J. Soc. Chem. Ind.* **54**, 413T.

BARANYI, J. and ROBERTS, T. A. (1993) *Mathematical and statistical tools in predicting bacterial growth*, IMA Conference on Modelling for Food Safety, Queen's Univ., Belfast, 14–16th. April, 1993.

BARBER, R. S., BRAUDE, R. and MITCHELL, K. G. (1955) *J. Agric. Sci.* **46**, 97.

BARDSLEY, A. J. and TAYLOR, A. MCM. (1962) *B.F.M.I.R.A. Res. Repts.*, No. 99.

BARLOW, D. J., EDWARDS, M. S. and THORNTON, J. M. (1986) *Nature, Lond.* **332**, 747.

BARLOW, J. and KITCHELL, A. G. (1966) *J. Appl. Bact.* **29**, 185.

BARNARD, R. D. (1937) *J. Biol. Chem.* **120**, 177.

BARNES, E. M. (1956) *Food Manuf.* **31**, 508.

BARNES, E. M. (1957) *J. Roy. Soc. Health* **77**, 446.

BARNES, E. M. (1958) *Brit. Vet. J.* **144**, 333.

BARNES, E. M. and INGRAM, M. (1956) *J. Appl. Bact.* **19**, 117.

BARNES, E. M. and KAMPELMACHER, E. H. (1966) *Proc. 1st Intl. Congr. Indust. Aliment. Agricol.*, Abidjan, p. 1093.

BARNES, M. J., CONSTABLE, B. J. and KODICEK, E. (1969) *Biochem. J.* **113**, 387.

BARNES, R. H., LUNDBERG, W. O., HANSON, H. T. and BURN, G. O. (1943) *J. Biol. Chem.* **149**, 313.

BARRETT, A. J. (1977) In *Proteinases in Mammalian Cells and Tissues* (Ed. A. J. BARRETT), p. 1, North Holland Publish. Co., Amsterdam.

BARRON, E. S. G. and DICKMAN, S (1949) *J. Gen. Physiol.* **32**, 595.

BARRON, E. S. G. and LYMAN, C. M. (1938) *J. Biol. Chem.* **123**, 229.

BARTON-GADE, P. (2002) *Meat Sci.* **62**, 353.

BARTON-GADE, P. and CHRISTENSEN, L. (1998) *Meat Sci.* **48**, 237.

BASKIN, R. J. and DEAMER, D. W. (1969) *J. Cell. Biol.* **43**, 610.

BASS, J. J. and CLARK, R. G. (1989) In *Meat Production and Processing* (Eds. R. W. PURCHAS, B. W. BUTLER-HOGG and A. S. DAVIES), p. 103. Occas. Publicn. No. 11 (N.Z. Soc. Animal Production).

BATE-SMITH, E. C. (1937a) *Ann. Reps. Fd. Invest. Bd., Lond.*, p. 15.

BATE-SMITH, E. C. (1937b) *Proc. Roy. Soc.* **B 124**, 136.

BATE-SMITH, E. C. (1938) *J. Physiol.* **92**, 336.

BATE-SMITH, E. C. (1948) *J. Soc. Chem. Ind.* **67**, 83.

BATE-SMITH, E. C. (1957) *J. Linn. Soc. (Bot.)* **55**, 669.

BATE-SMITH, E. C. (1961) *New Scientist* **11**, 329.

BATE-SMITH, E. C. and BENDALL, J. R. (1947) *J. Physiol.* **106**, 177.

BATE-SMITH, E. C. and BENDALL, J. R. (1949) *J. Physiol.* **110**, 47.

BATZER, O. F. and DOTY, D. M. (1955) *J. Agric. Food Chem.* **3**, 64.

BATZER, O. F., SANTORO, A. T. and LANDMANN, W. A. (1962) *J. Agric. Food Chem.* **10**, 94.

BATZER, O. F., SANTORO, A. T., TAN, M. C., LANDMANN, W. A. and SCHWEIGERT, B. S. (1960) *J. Agric. Food Chem.* **8**, 498.

BATZER, O. F., SLIWINSKI, R. A., CHANG, L., PIH, K., FOX, J. B. JR., DOTY, D. M., PEARSON, A. M. and SPOONER, M. R. (1959a) *Food Tech.* **13**, 501.

BATZER, O. F., SRIBNEY, M., DOTY, D. M. and SCHWEIGERT, B. S. (1959b) *J. Agric. Food Chem.* **5**, 700.

BAUERNFEIND, J. C. (1953) *Adv. Fd. Res.* **4**, 359.

BEAR, R. S. (1952) *Adv. Prot. Chem.* **7**, 69.

BEARD, F. J. (1924) cited hy WANDERSTOCK, J. F. and MILLER, J. I. (1948) *Food Res.* **13**, 291.

BEATTIE, W. A. (1956) *A Survey of the Beef Cattle Industry of Australia*, Bulletin No. 278, C.S.I.R.O., Melbourne.

BEATTY, C. H. and BOCEK, R. M. (1970) In *The Physiology and Biochemistry of Muscle as a Food*, vol. II, p. 155 (Eds. E. J. BRISKEY, R. G. CASSENS and B. B. MARSH), Univ. Wisconsin Press.

BECHTIL, P. (1979) *J. Biol. Chem.* **254**, 1755.

BEECHER, G. R, KASTENSCHMIDT, L. L., CASSENS, R. G., HOEKSTRA, W. G. and BRISKEY, E. J. (1968) *J. Food. Sci.* **33**, 84.

BEERMAN, D. H., WANG, S. V., ARMBRUSTER, G., DICKSON, H. W., PICKER, E. L. and LARSON, J. G. (1990) *Proc. 42nd Ann. Recip. Meat Conf.*, p. 54 (Natl. Livestock & Meat Bd.: Chicago).

BEHNKE, R. J., FENNEMA, O. and CASSENS, R. G. (1973) *J. Agric. Food Chem.* **21**, 5.

BEJERHOLM, C. and BARTON-GADE, P. (1986) *Proc. 32nd Meeting Meat Res. Workers*, Ghent. Vol. ll, p. 389.

BELEW, J. B., BROOKS, J. C., MCKENNA, D. R. and SAVELL, J. W. (2003) *Meat Sci.* **64**, 507.

BELKIN, A. M. and STEPP, M. A. (2000) *Microsc. Res. Techniques* **51**, 280.

BELL, R. G., PENNEY, N., GILBERT, K. V., MOORHEAD, S. M. and SCOTT, S. M. (1996) *Meat Sci.* **42**, 371.

BELLATTI, M. and PAROLARI, G. (1982) *Meat Sci.* **7**, 59.

BENDALL, J. R. (1946) *J. Soc. Chem. Ind.* **65**, 226.

BENDALL, J. R. (1947) *Proc. Roy. Soc.* **B 134**, 272.

BENDALL, J. K. (1951) *J. Physiol.* **114**, 71.

BENDALL, J. R. (1954) *J. Sci. Fd. Agric.* **5**, 468.

BENDALL, J. R. (1960) In *The Structure and Function of Muscle*, Vol. III, Ed. G. H. BOURNE, p. 227 (Academic Press: New York).

BENDALL, J. R. (1962) *Recent Advances in Food Science*, vol. 1, p. 58, Butterworths, London.

BENDALL, J. R. (1963) *Proc. Meat Tenderness Symp.*, p. 33, Campbell Soup Co., Camden, NJ.

BENDALL, J. R. (1966) *J. Sci. Fd. Agric.* **17**, 334.

BENDALL, J. R. (1967) *J. Sci. Fd. Agric.* **18**, 553.

BENDALL, J. R. (1969) *Muscles, Molecules and Movement*, p. 22, Heinemann Educ. Books Ltd., London.

BENDALL, J. R. (1973) *The Structure and Function of Muscle*, vol. 12 (Ed. G. H. BOURNE), p. 244, Academic Press, New York.

BENDALL, J. R. (1975) *J. Sci. Fd. Agric.* **26**, 55.

BENDALL, J. R. (1976) *J. Sci. Fd. Agric.* **27**, 819.

BENDALL, J. R. (1977) *Technol. mesa* (Yugl.), **18**, 34.

BENDALL, J. R. (1978a) *Proc. 24th Europ. Meeting Meat Res. Workers*, Kulmbach. E 1. 2.

BENDALL, J. R. (1978b) *Meat Sci.* **2**, 91.

BENDALL, J. R. (1980) In *Developments in Meat Science*, Vol. I (Ed. R. A. LAWRIE), p. 37 (Applied Sci.: Lond.).

BENDALL, J. R. and DAVEY, C. L. (1957) *Biochim. Biophys. Acta*, **26**, 93.

BENDALL, J. R, HALLUND, O. and WISMER-PEDERSEN, J. (1963) *J. Food Sci.* **28**, 156.

BENDALL, J. R., KETTERIDGE, C. C. and GEORGE, A. R. (1976) *J. Sci. Ed. Agric.* **27**, 1123.

BENDALL, J. R. and LAWRIE, R. A. (1962) *J. Comp. Path.* **72**, 118.

BENDALL, J. R. and LAWRIE, R. A. (1964) *Anim. Breed Abs.* **32**, 1.

BENDALL, J. K. and MARSH, B. B. (1951) *Proc. 8th Int. Congr. Refrig., Lond.*, p. 351.

BENDALL, J. R. and RESTALL, D. J. (1983) *Meat Sci.* **8**, 93.

BENDALL, J. R. and SWATLAND, H. (1988) *Meat Sci.* **24**, 85.

BENDALL, J. R. and WISMER-PEDERSEN, J. (1962) *J. Food Sci.* **27**, 144.

BENDER, A. E. (1966) *J. Food Technol.* **1**, 261.

BENDER, A. E. (1975) In *Meat* (Eds. D. J. A. COLE and R. A. LAWRIE) Butterworth, London, p. 433.

BENDER, A. E. and BALLANCE, P. E. (1961) *J. Sci. Fd. Agric.* **12**, 683.

BENDER, A. E. and HUSAINI (1976) *J. Fd. Technol.* **11**, 499.

BENDER, A. E., WOOD, T. and PALGRAVE, J. A. (1958) *J. Sci. Fd. Agric.* **9**, 812.

BENDER, A. E. and ZIA, M. (1976) *J. Fd. Technol.* **11**, 495.

BENDIXEN, C. (2005) *Meat Sci.* **71**, 138.

BENDIXEN, C., HEDEGAARD, J. and HORN, P. (2005) *Meat Sci.* **71**, 128.

BENEDICT, F. G. (1958) *Vital Energetics: A Study in Comparative Basal Metabolism* (Carnegie Inst.: Washington) Puhl. No. 503.

BENNETT, H. S. (1960) *The Structure and Function of Muscle*, vol. 1 (Ed. G. H. BOURNE), p. 137, Academic Press, New York.

BENNET, M., UAUY, R. and GRUNDY, S. M. (1987) *Amer. J. Path.* **126**, 103.

BERDAGUÉ, J. L., MONTEIL, P., MOMTEL, M. C. and TALON, R. (1993) *Meat Sci.* **35**, 275.

BERG, R. T. and BUTTERFIELD, R. M. (1975) In *Meat* (Eds. D. J. A. COLE and R. A. LAWRIE) Butterworths, London, p. 19.

BERG, R. T. and BUTTERFIELD, R. M. (1976) *New Concepts of Cattle Growth* (John Wiley: New York).

BERGE, P., CULIOLI, J. and OUALI, A. (1993) *Meat Sci.* **33**, 191.

BERGE, P., ERTBJERG, P., LARSEN, M. L., ASTRUC, T., VIGNON, X. and MØLLER, A. J. (2001) *Meat Sci.* **57**, 347.

BERGMANN, A. (1847) *Göttnger, Studien* **1**, 595, cited by N. C. WRIGHT (1954) *loc. cit.*

BERMAN, M. (1961) *J. Food Sci.* **26**, 422.

BERNARD, C. (1877) *Leçons sur la Diabète et le Glycogénèse animale*, p. 426, Baillière, Paris.

BERNOFSKY, C., FOX, J. B. JR. and SCHWEIGERT, B. S. (1959) *Food Res.* **24**, 339.

BERNOTHY, J. M. (1963) *Nature, Lond.* **200**, 86.

BERTRAM, H. C., ANDERSEN, H. J. and KARLSSON, A. H. (2001) *Meat Sci.* **57**, 125.

BERTRAM, H. C., SCHÄFER, A. F., ROSENVOLD, K. and ANDERSEN, H. J. (2004a) *Meat Sci.* **66**, 915.

BERTRAM, H. C., STØDKILDE-JØRGENSEN, H., KARLSSON, A. H. and ANDERSEN, H. J. (2002) *Meat Sci.* **62**, 113.

BERTRAM, H. C., WHITTAKER, A. K., SHORTHOSE, W. R., ANDERSEN, H. T. and KARLSSON, A. H. (2004b) *Meat Sci.* **66**, 301.

BETTS, A. O. (1961) *Vet. Rec.* **73**, 1349.

BEUK, J. F., CHORNOCK, F. W. and RICE, E. E. (1948) *J. Biol. Chem.* **175**, 291.

BEUK, J. F., CHORNOCK, F. W. and RICE, E. E. (1949) *J. Biol. Chem.* **180**, 1243.

BEUK, J. F., SAVICH, A. L, GOESER, P. A. and HOGAN, J. M. (1959) U.S. Pat. No. 2,903,362.

BICKERSTAFFE, R., BEKHIT, A. E. D., ROBERTSON, L. J., ROBERTS, N. and GEESINK, G. H. (2001) *Meat Sci.* **59**, 303.

BIGALKE, R. C. (1964) *New Scientist* **14**, 141.

BILLETER, R, HEIZMANN, C. W., REIST, V., HOWALD, H. and JENNY, E. (1982) *FEBS Lett.* **139**, 45.

BING, R. (1902) *Virchows Arch. Path. Anat.* **170**, 171, cited by ADAMS *et al.* (1962).

BIÖRCK, G. (1949) *Acta Med. Scand.* **133**, Sup. 226.

BIRD, J. W. C., SCHWARTZ, W. N. and SPAMER, A. M. (1977) *Acta Biol. Med Germ.* **36**, 1587.

BISHOP, M. D., KAPPES, S. M., KEELE, J. M. W., STONE, R. T., GUNDEN, C. L. F., HAWKINS, G. A., TOLDO, S. S., FRIES, R., GROZ, M. D., YOU, J. and BEATTIE, C. W. (1994) *Genetics* **136**, 231.

BISHOP, M. D., KOOHMARAIE, M., KILLEFER, J. and KAPPES, S. (1993) *J. Anim. Sci.* **71**, 2277.

BJERRE, J. (1956) *The Last Cannibals*, Michael Joseph, London.

BLACKMORE, D. K. and NEWHOOK, J. C. (1982) *Meat Sci.* **7**, 19.

BLAKESLEE, A. F. (1932) *Proc. Nat. Acad. Sci., U.S.A.* **18**, 120.

BLANCHAER, M. and VAN WIJHE, M. (1962) *Nature, Lond.* **193**, 877.

BLAXTER, K L. (1962) *Vits. Hormones* **20**, 633.

BLAXTER, K. L. (1968) *Science J.* **4**, 53.

BLAXTER, K. L. (1971–72) *Scot. Agric.* **51**, 225.

BLAXTER, K. L. and MCGILL, R. F. (1955) *Vet. Rev.* **1**, 91.

BLAXTER, K. L. and WOOD, W. A. (1952) *Brit. J. Nutr.* **6**, 144.

BLOMQUIST, S. M. (1957) *Food Manuf.* **32**, 227.

BLOMQUIST, S. M. (1958) *Food Manuf.* **33**, 491.

BLOMQUIST, S. M. (1959) *Food Manuf.* **34**, 21.

BOARDMAN, N. K. and ADAIR, G. S. (1956) *Nature, Lond.* **170**, 679.

BOCCARD, K. (1981) In *Developments in Meat Science*, Vol. II (Ed. R. A. LAWRIE), p. 1 (Applied Sci.: Lond.).

BOCCARD, R. and DUMONT, B. L. (1974) *Annls. Génét. Sell Anim.* **6**, 177.

BOCEK, R. M., BASINGER, G. M. and BEATTY, C. H. (1966) *Amer. J. Physiol.* **210**, 1108.

BODWELL, C. E. and MCLAIN, P. E. (1971) In *The Science of Meat and Meat Products* (2nd ed.) (Eds. J. E. PRICE and B. S. SCHWEIGERT), W. H. Freeman, San Francisco, p. 78.

BOGGS, D., MERKEL, R. A. and DOURMIT, M. E. (1998) 'Livestock and Carcasses' (Dubuque, Iowa: Kendall/ Hunt. Publ. Co.).

BOLEMAN, S. J., BOLEMAN, S. L., BIDNER, T. D., MCMILLIN, K. W. and MONLEZUM, C. J. (1995) *Meat Sci.* **39**, 35.

BOLTON, F. J., DAWKINS, H. C. and ROBERTSON, L. (1982) *J. Inf.* **4**, 243.

BONNEAU, M., KEMPSTER, A. J., CLAUS, R., CLAUDI-MAGNUSSEN, C., DIESTRE, A., TORNBERG, E., WALSTRA, P., CHEVILLON, P., WEILER, U. and COOK, G. L. (2000a) *Meat Sci.* **54**, 251.

BONNEAU, M., WALSTRA, P., CLAUDI-MAGNUSSEN, C., KEMPSTER, A. J., TORNBERG, E., FISCHER, K., DIESTRE, A., SIRET, F., CHEVILLON, P., CLAUS, R., DIJKSTERHUIS, G., PUNTER, P., MATHEWS, K. R., AGERHEM, H., BEAGUE, M. P., OLIVER, M. A., GISPERT, M., WEILER, U., VON SETH, G., LEASK, H., FONT I FURNOLS, M., HOMER, D. B. and COOK, G. L. (2000b) *Meat Sci.* **54**, 285.

BOREK, E. and WAELSCH, H. (1951) *J. Biol. Chem.* **190**, 191.

BORTON, R. J., BRATZLER, L. J. and PRICE, J. F. (1970) *J. Fd. Sci.* **35**, 783.

BOUCEK, R. J, NOBLE, N. L. and MARKS, A. (1961) *J. Gerontol.* **5**, 150.

BOULEY, C., CHAMBON, C., DESMET, S., HOCQUETTE, J. H. and PICARD, B. (2005) *Proteomics*, **5**, 450.

BOULEY, M. (1874) *C.R. Acad Sci. Fr.* **79**, 739.

BOUTON, P. E., BROWN, A. D. and HOWARD, A. (1954) *C.S.I.R.O. Food Pres. Quart.* **14**, 62.

BOUTON, P. E., CARROL, F. D., FISHER, A. L., HARRIS, P. V. and SHORTHOSE, W. R. (1973b) *J. Fd. Sci.* **38**, 816.

BOUTON, P. E., CARROL, E. D., HARRIS, P. V. and SHORTHOSE, W. R. (1973c) *J. Fd. Sci.* **38**, 401.

BOUTON, P. E., ELLIS, R. W., HARRIS, P. V. and SHORTHOSE, W. R. (1976) *Meat Res. C.S.I.R.O.*, 42.

BOUTON, P. E., FISHER, A. L., HARRIS, P. V. and BAXTER, R. I. (1973a) *J. Fd. Technol.* **8**, 39.

BOUTON, P. E., FORD, A. L., HARRIS, P. V., MACFARLANE, J. J. and O'SHEA, J. M. (1977a) *J. Fd. Sci.* **42**, 132.

BOUTON, P. E., FORD, A. L., HARRIS, P. V. and SHAW, F. D. (1980a) *Meat Sci.* **4**, 145.

BOUTON, P. E., FORD, A. L., HARRIS, P. V., SHORTHOSE, W. R., RATCLIFF, D. and MORGAN, J. H. L. (1978a) *Meat Sci.* **2**, 301.

BOUTON, P. E., HARRIS, P. V., MACFARLANE, J. J. and O SHEA, J. M. (1977b) *Meat Sci.* **1**, 307.

BOUTON, P. E., HARRIS, P. V., RATCLIFF, D. and ROBERTS, D. W. (1978b) *J. Fd. Sci.* **43**, 1038.

BOUTON, P. E., HARRIS, P. V. and SHORTHOSE, W. R. (1975) *J. Text. Stud.* **6**, 297.

BOUTON, P. E., HARRIS, P. E., SHORTHOSE, W. R. and ELLIS, R. W. (1978c) *Meat Sci.* **2**, 161.

BOUTON, P. E., HARRIS, P. V., SHORTHOSE, W. R. and SMITH, M. G. (1974) *J. Fd. Technol.* **9**, 31.

BOUTON, P. E. and HOWARD, A. (1956) *C.S.I.R.O. Food Pres. Quart.* **16**, 50.

BOUTON, P. E. and HOWARD, A. (1960) *Proc. 6th Meeting European Meat Res. Workers, Utrecht.* Pap. No. 24.

BOUTON, P. E., HOWARD, A. and LAWRIE, R. A. (1957) *Spec. Rept. Fd. Invest. Bd., Lond.*, No. 66.

BOUTON, P. E., HOWARD, A. and LAWRIE, R. A. (1958) *Spec. Rept. Fd. Invest. Bd., Lond.*, No. 67.

BOUTON, P. E. and SHORTHOSE, W. R. (1969) *Proc. 15th Meeting European Meat Res. Workers, Helsinki*, p. 78.

BOUTON, P. E., WESTE, R. R. and SHAW, F. D. (1980b) *J. Fd. Sci.* **45**, 148.

BOVARD, K. P. and HAZEL, L. N. (1963) *J. Anim. Sci.* **22**, 188.

BOWES, J. H., ELLIOT, R. G. and MOSS, J. A. (1957) *J. Soc. Leather Trades Chemists* **41**, 249.

BOWLING, R. and CLAYTON, R. P. (1992) U.S. Pat. No. 5,149, 295.

BOYER, R. A. (1954) U.S. Pat. No. 2,560,621.

BOYLE, M. P. (1987) *Proc. 39th Ann. Recip. Meat Conf.*, p. 7 (Natnl. Livestock 8c Meat Bd.: Chicago).

BRADEN, A. W. H., SOUTHCOTT, W. H. and MOULE, G. R. (1964) *Aust. J. Agric. Res.* **15**, 142.

BRAMBLETT, V. D., HOSTETLER, R. L, VAIL, G. E. and DRAUDT, H. N. (1959) *Food Tech.* **13**, 707.

BRAMBLETT, V. D. and VAIL, G. E. (1964) *Food Tech.* **18**, 123.

BRANDT, N. R., CASWELL, A. H., BRANDT, T., BREW, K. and MELLGREN, R. L. (1992) *Membrane Biol.* **127**, 25.

BRANNAN, R. G. and DECKER, E. A. (2002) *Meat Sci.* **62**, 229.

BRASCH, A. and HUBER, W. (1947) *Science* **105**, 112.

BRAUDE, R. (1950) *Brit. J. Nutr.* **4**, 138.

BRAUDE, R. (1967) *Proc. C.I.C.R.A. Conf., Dublin*, p. 21.

BRAZEAU, P., VALE, W., BURGUS, R., LING, N., BUTCHER, M., RIVIER, J. and GUILLEMIN, R. (1973) *Science*, **179**, 177.

BRENNAND, C. P. and LINDSAY, R. C. (1992) *Meat Sci.* **31**, 411.

BRENNER, S. (2002) *Nature, Lond.* **425**, 285.

BREWER, S. (2004) *Meat Sci.* **68**, 1.

BRIGELOW, W. D. and ESTY, J. R. (1920) *J. Infect. Dis.* **27**, 602.

BRISKEY, E. J. (1964) *Adv. Food Res.* **13**, 90.

BRISKEY, E. J. (1969) In *Recent Points of View on the Condition and Meat Quality of Pigs for Slaughter*, p. 41 (Eds. W. SYBESMA, P. G. VAN DER WALS and P. WALSTRON), Res. Inst. Animal Husbandry, Zeist.

BRISKEY, E. J. and WISMER-PEDERSEN, J. (1961) *J. Food Sci.* **26**, 297.

BRISSEY, G. E. (1952) U.S. Pat. No. 2,596,067.

BROCKMANN, M. C. (1969) *Proc. 15th Meeting European Meat Res. Workers, Helsinki*, p. 468.

BROCKMANN, M. C. (1970) *Food Technol.*, Champaign, **24**, 896.

BRODMANN, P. D. and MOOR, D. (2003) *Meat Sci.* **65**, 599.

BRODY, S. (1927) *Mo. Agric. Expt. Sta. Res. Bull.*, Nos. 97, 98, 101, 104.

BRØNDUM, J., BYRNE, D. V., BAK, L. S., BERTELSEN, G. and ENGELSEN, S. B. (2000) *Meat Sci.* **54**, 83.

BROOKS, J. (1929a) *Ann. Rept. Fd. Invest. Bd. Lond.*, p. 29.

BROOKS, J. (1929b) *Biochem. J.* **23**, 1391.

BROOKS, J. (1930) *Biochem. J.* **24**, 1379.

BROOKS, J. (1931) *Ann. Rept. Fd. Invest. Bd., Lond.*, p. 36.

BROOKS, J. (1933) *J. Soc. Chem. Ind.* **52**, 17T.

BROOKS, J. (1935) *Proc. Roy. Soc.* **B 118**, 560.

BROOKS, J. (1937) *Proc. Roy. Soc.* **B 123**, 368.

BROOKS, J. (1938) *Food Res.* **3**, 75.

BROOKS, J., HAINES, R. B., MORAN, T. and PACE, J. (1940) *Spec. Rept. Fd. Invest. Bd., London*, No. 49.

BROOKS, J. C. and SAVELL, J. W. (2004) *Meat Sci.* **67**, 329.

BROSTROM, C. O., HUNKELWR, E. L. and KREBS, E G. (1971) *J. Biol. Chem.* **246**, 1961.

BROTHWELL, D. and BROTHWELL, P. (1969) *Food in Antiquity*, p. 195 (Thames & Hudson: Lond.)

BROWN, A. D., COOTE, G. G. and MEANEY, M. F. (1957) *J. Appl. Bact.* **20**, 75.

BROWN, H. (2000) *HACCP in the Meat Industry* (Woodhead Publ. Cambridge).

BROWN, R. H., BLASER, R. E. and FONTENOT, J. P. (1963) *J. Anim. Sci.* **22**, 1038.

BROWN, S. N., BEVIS, E. A. and WARRISS, P. D. (1990) *Meat Sci.* **27**, 249.

BROWN, T., CHOUROUZIDIS, K. N. and GIGIEL, A. J. (1993) *Meat Sci.* **34**, 311.

BROWN, T., GIGIEL, A. J., VERONICA, M., SWAIN, L. and HIGGINS, J. A. (1988) *Meat Sci.* **22**, 173.

BRUMBY, P. J. (1959) *N.Z. J. Agric. Res.* **2**, 683.

BRYCE-JONES, K. (1962) *Inst. Meat Bull.*, April, p. 2.

BRYCE-JONES, K. (1969) *Inst. Meat Bull.*, No. 65, p. 3.

BRYCE-JONES, K., HOUSTON, T. W. and HARRIES, J. M. (1963) *J. Sci. Fd. Agric.* **14**, 637.

BRYCE-JONES, K., HOUSTON, T. W. and HARRIES, J. M. (1964) *J. Sci. Fd. Agric.* **16**, 790.

BRYNJOLFSSON, A. (1980) *Proc. 26th Meeting Eur. Meat Res. Workers, Colorado Springs*, **1**, 172.

BRYNKO, C. and SMITHIES, W. R. (1956) *Food in Canada* **16** (10), 26.

BUCHANAN, R. L., STAHL, H. G. and ARCHER, D. L. (1987) *Food Microbiol.* **44**, 269.

BUEGGE, D. R. and MARSH, B. B. (1975) *Biochem. Biophys. Res. Commun.* **65**, 478.

BURK, R. F. (1997) In *Comprehensive Toxicology*, (Ed. F. P. GUENGERICH) (London: Pergamon), p. 229.

BURLEIGH, L G. (1974) *Biol. Rev.* **49**, 267.

BURN, C. G. and BURKET, L W. (1938) *Arch. Path.* **25**, 643.

BURRIS, M. J., BOGARD, R., OLIVER, A. W., MCKEY, A. O. and OLDFIELD, J. E. (1954) *Ore. Agric. Expt. Sta. Tech. Bull.*, No. 31.

BURTON, V. L. (1949) *J. Am. Chem. Soc.* **71**, 4117.

BUSCH, W. A., GOLL, D. E. and PARRISH, F. C., JR. (1972b) *J. Fd. Sci.* **37**, 289.

BUSCH, W. A., PARRISH, F. C. and GOLL, D. E. (1967) *J. Fd. Sci.* **32**, 390.

BUSCH, W. A., STROMER, M. H., GOLL, D. E. and SUZUKI, A. (1972a) *J. Cell. Biol.* 52, 167.

BUTLER-HOGG, B. W. and CRUICKSHANK, G. J. (1989) In *Meat Production and Processing* (Eds. R. W. PURCHAS, B. W. BUTLER-HOGG and A. S. DAVIES), p. 87. Occas. Publicn. No. 11 (N.Z. Soc. Animal Production).

BUTTERFIELD, R. M. and BERG, R. T. (1966) *Aust. Vet. Sci.* **7**, 389.

BUTTERMAN, L. B. (1979) *J. Fd. Protect.* **42**, 65.

BUTTERY, P. J. (1983) *Proc. Nutr. Soc.* **42**, 137.

BUTTERY, R. G., LING, L. C., TERENISHI, R. and MON, T. R. (1977) *J. Agric. Fd. Chem.* **25**, 1227.

BYRNE, D. V., BREDIE, W. L. P., BAK, L. S., BERTELSEN, G., MARTENS, H. and MARTENS, M. (2001) *Meat Sci.* **59**, 29.

BYWATERS, E. G. L. (1944) *J. Am. Med Ass.* **124**, 1103.

CAIN, R. F., ANGLEMIER, A. F., SATHER, L A., BAUTISTA, F. R. and THOMPSON, R. H. (1958) *Food Res.* **23**, 603.

CALLOW, E. H. (1925–26) *Ann. Rept. Fd. Invest. Bd., Lond.*, p. 17.

CALLOW, E. H. (1927a) *Ann. Rept. Fd. Invest. Bd., Lond.*, p. 17.

CALLOW, E. H. (1927b) *Analyst* **52**, 391.

CALLOW, E. H. (1930) *Ann. Rept. Fd. Invest. Bd., Lond.*, p. 71.

CALLOW, E. H. (1931) *Ann. Rept. Fd. Invest. Bd., Lond.*, p. 134.

CALLOW, E. H. (1932) *Ann. Rept. Fd. Invest. Bd., Lond.*, p. 97.

CALLOW, E. H. (1933) *Ann. Rept. Fd. Invest. Bd., Lond.*, p. 100.

CALLOW, E. H. (1934) *Food Invest. Bd., Lond.*, Leafl. No. 5.

CALLOW, E. H. (1935) *Ann. Rept. Fd. Invest. Bd., Lond.*, p. 57.

CALLOW, E. H. (1936) *Ann. Rept. Fd. Invest. Bd., Lond.*, pp. 75, 81.

CALLOW, E. H. (1937) *Ann. Rept. Fd. Invest. Bd., Lond.*, pp. 34, 49.

CALLOW, E. H. (1938a) Transit Shrinkage in Easting Pigs, *Bacon Development Board Bull.*, No. 3, London.

CALLOW, E. H. (1938b) *Ann. Rept. Fd. Invest. Bd., Lond.*, p. 54.

CALLOW, E. H. (1939) *Ann. Rept. Fd. Invest. Bd., Lond.*, p. 29.

CALLOW, E. H. (1947) *J. Agric. Sci.* **37**, H3.

CALLOW, E. H. (1948) *J. Agric. Sci.* **38**, 174.

CALLOW, E. H. (1954) *Ann. Rept. Fd. Invest. Bd., Lond.*, p. 28.

CALLOW, E. H. (1955a) *Inst. Meat Bull.*, March, p. 2.

CALLOW, E. H. (1955b) *Ann. Rept. Fd. Invest. Bd., Lond.*, p. 16.

CALLOW, E. H. (1956) *J. Sci. Fd. Agric.* **7**, 173.

CALLOW, E. H. (1958) *J. Agric. Sci.* **51**, 361.

CALLOW, E. H. (1961) *J. Agric. Sci.* **56**, 265.

CALLOW, E. H. (1962) *J. Agric. Sci.* **58**, 295.

CALLOW, E. H. and BOAZ, T. G. (1937) *Ann. Rept. Fd. Invest. Bd., Lond.*, p. 51.

CALLOW, E. H. and INGRAM, M. (1955) *Food* **24**, 52.

CALLOW, E. H. and SEARLE, R. L. (1956) *J. Agric. Sci.* **48**, 61.

CALVALO, A. (1981) In *Developments in Meat Science*, Vol. 2. (Ed. R. A. LAWRIE) p. 125 (Appl. Sci.: Lond.)

CAMBERO, M. I., DE LA HOZ, L., SANZ, B. and ORDOÑEZ, J. A. (1991) *Meat Sci.* **29**, 153.

CAMERON, H. S., GREGORY, P. W. and HUGHES, E. H. (1943) *J. Vet. Res.* **4**, 387.

CAMERON, N. D. and ENSER, M. (1991) *Meat Sci.* **29**, 293.

CAMERON, N. D., ENSER, M., NUTE, G. R., WHITTINGTON, F. M., PENMAN, J. C., FISKEN, A. C., PERRY, A. M. and WOOD, J. D. (2000a) *Meat Sci.* **55**, 187.

CAMERON, N. D., PENMAN, J. C., FISKEN, A. C., NUTE, G. R., PERRY, A. M. and WHITTINGTON, F. W. (2000b) *Meat Sci.* **54**, 147.

CAMPBELL, K. P., LEUNG, A. T. and SHARP, A. H. (1988) *Trends Neurosci.* **11**, 425.

CAMPBELL, K. H. S., MCWHIR, J., RITCHIE, W. A. and WILMUT, I. (1996) *Nature, Lond.* **380**, 64.

CAMPBELL, R. G. and KING, R. H. (1982) *Anim. Prod.* **35**, 177.

CAMPBELL, R. G., TAVERNER, M. R. and CURIC, D. M. (1984) *Anim. Prod.* **38**, 233.

CAMPBELL-PLATT, G. and COOK, P. E. (Eds) (1995) *Fermented Meats* (Blackie: Glasgow).

CAMPION, D. R. (1987) *Proc. 40th Ann. Recip. Meat Conf.*, 8t. Paul, p. 75.

CAMPO, M. M., NUTE, G. R., WOOD, J. D., ELMORE, S. J., MOTTRAM, D. J. and ENSER, M. (2003) *Meat Sci.* **63**, 367.

CANONICO, P. G. and BIRD, J. W. G. (1970) *J. Cell. Biol.* **45**, 321.

CARBEN, E., FUCHS, E. and KNAPPEIS, G. G. (1965) *J. Cell. Biol.* **27**, 35.

CARLEZ, A., ROSEC, J. P., RICHARD, N. and CHEFTEL, J. C. (1994) *Lebensm. Wiss. Technol.* **27**, 48.

CARLEZ, A., VECIANA-NOGUES, T. and CHEFTEL, J. C. (1995) *Lebens. Wiss. Technol.* **28**, 258.

CARMACK, C. F., KASTNER, C. L., DIKEMAN, M. E., SCHWENKE, J. R., GARCIA, R. and ZEPEDA, C. M. (1995) *Meat Sci.* **39**, 143.

CARMICHAEL, D. J. and LAWRIE, R. A. (1967a) *J. Food Technol.* **2**, 299

CARMICHAEL, D. J. and LAWRIE, R. A. (1967b) *J. Food Technol.* **2**, 313.

CARNEGIE, P. R, COLLINS, M. C. and ILIC, M. Z. (1984) *Meat Sci.* **10**, 145.

CARPENTER, J. W., PALMER, A. Z., KIRK, W. G., PEACOCK, E. M. and KOGER, M. (1955) *J. Anim. Sci.* **14**, 1228.

CARSE, W. A. (1973) *J. Fd. Technol.* **8**, 163.

CARSE, W. A. and LOCKER, R. H. (1974) *J. Sci. Fd. Agric.* **25**, 1529.

CARSON, N. (1970) In *Proteins as Human Food* (Ed. R. A. LAWRIE), Butterworths, London, p. 458.

CARSTEN, M. E. (1968) *Biochemistry* **7**, 960.

CARTWRIGHT, T. C., BUTLER, O. D. and COVER, S. (1958) *Proc. 10th Res. Conf., Amer. Meat Inst. Found.*, p. 75.

CASEY, J. C., CROSSLAND, A. R. and PATTERSON, R. L. S. (1985) *Meat Sci.* **12**, 189.

CASSELLA, J. F., CRAIG, S. W., MAACK, D. J. and BROWN, A. E. (1987) *J. Cell Biol.* **106**, 371.

CASSENS, R. G. (1970) In *The Physiology and Biochemistry of Muscle as a Food*, Vol. 11 (Eds. E. J. BRISKEY, R. G. CASSENS and B. B. MARSH), Univ. Wisconsin Press, Madison, p. 679.

CASSENS, R. G., COOPER, C. C. and BRISKEY, E. J. (1969) *Acta neuropath. (Berl.)* **12**, 300.

CASSENS, R. G. and NEWBOLD, R. P. (1966) *J. Sci. Fd. Agric.* **17**, 254.

CASSENS, R. G. and NEWBOLD, R. P. (1967) *J. Food Sci.* **32**, 269.

CASTELLANI, A. G. and NIVEN, C. E. JR. (1955) *Appl. Microbiol.* **3**, 154.

CATCHPOLE, C., HORSFIELD, C. and LAWRIE, R. A. (1970) *Proc. 16th Europ. Meeting Meat Res.*, Workers, Varna, p. 214.

CAUSEY, M., HAUSRATH, M. E., RAMSTAD, P. E. and PENTON, E. (1950) *Food Res.* **15**, 237, 249, 256.

CAVALIER-SMITH, T. (1980) *Nature, Lond.* **285**, 619.

CAVETT, J. J. (1962) *J. Appl. Bact.* **25**, 282.

CENA, P., JAIMÉ, I., BELTRAN, A. and RONCALES, P. (1992) *J. Muscle Foods* **3**, 253.

CHADWICK, J. P. and KEMPSTER, A. J. (1983) *Meat Sci.* **9**, 101.

CHAMBERS, R. and HALE, H. P. (1932) *Proc. Roy. Soc.* **B 110**, 336.

CHAMBON, P. (1981) *Sci. Amer.* **244** (5), 48.

CHAMPAGNAT, A. (1966) *7th Intl. Congr. Nutrition, Hamburg*.

CHAMPION, A., PARSONS, A. L. and LAWRIE, R. A. (1970) *J. Sci. Fd. Agric.* **21**, 7.

CHANG, K. C., DA COSTA, N., BLACKLEY, R., SOUTHWOOD, O., EVANS, G., PLASTOW, G., WOOD, J. and RICHARDSON, R. (2003) *Meat Sci.* **64**, 93.

CHANG, S. S., HIRAI, C., REDDY, B. R., HERZ, K. O., KATO, A. and SIPMA, G. (1978) *Chem. Ind.*, p. 1639.

CHANNON, H. A., PAYNE, A. M. and WARNER, R. D. (2002) *Meat Sci.* **60**, 63.

CHAPPEL, J. B. and PERRY, S. V. (1953) *Biochem. J.* **55**, 586.

CHARLES, L. M. T. and NICOL, T. (1961) *Nature, Lond.* **192**, 565.

CHEAH, A. M. (1981) *Biochim. Biophys. Acta* **648**, 113.

CHEAH, A. M., CHEAH, K. S., LAHUCKY, R., KOVAC, L., KRAMER, H. L. and MCPHEE, C. P. (1994) *Meat Sci.* **38**, 375.

CHEAH, K. S. (1971) *Fed. Eu. Biochem. Soc. Letters*, **19**, 105.

CHEAH, K. S. (1973) *J. Sci. Fd. Agric.* **24**, 51.

CHEAH, K. S. (1976) *J. Fd. Technol.* **11**, 181.

CHEAH, K. S. and CHEAH, A. M. (1976) *J. Sci. Fd. Agric.* **27**, 1137.

CHEAH, K. S. and CHEAH, A. M. (1981a) *Biochim. Biophys. Acta* **634**, 70.

CHEAH, K. S. and CHEAH, A. M. (1981b) *Biochim. Biophys. Acta* **638**, 40.

CHEAH, K. S., CHEAH, A. M., CROSLAND, A. R. and CASEY, J. C. (1984) *Meat Sci.* **10**, 117.

CHEAH, K. S., CHEAH, A. M. and WARING, J. C. (1986) *Meat Sci.* **17**, 37.

CHEAH, P. B. and LEDWARD, D. A. (1995) *J.A.O.C.S.* **72**, 1059.

CHEAH, P. B. and LEDWARD, D. A. (1996) *Meat Sci.* **43**, 123.

CHEAH, P. B. and LEDWARD, D. A. (1997) *Meat Sci.* **45**, 411.

CHEFTEL, J. C. (1995) *Food Sci. Technol. Internat.* **1**, 75.

CHEFTEL, J. C. and CULIOLI, J. (1997) *Meat Sci.* **46**, 211.

CHEN, F. C., HSIEH, Y. H. P. and BRIDGMAN, R. C. (2002) *Meat Sci.* **62**, 405.

CHESNEY, M. S., MANDIGO, R. W. and CAMPBELL, J. E. (1978) *J. Fd. Sci.* **43**, 1535.

CHIKUNI, K., TANABE, R., MUROYA, S. and NAKAJIMA, I. (2001) *Meat Sci.* **57**, 311.

CHRISTIAN, J. H. B. and SCOTT, W. J. (1953) *Aust. J. Biol. Sci.* **6**, 565.

CHRISTENSEN, L. B. (2003) *Meat Sci.* **63**, 469.

CHRISTIANSEN, M., HENCKEL, P. and PURSLOW, P. P. (2004) *Meat Sci.* **66**, 595.

CHRYSTALL, B. B. and DEVINE, C. E. (1976–77) *Ann. Res. Rept.*, Meat Ind. Res. Inst. N.Z. Inc., p. 32.

CHRYSTALL, B. B. and DEVINE, C. E. (1978) *Meat Sci.* **2**, 49.

CHRYSTALL, B. B., DEVINE, C. E. and DAVEY, C. L. (1980) *Meat Sci.* **4**, 69.

CHRYSTALL, B. B., DEVINE, C. E. and NEWTON, K. G. (1980–81) *Meat Sci.* **5**, 339.

CHRYSTALL, B. B. and HAGYARD, C. J. (1976) *N.Z. J. Agric. Res.* **19**, 7.

CITOLER, P., BENISTY, and MAURERR, W. (1966) *Exp. Cell. Res.* **45**, 195.

CLARKE, F. M., MORTON, D. J. and SHAW, F. D. (1980) *Biochem. J.* **186**, 105.

CLAUS, R. (1975) In *Immunization with Hormones in Reproduction Research.* (Ed. E. NIESCHLAG), p. 189 (North-Holland: Amsterdam).

CLAUS, R., WEILER, U. and HERZOG, A. (1994) *Meat Sci.* **38**, 289.

CLAUSEN, H. (1965) *Wld Rev. Anim. Prod* **1**, 28.

CLAYTON, W. (1932) *Food Manuf.* **7**, 109.

CLELAND, K. W. and SLATER, E. C. (1953) *Biochem. J.* **53**, 547.

CLOSE, R. (1967) *J. Physiol.* **193**, 45.

CLOUSTON, J. G. and WILLS, P. A. (1969) *J. Bact.* **97**, 684.

CLUTTON-BROCK, J. (1981) *Domesticated Animals from Early Times* (Heinemann: British Museum (Natural History): Lond.).

COATES, M. E. (1987) *B. N. F. Nutr. Bull.* **12**, 87.

COCKS, D. H., DENNIS, P. O. and NELSON, T. H. (1964) *Nature, Lond.* **202**, 184.

COGNIE, Y., HERNANDEZ-BARRETI, M. and SAUNAUDE, J. (1975) *Ann. Biol. Anim. Bioch. Biophys.* **15**, 329.

COHEN, B. and CLARK, W. M. (1919) *J. Bact.* **4**, 409.

COHEN, D. M. and MURPHY, R. A. (1979) *Circ. Res.* **45**, 661.

COLDITZ, P. J. and KELLAWAY, R. C. (1972) *Aust. J. Agric. Res.* **23**, 717.

COLE, D. J. A., WHITE, M. R., HARDY, B. and CARR, J. R. (1976) *Anim. Prod.* **22**, 341.

COLE, H. A. and PERRY, S. V. (1975) *Biochem. J.* **119**, 525.

COLEBY, B. (1959) *Intl. J. Appl. Radn. Isotopes* **6**, 71.

COLEBY, B., INGRAM, M., RHODES, D. N. and SHEPHERD, H. J. (1962) *J. Sci. Fd. Agric.* **13**, 628.

COLEBY, B., INGRAM, M. and SHEPHERD, H. J. (1961) *J. Sci. Fd. Agric.* **12**, 417.

COLEMAN, H. M. and STEFFEN, H. H. (1949) U.S. Pat. 2,491,646.

COLEMAN, R. and CRESWICK, N. (1966) U.S. Pat. No. 3,253,931.

COLLEY, N. J., TOKUYASU, K. T. and SINGER, S. J. (1990) *J. Cell Res.* **95**, 11.

COMMITTEE ON MEDICAL ASPECTS OF FOOD POLICY (1984) 'Diet and Cardiovascular Disease: Report on Meat and Social Subjects' (H.M.S.O.: Lond.).

COMMONWEALTH ECONOMIC COMMITTEE (1961) Report.

CONCEPCION-ARISTOY, M. and TOLDRA, F. (2004) *Meat Sci.* **67**, 211.

CONNOLLY, B. J. and DECKER, E. A. (2004) *Meat Sci.* **66**, 499.

COOK, C. F. and LANGSWORTH, R. F. (1966) *J. Food Sci.* **31**, 497.

COOK, C. F., MAYER, E. W., CATSIMPOULAS, N. and SIPOS, E. F. (1969) *Proc. 15th Meeting European Meat Res. Workers*, Helsinki, p. 381.

COOK, C. J., DEVINE, C. E., GILBERT, K. V., SMITH, D. D. and MAASLAND, S. A. (1995) *Meat Sci.* **40**, 137.

COOK, D. J. (2003) *J. Inst. Fd. Sci. Technol.* **17**(1), 26.

COOK, G. A., LOVE, E. F. G., VICKERY, J. R. and YOUNG, W. G. (1926) *Aust. J. Exp. Biol. Med. Sci.* **3**, 15.

COOK, J. D. and MONSEN, E. R. (1976) *Am. J. Clin. Nutr.* **29**, 859.

COOK, J. W. (1933) *Proc. Roy. Soc.* **B 113**, 277.

COOK, J. W. (1934) *Nature, Lond.* **134**, 758.

COOK, L. J., SCOTT, T. W., FERGUSON, K. A. and MCDONALD, I. (1970), *Nature, Lond.* **228**, 178.

COOPER, M. J. (1974) *Vet. Rec.* **94**, 161.

COPPOCK, B. (1975) Ph.D. Dissertation, Univ. London.

COPPOCK, B. M. and MACLEOD, G. (1977) *J. Sci. Fd. Agri.* **28**, 206.

CORI, C. F. (1956) *Enzymes: Units of Biological Structure and Function* (Ed. O. H. GAEBLER), p. 573, Academic Press, New York.

CORI, C. F. and CORI, G. T. (1928) *J. Biol Chem.* **79**, 309.

CORI, G. T. (1957) *Mod. Probl. Paediat.* 344.

COSMOS, E. (1966) *Develop. Biol.* **13**, 163.

COSNETT, K. S., HOGAN, D. J., LAW, N. H. and MARSH, B. B. (1956) *J. Sci. Fd. Agric.* **7**, 546.

COULSTON, F. and WILLS, J. H. (1976) In *Anabolic Agents in Animal Production* (Eds. F. E. LU and J. RENDEL), p. 238 (Thieme: Stuttgart).

COVER, S. and HOSTETLER, R. L. (1960) *Texas Agric. Expt. Sta. Bull.*, No. 947.

CRAIG, H. B., PEARSON, A. M. and WEBB, N. B. (1962) *Food Res.* **27**, 29.

CRAWFORD, M. A. (1968) *Proc. Nutr. Soc.* **27**, 163.

CRAWFORD, M. A. (1975) In *Meat* (Eds. D. J. A. COLE and R. A. LAWRIE), Butterworths, London, p. 451.

CRAWFORD, M. A., GALE, M. M., WOODFORD, M. H. and GASPED, N. M. (1970) *Intl. J. Biochem.* **1**, 295.

CRICHTON, D. B. (1972) *J. Physiol.* **226**, 68.

CRICHTON, D. B. (1980) In *Developments in Meat Science* Vol. I (Ed. R. A. LAWRIE), p. 1 (Applied Sci.: Lond.).

CRICK, F. (1979) *Science*, **204**, 264.

CRICK, F. H. C., BARNETT, L., BRENNER, S. and WATTS-TOBIN, R. J. (1961) *Nature, Lond.* **192**, 1227.

CRITCHELL, J. T. and RAYMOND, J. (1912) *A History of the Frozen Meat Trade*, 2nd ed., Constable & Co. Ltd., London.

CROALL, D. E. and DEMARTINO, G. N. (1991) *Physiol. Rev.* **71**, 813.

CROCKER, E. C. (1945) *Flavor*, McGraw-Hill, New York.

CROCKER, E. C. (1948) *Food Res.* **13**, 179.

CROMBACH, J. J. M. L., DE ROVER, W. and DE GROOTE, B. (1956) *Proc. 3rd Intl. Congr. Anim. Prod., Cambridge*, sect. III, p. 80.

CROSS, C. K. and ZIEGLER, P. (1965) *J. Fd. Sci.* **30**, 610.

CROSS, H. R. and STANSFIELD, M. S. (1976) *J. Fd. Sci.* **41**, 1257.

CROSSLAND, A. R., PATTERSON, R. L. S., HIGMAN, R. C., STEWART, C. A. and HARGIN, K. D. (1995) *Meat Sci.* **40**, 289.

CRUFT, P. G. (1957) *Meat Hygiene*, F.A.O. Agric. Series. No. 34, p. 147.

CUFF, P. W. W., MADDOCK, H. M., SPUR, V. C. and CATRON, D. V. (1951) *Iowa State Coll. Sci. J.* **25**, 575.

CUMMINS, P. and PERRY, S. V. (1973) *Biochem. J.* **133**, 765.

CURLEY, A., SEDLAK, V. A., GIRLING, E. F., HAWK, K. E., BARTHEL, W. F., PIERCE, T. E. and LIKOSKY, W. A. (1971) *Science*, **172**, 65.

CURRIE, R. W. and WOLFE, F. H (1980) *Meat Sci.* **4**, 123.

CURTIS, B. M. and CATTERALL, W. A. (1988) *Proc. Nat. Acad. Sci., U.S.A.* **82**, 2538.

CUTHBERTSON, A. and POMEROY, R. W. (1962) *J. Agric. Sci.* **59**, 207, 215.

CUTHBERTSON, A. and POMEROY, R. W. (1970) *Anim. Prod.* **12**, 37.

CUTTER, C. N., DORSA, W. J., HARDIE, A., RODRÍGUEZ-MORALES, S., ZHOU, X., BREEN, P. J. and COMPADRE, C. M. (2000) *J. Fd. Prot.* **63**, 593.

CUTTING, C. L. and BANNERMAN, A. (1951) *Fd. Invest. Bd., Lond.*, Leafl. No. 14.

CUTTING, C. L and MALTON, R. (1971) *Meat Res. Inst.*, Meino No. 2.

CZOK, R. and BÜCHER, TH. (1960) *Adv. Prot. Chem.* **15**, 315.

DAHL, O. (1958a) *Svensk kem. Tidskrift.* **70**, 43.

DAHL, O (1958b) *J. Refrig.* **1**, 170.

DAHL, O. (1965) *J. Sci. Fd. Agric.* **16**, 619.

DAINTON, F. S. (1948) *Rep. Progr. Chem.* **45**, 5.

DAINTY, R. H., SHAW, B. G., DE BOER, K. A. and SCHEPS, E. S. J. (1975) *J. Appl. Bact.* **139**, 73.

DALE, W. M. and DAVIES, J. V. (1951) *Biochem. J.* **48**, 129.

DANNERT, R. D. and PEARSON, A. M. (1967) *J. Fd. Sci.* **32**, 49.

DASTOLI F. R., LOPIEKES, D. V. and DOIG, A. R. (1968) *Nature, Lond.* **218**, 884.

DAUGHADAY, W. H., HALL, K., RABEN, M. S., SALMEN, W. D., VAN DEN BRAUDE, J. L and VAN WYK, J. J. (1972) *Nature, Lond.* **235**, 107.

DAUN, C. and ÅKESSON, B. (2004) *Meat Sci.* **66**, 801.

DAUPHIN, J. F. and SAINT-LEBE, L. R. (1977) In *Radiation Chemistry of Major Food Components* (Eds. P. S. ELIAS and A. J. COHEN) (Amsterdam: Elsevier), p. 131.

DAVEY, C. L. (1960) *Arch. Biochem. Biophys.* **89**, 303.

DAVEY, C. L. (1961) *Arch. Biochem. Biophys.* **95**, 296.

DAVEY, C L. (1964) *Ann. Rept. Meat Ind. Res. Inst., N.Z.*, p. 17.

DAVEY, C. L. (1970) *Ann. Rept. Meat Ind. Res. Inst., N.Z.*, p. 26.

DAVEY, C. L. (1980) *C.S.I.R.O. Fd. Pres. Quart.*, **40**, 51.

DAVEY, C. L. and CURSON, P. (1971) *Meat Ind. Res. Inst., N.Z.*, Rept. No. 215.

DAVEY, C. L. and DICKSON, M. R. (1970) *J. Food Sci.* **35**, 56.

DAVEY, C. L. and GARNETT, K. J. (1980) *Meat Sci.* **4**, 319.

DAVEY, C. L. and GILBERT, K. V. (1967) *J. Food Technol.* **2**, 57.

DAVEY, C. L. and GILBERT, K. V. (1969) *J. Food Sci.* **34**, 69.

DAVEY, C. L. and GILBERT, K. V. (1974) *J. Fd. Technol.* **9**, 51.

DAVEY, C. L. and GILBERT, K. V. (1975a) *J. Sci. Fd. Agric.* **26**, 755.

DAVEY, C. L. and GILBERT, K. V. (1975b) *J. Fd. Technol.* **10**, 333.

DAVEY, C. L. and GILBERT, K. V. (1975c) *J. Sci. Fd. Agric.* **26**, 953.

DAVEY, C. L. and GILBERT, K. V. (1976a) *J. Sci. Fd. Agric.* **27**, 1085.

DAVEY, C. L. and GILBERT, K. V. (1976b) *J. Sci. Fd. Agric.* **27**, 244.

DAVEY, C. L., GILBERT, K. V. and CURSON, P. (1971) *Ann. Rept. Meat Ind. Res. Inst., N.Z.*, p. 39.

DAVEY, C. L. and GRAAFHUIS, A. E (1976a) *Experentia*, **32**, 32.

DAVEY, C. L and GRAAFHUIS, A. E. (1976b) *J. Sci. Fd. Agric.* **27**, 301.

DAVEY, C. L., NIEDERER, A. F. and GRAAFHUIS, A. E. (1976) *J. Sci. Fd. Agric.* **27**, 251.

DAVEY, C. L., KUTTEL, H. and GILBERT, K. V. (1967) *J. Food Technol.* **2**, 53.

DAVEY, C. L. and WINGER, R. J. (1980) In *Fibrous Proteins: Scientific, Industrial & Medical Aspects.* Vol II. (Eds. D. A. D. PARRY and L. K. CREAMER), p. 125. (Acad. Press: New York).

DAVIDSON, H. R. (1953) *The Production and Marketing of Pigs*, 2nd ed., Longmans, London.

DAVIDSON, J. M., LINFORTH, R. S. T., HOLLOWOOD, T. A. and TAYLOR, A. J. (2000) In *Flavour Release* (Eds. D. D. ROBERTS and A. J. TAYLOR), ACS, Washington, DC, pp. 99–111.

DAVIDSON, M. H., HUNNINGHAKE, D., MSKI, K. C., KWITEROVITCH, P. and KAFONEK, S. (1999) *Arch. Intern. Med.* **159**, 1331.

DAVIES, A. S. (1974) *Anim. Prod.* **19**, 367.

DAVIES, A. S. (1989) In *Meat Production and Processing* (Eds. R. W. PURCHAS, B. W. BUTLER-HOGG and A. S. DAVIES), p. 43. Occas. Public. No. 11 (N.Z. Soc. Animal Production).

DAVIES, R. E. (1963) *Nature, Lond.* **199**, 1068.

DAVIS, S. R, BARRY, T. N. and HUGHSON, G. A. (1981) *Brit. J. Nutr.* **46**, 409.

DAWSON, D. M., GOODFRIEND, L. and KAPLAN, N. O. (1964) *Science* **143**, 929.

DAWSON, E. H., BRAGDON, J. L. and MCMANUS, S. (1963) *Food Tech.* **17**, Nos. 9, 45, 51; Nos. 10, 39, 43.

DAWSON, J. M., BUTTERY, P. J., GILL, M. and BEEVER, D. E. (1990) *Meat Sci.* **28**, 289.

DAWSON, M. J., GADIAN, D. G. and WILKIE, D. R. (1978) *Nature, Lond.* **274**, 861.

DAYTON, S. M., PEARCE, M. L., HASHIMOTO, S., DIXON, W. J. and TOMIYASHI, U. (1969) *Circulation*, **40**, Suppl. II.

DAYTON, W. R., GOLL, D. E., ZEECE, M. G., ROBSON, R. M. and REVILLE, W. J. (1976) *Biochemistry*, **15**, 2150.

DEAN, R. W. and BALL, C. O. (1960) *Food Tech.* **14**, 222, 271.

DEAR, T. N., MATENA, K. and VINGSON, M. (1997) *Genomics*, **45**, 175.

DEATHERAGE, F. E. (1955) *1st Intl. Conf. on Antibiotics in Agriculture, Washington*, Nat. Acad. Sci., Nat. Res. Counc. Publ. No. 397, p. 211.

DEATHERAGE, F. E. (1957) U.S. Pat. No. 2,786,768.

DEATHERAGE, F. E. (1965) *Proc. 1st Intl. Congr. Food Sci. Tech.* **2**, 65.

DEATHERAGE, F. E. and FUJIMAKI, M. (1964) *J. Food Sci.* **29**, 316.

DE DUVE, C. (1959a) *Subcellular Particles*, p. 128, Ronald Press, New York.

DE DUVE, C. (1959b) *Exp. Cell. Res., Suppl.* **7**, 169.

DE DUVE, C. and BEAUFAY, H. (1959) *Biochem. J.* **73**, 610.

DE DUVE, C. and BAUDHUIN, P. (1966) *Physiol. Rev.* **46**, 323.

DEEBLE, D. J., JAHIR, A. W., PARSONS, J. B., SMITH, C. J. and WHEATLEY, P. (1990) In *Food Irradiation and the Chemist* (Eds. D. E. JOHNSON and M. H. STEVENSON) p. 57 (Roy. Soc. Chem.: Lond.).

DEIBEL, R. H. and NIVEN, C. F., JR. (1959) *Appl. Microbiol.* **7**, 138.

DEISINGE, A. K., STONE, C. and THOMPSON, M. (2004) *Intl. J. Fd. Sci. Tech.* **39**, 587.

DEMOS, B. P. and MANDIGO, R. W. (1996) *Meat Sci.* **42**, 415.

DENG, Y., ROSENVOLD, K., KARLSSON, A. H., HORN, P., HEDEGAARD, J., STEFFENSEN, C. L. and ANDERSEN, H. J. (2002) *J. Fd. Sci.* **67**(5), 1642.

DENMEAD, C. F., WILKINSON, B. H. P., MORRIS, M. A., BAIN BRIDGE, A. P. and ROSE, L. J. (1973–74) *Ann. Res. Rept. Meat Ind Res. Inst. N.Z.*, Inc. p. 52.

DENNY-BROWN, D. (1929) *Proc. Roy. Soc.* **B 104**, 3 71.

DENNY-BROWN, D. (1961) *Neuromuscular Disorders* **38**, 147.

DENOYELLE, C. and LEBIHAN, E. (2003) *Meat Sci.* **66**, 241.

DEPREUX, F. F. S., OKAMURA, C. S., SWARTZ, D. R., GRANT, A. L., BRANDSTETTER, A. N. and GERRARD, D. E. (2000) *Meat Sci.* **56**, 261.

DESOUBRY, M. A. and PORCHER, M. C. M. C. (1895) *C.R. Soc. Biol., Paris* **47**, 101.

DESROSIER, N. W. (1959) *The Technology of Food Preservation*, AVI Publishing Co., Westport, Conn.

DEVEAUX, V., CASSAR-MALEK, F. and PICARD, B. (2000) *Comp. Biochem. Physiol.* **131**A, 21.

DEVEAUX, V., PICARD, B., BOULEY, J. and CASSAR-KALEK, I. (2003) *Rep. Nut. Develop.* **43**, 527.

DEVINE, C. E., ELLERY, S. and AVERILL, S. (1984) *Meat Sci.* **10**, 35.

DEVINE, C. E., GILBERT, K. V., GRAAFHUIS, A. E., TAVERNER, A., REED, H. and LEIGH, P. (1986) *Meat Sci.* **17**, 267.

DEVINE, C. E. and GRAAFHUIS, A. E. (1995) *Meat Sci.* **39**, 285.

DEVINE, C. E., GRAAFHUIS, A. E., MUIR, P. D. and CHRYSTALL, B. B. (1993) *Meat Sci.* **35**, 63.

DE VRIES, A. G., SOSNICKI, A., GARNIER, J. P. and PLASTOW, G. S. (1998) *Meat Sci.* **49**, S 245.

DE VRIES, A. L. (1998) *Comp. Biochem. Phys.* **90b**, 611.

DHOOT, G. K. and PERRY, S. V. (1979) Nature, Lond. **278**, 714.

DHOOT, G. K. and PERRY, S. V. (1980) Exp. Cell Res. **127**, 75.

DHOOT, G. K., PERRY, S. V. and URBOVA, G. (1981) Exp. Neurol. **72**, 513.

DICKINSON, A. G., HANCOCK, J. L., HOVELL, J. R. and TAYLER, ST. C. S. (1962) Anim. Prod. **4**, 64.

DICKINSON, M. R. (1969) Ann. Rept. Meat Ind. Res. Inst., N.Z., p. 17.

DICKSON, M. R., MARSH, B. B. and LEET, N. G. (1970) Ann. Rept. Meat Ind Res. Inst., N.Z., p. 31.

DIEHL, J. F. (1966) Proc. 2nd Intl. Congr. Fd. Sci. Tech., Warsaw, C.3.9.

DILL, C. W. (1976) Proc. 29th Ann. Recip. Meat Conf., Provo, Utah. p. 162. (Natl. Livestock & Meat Bd.: Chicago).

DIXSON, M. and WEBB, E. C. (1958) Enzymes, p. 152, Longmans, London.

DODD, C. E. R., CHAFFEY, B. J. and WAITES, W. M. (1988a) Appl. Envir. Microbiol. **54**, 1541.

DODD, C. E. R., MEAD, G. C. and WAITES, W. M. (1988b) Letters Appl. Microbiol. **7**, 63.

DOLATA, W., PIETROWSKA, E., WAJDZIK, J. and TRITT-GOX, J. (2004) Meat Sci. **67**, 25.

DOLMAN, C E. (1957) Meat Hygiene, F.A.O. Agric. Series, No. 34, p. 11.

DORSA, W. J. (1996) Proc. 49th Ann. Recip. Meat Conf., Provo, Utah, p. 114.

DOS REMEDIOS, C. G. and MOENS, P. D. J. (1995) Biochim. Biophys. Acta **1228**, 99.

DOTY, D. M (1960) The Science of Meat and Meat Products, Amer. Meat Inst. Found., p. 288, Reinhold Publishing Co., New York.

DOTY, D. M. and PIERCE, J. C. (1961) U.S. Dept. Agric. Tech. Bull., No. 1231.

DOTY, D. M. and WACHTER, J. P. (1955) J. Agric. Food Chem. **3**, 61.

DOUGLAS, J., CIVELLI, O. and HERBERT, F. (1984) Ann. Rev. Biochem. **53**, 665.

DOWBEN, R. M and ZUCKERMAN, L. (1963) Nature, Lond. **197**, 400.

DOYLE, M. (1998) Proc. Beef Safety Sympos., Chicago, Ill. (AMIF: Chicago) p. 42.

DOZIER, C. C. (1924) J. Infect. Dis. **35**, 134.

DRAKE, M. P., GERNON, G. D. and KRAUS, F. J. (1961) J. Food Sci. **26**, 156.

DRANSFIELD, E. (1983) Proc. 2nd Conf. Food Chem., Rome. In press.

DRANSFIELD, E. (1992) Meat Sci. **31**, 85.

DRANSFIELD, E. (1993) Meat Sci. **34**, 629.

DRANSFIELD, E. (1994) Meat Sci. **37**, 391.

DRANSFIELD, E. (1996) Meat Sci. **43**, 311.

DRANSFIELD, E., ETHERINGTON, D. J. and TAYLOR, M. A. J. (1992) Meat Sci. **31**, 75.

DRANSFIELD, E., JONES, R. C. D. and MCFIE, H. J. H. (1981) Meat Sci. **5**, 131.

DRANSFIELD, E, LEDWITH, M. J. and TAYLOR, A. A. (1991) Meat Sci. **29**, 129.

DRANSFIELD, E. and LOCKYER, D. K. (1985) Meat Sci. **13**, 19.

DRANSFIELD, E., LOCKYER, D. K. and PRABAAKARNA, P. (1986) Meat Sci. **16**, 127.

DRANSFIELD, E., NUTE, G. R., ROBERTS, T. A., BOCCARD, R., TOURAILLE, C., BUCHTER, L., CASTEELS, M., COSENTINO, E., HOOD, D. E., JOSEPH, R. L., SCHON, L. and PAARDEKOOPER, E. J. C. (1984) Meat Sci. **10**, 1.

DREYFUS, J. C., SCHAPIRA, G. and SCHAPIRA, F. (1954) J. Clin. Invest. **33**, 794.

DRUMMOND, J. C. and MACARA, T. (1938) Chem. Ind., p. 828.

DUBE, G., BRAMBLETT, V. D., JUDGE, M. D. and HARRINGTON, R. B. (1972) J. Fd. Sci. **37**, 13.

DUBOWITZ, V. (1963) Nature, Lond. **197**, 1215.

DUBOWITZ, V. (1966) Nature, Lond. **211**, 884.

DUBOWITZ, V. and PEARSE, A. S. E. (1961) J. Path. Bact. **81**, 365.

DUESBERG, J. (1909) Arch. Zellforsch. mikr. Anat. **4**, 602, cited by ADAMS, DENNY-BROWN and PEARSON (1962), loc. cit.

DUFF, J. T., WRIGHT, G. G. and YANNSKY, A. (1956) J. Bact. **72**, 455.

DUGAN, L. R. (1957) Amer. Meat Inst. Found. Circ., No. 36.

DUGAN, L. R. and LANDIS, P. W. (1956) J. Amer. Oil Chem. Soc. **33**, 152.

DUMONT, B. L. and BOCCARD, R. (1967) ATTI 2nd Sympos. Int. Zoot., Milano., p. 149.

DUNCAN, W. R. H. and GARTON, C. A. (1967) J. Sci. Fd. Agric. **18**, 99.

DUNKER, C. F., BERMAN, M., SNIDER, G. G. and TUBIASH, H. S. (1953) Food Tech. **7**, 288.

DUNKER, C. F. and HANKINS, O. G. (1955) Food Tech. **7**, 505.

DUNKER, C. F., HANKINS, O. G. and BENNETT, O. L. (1945) Food Res. **10**, 445.

DUTSON, T. (1976) Proc. 29th Ann. Recip. Meat Conf., Provo, Utah, p. 336.

DUTSON, T. (1977) Proc. Ann. Recip. Meat Conf., p. 79.

DUTSON, T. and LAWRIE, R. A. (1974) J. Fd. Technol. **9**, 43.

DUTSON, T. R., SAVELL, J. W. and SMITH, G. C. (1982) Meat Sci. **6**, 159.

DUTSON, T. R., SMITH, G. C. and CARPENTER, Z. L. (1980) J. Fd. Sci. **45**, 1097.

DVORAK, Z and VOGNAROVA, I. (1965) J. Sci. Fd. Agric. **16**, 305.

DYERBERG, J. and BANG, H. O. (1979) Lancet **ii**, 433.

DYETT, E. J. (1969) Proc. 15th Meeting European Meat Res. Workers, Helsinki, p. 509.

DYKHUISEN, R. S., FRAZER, R., DUNCAN, C., SMITH, C. C., GOLDEN, M., BENJAMIN, N. and SEIFERT, C. (1996) Antimicrobial Agents Chemother. **40**, 1425.

EARLE, R. L. and FLEMING, A. K. (1967) J. Food Technol. **21**, 79.

EBASHI, S. and EBASHI, F. (1964) Jap. J. Biochem. **55**, 604.

EBASHI, S. and ENDO, M. (1968) *Prog. Biophys. Molec. Biol.* **18**, 125.

EBBEHØJ, K. F. and THOMSEN, P. D. (1991) *Meat Sci.* **30**, 359.

EDDY, B. P. (1958) *The Microbiology of Fish and Meat Curing Brines*, p. 87, H.M.S.O., London.

EDDY, B. P., GATHERUM, D. P. and KITCHELL, A. G. (1960) *J. Sci. Fd. Agric.* **11**, 727.

EDDY, B. P. and INGRAM, M. (1956) *J. Appl. Bact.* **19**, 62.

EDDY, B. P. and INGRAM, M. (1965) *Proc. 1st Intl. Congr. Food Sci. Tech.* **2**, 405.

EDDY, B. P. and KITCHELL, A. G. (1959) *J. Appl. Bact.* **22**, 57.

EDDY, B. P. and KITCHELL, A. G. (1961) *J. Sci. Fd. Agric.* **12**, 146.

EDWARDS, R. A. and DAINTY, R. H. (1987) *J. Sci. Fd. Agric.* **38**, 57.

EDWARDS, R. A., DAINTY, K. H. and HIBBARD, C. M. (1985) *J. Appl. Bact.* **58**, 13.

EGAN, A. F. and SHAY, B. J. (1988) *Proc. 34th Intl. Congr. Meat Sci. Technol.*, Brisbane, p. 476.

EGELANDSDAL, B., MARTENSEN, B. and AUTIO, K. (1995) *Meat Sci.* **39**, 97.

EGGEN, A. and HOCQUETTE, J.-F. (2003) *Meat Sci.* **66**, 1.

EHRENBERG, A. S. C. and SHEWAN, J. M. (1953) *J. Sci. Fd. Agric.* **10**, 482.

EIKE, H., KOCH, R., FELDHUSEN, F. and SEIFERT, H. (2005) *Meat Sci.* **69**, 603.

EIKELENBOOM, G. and NANNI COSTA, L. (1988) *Meat Sci.* **23**, 9.

EILERS, J. D., MORGAN, J. B., MARTIN, A. M., MILLER, R. K., HALE, D. S., ACUFF, G. R. and SAVELL, J. V. (1994) *Meat Sci.* **38**, 443.

EISEN, A. Z., BAUER, E. A. and JEFFREY, J. H. (1971) *Proc. Natl. Acad Sci. U.S.A.* **68**, 248.

ELDRIDGE, G. A. (1988) *Proc. 34th. Intl. Congr. Meat Sci. Technol.*, Brisbane, p. 143.

ELGASIM, E. H. and KENNICK, W. H. (1980) *J. Fd. Sci.* **45**, 1122.

ELIAS, P. S. (1985) In *Developments in Meat Science* Vol. 4 (Ed. R. A. LAWRIE), p. 115 (Elsevier Appl. Sci. Publshrs.: Lond.).

ELLIOT, G. F. (1968) *J. Theoret. Biol.* **21**, 71.

ELLISON, N. S., GAULT, N. F. S and LAWRIE, R. A. (1980) *Meat Sci.* **4**, 77.

ELMORE, J. S., MOTTRAM, D. S., ENSER, M. and WOOD, J. D. (2000) *Meat Sci.* **55**, 149.

ELTON, C. (1927) *Animal Ecology*, Sidgwick & Johnson, London.

EMERSON, C., BRADY, D. E. and TUCKER, L. N. (1955) *Univ. Missouri Coll. Agric. Res. Bull.*, No. 470.

EMERY, A. E. H. (1964) *Nature, Lond.* **201**, 1044.

EMMENS, C. W. (1959) *Progress in the Physiology of Farm Animals*, vol. 3 (Ed. J. HAMMOND) p. 1047, Butterworths, London.

EMMENS, C. W. and BLACKSHAW, A. W. (1956) *Physiol. Rev.* **36**, 2.

EMPEY, W. A. (1933) *J. Soc. Chem. Ind.* **52**, 230 T.

EMPEY, W. A. and MONTGOMERY, W. A. (1939) *C.S.I.R.O. Food Pres. Quart.* **19**, 30.

EMPEY, W. A. and SCOTT, W. J. (1939) *C.S.I.R.O. Bull.*, No. 126.

EMPEY, W. A. and VICKERY, J. R. (1933) *J. Coun. Sci. Industr. Res. Aust.* **6**, 233.

ENGEL, W. K. (1963) *Nature, Lond.* **200**, 588.

ENGELHARDT, V. A. (1946) *Adv. Enzymol.* **6**, 147.

ENGESETH, N. J., GRAY, J. I., BOUREN, A. M. and ASGHAR, A. (1993) *Meat Sci.* **35**, 1.

ENSER, M., HALLETT, K., HEWITT, B., FURSEY, G. A. J. and WOOD, J. D. (1996) *Meat Sci.* **42**, 443.

ENSER, M., HALLETT, K. G., HEWETT, B., FURSEY, G. A. J., WOOD, J. D. and HARRINGTON, G. (1998) *Meat Sci.* **49**, 329.

ENSER, M., SCOLLAN, N., GULSTI, S., RICHARDSON, I., NUTE, G. and WOOD, J. D. (2001) *Proc. 47th Int. Congr. Meat Sci. Technol.* **1**, 12.

ERDÖS, T. (1943) *Stud. Inst. Med. Chem. Univ. Szeged* **3**, 51.

ESSEN-GUSTAVSSÖN, B., KARLSTRÖ, K. and LUNDSTRÖM, K. (1992) *Meat Sci.* **31**, 1.

ESSEN-GUSTAVSEN, B., KARLSSON, A., LUNDSTRÖM, K. and ENFÄLT, A-C. (1994) *Meat Sci.* **38**, 269.

ETHERINGTON, D. J. (1971) *Proc. 17th Meeting Meat Res. Insts.*, Bristol, p. 632.

ETHERINGTON, D. J. (1972) *Biochem. J.* **127**, 685.

ETHERINGTON, D. J. (1973) *Eur. J. Biochem.* **32**, 126.

ETHERINGTON, D. J. (1974) *Biochemistry* **137**, 547.

ETHERINGTON, D. J. (1984) *J. Anim. Sci.* **59**, 1644.

ETHERINGTON, D. J, TAYLOR, M. A. J. and DRANSFIELD, E. (1987) *Meat Sci.* **20**, 1.

EUSTACE, I. J. (1984) *C.S.I.R.O. Food Res. Quart.* **44**, 60.

EVANS, C. D. (1961) *Proc. Flavour Chem. Symp.*, p. 123. Campbell Soup Co., Camden, N.J.

EYRE, D. R., PAZ, M. A. and GALLOP, P. M. (1984) *Ann. Rev. Biochem.* **53**, 717.

FABIANSSON, S., LASER REUTERSWARD, H. and LIBELLIUS, R. (1985) *Meat Sci.* **12**, 177.

FAGERSON, I. S. (1969) *J. Agric. Food Chem.* **17**, 747.

FAIRBROTHER, K. S., HOPWOOD, A. J., LOCKLEY, A. K. and BARDSLEY, R. G. (1998) *Meat Sci.* **50**, 105.

FAO/IAEA/WHO (1981) *Report of the Working Party on Irradiated Food.* WHO Tech. Rept. Ser. No. 659.

FAROUK, M. M. and PRICE, J. F. (1994) *Meat Sci.* **38**, 477.

FAROUK, M. M., WIELICZKO, K., LIM, R., TURWALD, S. and MACDONALD, G. A. (2002) *Meat Sci.* **61**, 85.

FAWCETT, D. N. and REVELL, J. P. (1961) *J. Biophys. Biochem. Cytol.* **10**, Suppl., p. 89.

FEARSON, W. R. and FOSTER, D. L. (1922) *Biochem. J.* **16**, 564.

FELDHUSEN, F., KIRSCHNER, T., KOCH, R., GIESE, W. and WENZEL, S. (1995) *Meat Sci.* **40**, 245.

FELL, H. B. and DINGLE, J. T. (1963) *Biochem. J.* **87**, 403.

FENTON, F., FLIGHT, I. T., ROBSON, D. S., BEAMER, K. C. and HOW, I. S. (1956) *Cornell Univ. Agric. Expt. Sta., Ithaca, N.Y. Mem.*, No. 341.

FERNAND, V. S. V. (1949) Ph.D. Thesis, Univ. London.

FERNANDEZ, X., MOUROT, T., MOUNIER, A. and ECOLAN, P. (1995) *Meat Sci.* **41**, 335.

FICKER, M. (1905) *Arch. Hyg. Berl.* **54**, 354.

FIDDLER, W., DOERR, R. C., ERTEL, J. R. and WASSERMAN, A. E. (1971) *J.O.A.C.* **54**, 1160.

FIDDLER, W., DOERR, R. C. and WASSERMAN, A. E. (1970) *Agric. Food Chem.* **18**, 310.

FIEDLER, I., ENDER, K., WICKE, M., MAAK, S., VON LENDERKEN, G. and MEYER, W. (1999) *Meat Sci.* **53**, 9.

FIELD, R. A. (1974) *Proc. Meat Ind. Res. Conf., Arlington*, p. 35. (Amer. Meat Inst. Chicago).

FIELD, R. A. (1976) *Wld. Rev. Anim. Prod.* **12**, 73.

FIEMS, L. O., BUTS, B., BOUCQUÉT, CH. V., DEMEYER, D. I. and COTTYN, B. G. (1989) *Meat Sci.* **27**, 29.

FINDLAY, J. D. (1950) *Bull. Hannah Dairy Res. Inst.*, No. 9.

FINDLAY, J. D. and BEAKLEY, W. R. (1954) *Progr. in the Physiology of Farm Animals* (Ed. J. HAMMOND), vol. 1, p. 252, Butterworths Scientific Publications, London.

FINN, D. B. (1932) *Proc. Roy. Soc.* **B 111**, 396.

FIRMAN, M. C., BACHMANN, H. J., HEYRICH, E. J. and HOPPER, P. F. (1959) *Food Tech.* **13**, 529.

FIRST, N. L. and SQUIRE, K. R. E. (1987) *Proc. 39th Ann. Recip. Meat Conf.* p. 41 (Natl. Livestock & Meat Bd.: Chicago).

FISCHER, R. G., BLAIR, J. M. and PETERSEN, M. S. (1954) *The Quality and Stability of Canned Meats*, Q.M. Food and Container Inst., Nat. Res. Coun., Washington.

FISHER, P., MELLETT, F. D. and HOFFMAN, L. C. (2000) *Meat Sci.* **54**, 107.

FISHER, A. V., POUROS, A., WOOD, J. D., YOUNG-BOONG, K. and SHEARD, P. R. (2000) *Meat Sci.* **56**, 127.

FJELKNER-MODIG, S. and TORNBERG, E. (1986) *Meat Sci.* **17**, 213.

FLAIN, R. (1964) *Food Tech.* **18**, 753.

FLAMANT, F., KOHLER, M. and ASCHIERO, R. (1976) *Helv. Chim. Acta* **59**, 2308.

FLAMANT, E., SONNAY, P. and OHLOFF, G. (1977) *Helv. Chim. Acta* **60**, 1872.

FLEMING, A. (1929) *Brit. J. Exp. Path.* **10**, 226.

FLEMING, A. K. (1969) *J. Food Technol.* **4**, 199.

FLEMING, D. W., COCHI, S. L., MACDONALD, K. L., BRONDUM, J., HAYES, P. S., PLIKAYTIS, B. D., HOLMS, M. B., AUDURIER, A., BROOME, C. V. and REINGOLD, A. L. (1985) *New England J. Med.* **312**, 404.

FLINT, D. J. (1987) *J. Endocr.* **115**, 365.

FLINT, F. O. and FIRTH, B. M. (1983) *Analyst,* **108**, 757.

FLUX, D. S., MUMFORD, R. E. and BARCLAY, P. C. (1961) *N.Z. J. Agric Res.* **4**, 328.

FOLEY, C. W., HEIDENREICH, C. J. and LASLEY, J. F. (1960) *J. Hered.* **51**, 278.

FOLLET, M. J. (1974) Ph.D. Dissertation, Univ. Nottingham.

FOLLET, M. J., NORMAN, G. A. and RATCLIFF, P. W. (1974) *J. Fd. Technol.* **9**, 509.

FOOD STANDARDS COMMITTEE (1975) *Report on Novel Proteins*, H.M.S.O., London.

FORD, A. L., HARRIS, P. V, MCFARLANE, J. J., PARK, R. J. and SHORTHOSE, N. R. (1974) *C.S.I.R.O. Meat Res.* Rept. No. 2174.

FORD, A. L and PARK, R. J. (1980) In *Developments in Meat Science* Vol. I. (Ed. R. A. LAWRIE), p. 219 (Applied Sci.: Lond.).

FORD, A. L., PARK, R. J., RATCLIFF, D. and MURRAY, K. E. (1975) *Meat Res. in C.S.I.R.O.* p. 21.

FORMANEK, Z., LYNCH, A., GALVIN, K., FARKAS, J. and KERRY, J. P. (2003) *Meat Sci.* **63**, 433.

FORTIN, A., ROBERTSON, W. M. and TONG, A. K. W. (2005) *Meat Sci.* **69**, 297.

FOSTER, W. W. and JASON, A. C. (1954) U.K. Pat. Applic. No. 24,329,154.

FOSTER, W. W. and SIMPSON, T. H. (1961) *J. Sci. Food Agric.* **12**, 363.

FOX, J. B. JR. (1962) *Proc. 14th Res. Conf. Amer. Meat Inst. Found Chicago*, p. 93.

FOX, J. B., JR., STREHLER, T., BERNOFSKY, C. and SCHWEIGERT, B. S. (1958) *J. Agric. Food Chem.* **6**, 692.

FOX, J. B., JR. and THOMSON, J. S. (1963) *Biochemistry* **2**, 465.

FREARSON, N. and PERRY, S. V. (1975) *Biochem. J.* **151**, 99.

FREARSON, N., SOLERO, R. J. and PERRY, S. V. (1976) *Nature, Lond.* **264**, 801.

FREARSON, N., TAYLOR, R. D. and PERRY, S. V. (1981) *Clin. Sci.* **61**, 141.

FRICKE, H. (1938) *Cold Spr. Harb. Symp. Quant. Biol.* **6**, 134.

FRITZ, J. D., MITCHELL, M. C., MARSH, B. B. and GREASER, M. (1993) *Meat Sci.*, **33**, 41.

FRONING, G. W. and JOHNSON, F. (1973) *J. Fd. Sci.* **38**, 279.

FRUHBECK, G., AGUADO, M. and MARTÍNEZ, J. A. (1997) *Biochem. Biophys. Res. Commun.* **240**, 590.

FUJII, J., ORTSU, K., ZORZATO, F., DELEON, S., KHAMANA, V. K., WEILER, J. E., O'BRIEN, P. J. and MACLENNAN, D. H. (1991) *Science* **253**, 448.

FUJIMAKI, M., KATO, S. and KANATA, T. (1969) *Agric. Biol. Chem.* **33**, 1144.

FURMINGER, I. G. S. (1964) *Nature, Lond.* **202**, 1332.

FURUTA, M., DOHMARU, T., KATAYAMA, T., TORATANI, H. and TAKEDA, A. (1992) *J. Agric. Food Chem.* **40**, 1099.

GADDIS, A. M., HANKINS, O. G. and HINER, R. L. (1950) *Food Tech.* **4**, 498.

GADIAN, D. G. (1980) In *Developments in Meat Science* Vol. I (Ed. R. A. LAWRIE), p. 89 (Applied Sci., Lond.).

GAGNIÈRE, H., PICARD, B., JURIE, C. and GEAY, Y. (1997) *Meat Sci.* **45**, 145.
GALBRAITH, N. S., BARRETT, N. and SOCKETT, P. N. (1987) *B.N.F. Nutr. Bull.* **12**, 21.
GALE, E. F. (1947) *The Chemical Activities of Bacteria*, Univ. Tut. Press, London.
GALT, A. M. (1981) Ph.D. Dissertation, Univ. London.
GALT, A. M. and MACLEOD, G. (1984) *J. Agric. Food Chem.* **32**, 59.
GALTON, M. M., LOWERY, W. D. and HARDY, A. V. (1954) *J. Infect. Dis.* **95**, 236.
GALVIN, K., LYNCH, A-H., KERRY, J. P., MORRISSEY, P. A. and BUCKLEY, D. J. (2000) *Meat Sci.* **55**, 7.
GAMMON, D. L, KEMP, J. D., EDNEY, J. M. and VARNEY, W. Y. (1968) *J. Food Sci.* **33**, 417.
GARDNER, G. A., CARSON, A. W. and PATTON, J. (1967) *J. Appl. Bact.* **30**, 321.
GARDNER, G. A. and PATTERSON, R. L S. (1975) *J. Appl. Bact.* **39**, 263.
GARDNER, G. A. and PATTON, J. (1969) *Proc. 15th Meeting European Meat Res. Workers*, Helsinki, p. 176.
GARNER, F. H. (1944) *The Cattle of Britain*, Longmans, London.
GARTON, G. A., DUNCAN, W. R. H. and MCEWAN, E. H. (1971) *Canad. J. Zool.* **49**, 1159.
GARVEN, H. S. D. (1925) *Brain* **48**, 380.
GASS, G. H., COATS, D. and GRAHAM, D. (1964) *J. Nat. Cancer Inst.* **33**, 971.
GAULT, N. F. S. (1978) Ph.D. Dissertation, Univ. Nottingham.
GAULT, N. F. S. (1985) *Meat Sci.* **15**, 15.
GAULT, N. F. S. (1991) In *Developments in Meat Science–5* (Ed. R. A. LAWRIE) p. 191 (Elsevier Appl. Sci.: Lond.).
GAUNT, R. A., BIRNIE, J. H. and EVERSOLE, W. J. (1949) *Physiol. Rev.* **29**, 281.
GAUTHIER, G. F. (1969) *Zellforsch.* **95**, 462.
GAUTHIER, G. F. (1970) In *The Physiology and Biochemistry of Muscle as a Food* (Eds. E. J. BRISKEY, R. G. GASSES and B. B. MARSH) p. 103 (Univ. Wisconsin Press: Madison).
GAUTHIER, G. F., LOWEY, S., BENFIELD, P. A. and HOBBS, A. W. (1982) *J. Cell Biol.* **92**, 471.
GAUTHIER, G. F. and PADYKULA, H. A. (1986) *J. Cell Biol.* **28**, 333.
GEESINK, G. H., MAREKO, M. H. D., MORTON, J. D. and BICKERSTAFFE, R. (2001) *Meat Sci.* **57**, 145.
GEIGER, B. (1979) *Cell* **18**, 193.
GEORGALA, D. L. and DAVIDSON, C. M. (1970) Brit. Pat. No. 1,199,998.
GEORGE, A. R., BENDALL, J. R. and JONES R. C. D. (1980) *Meat Sci.* **4**, 51.
GEORGE, J. C. and NAIK, R. M. (1958) *Nature, Lond.* **181**, 782.
GEORGE, J. C. and SCARIA, K. S. (1958) *Nature, Lond.* **181**, 783.
GERELT, B., IKEUCHI, Y., NISHIUMI, T. and SUZUKI, A. (2002) *Meat Sci.* **60**, 237.
GERELT, B., IKEUCHI, I. and SUZUKI, A. (2000) *Meat Sci.* **56**, 211.
GERGELY, J. (1970) In *The Physiology and Biochemistry of Muscle as a Food*, vol. II (Eds. E. J. BRISKEY, K. G. CASSENS and B. B. MARSH), Univ. Wisconsin Press, Madison, p. 349.
GERGELY, J., PRAGEY, D., SCHOLTZ, A. F., SEIDEL, J. C., SRETER, F. A. and THOMPSON, M. M. (1965) In *Molecular Biology of Muscular Contraction*, p. 145, Igaku Shoin Ltd., Tokyo.
GERRARD, F. (1935) *Sausage and Small Goods Manufacture*, 1st ed., Leonard Hill, London.
GERRARD, F. (1951) *Meat Technology*, 2nd ed., Leonard Hill, London.
GERSHMAN, L. C., STRACHER, A. and DREIZEN, P. (1969) *J. Biol. Chem.* **244**, 2726.
GIACINO, C. (1968) U.S. Pat. No. 394,017.
GIBBONS, N. E. (1958) *The Microbiology of Fish and Meat Curing Brines* (Ed. B. P. EDDY), p. 69, H.M.S.O., London.
GIBBONS, N. E. and ROSE, D. (1950) *Canad. J. Res.* **28**, 438.
GIBSON, A. M., ROBERTS, T. A. and ROBINSON, A. (1982) *J. Fd. Technol.* **17**, 471.
GIBSON, Q. H. (1948) *Biochem. J.* **42**, 13.
GIFFEE, J. W., MADISON, H. L. and LANDMANN, W. A. (1963) *Proc. 9th Meeting European Meat Res. Workers*, Paper No. 28.
GIGIEL, A. J. (1985) *Int. J. Refrig.* **8**(2), 91.
GIGIEL, A. J. and JAMES, S. J. (1984) *Meat Sci.* **11**, 1.
GILBERT, K. V. and DAVEY, C. L. (1976) *N.Z. J. Agric. Res.* **19**, 1.
GILBERT, K. V. and DEVINE, C. E. (1982) *Meat Sci.* **7**, 197.
GILES, B. G. (1969) *Proc. 15th Meeting European Meat Res. Workers*, Helsinki, p. 289.
GILL, C. O. (1987) *Proc. 4th A.A.A.P. Anim. Sci. Conf.*, Hamilton, N.Z. p. 388.
GILL, C. O. (1988) *Meat Sci.* **22**, 65.
GILL, C. O. and JONES, T. (1994) *Meat Sci.* **38**, 385.
GILL, C. O. and JONES, T. (1996) *Meat Sci.* **42**, 203.
GILL, C. O., LEET, N. G. and PENNEY, N. (1984) *Meat Sci.* **10**, 265.
GILL, C. O. and MCGINNIS, J. C. (1995) *Meat Sci.* **39**, 387.
GILL, C. O. and MCGINNIS, J. C. (2004) *Meat Sci.* **68**, 333.
GILL, C. O. and NEWTON, K. G. (1978) *Meat Sci.* **2**, 207.
GILL, C. O. and PENNEY, N. (1986) *Meat Sci.* **18**, 41.
GILL, C. O., PENNEY, N. and NOTTINGHAM, P. M. (1976) *Appl. Envir. Microbiol.* **31**, 465.
GILL, C. O. and TAN, K. H. (1980) *Appl. Environ. Microbiol.* **39**, 317.
GILLESPIE, C. A., SIMPSON, D. R. and EDGERTON, V. R. (1970) *J. Histochem. Cytochem.* **18**, 552.

GINDLIN, I., FRID, N. and YAKEVLEV, N. (1958) *Bull. Inst. Froid. Annexe*, 1958–2, p. 153.

GINGER, B. and WEIR, C. E. (1958) *Food Res.* **23**, 662.

GINGER, I. D., LEWIS, U. J. and SCHWEIGERT, B. S. (1955) *J. Agric. Food Chem.* **3**, 156.

GINSBERG, A., REID, M., GRIEVE, J. M. and OGONOWSKI, K. (1958) *Vet. Rec.* **70**, 700.

GOLDBERG, H. S. (1962) *Antibiotics in Agriculture* (Ed. M. WOODBINE), p. 289, Butterworths, London.

GOLDBERG, H. S., WEISER, H. H. and DEATHERAGE, F. E. (1953) *Food Tech.* **7**, 165.

GOLDSCHMIDT, V. M. (1922) Naturwiss. **10**, 918; *Chem. Abs.* **17**, 3665.

GOLDSCHMIDT, V. M. (1923) *Videnskorps. Skrift.*, *Mat-Nat. Kl.* No. 23; *Chem. Abs.* **17**, 3665.

GOLDSMITH, E., SPRANG, S. and FLATTERICK, R. (1982) *J. Mol. Biol.* **156**, 411.

GOLDSPINK, G. (1962a) *Comp. Biochem. Physiol.* **7**, 157.

GOLDSPINK, G. (1962b) *Proc. Roy. Irish Acad.* **62B**, 135.

GOLDSPINK, G. (1962c) Ph.D. Thesis, Univ. of Dublin.

GOLDSPINK, G. (1970) In *The Physiology and Biochemistry of Muscle as a Food*, vol. II, p. 521 (Eds. E. J. BRISKEY, R. G. CASSENS and B. B. MARSH), Univ. Wisconsin Press, Madison.

GOLDSPINK, G. and MCLOUGHLIN, J. V. (1964) *Irish J. Agric. Res.* **3**, 9.

GOLL, D. E. (1968) *Proc. 21st Ann. Recip. Meat Conf.*, *Athens, Georgia*, p. 16.

GOLL, D. E. (1970) In *The Physiology and Biochemistry of Muscle as a Food*, vol. II, p. 255 (Eds. E. J. BRISKEY, R. G. CASSENS and B. B. MARSH), Univ. Wisconsin Press, Madison.

GOLL, D. E., BRAY, R. W. and HOEKSTRA, W. G. (1963) *J. Food Sci.* **28**, 503.

GOLL, D. E., STROMER, M. H., ROBSON, R. M., TEMPLE, J., EASON, B. A. and BUSCH, W. H. (1974) *J. Anim. Sci.* **33**, 963.

GOLL, D. E., THOMPSON, V. F., LI, H., WEI, W. and CONG, J. (2003) *Physiol. Rev.* **83**, 731.

GOLL, D. E., THOMPSON, V. F., TAYLOR, R. G. and CHRISTIANSEN, J. A. (1992) *Biochemie* **74**, 225.

GONZALEZ, C. B., SALILTO, V. A., CARDUZA, F. T., PAZOS, A. A. and LASTA, J. A. (2001) *Meat Sci.* **57**, 251.

GOODENOUGH, P. W. (1998) *Int. J. Fd. Sci. Technol.* **33**, 63.

GORBATOV, V. M. and LYASKOVSKAYA, Y. N. (1980) *Meat Sci.* **4**, 209.

GORDON, L. and BOLAND, M. P. (1978) *Wld. Rev. Anim. Prod.* **14**, 2, 9.

GORDON, L., WILLIAMS, G. and EDWARDS, J. (1962) *J. Agric. Sci.* **59**, 143.

GORDON, R. A. and MURRELL, W. G. (1967) *C.S.I.R.O. Food Pres. Quart.* **27**, 6.

GOTTSCHALL, R. A. and KIES, M. W. (1942) *Food Res.* **7**, 373.

GOULD, S. E., GOMBERG, H. J. and BETHALL, F. H. (1953) *Amer. J. Publ. Hlth.* **43**, 1550.

GOUTEFONGEA, R., RAMPSON, V., NICOLAS, N. and DUMONT, J. D. (1995) *Proc. 41st Int. Congr. Meat Sci. Technol.*, Vol, II, p. 384.

GRAHAM, A. and HUSBAND, P. M. (1976) *Meat Res. C.S.I.R.O.*, p. 10.

GRAND, R. J. A., PERRY, S. V. and WEEKS, R. A. (1979) *Biochem. J.* **177**, 521.

GRANDE, F. (1975) *Proc. 9th Intl. Congr. Nutr.*, *Mexico.* **1**, 346.

GRANDIN, T. (1993) *Livestock Handling & Transport* (CAB International: Wallingford).

GRANER, M., CAHILL, V. R. and OCKERMAN, H. (1969) *Food Technol.* **23**, 94.

GRANGER, B. L. and LAZARIDES, E. (1978) *Cell*, **15**, 1253.

GRANT, R. (1974) *Process Biochem.* **9**, 11.

GRAU, F. (1988) *J. Food Protect.* **51**, 857.

GRAU, F., EUSTACE, L. J. and BILL, B. A. (1985) *J. Fd. Sci.* **50**, 482.

GRAU, F. and HAMM, R. (1953) *Naturwiss.* **40**, 29.

GRAU, F. and SMITH, M. G. (1974) *J. Appl. Bact.* **137**, 111.

GRAU, R. (1952) *Fleischwirts.* **4**, 83.

GRAU, R. and FRIESS-SCHULTHEISS, A. (1962) *Fleischwirts.* **14**, 207.

GRAU, R. and HAMM, R. (1952) *Fleischwirts.* **4**, 295.

GRAY, J. H. (1966) *The Winter Years*, p. 180 (Laurentian Library: Canada).

GRAY, J. I. and PEARSON, A. M. (1987) In *Advances in Meat Research, Vol. 3. Restructured Meat and Poultry Products* (Eds. A. M. PEARSON and T. R. DUTSON) p. 222 (Van Nostrand Rheinhold Co.: New York).

GREASER, M. L (1974) *Proc. 27th Ann. Recip. Meat Conf.* (Meat Livestock and Meat Bd., Chicago), p. 337.

GREASER, M. L. (1977) *Proc. 30th Ann. Recip. Meat Conf.* p. 149 (Natl. Meat & Livestock Bd.: Chicago).

GREASER, M. L., CASSENS, R. G., HOEKSTRA, W. G. and BRISKEY, E. J. (1969a) *J. Food Sci.* **34**, 633.

GREASER, M. L, CASSENS, R. G., HOEKSTRA, W. G., BRISKEY, E. J., SCHMIDT, G. R., CARR, S. D. and GALLOWAY, D. E. (1969b) *J. Anim. Sci.* **28**, 589.

GREASER, M. L. and GERGELY, J. (1971) *J. Biol. Chem.* **248**, 2128.

GREASER, M. L., WANG, S. M. and LEMANSKI, L. F. (1981) *Proc. 34th Ann. Recip. Meat Conf.*, **34**, 12.

GREAVES, R. L. N. (1960) *Recent Research in Freezing and Drying* (Eds. A. S. PARICES and A. U. SMITH), p. 203, Blackwell, Oxford.

GREEN, F. and BRONLEE, L. E. (1965) *Aust. Vet. J.* **41**, 321.

GREENBAUM, A. L. and YOUNG, F. G. (1953) *J. Endocrin.* **9**, 127.

GREENBERG, L. A., LESTER, D. and HAGGARD, H. W. (1943) *J. Biol. Chem.* **151**, 665.

GREENBERG, R. A. (1972) *Proc. Meat Ind. Res. Conf. A.M.I.F.*, Chicago, p. 25.

GREENBERG, R. A. (1975) *Proc. Meat Ind. Res. Conf. A.M.I.F.*, Chicago, p. 71.

GREENE, H. (1956) In RUSSELL, F. C. and DUNCAN, D. L. (1956) *loc. cit.*

GREENE, L. E. (1981) *Biochem.* **20**, 2120.

GREGORY, H. D., TRUSCOTT, T. G. and WOOD, J. D. (1980) *Proc. Nutr. Soc.* **39**, 74.

GREER, G. G. and JONES, S. D. M. (1997) *Meat Sci.* **45**, 61.

GREGORY, K. E. and DICKERSON, G. E. (1952) *Mo. Agric. Exp. Sta. Res. Bull.*, No. 493.

GREGORY, N. G. (1994) *Meat Sci.* **36**, 45.

GREGORY, P. W. and CASTLE, W. E. (1931) *J. Exp. Zool.* **59**, 199.

GREVER, A. B. G. (1955) *Ann. Inst. Pasteur, Lille* **7**, 24.

GRIFFITHS, E., VICKERY, J. R. and HOLMES, N. E. (1932) *Spec. Rept. Fd. Invest. Bd., Lond.*, No. 41.

GRIFFITHS, N. M. and PATTERSON, R. L. S. (1970) *J. Sci. Food Agric.* **21,** 4.

GRONERT, G. A. (1980) *Anaesthesiology*, **53**, 35.

GRONINGER, H. S., TAPPEL, A. L. and KNAPP, F. W. (1956) *Food Res.* **21**, 555.

GROSS, J. (1958) *J. Exp. Med.* **107**, 265.

GROSS, J. (1961) *Sci. Amer.* May, p. 120.

GROSS, J. (1970) In *Chemistry and Molecular Biology of the Intracellular Matrix* (Ed. E. A. BALEZS), Vol. 3, p. 1623. Acad. Press, N.Y.

GROSS, M. (2002) *Chem. Britain* **38**(4), 27.

GRUHN, H. (1965) *Sond. Nahrung*, **9**, 325.

GUERRERO, I., MENDIOLA, R. L. S., PONCE, E. and ANATO, H. (1995) *Meat Sci.* **40**, 397.

GULBRANDSEN, L. F. (1935) *Amer. J. Hyg.* **22**, 257.

GUSTAVSEN, K. H. (1956) *The Chemistry and Reactivity of Collagen*, p. 174, Academic Press, New York.

GUTHRIE, H. D. and POLGE, C. (1976) *J. Reprod Fertil.* **48**, 423.

GUYON, R., DOREY, F., MALAS, J. P., GRIMONT, F., FORET, J., ROUVIERE, B. and COLLOBERT, J. F. (2001) *Meat Sci.* **58**, 329.

HAAS, M. C. and BRATZLER, L. J. (1965) *J. Food Sci.* **30**, 64.

HADLOCK, R. (1969) *Proc. 15th Meeting European Meat Res. Workers, Helsinki*, p. 149.

HAFEZ, E. S. E. (1961) *Cornell Vet.* **51**, 299.

HAGAN, H. F. (1954) *Proc. 6th Res. Conf., Amer. Meat Res. Inst. Found., Chicago*, p. 59.

HAGYARD, C. J. and HAND R. J. (1976 77) *Ann. Res. Rept., Meat Ind. Res. Inst. N.Z. Inc.*, p. 31.

HAINES, R. B. (1931) *J. Soc. Chem. Ind.* **50**, 223T.

HAINES, R. B. (1933) *J. Soc. Chem. Ind.* **52**, 13T.

HAINES, R. B. (1934) *J. Hyg., Camb.* **34**, 277.

HAINES, R. B. (1937) *Spec. Rept. Food Invest. Bd., Lond.*, No. 45.

HAINES, R. B. and SMITH, E. C. (1933) *Spec. Rept. Food Invest. Bd., Lond.*, No. 43.

HALL, G. O. (1950) U.S. Pat. No. 2513094.

HALLECK, F. E., BALL, C. O. and STIER, E. F. (1958) *Food Tech.* **12**, 197, 654.

HALLIWELL, B., MURCIA, M. A., CHIRICO, S. and ARUOMA, O. I. (1995) *Crit. Rev. Fd. Sci. Nutr.* **35**, (1/2).

HALLMARK, E. L. and VAN DUYNE, F. U. (1961) *J. Amer. Diet. Ass.* **45**, 139.

HALLUND, O. and BENDALL, J. R. (1965) *J. Food Sci.* **30**, 296.

HALVORSON, H. O. (1955) *Ann. Inst. Pasteur, Lille* **7**, 53.

HAMDY, M. K., MAY, K. N. and POWERS, J. J. (1961) *Proc. Soc. Exp. Biol. Med.* **108**, 185.

HAMILTON, D. N., ELLIS, M., MCKEITH, F. K. and EGGERT, J. J. (2003) *Meat Sci.* **65**, 853.

HAMILTON, R. G. and RICHERT, S. H. (1976) *C.S.I.R.O. Div. Fd. Res.*, Meat Res. Lab., Pap. No. 12.

HAMM, R. (1953) *Deut. Lebensm. Rundschau* **49**, 153.

HAMM, R. (1955) *Fleischwirts.* **7**, 196.

HAMM, R. (1957) *Z. Lebensmitt. Untersuch.* **106**, 281.

HAMM, R. (1959) *Z. Lebensmitt. Untersuch.* **110**, 95, 227.

HAMM, R. (1960) *Adv. Fd. Res.* **10**, 356.

HAMM, R. (1966) *Fleischwirts.* **18**, 856.

HAMM, R. (1975) In *Meat* (Eds. D. J. A. COLE and R. A. LAWRIE), Butterworths, London, p. 321.

HAMM, R. (1981) In *Developments in Meat Science* Vol. II. (Ed. R. A. LAWRIE), p. 93. (Applied Sci., Lond).

HAMM, R. and DEATHERAGE, F. E (1960a) *Food Res.* **25**, 387.

HAMM, R. and DEATHERAGE, F. E. (1960b) *Food Res.* **25**, 573.

HAMM, R. and GRAU, R. (1958) *Z. Lebensmitt. Untersuch.* **108**, 280.

HAMM, R. and KORMENDY, L. (1966) *Fleischwirts.* **46**, 615.

HAMM, R. and HOFMAN, K (1965) *Nature, Lond.* **207**, 1269.

HAMMAN, A. and MATTHAEI, S. (1996) *Exp. Clin. Endocrinol. Diabetes* **104**, 293.

HAMMES, W. P. and HERTEL, C. (1998) *Meat Sci.* **49**, S 125.

HAMMOND, J. (1932a) *Growth and Development of Mutton Qualities in the Sheep*, Oliver & Boyd, London.

HAMMOND, J. (1932b) *J. Roy. Agric. Soc.* **93**, 131.

HAMMOND, J. (1933–4) *Pig Breeders Annual*, p. 28, Nat. Pig Breeders Assoc., London.

HAMMOND, J. (1936) *Festschrift Prof. Duerst Berne.*

HAMMOND, J. (1940) *Farm Animals: Their Breeding, Growth and Inheritance,* 1st ed., Edward Arnold, London.

HAMMOND, J. (1944) *Proc. Nutr. Soc.* **2**, 8.

HAMMOND, J. (1949) *Brit. J. Nutr.* **3**, 79.

HAMMOND, J. (1957) *Outlook in Agriculture* **1**, 230.

HAMMOND, J. (1963a) Personal communication.

HAMMOND, J. (1963b) *Meat Trades J.* **190**, 439.

HAMMOND, J., JR and BATTACHANYHA, P. (1944) *J. Agric. Sci.* **34**, 1.

HAMMOND, J., JR., HAMMOND, J. and PARKES, A. S. (1942) *J. Agric. Sci.* **32**, 308.

HAMOIR, G. and LASZT, L. (1962) *Biochim. Biophys. Acta* **59**, 365.

HAN, G. D., MATSUMO, M., ITO, G., IKEUCHI, Y. and SUZUKI, A. (2000) *Biosci. Biotechnol. Biochem.* **64**, 1887.

HANKINS, O. G., SULZBACHER, W. L., KAUFFMAN, W. R. and MAYO, M. E. (1950) *Food Tech.* **4**, 33.

HANNAN, R. S. (1955) *Spec. Rept. Fd. Invest. Bd., Lond.*, No. 61.

HANSEN, D. and RIEMANN, H. (1958) *Slagteriernes Forskningsinstitut, Denmark,* No. 84.

HANSEN, E., ANDERSEN, M. L. and SKIBSTED, L. H. (2003) *Meat Sci.* **63**, 63.

HANSEN, N. H. (1960) *Slagteriernes Forshningsinstitut,* Denmark, No. 28.

HANSEN, S., HANSEN, T., AASLING, M. D. and BYRNE, D. V. (2002) *Meat Sci.* **68**, 611.

HANSET, R. and MICHAUX, C. (1985) *Génét. Sél. Evol.* **17**, 359.

HANSON, J. and HUXLEY, H. E. (1953) *Nature, Lond.* **172**, 530.

HANSON, J. and HUXLEY, H. E. (1955) *Symp. Soc. Exp. Biol.* **9**, 228.

HANSON, J. and LOWY, J. (1963) *J. Mol. Biol.* **6**, 46.

HANSON, S. W. F. (1961) *The Accelerated Freeze-drying (AFD) Method of Food Preservation*, H.M.S.O., London.

HARDING, S. S. and BARDSLEY, R. G. (1986) *Macromolecular Preprints*, 146.

HARDY, M. F., HARRIS, C. J., PERRY, S. V. and STONE, D. (1970) *Biochem. J.* **120**, 643.

HARDY, W. V., DOWNING, H. E., REYNOLDS, W. M. and LUTFER, H. G. (1953–4) *Antibiotics Annual*, p. 372, Medical Encyclopaedia Inc., New York.

HARESIGN, W. (1976) *Anim. Prod.* **22**, 137.

HARGIN, K. D. (1996) *Meat Sci.* **43**, Suppl., S277.

HARPER, J. M. (1979) *CRC Crit. Rev. Fd. Sci. Nutr.* **11**, 1551.

HARPER, J. M. M. and BUTTERY, P. J. (1988) In *Developments in Meat Science* Vol. 4 (Ed. R. A. LAWRIE), p. L (Elsevier Appl. Sci., Lond.).

HARRIES, J. M., BRYCE-JONES, K., HOUSTON, T. W. and ROHERTSON, J. (1963) *J. Sci. Fd. Agric.* **14**, 501.

HARRIES, J. M., HUBBARD A. W., ALDER, F. E., KAY, M. and WILLIAMS, D. R. (1968) *Brit. J. Nutr.* **22**, 21.

HARRINGTON, G. (1962) *Outlook in Agriculture* **3**, 180.

HARRIS, G. W., REED, M. and FAWCETT, C. P. (1966) *Brit. Med. Bull.* **22**, 266.

HARRIS, P. V. and MCFARLANE, J. J. (1971) *Proc. 17th Meeting Res. Inst.*, Bristol, p. 102.

HARRIS, P. V. and SHORTHOSE, R. (1987) In *Developments in Meat Science* Vol. 4 (Ed. R. A. LAWRIE), p. 245.

HARRIS, R. E. and VON LOESECKE, H. (1960) *Nutritional Evaluation of Food Processing*, Wiley, New York.

HARRIS, R. C., VIRU, M., GREENHAFF, P. L. and HULTMAN, E. (1993) *J. Physiol.* **467**, 74P.

HARRISON, D. L. (1948) *Iowa Sta. Coll. J. Sci.* **23**, 36.

HARRISON, W. C. (1967) *Health Phys.* **13**, 383.

HARSHAM, A. and DEATHERAGE, F. E. (1951) U.S. Pat No. 2,544,681

HATEM, I., TAN, J. and GERRARD, D. E. (2003) *Meat Sci.* **65**, 999.

HATTON, M. W. C., LAWRIE, R. A., RATCLIFF, P. W. and WAYNE, N. (1972) *J. Fd. Technol.* **7**, 443.

HATTORI, A. and TAKAHASHI, K. (1988) *J. Biochem.* **103**, 809.

HAUGHEY, D. P. and MARER, J. M. (1971) *J. Fd. Technol.* **6**, 119.

HAUROWITZ, F. (1950) *Chemistry and Biology of Proteins*, Academic Press, New York.

HAWLEY, H. B. (1962) *Antibiotics in Agriculture* (Ed. M. WOODBINE), p. 272, Butterworths, London.

HAY, P. H., HARRISON, D. L. and VAIL, G. E. (1953) *Food. Tech.* **7**, 217.

HAZELL, T., LEDWARD, D. A., NEALE, R. J. and ROOT, I. C. (1981) *Meat Sci.*, **5**, 397.

HAZLEWOOD, C. F., NICHOLS, B. L. and CHAMBERLAIN, N. F. (1969) *Nature, Lond.* **222**, 747.

HEAD, J. F. and PERRY, S. V. (1974) *Biochem. J.* **137**, 145.

HEAD, J. F., WEEKS, R. A. and PERRY, S. V. (1977) *Biochem. J.* **161**, 465.

HEADIN, P. A., KURTZ, G. W. and KOCH, R. B. (1961) *J. Food Sci.* **26**, 112, 212.

HECTOR, D. A., BREW-GRAVES, C., HASSEN, N. and LEDWARD, D. A. (1992) *Meat Sci.* **31**, 299.

HEDRICK, H. B., BAILEY, M. E., DUPUY, H. P. and LEGENDRE, M. G. (1980) *Proc. 26th Europ. Meeting Meat Res. Workers, Colorado Springs*, **1**, 307.

HEGARTY, G. R., BRATZLER, L. J. and PEARSON, A. M. (1963) *J. Food Sci.* **28**, 525.

HEIDEMANN, F. H. and WISMER-PEDERSEN, J. (1976) *Proc. 22nd Europ. Meeting Meat Res. Workers, Stockholm*, **1**, A6:1.

HEIDEMANN, F. H. and WISMER-PEDERSEN, J. (1977) *Proc. 23rd Europ. Meeting Meat Res. Workers, Moscow.*

HEIDENHAIN, M. (1913) *Arch. Mikr. Anat.* **83**, 427.

HEIDTMANN, R. (1959) *Fleischwirts.* **11**, 199.

HEITZMAN, R. J. (1976) In *Anabolic Agents in Animal Production* (Eds. F. C. LU and J. RENDEL), p. 9. (Thieme: Stuttgart).

HELANDER, E. (1957) *Acta Physiol. Scand.*, Suppl. **41**, 141.

HELANDER, E. (1958) *Nature, Lond.* **182**, 1035.

HELANDER, E. (1959) *Acta Morph. Neer-scand.* **22**, 230.

HELLENDOORN, E. W. (1962) *Food Tech.* **16**, 119.

HENCKEL, P., JØRGENSEN, P. F. and JENSEN, P. (1992) *Meat Sci.* **32**, 131.

HENDERSON, D. W., GOLL, D. E. and STROMER, M. H. (1970) *Amer. J. Anat.* **128**, 117.

HENRICKSON, R. L., BRADY, D. E., GEHRKE, C. W. and BROOKS, R. F. (1955) *Food Tech.* **9**, 290.

HENRICKSON, R. L., PARR, A. F., CAGLE, E. D., ARGANOSA, F. C. and JOHNSON, R. G. (1969) *Proc. 15th Meeting European Meat Res. Workers, Helsinki,* p. 23.

HENRY, M., ROMANI, J. D. and JOUBERT, L. (1958) *Rev. Path. Gén Physiol. Clin.* **696**, 355.

HERNANDEZ, P., PARK, D. and RHEE, K. S. (2002) *Meat Sci.* **61**, 405.

HERRING, H. K. (1968) *Proc. 21st Ann. Recip. Meat Conf., Athens, Georgia,* p. 47.

HERRING, H. K., CASSENS, R. G. and BRISKEY, E. J. (1965) *J. Food Sci.* **30**, 1049.

HERSCHDOERFER, S. M. and DYETT, E. J. (1959) *5th Meeting European Meat Res. Workers, Paris,* Paper No. 27.

HERTER, M. and WILSDORF, G. (1914) *Deut. Landwirtschafts-Gesellschaft,* Heft 270, Berlin.

HERZ, K. O. and CHANG, S. S. (1970) *Adv. Food Res.* **18**, 2.

HESS, E. (1928) *J. Bact.* **15**, 33.

HEYWOOD, S. M. (1970) in *The Physiology and Biochemistry of Muscle as a Food,* vol. II, p. 13 (Eds. E. J. BRISKEY, R. G. CASSENS and B. B. MARSH), Madison, Univ. Wisconsin Press.

HEYWOOD, S. M. and RICH, A. (1968) *Proc. Nat. Acad. Sci., U.S.A.* **59**, 590.

HIBBERT, I. (1973) Ph.D. Dissertation, Univ. Nottingham.

HIBBERT, I. and LAWRIE, R. A. (1972) *J. Fd. Technol.* **7**, 326.

HICKS, E. W., SCOTT, W. J. and VICKERY, J. R. (1956) *Food Pres. Quart.* **16**, 72.

HIGGS, A. R., NORRIS, R. T. and RICHARDS, R. (1991) *Aust. J. Agric. Res.* **42**, 205.

HIGGS, J. D. (2000) *Food Sci. Technol. Today* **14**(1), 22.

HIGLEY, N. A., TAYLOR, S. L., HERIAN, A. M., LEE, K. and MARSH, B. B. (1986) *Meat Sci.* **16**, 175.

HILL, A. V. (1930) *Proc. Roy. Soc.* **B 106**, 477.

HILL, H. J. and HUGHES, C. E. (1959) *Armours Analysis* **8**, 1.

HIMWICH, H. E (1955) *Sci. Amer.* **195**, 74.

HINER, R. L. (1951) *Proc. 2nd Res. Conf., Amer. Meat. Inst. Found,* p. 92.

HINER, R. L., ANDERSON, E. E. and FELLERS, C. R. (1955) *Food Tech.* **9**, 80.

HINER, R. L., GADDIS, A. M. and HANKINS, O. G. (1951) *Food Tech.* **5**, 223.

HINER, R. L., and HANKINS, O. G. (1950) *J. Anim. Sci.* **9**, 347.

HINER, K. L., HANKINS, O. G., SLOANE, H. S., FELLERS, C. R. and ANDERSON, E. E. (1953) *Food Res.* **18**, 364.

HINER, R. L., MADSEN, L. L and HANKINS, O. G. (1945) *Food Res.* **10**, 312.

HINTON, M. H., RIXSON, P. D., ALLEN, V. and LINTON, A. H. (1984) *J. Hyg. Camb.* **93**, 547.

HINTZ, C. S., CHI, M. M. H., FELL, R. D., IVY, J. L., KAISER, K K, LOWRY, C. V. and LOWRY, O. H. (1982) *Amer. J. Physiol.* **242**, C218.

HITCHCOCK, C. H. S. and CRIMES, A. A. (1985) *Meat Sci.* **15**, 215.

HO, C. T. (1980) *Proc. Meat Ind Res. Conf.,* p. 41 (Amer. Meat Inst., Chicago).

HOAGLAND, R., MCBRYDE, C. N. and POWICK, W. C. (1917) *U.S. Dept. Agric. Bull.,* No. 433.

HOBBS, B. C. (1955) *Ann. Inst. Pasteur, Lille* **7**, 190.

HOBBS, B. C. (1960) *6th Meeting European Meat Res. Workers, Utrecht,* Paper No. 32.

HOCK, R. A. and MILLER, A. D. (1986) *Nature, Lond.* **320**, 275.

HOERSCH, T. M. (1967) U.S. Pat. Nos. 3,316,099 and 3,316,092.

HOET, J. P. and MARKS, H. P. (1926) *Proc. Roy. Soc.* **B 100**, 72.

HOFMAN, K. and HAMM, R. (1969) *Fleischwirts.* **49**, 1180.

HOFFMAN, B. and KUNG, H. (1976) In *Anabolic Agents in Animal Production* (Eds. F. C. LU and J. RENDEL), p. 181, (Thieme: Stuttgart).

HOFFMAN, L. C., KRITZINGER, B. and FERRIERA, A. V. (2005) *Meat Sci.* **69**, 551.

HOFLUND, S., HOLMBERG, J. and SELLMAN, G. (1956) *Cornell Vet.* **46**, 51.

HÖGBERG, A., PICKOVA, J., BABOL, J., ANDERSSON, K. and DUTTA, P. C. (2002) *Meat Sci.* **60**, 411.

HOGUE, A., AKKINA, J., ANGULA, F., JOHNSON, R., PEDERSEN, K., SAINI, P. and SCHLOSSER, W. (1998) *Proc. Beef Safety Sympos.,* Chicago, III. (AMIF: Chicago) p. 13.

HOLLENBECK, C. M. and MARINELLI, L. J. (1963) *Proc. 15th Res. Conf., Amer. Meat Inst. Found, Chicago,* p. 67.

HOLLIS, F., KAPLOW, M., KLOSE, R. and HALIK, J. (1968) *Tech. Rept.* 62–26 FL. U.S. Army, Natick.

HOLMES, A. W. (1960) Unilever, U.K. Pat. No. 848,014.

HOLTFRETER, J. (1934) *Archiv. f. Entwicklungsmechanik* **132**, 307, cited by J. NEEDHAM (1942) *loc. cit.*

HOLTZER, H., MARSHALL, J. M., JR and FINOK, H. (1957) *J. Biophys. Biochem. Cytol.* **3**, 705.

HOMAN, K. and HAMM, R. (1978) *Ad. Fd. Res.* **24**, 2.

HOMER, D. B. and MATTHEWS, K. R. (1998) *Meat Sci.* **49**, 425.

HOMMA, N., IKEUCHI, Y. and SUZUKI, A. (1994) *Meat Sci.* **38**, 219.

HOMMA, N., IKEUCHI, Y. and SUZUKI, A. (1995) *Meat Sci.* **41**, 251.

HONIKEL, K. O. (1998) *Meat Sci.* **49**, 447.

HONIKEL, K. O., FISCHER, C., HAMID, A. and HAMM, R. (1981) *J. Fd. Sci.* **46**, 1.

HONIKEL, K. O. and HAMM, R. (1978) *Meat Sci.* **2**, 181.

HONIKEL, K. O., KIM, C. J. and HAMM, R. (1986) *Meat Sci.* **16**, 267.

HONIKEL, K. O., RONCALES, P. and HAMM, R. (1983) *Meat Sci.* **8**, 221.

HOOD, D. E. (1971) *Proc. 17th Meeting European Meat Res. Workers, Bristol*, p. 677.

HOOD, D. E. (1975) *J. Sci. Fd. Agric.* **26**, 85.

HOOD, R. L. and ALLEN, E. (1971) *J. Fd. Sci.* **36**, 786.

HOPKINS, D. L. and THOMPSON, J. M. (2001a) *Meat Sci.* **57**, 1.

HOPKINS, D. L. and THOMPSON, J. M. (2001b) *Meat Sci.* **59**, 199.

HORGAN, D. J. (1980–81) *Meat Sci.* **5**, 297.

HORGAN, D. J. (1987) *Meat Sci.* **19**, 285.

HORGAN, D. J. (1991) *Meat Sci.* **29**, 243.

HORGAN, D. J., JONES, P. N., KING, N. L., KURTH, L. B. and KUYPERS, R. (1991) *Meat Sci.* **29**, 251.

HORGAN, D. J. and KUYPERS, R. (1983) *Meat Sci.* **8**, 65.

HORNSEY, H. C. (1959) *J. Sci. Fd. Agric.* **10**, 114.

HORNSTEIN L., CROWE, P. F. and HEIMBERG, M. J. (1961) *Food Res.* **26**, 581.

HORNSTEIN, L., CROWE, P. F. and HINER, R. L. (1968) *J. Food Sci.* **32**, 650.

HORNSTEIN, L., CROWE, P. F. and SULZBACHER, W. L. (1960) *J. Agric. Food Chem.* **8**, 65.

HORNSTEIN, L., CROWE, P. F. and SULZBACHER, W. L. (1963) *Nature, Lond.* **199**, 1252.

HOROWITS, R., KEMPNER, E. S., BISHER, M. E and PODLOSKY, R. H. (1986) *Nature, Lond.* **323**, 160.

HOSTETLER, R. L. (1970) *Proc. 16th Meeting European Meat Res. Workers, Varna*, p. 100.

HOUBEN, J. H. and KROL, B. (1985) *Meat Sci.* **13**, 193.

HOUDE, A. and POMMIER, S. A. (1993) *Meat Sci.* **33**, 349.

HOUGHAM, D. and WATTS, B. M. (1958) *Food Tech.* **12**, 681.

HOUSE OF LORDS (1985–86) Select Committee on the European Community. *Examination of Animals and Fresh Meat for the Presence of Residues* (HMSO: Lond.).

HOUTHIUS, M. J. J. (1957) Meat Hygiene, *F.A.O. Agricultural Series*, No. 34, p. 111

HOWARD, A. (1956) C.S.I.R.O. *Food Pres. Quart.* **16**, 26.

HOWARD, A. and LAWRIE, R. A. (1956) *Spec. Rept. Fd. Invest. Bd., Lond.*, No. 63.

HOWARD, A. and LAWRIE, R. A. (1957a) *Spec. Rept. Fd. Invest. Bd., Lond.*, No. 65.

HOWARD, A. and LAWRIE, R. A. (1957b) *Spec. Rept. Fd. Invest. Bd., Lond.*, No. 64.

HOWARD, A., LAWRIE, R. A. and LEE, C. A. (1960a) *Spec. Rept. Fd. Invest. Bd., Lond.*, No. 68.

HOWARD, A., LEE, C. A. and WEBSTER, H. L. (1960b) *C.S.I.R.O. Div. Food Pres. Tech. Paper*, No. 21.

HOWARD, A. J. (1949) *Canning Technology*, J. and A. Churchill, London.

HOWE, P. E. and BARBELLA, N. G. (1937) *Food Res.* **2**, 197.

HOWELL, N. (1981) Ph.D. Dissertation, Univ. Nottingham.

HOWELL, N. and LAWRIE R. A. (1981) *Proc. 27th Meeting Europ. Meat Res. Workers, Vienna*, p. 383.

HOWLAND, J. L. (2001) *Biologist* **48**(6), 278.

HSIAO, K., CHENG, C-H., FERNANDES, I. F., DETRICH, J. and DE VRIES, H. W. A. L. (1990) *Proc. Natl. Acad. Sci.* **87**, 9265.

HUBER, W., BRASCH, A. and WALY, A. (1953) *Food Tech.* **7**, 109.

HUFFMAN, D. L. (1981) *J. Fd. Sci.* **46**, 1563.

HUFFMAN, D. L. and CORDRAY, J. C. (1983) In *International Sympos.: Meat Science & Technology*, p. 229 (Natl. Meat & Livestock Bd.: Chicago).

HUFFMAN, D. L., PALMER, A. Z., CARPENTER, J. W., HARGROVE, D. D. and KOGER, M. (1967) *J. Anim. Sci.* **26**, 290.

HUFFMAN, R. D. (2002) *Meat Sci.* **62**, 285.

HUGAS, M., GARRIGA, M. and MONFORT, J. J. (2002) *Meat Sci.* **62**, 359.

HUGHES, M. C., HEALY, A., MCSWEENEY, P. L. H. and O'NEILL, E. E. (2000) *Meat Sci.* **56**, 165.

HUGHES, S. M., CHI, M. M., LOWRY, O. H. and GUNDERSEN, K. (1999) *J. Cell. Biol.* **145**, 633.

HUNT, D., PARKES, H. C. and DAVIES, I. D. (1997) *Food Chem.* **60**, 437.

HUNT, M. C. and HEDRICK, H. B. (1977) *J. Fd. Sci.* **42**, 573.

HUNT, S. and MATHESON, N. (1958) *Nature, Lond.* **181**, 472.

HUNTER, W. (1794) *A Treatise on Blood, Inflammation and Gunshot Wounds*, p. 88.

HUXLEY, H. E. (1960) *The Cell* (Eds. J. BRACHET and A. E. MIRSKY), vol. 4, p. 365, Academic Press, New York.

HUXLEY, H. E. (1963) *J. Mol. Biol.* **7**, 281.

HUXLEY, H. E. (1969) *Science* **164**, 1356.

HUXLEY, H. E. (1971) *Proc. Roy. Soc.* **B 178**, 131.

HUXLEY, H. E. and HANSON, J. (1957) *Biochim. Biophys. Acta* **23**, 229, 250.

HUXLEY, H. E. and HANSON, J. (1960) *The Structure and Function Of Muscle* (Ed. G. H. BOURNE), vol. 1, p. 183, Academic Press, New York.

HWANG, K. and IVY, A. C. (1951) *Ann. N. Y. Acad. Sci.* **54**, 143.

HYNES, R. O. (1992) *Cell* **69**, 11.

IGENE, J. O., KING, J. A., PEARSON, A. M. and GRAY, J. L. (1979) *J. Agric. Fd. Chem.* **27**, 838.

IKKAL, T. and OOI, T. (1966) *Biochem.* **5**, 1551.

IKKAI, T. and OOI, T. (1969) *Biochem.* **8**, 2615.

ILIAN, M. A., BECKIT, A. E.-D. and BICKERSTAFFE, A. (2004) *Meat Sci.* **66**, 317, 387.

IMESON, A. P., LEDWARD, D. A. and MITCHELL, J. R. (1979) *Meat Sci.* **3**, 287.

IMMONEN, K., RUUSUMEN, E. and PUOLANNE, E. (2000) *Meat Sci.* **55**, 33.

INGRAM, D. L. (1974) In *Heat Loss in Animals and Man* (Eds. J. L. MONTEITH and L. E. MOUNT), p. 233, Butterworths, London.

INGRAM, G. C. (1971) *Ann. Rept. A.R.C. Meat Res. Inst.*, 1970–71, p. 71.

INGRAM, M. (1948) *Ann. Inst. Pasteur, Lille* **75**, 139.

INGRAM, M. (1952) *J. Hyg. Camb.* **50**, 165.

INGRAM, M. (1958) *The Chemistry and Biology of Yeasts,* p. 603, Academic Press, New York.

INGRAM, M. (1959) *Intl. J. Appl. Radn. Isotopes* **6**, 105.

INGRAM, M. (1960) *J. Appl. Bact.* **23**, 206.

INGRAM, M. (1962) *J. Appl. Bact.* **25**, 259.

INGRAM, M. (1969a) *J. Soc. Chem. Ind.* **68**, 356.

INGRAM, M. (1969b) *Food Manuf.* **24**, 292.

INGRAM, M., HAWTHORNE, J. R. and GATHERUM, D. P. (1947) *Food Manuf.* **10**, 457, 506, 543.

INGRAM, M. and HOBBS, B. C. (1954) *J. Roy. Sanit. Inst.* **74**, 12.

INGRAM, M. and RHODES, D. N. (1962) *Food Manuf.* **37**, 318.

IRONSIDE, J. W. (1999) *Biologist* **46**, 172.

IRVING, J. (1956) *J. Appl. Physiol.* **9**, 414.

IRVING, L. (1951) *Fed. Proc.* **10**, 543.

ISKANDARYAN, A. K. (1958) *Proc. 4th Meeting European Meat Research Workers, Camb.*, Paper No. 25.

IYENGAR, M. R. and GOLDSPINK, G. (1971) Cited by G. Goldspink in *The Structure and Function of Muscle,* 2nd Ed. Vol. 1, p. 181. (Ed. G. H. BOURNE). (Acad. Press: New York).

JACKSON, R. L, MERRISET, J. D. and GOTTI, A. M. J. (1975) In *Hyperlipidaemia: Diagnosis and Therapy.* (Eds. B. M. RIFKIND and R. I. LEVY), p. 1. (Grune & Stratton: New York).

JACKSON, T. R, ACUFF G. R, VANDERZOUT, C., SHARP, T. R. and SAVELL, J. W. (1992) *Meat Sci.* **31**, 175.

JACOBSON M. WELLER, M., GALGON, M. W. and RUPNOW, E. H. (1962) *Factors in the Flavour and Tenderness of Lamb, Beef and Pork,* Washington Agric. Exp. Sta.

JAKOBBSON, B. and BERGTSSON, N. E. (1969) *Proc. 15th Meeting European Meat Res. Workers, Helsinki,* p. 482.

JALANGO, J. W., SAUL, G. L. and LAWRIE, R. A. (1987) *Meat Sci.* **21**, 73.

JAMES, S. M., FANNIN, S. L., HALL, B., PARKER, E., VOGT, J., RUN, G. and WILHEMS, D. L. (1985) *M.M.W.R.*, **34**, 357.

JANICKI M. A., KOLACZYK, S. and KORTZ, J. (1963) *Proc. 9th Meeting European Meat Res. Workers, Budapest,* Paper No. 2.

JANSEN, M. M. (1966) *Acta Vet. Scand.* **7**, 394.

JASPER, T. W. (1958) *Bull. Int. Froid. Annexe* 1958–2, p. 475.

JAULHAC, B., BES, M., BORNSTEIN, N., PIERMONT, Y., BRUN, Y. and FLEURETTE, J. (1992) *J. Appl. Bact.* **72**, 386.

JEACOCKE, R. E. (1984) *Meat Sci.* **11**, 237.

JEFFERSON, N. C., ARAI, T., GEISEL, T. and NECHELES, H. (1964) *Science,* **144**, 58.

JEFFREYS, A. J., WILSON, V. and THEIN, S. L. (1985) *Nature, Lond.* **314**, 67.

JENSEN, C., GUIDERA, J., SKOVGAARD, I. M., STAUN, H., SKIBSTED, L. H., JENSEN, S. K., MØLLER, P. J., BUCKLEY, J. and BERTELSEN, G. (1997) *Meat Sci.* **45**, 491.

JENSEN, L B. (1945) *Microbiology of Meats,* 2nd ed., Garrard Press, Champaign, Ill.

JENSEN, L. B. (1949) *Meat and Meat Foods,* p. 47, Ronald Press, New York.

JENSEN, L. B. and HESS, W. R. (1941) *Food Res.* **6**, 273.

JEREMIAH, L. E., DUGAN, M. E. R., AALHUS, J. L. and GIBSON, L. L. (2003b) *Meat Sci.* **65**, 985.

JEREMIAH, L. E., DUGAN, M. E. R., AALHUS, J. L. and GIBSON, L. L. (2003c) *Meat Sci.* **65**, 1013.

JEREMIAH, L. E., GIBSON, L. L., AALHUS, J. L. and DUGAN, M. E. R. (2003a) *Meat Sci.* **65**, 949.

JEREMIAH, L E., TONG, A. K. W. and GIBSON, L. L. (1991) *Meat Sci.* **30**, 97.

JIMINEZ-COLMENERO, F., CARBALLO, J. and COFRADES, S. (2001) *Meat Sci.* **59**, 5.

JOBLING, A. and HUGHES, R. G. (1977) 'Recovery and Utilization of Edible Collagen'. Meat Panel Res. Pap. SCI Food Group.

JOBLING, A. and JOBLING, C. A. (1983) In *Upgrading Wastefor Feeds and Food.* (Eds. D. A. LEDWARD, A. J. TAYLOR and R. A. LAWRIE), p. 183 (Butterworths, Lond.).

JOCSIMOVIC, J. (1969) *J. Sci. Agric. Res., Beograd* **22**, 109.

JOHNS, A. T., MANAGAN, J. L. and REID, C. S. W. (1957) *N. Z. Vet. J.* **5**, 115.

JOHNSON, D. W. and PALMER, L S. (1939) *J. Agric. Res.* **58**, 929.

JOHNSON, P. and PERRY, S. V. (1970) *Biochem. J.* **119**, 293.

JOHNSON, S. K and LAWRIE, R. A. (1988) *Meat Sci.* **22**, 303.

JOHNSON, S. K., WHITE, W. J. P. and LAWRIE, R. A. (1986) *Meat Sci.* **18**, 235.

JOHNSON, T. H. and BANCROFT, M. J. (1918) *Proc. Roy. Soc., Queensland,* **30**, 219.

JOLLEY, P. D. (1979) *J. Fd. Technol.* **14**, 81.

JONES, A. D., HOMAN, A. C., FAVELL, D. J., HITCHCOCK, C. H. S., BERRYMAN, P. M., GRIFFITHS, N. M. and BILLINGTON M. J. (1985) *Meat Sci.* **15**, 137.

JONES, A. D., SHORLEY, D. and HITCHCOCK, C. H. S. (1982) *J. Sci. Fd. Agric.* **33**, 677.

JONES, I. (2002) *Chem. Britain* **38**(4), 34.

JONES, R. W., EASTER, R. A., MCKEITH, E. K., DALRYMPLE, R. H., MADDOCK, H. M. and BECHTEL, P. J. (1985) *J. Anim. Sci.* **61**, 905.

JONES, S. D. M., GREER, G. G., JEREMIAH, L. E., MURRAY, A. C. and ROBERTSON, W. M. (1991) *Meat Sci.* **29**, 1.

JONES, S. D. M., JEREMIAH, L. E. and ROBERTSON, W. M. (1993) *Meat Sci.* **34**, 351.

JONES, S. D. M., SCHAEFER, A. L, ROBERTSON, W. M. and VINCENT, B. C. (1990) *Meat Sci.* **28**, 131.

JONES, S. J., CALKINS, C. R. and PODANY, K. (2001) *J. Anim. Sci.* **79**, Suppl. 1.

JONES, W. T. and MANGAN, J. L. (1977) *J. Sci. Food Agric.* **28**, 126.

JØRGENSEN, T. W. (1963) *Proc. 9th Meeting European Meat Research Workers, Budapest,* Paper No. 3.

JOSELL, Å., MARTINSSON, I. and TORNBERG, E. (2003) *Meat Sci.* **64**, 341.

JOSEPH, A. L., DUTSON, T. R. and CARPENTER, Z. L. (1980) *Proc. 26th Europ. Meeting Meat Res. Workers, Colorado Springs,* **2**, 77.

JOSEPH, R. L. (1996) *Meat Sci.* **43**, Suppl., S217.

JØSEPHS, R. and HARRINGTON, W. F. (1968) *Biochem.* **7**, 2834.

JOUBERT, D. H. (1956) *J. Agric. Sci.* **47**, 59.

JUDGE, M. D. and ABERLE, E. D. (1986) *J. Fd. Sci.* **45**, 1736.

JUDGE, M. D., REEVES, E. S. and ABERLE, E. D. (1980) Proc. 26th Europ. Meeting Meat Res. Workers, Colorado Springs, **2**, 74.

JUDGE, M. D. and STOB, M. (1963) *J. Anim. Sci.* **22**, 1059.

JUHÁSZ, A., BERNDORFER-KRASZNER, E., KORMENDY, L. and GABOR, T. (1976) *Acta Aliment.* **5**, 23.

JUKES, T. H. (1976) *Bioscience* **26**, 544.

JUL, M. (1957) *Food Manuf.* **32**, 259.

JUL, M., NEILSON, H. and PETERSEN, H. (1958) *Fleischwirts.* **10**, 840.

KAESS, G. (1956) *Aust. J. Appl. Sci.* **7**, 242.

KAESS, G. and WEIDEMANN, J. F. (1961) *Food Tech.* **15**, 129.

KAESS, G. and WEIDEMANN, J. F. (1962) *Food Tech.* **16**, 125.

KAESS, G. and WEIDEMANN, J. F. (1963) *Proc. 9th Meeting European Meat Res. Workers, Budapest,* Paper No. 58.

KAESS, G. and WEIDEMANN, J. F. (1973) *J. Fd. Technol.* **8**, 59.

KAMSTRA, L. D. and SAFFLE, R. L. (1 959) *Food Tech.* **13**, 11, 652.

KANNEL, W. B. and GORDON, T. (1970) *Some Characteristics Related to the Incidence of Cardio-vascular Disease and Death. Framlingham Study 16-Year Follow up* (US Govt. Printing Office: Washington).

KASSAI, D. and KÁRPÁTI, GY. (1963) *Proc. 9th Meeting European Meat Res. Workers, Budapest,* Paper No. 59.

KASTENSCHMIDT, L. L. (1970) In *The Physiology and Biochemistry of Muscle as a Food,* vol. II, p. 735 (Eds. E. J. BRISKEY, R. G. CASSENS and B. B. MARSH), Univ. Wisconsin Press, Madison.

KATO, S., KURATA, T. O. and FUJIMAKI, M. (1971) *Agric. Biol. Chem.* **35**, 2106.

KATO, S., KURATA, T. O., ISHIGURO, S. and FUJIMAKI, M. (1973) *Agric. Biol. Chem.* **37**, 1759.

KAUFFMAN, R. G., CARPENTER, Z L., BRAY, R. W. and HOEKSTRA, W. G. (1964) *J. Food Sci.* **29**, 65.

KAY, C. M. and PABST, H. F. (1962) *J. Biol. Chem.* **237**, 727.

KAY, M. and HOUSEMAN, R. (1975) In Meat (Eds. D. J. A. COLE and R. A. LAWRIE), Butterworths, London, p. 85.

KEAST, D., ARSTEIN, D. L., HANSEN, W., FRAY, R. W. and MORTON, A. R. (1995) *Med. J. Aust.* **162**, 15.

KEFFORD, J. F. (1948) *J. Coun. Sci. Industr. Res. Aust.* **21**, 116.

KELLER, H. and HEIDTMANN, H. H. (1955) *Fleischwirts.* **7**, 502.

KEKOMAKI, M., VISAKORPI, J. K., PERHEENTUPA, J. and SAXEN, L. (1967) *Acta Paediat. Scand.* **56**, 617.

KEMPE, L. L. (1955) *Appl. Microbiol.* **3**, 346.

KEMPSTER, A. J. (1979) *Meat Sci.* **3**, 199.

KEMPSTER, A. J., COOK, G. L. and GRANTLEYSMITH, M. (1986) *Meat Sci.* **17**, 107.

KEMPSTER, A. J. and HARRINGTON, G. (1979) *Meat Sci.* **3**, 53.

KENDRICK, J. L. and WATTS, B. M. (1969) *J. Food Sci.* **34**, 292.

KENNICK, W. H. and ELGASIM, E. H. (1981) *Proc. 34th. Ann. Recip. Meat Conf.* p. 68 (Amer. Meat Sci. Assoc.: Chicago).

KESTIN, S. R., KENNEDY, E., TONNER, M., KIERMAN, A., CRYER, H., GRIFFIN, S., BUTTERWORTH, S., RHIND, S. and FLINT, D. (1993) *J. Anim. Sci.* **71**, 1486.

KEVEL, E. and KOGMA, E. (1976) *Nahrung,* **20**, 243.

KEYS, A., ANDERSON, J. T. and GRANDE, F. (1960) *J. Nutr.* **70**, 257.

KIDNEY, A. J. (1967) *Proc. 13th Meeting European Meat Res. Workers, Rotterdam,* Paper No. A 7.

KIDWELL, J. F. (1952) *J. Hered.* **43**, 157.

KIERNAT, B. H., JOHNSON, J. A. and SIEDLER, A. J. (1964) 'A Survey of the Nutritional Content of Meat', *Amer. Meat Inst. Foundn. Bull. No.* 47.

KIESSLING, K. H. and KEISSLING, A. (1984) *Comp. Biochem. Physiol.* **77A**, 75.

KILLDAY, K. B., TEMPESTA, M. S., BAILEY, H. E. and METRAL, C. J. (1988) *J. Agric. Food Chem.* **36**, 909.

KIM, K., IKEUCHI, Y. and SUZUKI, A. (1992) *Meat Sci.* **32**, 237.

KING, N. L. and HARRIS, P. V. (1982) *Meat Sci.* **6**, 137.

KING, N. L. (1984) *Meat Sci.* **11**, 27, 59.

KING, N. L. R. (1978) *Meat Sci.* **2**, 313.

KINSELLA, J. E. (1978) *CRC Crit. Rev. Fd. Sci. Nutr.* **10**, 147.

KIRBY, L. T. (1992) *DNA Fingerprinting: an Introduction* (W. H. Freeman: New York).

KIRTON, A. H., BISHOP, W. H., MULLORD, M. M. and FRAZER HURST, L. F. (1978) *Meat Sci.* **2**, 199.

KIRTON, A. H., CARTER, A. H., SINCLAIR, D. P. and JURY, K. E. (1974) *Proc. Ruakura Farmers Confrc.*, N.Z. p. 29.

KIRTON, A. H., FRAZERHURST, L. F., BISHOP, W. H. and WINN, G. W. (1981a) *Meat Sci.* **5**, 407.

KIRTON, A. H., FRAZERHURST, L. F., WOODS, E. G. and CHRYSTALL, B. B. (1981b) *Meat Sci.* **5**, 347.

KITCHELL, A. G. (1958) *2nd Intl. Symp. Fd. Microbiol., Camb.* (Ed. B. P. EDDY), p. 191, H.M.S.O., London.

KITCHELL, A. G. (1959) *Proc. 10th Intl. Congr. Refrig., Copenhagen* **3**, 65.

KITCHELL, A. G. (1967) *Inst. Meat Bull.*, Jan., p. 7.

KITCHELL, A. G. (1971) *Proc. 17th Meeting Meat Res. Inst., Bristol,* p. 194.

KITCHELL, A. G. and INGRAM, M. (1956) *Ann. Inst. Pasteur, Lille* **8**, 121.

KITCHELL, A. G. and INGRAM, M. (1963) *Food Process. Packag.* **32**, 3.

KIVIRIKKO, K. I. (1963) *Nature, Lond.* **197**, 385.

KJØLBERG, O., MANNERS, D. J. and LAWRIE, R. A. (1963) *Biochem. J.* **87**, 351.

KLICKA, M. W. (1969) *Proc. Sympos. Feeding the Military Man*, U.S. Army, Natick, p. 63.

KLONT, R. E., BARNIER, V. M. H., SMULDERS, F. J. M., VAN DIJK, A., HOVING-BOLINK, A. H. and EIKELENBOOM, G. (1999) *Meat Sci.* **53**, 195.

KNAPPEIS, G. G. and CARLSEN F. (1968) *J. Cell Biol.* **38**, 202.

KNIGHT, B. C. J. G. and FILDES, P. (1930) *Biochem. J.* **24**, 1496.

KNIGHT, M. K. (1992) In *The Chemistry of Muscle-based Foods* (Eds. D. E. JOHNSON, M. K. KNIGHT and D. E. LEDWARD) p. 222 (Roy. Soc. Chem.: Lond.).

KNIGHT, P. J., ELSEY, J. and HEDGES, N. (1989) *Meat Sci.* **26**, 209.

KNIGHT, P. J. and PARSONS, N. (1988) *Meat Sci.* **24**, 275.

KNOESS, K. H. (1977) *Wld. Anim. Rev.* **22**, 3.

KNUDSEN, K. A., MYERS, L. and MCELWEE, S. A. (1970) *Exp. Cell. Res.* **188**, 175.

KOCH, R. M., CUNDIFF, L. V., GREGORY, K. E. and DIKEMAN, M. E. (1982) *Beef Res. Progr. Rept.* No. 1, p. 13 (Agric. Rev. Manuals: USDA: ARM-NC–21).

KOCHAKIAN, C. D. and TILLOTSON, C. (1957) *J. Endocrin.* **60**, 607.

KODAMA, S. (1913) *J. Tokyo Chem. Soc.* **34**, 751.

KOEHLER, P. E. and ODELL, G. V. (1970) *J. Agric. Fd. Chem.* **17**, 393.

KOEPPE, S. (1954) *Rev. de conserve* **9**, 83.

KOFRANYI, E. and JEKAT, F. K. (1969) *Hoppe-Seyl. Z. physiol. Chem.* **350**, 1405.

KOHLBRUGGE, J. H. F. (1901) *Zbl. Bakt.* 1, Abt. **29**, 571, cited by R. B. HAINES (1937) *loc. cit.*

KOLEDIN, L G. (1963) *Proc. 9th Meeting European Meat Res. Workers, Budapest,* Paper No. 6.

KOLLE, B. K., MCKENNA, D. R. and SAVELL, J. W. (2004) *Meat Sci.* **68**, 145.

KOKOSKY, K. (1998) *Stress Waves in Solids* (Dover Publishing Inc.: New York).

KOOHMARIAE, M., BABIKER, A. S., MERKEL, R. A. and DUTSON T. R. (1988) *J. Fd. Sci.* **53**, 1253.

KOOHMARAIE, M., SEIDEMAN, S. C., SCHOLLMEYER, J. E., DUTSON, T. R. and GROUSE, J. D. (1987) *Meat Sci.* **19**, 187.

KOOHMARAIE, M., WHIPPLE, G., KRETCHMAR, D. H., CROUSE, D. and MARSMANN, H. J. (1991) *J. Anim. Sci.* **69**, 617.

KOPPENOL, W. H. (1995) *Metal Ions Biol. Syst.* **36**, 597.

KÖRMENDY, L. (1955) *Elelmezési Ipar.* **8**, 172.

KÖRMENDY, L. and GANTNER, G. Y. (1958) *Proc. 4th Meeting European Meat Res. Workers, Cambridge,* Paper No. 21.

KORNER, A. (1963) *Biochem. J.* **89**, 14.

KOTERA, A., YOKOYAMA, M., YAMAGUCHI, M. and MIYAZAWA, Y. (1969) *Biopolymers* **7**, 99.

KOU, P. C. and SCHROEDER, R. A. (1995) *Annals Surgery* **221**, 220.

KRAFT, A. A. and AYRES, J. C. (1952) *Food Tech.* **6**, 8.

KRAMER, A. (1957) *Food Eng.* **29**, 57.

KRAMER, H. and LITTLE, K. (1955) *Nature and Structure of Collagen* (Ed. J. T. RANDALL), p. 33, Butterworths, London.

KRAMLICH, W. E and PEARSON, A. M. (1958) *Food Res.* **23**, 567.

KRISTENSEN, L. and PURSLOW, P. P. (2001) *Meat Sci.* **58**, 17.

KRITCHEVSKY, D. (2000) *BNF Bull.* **25**, 25.

KROGH, P. (1977) *Nord Veterinuermed.* **29**, 402.

KROPF, D. H. (1980) *Proc. Ann. Recip. Meat Conf.* (AMIF: Chicago), Vol. 33, p. 15.

KRYLOVA, N. N., BAZAROVA, K. I. and KAZNETSNOVA, V. V. (1962) *Proc. 8th Meeting European Meat Res. Workers, Moscow,* Paper No. 38.

KUBO, T., GERELT, B., HAN, G. D., SUGIYAMA, T., NASHIUMI, T. and SUZUKI, A. (2002) *Meat Sci.* **61**, 415.

KÜCHENMEISTER, U., KUHN, G. and ENDER, K. (2000) *Meat Sci.* **55**, 239.

KUKOWSKI, A. C., WULF, D. M., SHANKS, B. C., PAGE, J. K. and MADDOCK, R. J. (2004) *Meat Sci.* **66**, 889.

KUNINAKA, A. (1981) In *Flavour Research: Recent Advances* (Eds. R. TERENISHI and L. A. FLAT), p. 305 (Marcel Dekker).

KUNKEL, H. O. (1961) *Biochemical and Fundamental Physiological Bases for Genetically Variable Growth of Animals,* College Station, Texas Agric. Exp. Sta.

KUPRIANOFF, J. (1956) *Proc. Int. Inst. Refrig.* **53**, 129.

KURICHARA, K. and WOOL, I. G. (1968) *Nature, Lond.* **219**, 721.

KYLÄ-PUHJU, M., RUUSUNEN, M. and PUOLANNE, E. (2005) *Meat Sci.* **69**, 143.

KYLE, R. (1972) *Meat Production in Africa: the Case for a New Domestic Species,* Univ. Press, Bristol.

LAAKKONEN, E., WELLINGTON, G. H. and SHERBON, J. W. (1970) *J. Fd. Sci.* **35**, 175, 178.

LACOURT, A. and TARRANT, P. V. (1985) *Meat Sci.* **15**, 85.

LAING, J. A. (1959) *Progress in the Physiology of Farm Animals* (Ed. J. HAMMOND), vol. 3, p. 760, Butterworths, London.

LAKNITZ, L., SIMENHOFF, M. L., DUNN, S. R. and FIDDLER, W. (1980) *Fd. Cosmet. Toxicol.* **18**, 77.

LAKRITZ, L., and MAERKER G. (1988) *Meat Sci.* **23**, 77.

LAMARE, M., TAYLOR, R. G., FAROUT, L., BRIAND, Y. and BRIAND, M. (2002) *Meat Sci.* **61**, 199.

LAMBOOIJ, E. and SCHATZMANN, U. (1994) *Meat Sci.* **37**, 381.

LAMBOOY, E. (1981) *Proc. 27th. Europ. Meeting Meat Res. Workers, Vienna,* B3.

LAMETSCH, R., KARLSSON, A. F. ROSENVOLD, K., ANDERSEN, H. J., ROEPSTORFF, P. and BENDIXEN, C. (2003) *J. Agric. Fd. Chem.* **51**, 6992.

LAMMING, G. E. (1956) *Rep. Sch. Agric.*, p. 66, Univ. Nottingham.

LAMPREY, H. F. (1963) *J. E. Afr. Wildl.* **1**, 63.

LANARI, M. C., CASSENS, R. G., SCHAEFER, D. M. and SCHELLER, K. K. (1994) *Meat Sci.* **38**, 3.

LANDMANN, W. A. (1963) *Proc. Meat Tenderness Sympos.*, p. 87, Campbell Soup Co., Camden, New Jersey.

LAPSHIN, I. (1962) *Proc. 8th Meeting European Meat Res. Workers, Moscow,* Paper No. 39.

LARA, J. A. F., SENIGALIA, S. W. B., OLIVIERA, T. C. R. M., DUTRA, I. S., PINTO, M. F. and SHIMOKOMAKI, M. (2003) *Meat Sci.* **65**, 609.

LASTA, J. A., PENSEL, N., MASANO, M., RICARDO-RODRIGUEZ, H. and GARCIA, P. T. (1995) *Meat Sci.* **39**, 149.

LAW, N. H. and VERE-JONES, N. W. (1955) D.S.I.R., *N.Z.*, Bull. No. 118.

LAWRENCE, T. E., DIKEMAN, M. E., STEPHENS, J. W., OBUZ, E. and DAVIS, J. R. (2004) *Meat Sci.* **66**, 69.

LAWRIE, R. A. (1950) *J. Agric. Sci.* **40**, 356.

LAWRIE R. A. (1952a) *Nature, Lond.* **170**, 122.

LAWRIE R. A. (1952b) Ph.D. Dissertation, Univ. Cambridge.

LAWRIE, R. A. (1953a) *Biochem. J.* **55**, 298.

LAWRIE, R. A. (1953b) *Biochem. J.* **55**, 305.

LAWRIE, R. A. (1953c) *J. Physiol.* **121**, 275.

LAWRIE, R. A. (1955) *Biochim. biophys. Acta* **17**, 282.

LAWRIE, R. A. (1958) *J. Sci. Fd. Agric.* **9**, 721.

LAWRIE, R. A. (1959) *J. Refrig.* **2**, 87.

LAWRIE, R. A. (1960a) *Brit. J. Nutr.* **14**, 255.

LAWRIE, R. A. (1960b) *J. Comp. Path.* **70**, 273.

LAWRIE, R. A. (1961) *Brit. J. Nutr.* **15**, 453.

LAWRIE, R. A. (1966) In *Physiology and Biochemistry of Muscle as a Food,* p. 137 (Eds. E. J. BRISKEY, R. G. CASSENS and J. C. TRAUTMAN), Univ. Wisconsin Press, Madison.

LAWRIE, R. A. (1968) *J. Food Technol.* **3**, 203.

LAWRIE, R. A. (1975) In *Meat* (Eds. D. J. A. COLE and R. A. LAWRIE), Butterworths, London, p. 249.

LAWRIE, R. A. and GATHERUM, D. P. (1962) *J. Agric. Sci.* **58**, 97.

LAWRIE, R. A. and GATHERUM, D. P. (1964) *J. Agric. Sci.* **62**, 381.

LAWRIE, R. A., GATHERUM, D. P. and HALE, H. P. (1958) *Nature, Lond.* **182**, 807.

LAWRIE, R. A. and LEDWARD, D. A. (1983) In *Upgrading Waste for Feeds and Food.* (Eds. D. A. LEDWARD, A. J. TAYLOR and R. A. LAWRIE), p. 163 (Butterworths, Lond.).

LAWRIE, R. A., MANNERS, D. J. and WRIGHT, A. (1959) *Biochem. J.* **73**, 485.

LAWRIE, R. A., PENNY, I. E., SCOPES, R. K. and VOYLE, C. A. (1963a) *Nature, Lond.* **200**, 673.

LAWRIE, R. A. and POMEROY, R. W. (1963c) *J. Agric. Sci.* **61**, 409.

LAWRIE, R. A., POMEROY, R. W. and CUTHBERTSON, A. (1963b) *J. Agric. Sci.* **60**, 195.

LAWRIE, R. A., POMEROY, R. W. and WILLIAMS, D. R. (1964) *J. Agric. Sci.* **62**, 89.

LAWRIE, R. A. and PORTREY, E. (1967) *Ann. Rept. Sch. Agric., Univ. Nottingham,* p. 135.

LAWRIE, R. A., SHARP, J. G., BENDALL, J. R. and COLEBY, B. (1961) *J. Sci. Fd. Agric.* **12**, 742.

LAWRIE, R. A. and VOYLE, C. A. (1962) *Ann. Rep. Low. Temp. Res. Sta.*, p. 16.

LAWSON, M. A. (2004) *Meat Sci.* **68**, 559.

LEA, C. H. (1931) *J. Soc. Chem. Ind.* **50**, 215 T.

LEA, C. H. (1937) *J. Soc. Chem. Ind.* **56**, 376 T.

LEA, C. H. (1938) *Spec. Rept. Fd. Invest. Bd., Lond.*, No. 46.

LEA, C. H. (1939) *Rancidity in Edible Fats*, Chemical Publishing Co., New York.

LEA, C. H. (1943) *J. Soc. Chem. Ind.* **62**, 200.

LEA, C. H. (1959) *Intl. J. Appl. Radn. Isotopes* **6**, 86.

LEA, C. H. (1962) *Recent Advances in Food Research* (Eds. J. HAWTHORN and J. M. LEITCH), vol. 1, p. 83, Butterworths, London.

LEA, C. H., HAINES, R. B. and COULSON, C. A. (1936) *Proc. Roy. Soc.* **B 120**, 47

LEA, C. H., HAINES, R. B. and COULSON, C. A. (1937) *Proc. Roy. Soc.* **B 123**, 1.

LEA, C. H. and HANNAN, R. S. (1950) *Nature, Lond.* **65**, 438.

LEA, C. H. and HANNAN, R. S. (1952) *Food Science* (Eds. T. N. MORRIS and E. C. BATE SMITH), p. 228, Cambridge Univ. Press.

LEA, C. H., MCFARLANE, J. J. and PARR, L. J. (1960) *J. Sci. Fd. Agric.* **11**, 690.

LEA, C. H., SWOBODA, P. A. T. and GATHERUM, D. P. (1969) *J. Sci. Fd. Agric.* **74**, 279.

LEACH, T. M. (1971) *Proc. 17th Meeting Meat Res. Inst., Bristol*, p. 161.

LEAF, A. and WEBER, P. C. (1988) *New England J. Med.* **318**, 549.

LEAKEY, R. (1981) *The Making of Mankind* (Book Club Association, Lond.).

LEBLANC, F. R., DEVLIN, K. A. and STRUMBO, D. R. (1953) *Food Tech.* **7**, 181.

LEDGER, H. P. (1959) *Nature, Lond.* **184**, 1405.

LEDWARD, D. A. (1971a) *J. Fd. Sci.* **36**, 138.

LEDWARD, D. A. (1971b) *J. Fd. Sci.* **36**, 883.

LEDWARD, D. A. (1972) *J. Fd. Sci.* **37**, 634.

LEDWARD, D. A., CHIZZOLINI, R. and LAWRIE, R. A. (1975) *J. Fd. Technol.* **10**, 349.

LEDWARD, D. A., DICKINSON, R. E., POWELL, V. H. and SHORTHOSE, W. R. (1986) *Meat Sci.* **16**, 245.

LEDWARD, D. A. and LAWRIE, R. A. (1975) *J. Sci. Fd. Agric.* **26**, 691.

LEDWARD, D. A. and MACKEY, B. M. (2002) *Int. Rev. Fd. Sci. Tech.* **1**, 78.

LEDWARD, D. A. and VARLEY, J. (1999) *Ind. Proteins Senter.* **7**, 17.

LEE, C. A. and WEBSTER, H. L. (1963) *C.S.I.R.O. Div. Food Pres. Tech. Paper*, No. 30.

LEE, H. and LANIER, T. C. (1995) *J. Muscle Foods* **6**, 125.

LEE, M.-Y., MCKEITH, F. K., NOVASKOFSKI, J. and BECHTEL, P. J. (1987) *J. Anim. Sci.* **57**, Suppl. 1, 283.

LEET, N. G., DEVINE, C. E. and GAVEY, A. B. (1977) *Meat Sci.* **1**, 229.

LEFEBVRE, N., THIBAULT, C. and CHARBONNEAU, R. (1992) *Meat Sci.* **32**, 203.

LEFEBVRE, N., THIBAULT, C., CHARBONNEAU, R. and PIETTE, J.-P.-G. (1994) *Meat Sci.* **36**, 370.

LEHMANN, K. B. (1904) *Z. Biol.* **45**, 324.

LEHMANN, K. B. (1907) *Arch. Hyg.* **63**, 134.

LEHMANN, B. T. and WATTS, B. M. (1951) *J. Am. Oil Chemists' Soc.* **28**, 475.

LEIGHTON, G. R. and DOUGLAS, L. M. (1910) *The Meat Industry and Meat Inspection*, vol. II, Educational Book Co., London.

LEISTNER, L. (1960) *Proc. 12th Res. Conf.*, p. 17 (Amer. Meat Inst. Found., Chicago).

LEISTNER, L. (1995) In *New Methods of Food Preservation* (Ed. G. W. GOULD), p. 1 (Blackie Acad. Press, Lond.).

LEMBERG, R. and LEGGE, J. W. (1949) *Haematin Compounds and Bile Pigments,* Interscience, London.

LEMBERG, R. and LEGGE, J. W. (1950) *Ann. Rev. Biochem.* **19**, 431.

LEPETIT, J., SALE, P., FAVIER, R. and DALLE, R. (2002) *Meat Sci.* **60**, 51.

LE ROY, P., NAVEAU, J., ELSEN, J. M. and SELLIER, P. (1990) *Génét. Res.* **55**, 33.

LESEIGNEUR, A. and GANDEMER, G. (1991) *Meat Sci.* **29**, 229.

LEVEILLE, G. A. (1980) *Proc. Meat. Ind. Res. Conf., Arlington.* p. 109 (Amer. Meat Inst.: Chicago).

LEVIEUX, D. and LEVIEUX, A. (1996) *Meat Sci.* **42**, 239.

LEVIN, E. (1970) *Food Technol.* **24**, 19.

LEVINE, R. and GOLDSTEIN, M. S. (1955) *Rec. Progr. Hormone Res.* **11**, 343.

LEWIS, J. and MARTIN, P. (1989) *Nature, Lond.* **342**, 134.

LEWIS, P. K., JR., BROWN, C. J. and HECK, M. C. (1962) *J. Anim. Sci.* **21**, 196.

LEWIS, P. K., JR., BROWN, C. J. and HECK, M. C. (1967) *Food Tech.* **21**, 75A.

LI, Y.-G., TANNER, G. and LARKIN, P. (1996) *J. Sci. Food Agric.* **90**, 89.

LIGHT, N. D. and BAILEY, A. J. (1989) *FEBS Lett.* **97**, 183.

LIGHT, N. D. and BAILEY, A. J. (1983) *Proc. 29th. Europ. Meeting Meat Res. Workers, Salsomaggiore*, p. 135.

LIGHT, N. D., CHAMPION, A. E., VOYLE, C. A. and BAILEY, A. J. (1985) *Meat Sci.* **13**, 137.

LIJINSKY, W. and EPSTEIN, S. S. (1970) *Nature, Lond.* **225**, 21.

LINDAHL, G., ENFÄLT, A.-C., VON SETH, G., JOSELL, Å., HEDEBRO-VELANDER, I., ANDEISEN BRAUNSCHWEIG, M., ANDERSON, L. and LUNDSTROM, K. (2004) *Meat Sci.* **66**, 609, 621.

LINDBERG, P. and ORSTADIUS, K. (1961) *Acta Vet. Scand.* **2**, 1.

LINEWEAVER, H., ANDERSON, J. D. and HANSON, H. L. (1952) *Food Tech.* **6**, 1.

LINEWEAVER, H. and HOOVER, S. R. (1941) *J. Biol. Chem.* **137**, 325.

LINDQUIST, S. and CRAIG, E. A (1988) *Ann. Rev. Genetics* **22**, 631.

LINFORTH, R. S. T. and TAYLOR, A. J. (1993) *Food Chem.* **48**, 115.

LINTON, A. H. and HINTON, M. H. (1984) In *Antimicrobials in Agriculture* (Ed. M. WOODBINE), p. 533 (Butterworths, Lond.).

LISTER, D. (1969) In *Recent Points of View on the Condition and Meat Quality of Pigs for Slaughter*, p. 123 (Eds. W. SYBESMA, P. G. VAN DER WALS and P. WALSTRON), Res. Inst. Animal Hushandry, Zeist.

LISTER, D., GREGORY, N. G. and WARRISS, P. D. (1981) In *Developments in Meat Science*, Vol. II. (Ed. R. A. LAWRIE), p. 61. (Applied Sci.: Lond.).

LISTER, D. and SPENCER, G. S. G. (1981) In *The Problem of Dark-Cutting in Beef.* (Eds. D. E. HOOD and P. V. TARRANT), p. 129 (Martinus Nijhoff: The Hague).

LOBLEY, G. E., PERRY, S. V. and STONE, D. (1971) Nature, *Lond.* **231**, 317.

LOCKER, R. H. (1956) *Biochim. Biophys. Acta* **20**, 514.

LOCKER, R. H. (1960a) *Food Res.* **25**, 304.

LOCKER, R. H. (1960b) *J. Sci. Fd. Agric.* **11**, 520.

LOCKER, R. H. (1968) *Proc. 10th Meat Ind Res. Conf., Hamilton, N.Z.*, p. 3.

LOCKER, R. H. (1973) *J. Fd. Technol.* **8**, 71.

LOCKER, R. H. (1976) *Proc. 18th Meat Res. Conf., Rotorua*, p. 1.

LOCKER, R. H. (1987) *Meat Sci.* **20**, 217.

LOCKER, R. H. (1989) In *Meat Production and Processing* (Eds. R. W. PURCHAS, B. W. BUTLER-HOGG and A. S. DAVIES) p. 173. Occas. Publcn. No. 11 (N.Z. Soc. Animal Prod.).

LOCKER, R. H. and DAINES, G. J. (1975) *J. Sci. Fd. Agric.* **26**, 1721.

LOCKER, R. H. and DAINES, G. J. (1976) *J. Sci. Fd. Agric.* **27**, 244.

LOCKER, R. H., DAINES, G. J., CARSE, W. A. and LEET, N. G. (1977) *Meat Sci.* **1**, 87.

LOCKER, R. H., DAVEY, C. L., NOTTINGHAM, P. M., HAUGHEY, D. P. and LAW, W. H. (1975) *Adv. Fd. Res.* **21**, 158.

LOCKER, R. H. and HAGYARD, C. J. (1963) *J. Sci. Food Agric.* **14**, 787.

LOCKER, R. H. and HAGYARD, C. J. (1968) *Arch. Biochem. Biophys.* **127**, 370.

LOCKER, R. H. and LEET, N. G. (1975) *J. Ultrastruct. Res.* **52**, 64.

LOCKER, R. H. and WILD, D. J. C. (1982) *Meat Sci.* **7**, 189.

LOCKER, R. H. and WILD, D. J. C. (1984) *Meat Sci.* **11**, 89.

LOCKER, R. H. and WILD, D. J. C. (1986) *J. Biochem.* **99**, 1473.

LOCKETT, C., SWIFT, C. E. and SULZBACHER, W. L. (1962) *J. Food Sci.* **27**, 36.

LOCKLEY, A. K. and BARDSLEY, R. G. (2000) *Trend Fd. Sci. Technol.* **71**, 67.

LOCKLEY, A. K. and BARDSLEY, R. G. (2002) *Meat Sci.* **61**, 163.

LODGE, G. A. (1970) In *Proteins as Human Food*, p. 141 (Ed. R. A. LAWRIE), Butterworths, London.

LONGDILL, G. R. (1989) In *Meat Production and Processing* (Eds. R. W. PURCHAS, B. W. BUTLER-HOGG and A. S. DAVIES), p. 191. Occas. Publcn. No. 11 (N.Z. Soc. Animal Production).

LONGDILL, G. R. (1994) *Meat Sci.* **36**, 277.

LORINCZ, E. and BIRO, G. (1963) *Acta Morph., Budapest* **12**, 15.

LOURDES-PEREZ, M., ESCALONA, H. and GUERERO, I. (1998) *Meat Sci.* **48**, 125.

LOVE, J. D. (1983) *Food Technol.* **37**, H7.

LOVE, R. M. (1955) *J. Sci. Fd. Agric.* **6**, 30.

LOVE, R. M. (1956) *Nature, Lond.* **178**, 198.

LOVE, R. M. (1958) *J. Sci. Fd. Agric.* **9**, 257, 262.

LOVE, R. M. and ELERIAN, M. K. (1963) *Proc. 11th Intl. Congr. Refr., Munich*, p. 887.

LOVE, R. M. and HARALDSSON, S. B. (1961) *J. Sci. Fd. Agric.* **12**, 442.

LOWE, B. (1934) M.Sc. Thesis, Univ. Chicago.

LOWE, J., MCDERMOTT, H., KENWARD, N., LANDON, M., MAYER, R. J., BRUCE, M., MCBRIDE, P., SOMERVILLE, R. A. and HOPE, J. (1990) *J. Path.* **162**, 61.

LOWE, J. and MAYER, R. J. (1990) *Neuropath. Appl. Neurobiol.* **16**, 281.

LOWEY, S. (1968) *Symposium on Fibrous Proteins*, Butterworths, London, p. 124.

LOWEY, S. and HOLT, J. L. (1972) *Cold. Spr. Harb. Sympos. Quant. Biol.* **37**, 19.

LOWEY, S., SLAYTER, H. S., WEEDS, A. G. and BAKER, H. (1969) *J. Molec. Biol.* **42**, 1.

LOWRY, O. H., GILLIGAN, D. R. and KATERSKY, E. M. (1941) *J. Biol. Chem.* **139**, 795.

LOWRY, P. D. and TIONG, L. (1988) *Proc. 34th Intl. Congr. Meat Sci. Technol.*, Brisbane, p. 528.

LOYD, E. J. and HINER, R. L. (1959) *J. Agric. Food Chem.* **7**, 860.

LUCCIA, A. DI., PICARIELLO, G., CACACE, G., SCALONI, A., FACCIA, M., LUIZZI, V. and ALVITA, G. (2005) *Meat Sci.* **69**, 479.

LÜCKE, F.-K. (2000) *Meat Sci.* **56**, 105.

LUDVIGSEN, L. (1954) *Beretn. Forsøgslab. Kbh.*, No. 272.

LUDVIGSEN, J. (1957) *Acta Endocrin., Copenhagen* **26**, 406.

LUECKE, F. K. (1990) In *Food Preservation* (Eds. G. W. GOULD and N. J. RUSSELLS) p. 181 (Blackie: Glasgow).

LUMLEY, I. (2001) *Fd. Sci. Technol.* **15**(2), 24.

LUNDBERG, P., VOGEL, H. J. and RUDERHUS, H. (1986) *Meat Sci.* **18**, 133.

LUNDGREN, H. P. (1945) *Textile Res. J.* **15**, 334.

LUNDSTROM, K., ESSEN-GUSTAVSSON, B., RUNDGREN, M., EDFORS-LILJA, I. and MALMFORS, G. (1989) *Meat Sci.* **25**, 251.

LUNDSTROM, K. and MALMFORS, G. (1985) *Meat Sci.* **15**, 203.

LUNNEN, J. E., FAULKNER, L. C., HOPWOOD, M. L. and PICKETT, B. N. (1974) *Biol. Reprod.* **10**, 453.

LUÑO, M., RONCALES, P., DJENNE, D. and BELTRAN, J. A. (2000) *Meat Sci.* **55**, 413.

LUPANDIN, Y. V. and POLESHCHUK, N. K. (1979) *Neirofiziologiya* **11**, 263.

LUSHBOUGH, C. H. and URBIN, M. C. (1963) *J. Nutr.* **81**, 99.

LUYERINK, J. H. and VAN BAAL, J. P. W. (1969) *Proc. 15th Meeting European Meat Res. Workers, Helsinki,* p. 41.

LUYET, B. J. (1961) *Proc. Symp. Low Temperature Biology*, p. 63, Campbell Soup Co., Camden, N.J.

LUYET, B. J. (1962) *Freeze-drying of Foods*, p. 194. Nat. Res. Counc., Nat. Acad. Sci., Washington, D.C.

LUYET, B. J., RAPATZ G. L. and GEHENIO, P. M. (1965) *Biodynamica* **9**, 283.

LYNCH, G. P., OLTJEN, R. R., THORNTON, J. W. and HINER, R. L. (1966) *J. Anim. Sci.* **25**, 1133.

LYTRAS, G. N., GEILESKY, A., KING, R. D. and LEDWARD, D. A. (1999) *Meat Sci.* **52**, 189.

MA, H.-J. and LEDWARD, D. A. (2004) *Meat Sci.* **68**, 347.

MABROUK, A. E., JARBOE, J. K and O CONNER E. M. (1969) *J. Agric. Fd. Chem.* **17**, 5.

MACARA, T. J. R. (1947) *Mod. Refrig.* **50**, 63.

MCARDLE, B. (1956) *Brit. Med Bull.* **12**, 226.

MCBRIDE, G. (1963) *Animal Behaviour* **11**, 53.

MCBRYDE, C. N. (1911) *J. Agric. Sci.* **33**, 761.

MCCANCE, R. A. and WIDDOWSON E. M. (1946) *The Chemical Composition of Food*, H.M.S.O., London.

MCCANCE, R. A. and WIDDOWSON, E. M. (1959) *J. Physiol.* **147**, 124.

MCCANCE, R. A. and WIDDOWSON, E. M. (1960) *M. R. C. Spec. Rept.* No. 297, H.M.S.O., London.

MCCARTHY, T. L., KERRY, J. P., KERRY, J. F., LYNCH, P. B. and BUCKLEY, D. J. (2001) *Meat Sci.* **58**, 45.

MCCAUGHEY, W. J. and DOW, C. (1977) *Vet. Rec.* **100**, 29.

MCCONNELL, P. (1902) The Elements of Agricultural Geology, Lockwood, London.

MCCRAE, S. E. and PAUL, P. C. (1974) *J. Fd. Sci.* **39**, 18.

MCCUTCHEON, S. N. (1989) In Meat Production and Processing (Eds. R. W. PURCHAS, B. W. BUTLER-HOGG and A. S. DAVIES), p. 113. Occas. Publcn. No. 11 (N.Z. Soc. Animal Production).

MCDONALD, I. W. (1968) *Aust. Vet. J.* **44**, 145.

MCDONALD, M. A. and SLEN, S. B. (1959) *Canad J. Anim. Sci.* **39**, 202.

MACDOUGALL, D. B. (1984) *Anal. Proc.* **21**, 494.

MACDOUGALL, D. B., SHAW, B. G., NUTE, G. R. and RHODES D. N. (1979) *J. Sci. Fd. Agric.* **30**, 1160.

MACDOUGALL, D. B. and TAYLOR, A. A. (1975) *J. Fd. Technol.* **10**, 339

MCELHINNEY, J., HANSON, A. O., BECKER, R. A., DUFFIELD, R. B. and DIVEN, B. C. (1949) *Phys. Rev.* **75**, 542.

MACFADYEN, A. and ROWLANDS, S. (1900) *Proc. Roy. Soc.* **B 66**, 488.

MACFARLANE, J. J. (1973) *J. Fd. Sci.* **38**, 294.

MACFARLANE, J. J. (1984) In *Developments in Meat Science* Vol. III (Ed. R. A. LAWRIE), p. 155. (Applied Sci., Lond.).

MACFARLANE, J. J. and MACKENZE, I. J. (1976) *J. Fd. Sci.* **41**, 1442.

MACFARLANE, J. J., MCKENZIE, I. J. and TURNER, R. H. (1986) *Meat Sci.* **17**, 161.

MACFARLANE, J. J., MCKENZIE, I. J., TURNER, R. H. and JONES, R. M. (1984) *Meat Sci.* **10**, 307.

MACFARLANE, J. J. and MORTON, D. J. (1978) *Meat Sci.* **2**, 281.

MACFARLANE, J. J., SCHMIDT, G. R. and TURNER, R. H. (1977) *J. Fd. Sci.* **42**, 1603.

MACHLIK, S. M. and DRAUDT, H. N. (1963) *J. Food Sci.* **28**, 711.

MACKEY. B. M., CROSS, D. and PARK, S. F. (1994) *J. Appl. Bact.*, in press.

MCINTOSH, E. N. (1965) *J. Food Sci.* **30**, 986.

MCKELLAR, J. C. (1960) *Vet. Rec.* **72**, 507.

MCKINSTRY, J. L. and ANIL, M. H. (2004) *Meat Sci.* **67**, 121.

MCLAIN, P. E., CREED, G. J., WILLEY, E. R. and HORNSTEIN, I. (1970) *J. Food Sci.* **35**, 258.

MCLAIN, P. E., PEARSON, A. M., BRUNNER, J. R. and CREVASSE, G. A. (1969) *J. Food Sci.* **34**, 115.

MCLAREN, A. and MICHIE, D. (1960) *Nature, Lond.* **187**, 363.

MCLEAN, R. A. and SULZBACHER, W. L (1953) *J. Bact.* **65**, 428.

MCLEAN, R. A. and SULZBACHER, W. L. (1959) *Appl. Microbiol.* **7**, 81.

MACLENNAN, D. H. (1974) *J. Biol. Chem.* **249**, 980.

MACLENNAN, D. H., DUFF, C., ZORZATO, F., FUJII, J., PHILLIPS, M., KORMELUK, R. G., FRODIS, W., BRITT, B. A. and WORTON, R. G. (1990) *Nature, Lond.* **343**, 559.

MACLENNAN, D. and WONC, P. T. S. (1971) *Proc. Nat. Acad. Sci., U.S.A.* **68**, 1231.

MACLEOD, G. (1984) *Proc. I.F.S.T. Meat and Poultry Sympos.* p. 184.

MACLEOD, G. (1986) In *Developments in Food Flavours* (Eds. G. G. BIRCH and M. G. LINDLEY), p. 191 (Elsevier Appl. Sci., Lond.).

MACLEOD, G. and COPPOCK, B. (1976) *J. Agric. Fd. Chem.* **24**, 835.

MACLEOD, G. and COPPOCK, B. (1977) *J. Agric. Fd. Chem.* **25**, 113.

MACLEOD, G. and SEYYEDAIN-ARDEBILI, M. (1981) *CRC Crit. Rev. Fd. Sci. Technol.* **14**, 309.

MCLEOD, K, GILBERT, K. V., WYBORN, R., WENHAM, L. M., DAVEY, C. L. and LOCKER, R. H. (1973) *J. Fd. Technol.* **8**, 71.

MCLUSKEY, R. T. and THOMAS, L (1958) *J. Exp. Med.* **108**, 371.

MCMAHAN, J. R., DOWNING, H. E., OTTKE, R. C., LUTHER, H. G. and WRENSHALL C. L. (1955–6) *Antibiotics Annual*, p. 727.

MCMEEKAN, C. P. (1940) *J. Agric. Sci.* **30**, 276, 287.

MCMEEKAN, C. P. (1941) *J. Agric. Sci.* **31**, 1.

MCNEIL, I. H., DIMICK, P. S. and MAST, M. G. (1973) *J. Fd. Sci.* **38**, 1077.

MACY, R. L., NAUMANN, H. D. and BAILEY, M. E. (1964) *J. Fd. Sci.* **29**, 136.

MACY, R. L, NAUMANN, H. D. and BAILEY, M. E. (1970) *J. Fd. Sci.* **33**, 78.

MADARENA, G., DAZZI, G., CAMPANINI, G. and MACCI, E. (1980) *Meat Sci.* **4**, 157.

MADDEN, R. H., ESPIE, W. E., MORAN, L., MCBRIDE, J. and SCATES, P. (2001) *Meat Sci.* **58**, 343.

MADSEN, J. (1942) *Beretn. Vet. Landbohøjskoles Slagterilaboratorium, Kbh.*, No. 11.

MAGRI, V. A., EWTON, D. A. and FLORINI, J. R. (1991) In *Molecular Biology and Physiology of Insulin and Insulin-like Growth Factors* (Eds. M. K. RAIZADA and D. LEROTH) p. 87 (Plenum Press: New York).

MAHDAVI, V., STREHLER, E. E., PERIASAMY, M., WIECZORAK, D. E., IZUMO, S. and NADAL GINARD, B. (1986) *Med. Sci. Sports Exercise*, **18**, 299.

MAHER, P. A., COX, C. F. and SINGER, S. J. (1985) *J. Cell. Biol.* **101**, 1871.

MAIJALA, R. and EEROLA, S. (1993) *Meat Sci.* **35**, 387.

MAIJALA, R., NURMI, E. and FISCHER, A. (1995) *Meat Sci.* **39**, 9.

MAILLET, M. (1955) *Ann. Inst. Pasteur, Lille* **7**, 9.

MAJOR, R. and WATTS, B. M. (1948) *J. Nutr.* **35**, 103.

MALTIN, C. A., WATKUP, C. C., MATTHEWS, K. R., GRANT, C. M., PORTER, A. D. and DELDAY, M. I. (1997) *Meat Sci.* **47**, 237.

MALTON, R. (1971) *Proc. 17th Meeting European Meat Res. Workers*, Bristol, p. 486.

MANDIGO, R. W. (1983) In *International Sympos.: Meat Science and Technology*, p. 235 (Natl. Livestock & Meat Bd.: Chicago).

MANGAN, J. L. (1959) *N.Z. J. Agric. Res.* **2**, 47.

MANI, R. S. and KAY, C. M. (1978) *Biochim. Biophys. Acta* **533**, 248.

MANN, J. (1999) *Chem. Britain*, Oct., p. 29.

MANTEL, M. C., MASON, F. and TALON, R. (1998) *Meat Sci.* **49**, Suppl. S 111.

MARKAKIS, P. C., GOLDBLITH, S. A. and PROCTOR, H. E. (1951) *Nucleonics* **9**, 71.

MARSH, B. B. (1952a) *Biochim. Biophys. Acta* **9**, 247.

MARSH, B. B. (1952b) *Biochim. Biophys. Acta* **9**, 127.

MARSH, B. B. (1954) *J. Sci. Fd. Agric.* **5**, 70.

MARSH, B. B. (1962) *4th Meat Industry Res. Conf., Hamilton*, Meat Industry Res. Inst. N.Z Inc. Publ. No. **55**, p. 32.

MARSH, B. B. (1964) *Ann. Rept. Meat Ind. Res. Inst., N.Z.*, p. 16.

MARSH, B. B. (1983) *Proc. 30th Ann. Recip. Meat Conf., Fargo, N. Dakota*, p. 131 (Natl. Livestock & Meat Bd.: Chicago).

MARSH, B. B. and LEET, N. G. (1966) *J. Food Sci.* **31**, 450.

MARSH, B. B., LEET, N. G. and DICKSON, M. R. (1974) *J. Fd. Technol.* **9**, 141.

MARSH, B. B., LOCHNER, J. V., TAKAHASHI, G. and KRAGNESS, D. D. (1981) *Meat Sci.* **5**, 479.

MARSH, B. B., RINGKOB, T. P., RUSSELL, R. L., SWARTZ, D. R. and PAGEL, L. A. (1987) *Meat Sci.* **21**, 241.

MARSH, B. B. and THOMPSON, J. T. (1958) *J. Sci. Fd. Agric.* **9**, 417.

MARSH, B. B., WOODHAMS, P. M. and LEET, N. G. (1968) *J. Food Sci.* **33**, 12.

MARSHALL, J. M., HOLTZER, H., FRICK, H. and PEPE, F. (1959) *Exp. Cell. Res.*, suppl. 7, p. 219.

MARTEHENKO, A. P. and KOZENYASHEVA, O. M. (1974) *Trudy VNIIMP* **29**, 87.

MARTENS, H., STABURSVIK, E. and MARTENS, M. (1982) *J. Text. Stud.* **13**, 291.

MARTIN, G. R., BYERS, P. H. and PIEZ, K. A. (1975) *Adv. Enzymol.* **42**, 167.

MARTIN, S., BATZER, O. F., LANDMANN, W. A. and SCHWEIGERT, B. S. (1962) *J. Agric. Food Chem.* **10**, 91.

MARTIN, W. M. (1948) Food Ind. **20**, 832.

MARTINO, M. N., OTERO, L., SANZ, P. D. and ZARITZKY, N. E. (1998) *Meat Sci.* **50**, 303.

MARTONOSI, A. N. and BEELER, T. (1983) In *Handbook of Physiology. Section 10. Skeletal Muscle* (Eds. L. D. PEACHEY, R. H. ADRIAN and S. R. GEIGER), p. 417 (Amer. Physiol. Soc.: Bethesda, Maryland).

MARUYAMA, K. (1970) In *Physiology and Biochemistry of Muscle as a Food*, vol. II (Eds. E. J. BRISKEY, R. G. CASSENS and B. B. MARSH), Univ. Wisconsin Press, Madison, p. 373.

MARUYAMA, K. (1986) *Int. Rev. Cytol.* **104**, 81.

MARUYAMA, K., KIMURA, S., OHASHI, K. and KUWANO, Y. (1981) *J. Biochem.* **89**, 701.

MARUYAMA, K, KIMURA, S., TOYOTA, N. and OHASHI, K. (1979) In *Fibrous Proteins: Scientific, Industrial and Medical Aspects.* Eds. D. A. D. PARRY and L. K. CREAMER. Vol. 2 (Acad. Press: New York).

MARUYAMA, K., KUNITOMO, S., KIMURA, S. and OHASHI, K. (1977a) *J. Biochem.* **81**, 243.

MARUYAMA, K., KURODA, M. and NONOMURA, Y. (1985) *Biochem. Biophys. Acta* **829**, 229.

MARUYAMA, K., MATSUBARA, S., NATORI, R., NONOMURA, Y. KIMURA, S., OHASHI, K, MURAKAMI, F., HANDA, S. and EGUCHI, G. (1977b) *J. Biochem.*, Tokyo **82**, 317.

MASON, L. L. (1951) *Tech. Bull. Commonwealth Bureau Animal Breeding and Genetics*, no. 8.

MATHESON, N. A. (1962) *Rec. Adv. Food Sci.* (Eds. J. HAWTHORN and J. M. LEITCH), vol. 2, p. 51, Butterworths, London.

MATSUISHI, M. A. and OKITANI, O. A. (2000) *Meat Sci.* **56**, 569.

MATSUKURA, V., OKITANI, A., NISHIMURO, T. and KATO, H. (1981) *Biochim. Biophys. Acta* **662**, 41.

MATSUO, H., BABA, Y., NAIR, R. M. G., ARMURA, A. and SCHALLY, A. V. (1971) *Biochem. Biophys. Res. Commun.* **43**, 1334.

MATTEY, M., PARSONS, A. L. and LAWRIE, R. A. (1970) *J. Fd. Technol.* **5**, 41.

MATTSON, F. H. and GRUNDY, S. M. (1985) *J. Lipid Res.* **26**, 197.

MAURER, F. (1894) *Gegensbaurs Jb.* **21**, 473, cited by PICKEN (1960) *loc. cit.*

MAW, S. J., FOWLER, V. R., HAMILTON, M. and PETCHEY, A. M. (2003) *Meat Sci.* **63**, 185.

MAXCY, R. B. and ROWLEY, D. B. (1978) In *Food Preservation by Irradiation* Vol. 1, p. 397 (Intl. Atomic Energy Assoc).

MAY, C. G. (1961) U.K. Pat. No. 858,333.

MAY, C. G. and MORTON, I. D. (1961) U.K. Pat. No. 858,660.

MEARUS, W. J. and SCHMIDT, G. R. (1986) *J. Fd. Sci.* **51**, 60.

MEAT & LIVESTOCK COMMISSION (1974–75) *Pig Improvement Yearbook.*

MEAT & LIVESTOCK COMMISSION (1988) *Sheep in Britain.*

MEAT RESEARCH CORPORATION (1994–95) *Project Guide* p. 106 (Sydney, Australia).

MEDREK, T. F. and BARNES, E. M. (1962) *J. Appl. Bac.* **25**, 159.

MEECH, M. V. and KIRK, R. S. (1986) *J. Assoc. Publ. Analysts* **24**, 13.

MEER, D. P. and EDDINGER, T. J. (1996) *Meat Sci.* **44**, 285.

MEIBURG, D. E., BEERY, K. E., BROWN, C. L. and SIMON, S. (1976) *Proc. Meat Ind Res. Conf.*, *Arlington*, p. 79 (Amer. Meat Inst.: Chicago).

MELLGREN, R. L., LANE, R. D. and MERICLE, M. T. (1989) *Biochem. Biophys. Acta* **999**, 71.

MELROSE, D. R. and GRACEY, W. (1975) In *Meat* (Eds. D. J. A. COLE and R. A. LAWRIE), Butterworths, London, p. 109.

MENZ, L. and LUYET, B. (1961) *Biodynamica* **8**, 261.

MERAT, P. (1990) *INRA Prod. Anim.* **3**, 151.

MERKEL, R. A., SCHROEDER, A. L., BURNETT, R. J., MALZET, P. D. and BERGEN, W. G. (1990) *Proc. 42nd Ann. Recip. Meat Conf.* p. 55 (Natl. Livestock and Meat Bd.: Chicago).

MERRITT, C., JR., ANGELINI, P. and GRAHAM, R. A. (1978) *J. Agric. Fd. Chem.* **26**, 29.

MERRITT, C., JR., BRESNICK, S. R, BAZINET, M. L., WALSH, J. T. and ANGELINI, P. (1959) *J. Agric. Food Chem.* **7**, 784.

MERRITT, C., JR. and ROBERTSON, D. H. (1967) *J. Gas Chrom.* **5**, 96.

MERYMAN, H. T. (1956) *Science* **124**, 515.

MESTRE-PRATES, J. A., GARCIA E COSTA, F. J. S., RIBEIRO, A. M. R. and DIAS CORREIA, A. A. (2002) *Meat Sci.* **61**, 103.

MEYER, E. W. (1967) *Proc. Intl. Conf. Soybean Protein Foods*, U.S.D.A. Pubn. ARS–7–35, p. 142.

MICKELSON, J. R. (1983) *Meat Sci.* **9**, 205.

MIDDLEHURST, J., PARKER, N. S. and COFFEY, M. F. (1969) C.S.I.R.O. *Food Pres. Quart.* **29**, 21.

MIKAMI, M., WHITING, A. H., TAYLOR, M. A. J., MACIEWICZ, R. A. and ETHERINGTON, D. J. (1987) *Meat Sci.* **21**, 81.

MILAN, D., JEON, T., LOFT, C., AMARGER, V., ROBIC, A., THELANDER, M., ROGEL-GAILLARD, C., PAUL, S., IANNACCELLI, N., RASK, L., RENNE, R., LUNDSTRÖM, K., REINSCH, N., GELLIN, J., KALM, E., ROY, P. L., CHARDON, P. and ANDERSSON, L. (2000) *Science* **288**, 1248.

MILER, K. (1963) *Proc. 9th Meeting European Meat Res. Workers, Budapest*, Paper No. 44.

MILES, C. L. and LAWRIE, R. A. (1970) *J. Fd. Technol.* **5**, 325.

MILLER, E. D., LONGNECKER, D. E. and PEACH, M. J. (1979) *Circ. Shock* **6**, 271.

MILLER, R. L. and UDENFRIEND, S. (1970) *Arch. Biochem. Biophys.* **139**, 104.

MILLER, S. J., MOSS, B. W. and STEVENSON, M. H. (2000) *Meat Sci.* **55**, 249.

MILLET, S., RAES, K., VAN DEN BROECK, W., DESLET, S. and JANSSENS, G. P. J. (2005) *Meat Sci.* **69**, 335.

MILLIKAN, G. A. (1939) *Physiol. Rev.* **19**, 503.

MILLMAN, B. M. (1981) *J. Physiol.* **320**, 118P.

MILLMAN, B. M. and NICKEL, B. G. (1980) *Biophys. J.* **32**, 49.

MILLS, C. F. (1979) *Phil. Trans. Roy. Soc., Lond.* **B. 288**, 51.

MIN, D. B. S., INA, K., PETERSON, R. J. and CHANG, S. S. (1979) *J. Fd. Sci.* **44**, 639.

MIR, Z., RUSHFELDT, M. L., MIR, P. S., PATERSON, L. J. and WESELAKE, R. J. (2003) *Small Rumin. Res.* **36**, 250.

MITCHELL, H. H., BEADLES, J. R. and KRUGER, J. H. (1927) *J. Biol. Chem.* **73**, 767.

MITCHELL, H. H. and HAMILTON, J. S. (1933) *J. Agric. Res.* **46**, 917.

MITCHELL, H. H., HAMILTON, J. S. and HAINES, W. T. (1928) *J. Nutr.* **1**, 165.

MITTAL, P. (1981) Ph.D. Dissertation, Univ. Nottingham.

MITTAL, P. and LAWRIE, R. A. (1984) *Meat Sci.* **10**, 101.

MIYADA, D. S. and TAPPEL, A. L. (1956) *Food Tech.* **10**, 142.

MIYAHARA, M., KISHI, K. and NODA, H. (1980) *J. Biochem.* **87**, 1341.

MOESGAARD, B., QUISTORFF, B., CHRISTENSEN, U. G., THERLESEN, I. and JØRGENSEN, P. G. (1995) *Meat Sci.* **39**, 43.

MOHR, V. and BENDALL, J. R. (1969) *Nature, Lond.* **223**, 404.

MOLINA-GARCIA, A. D., OTERO, L., MARTINO, M. N., ZARITZKY, N. E., ARABAS, J., SZCZOPEK, J. and SANZ, P. D. (2004) *Meat Sci.* **66**, 709.

MØLLER, A. J. and VESTERGAARD, T. (1986) *Meat Sci.* **18**, 78.

MONAHAN, F. J., BUCKLEY, D. J., MORISSEY, P. A., LYNCH, P. B. and GRAY, J. I. (1992) *Meat Sci.* **31**, 229.

MONCRIEFF, R. W. (1951) *The Chemical Senses,* 2nd ed., Leonard Hill, London.

MONIN, G., LAMBOOY, E. and KLONT, R. (1995) *Meat Sci.* **40**, 149.

MONIN, G. and OUALI, A. (1991) In *Developments in Meat Science–5* (Ed. R. A. LAWRIE), p. 89 (Elsevier Appl. Sci.: Lond.).

MONIN, G. and SELLIER. P. (1985) *Meat Sci.* **13**, 49.

MONTERO, P. and GOMÉZ-GUILLÉN, C. (2002) In *Symposium on Emerging Technologies*, Madrid, Spain, p. 291.

MOORE, J. (2004) *Meat Sci.* **67**, 565.

MOORE, V. (1990) *Meat Sci.* **27**, 91.

MOORE, V. J., LOCKER, R. H. and DAINES, C. J. (1976–77) *Ann. Res. Rept.*, Meat Ind. Res. Inst. N.Z. Inc. p. 38.

MOORE, S. S., BARENDSE, W., BERGER, K. T., ARMITAGE, S. M. and HETZEL, D. J. S. (1992) *Amer. Genet.* **23**, 463.

MORAN, T. (1927) *Proc. Roy. Soc.* **B 105**, 177.

MORAN, T. (1930) *Proc. Roy. Soc.* **B 107**, 182.

MORAN, T. (1932) *J. Soc. Chem. Ind.* **51**, 16T, 20T.

MORAN, T. and SMITH, E. C. (1929) *Spec. Rept. Fd. Invest. Bd., Lond.*, No. 36.

MORGAN, B. (1957) *Food Process.* **18**, 24.

MORGAN, M., PERRY, S. V. and OTTAWAY, J. (1976) *Biochem. J.* **157**, 687.

MORITA, S., IWAMOTO, H., FUKUMITSU, Y., GOTOH, T., NISHIMURA, S. and ONO, Y. (2000) *Meat Sci.* **54**, 59.

MORLEY, M. J. (1966) *J. Fd. Technol.* **1**, 303.

MORPURGO, B. (1897) Cited by DENNY-BROWN (1961) *Arch. Path. Anat.* **150**, 522.

MORRISON, C. A. (1986) *J. Endocr.* **111**, Suppl. Abs. 14.

MORRISSEY, P. A., SHEEHY, P. J. A., GALVIN, K., KERRY, J. P. and BUCKLEY, D. J. (1998) *Meat Sci.* **49**, S 73.

MORRISSEY, P. A. and TICHIVANGANA, J. Z. (1985) *Meat Sci.* **14**, 175.

MORTON, D. J., WEIDEMANN, J. F. and CLARKE, F. M. (1988) In *Developments in Meat Science*, Vol. 4 (Ed. R. A. LAWRIE), p. 37 (Elsevier Appl. Sci., Lond.).

MORTON, H. C. and NEWBOLD, R. P. (1982) *Meat Sci.* **7**, 285.

MORTON, I. D., ARKROYD, P. and MAY, C. G. (1960) U.K. Pat. No. 836,694.

MORZEL, M., CHAMBON, C., HAMELIN, M., SANTÉ-LLOUTELLIER, V., SAYD, T. and MONIN, G. (2004) *Meat Sci.* **67**, 689.

MOSSEL, D. A. A. (1955) *Ann. Inst. Pasteur, Lille* **7**, 171.

MOSSEL, D. A. A., DIIKMANN, K. E. and SNŸDERS, J. M. A. (1975) In *Meat* (Eds. D. J. A. COLE and R. A. LAWRIE), Butterworths, London, p. 223.

MOTTRAM, D. S., CROFT, S. E. and PATTERSON, R. L. S. (1984) *J. Sci. Fd. Agric.*, **35**, 233.

MOTTRAM, D. S. and EDWARDS, R. A. (1983) *J. Sci. Fd. Agric.* **34**, 517.

MOTTRAM, D. S., EDWARDS, R. A. and MACFIE, H. J. H. (1982) *J. Sci. Fd. Agric.* **33**, 934.

MÜLLER, E. A. (1957) *J. Amer. Physiol. Ment. Rehabilitation* **11**, 41.

MURACHI, T., TANAKA, K., HATANAKA, M. and MURAKAMI, T. (1981) *Adv. Enzym. Regul.* **19**, 405.

MURAMOTO, T., NAKANISHI, N., SHIBATO, M. and AIKAWA, K. (2003) *Meat Sci.* **63**, 39.

NADAL-GINARD, B., MEDFORD, R. M., NGUYEN, H. T., PERIASAMY, M., WYDKO, R. M., HORNIG, D., GABITS, R., GARFINKEL, L. I., WECZORAK, D., BEKESI, E. and MAHDAVI, V. (1982) In *Muscle Development: Molecular and Cellular Control* (Eds. M. L. PEARSON and H. F. EPSTEIN) p. 143 (Cold Spr. Harb. Laboratory, Cold Spring Harbor, N. Y.).

NAIRN, A. C. and PERRY, S. V. (1979) *Biochem. J.* **179**, 89.

NAKAMURA, Y-N., IWAMOTO, H., ONO, Y., SHIBA, N., NISHIMURA, S. and TABATA, S. (2003) *Meat Sci.* **64**, 43.

NAM, K. C., DU, M., JO, C. and AHN, D. U. (2001) *Meat Sci.* **58**, 431.

NASSER, A. H. and HU, C. Y. (1991) *J. Anim. Sci.* **69**, 578.

NATIONAL RESEARCH COUNCIL (1981) *The Health Effects of Nitrate, Nitrite and N-Nitroso Compounds* (Net. Acad. Press: Washington) p. 51.

NEEDHAM, D. M. (1926) *Physiol. Rev.* **6**, 1.

NEEDHAM, D. M. (1960) *The Structure and Function of Muscle* (Ed. G. H. BOURNE), vol. 2, p. 55, Academic Press, New York.

NEEDHAM, J. (1931) *Chemical Embryology*, p. 1577, Cambridge Univ. Press.

NEEDHAM, J. (1942) *Biochemistry and Morphogenesis*, Cambridge Univ. Press.

NEGISHI, H., YAMAMOTO, E. and KUWATA, T. (1996) *Meat Sci.* **42**, 289.

NESBAKKEN, T. and SKJERVE, E. (1996) *Meat Sci.* **43**, Suppl., S47.

NESTOROV, N., GEORGIEVA, R. and GROSDANOV, A. (1970) *Proc. 16th Meeting European Meat Res. Workers*, *Varna*, p. 771.

NEUMANN, R. E and LOGAN, M. A. (1950) *J. Biol. Chem.* **184**, 299.

NEUVY, A. and VISSAC, B. (1962) *Contribution á l'étude du phénomène culard*, Union Nationale des Livres Génèalogiques, Paris.

NEWBERRIE, P. M. (1979) *Science*, **204**, 1079.

NEWBOLD, R. P. (1980) *Proc. 32nd. Ann. Recip. Meat Conf.*, p. 70 (Amer. Meat Inst.: Chicago).

NEWBOLD, R. P. and SCOPES, R. K. (1967) *Biochem. J.* **105**, 127.

NEWBOLD, R. P. and SMALL, L. M. (1985) *Meat Sci.* **12**, 1.

NEWBOLD, R. P. and TUME, R. K. (1976) *Meat Res. C.S.I.R.O.* p. 37.

NEWBOLD, R. P. and TUME, R. K. (1977) *Aust. J. Biol. Sci.* **30**, 519.

NEWHOOK, J. C. and BLACKMORE, D. K. (1982a) *Meat Sci.* **6**, 221.

NEWHOOK, J. C. and BLACKMORE, D. K. (1982b) *Meat Sci.* **6**, 295.

NEWMAN, P. B. (1980–81) *Meat Sci.* **5**, 17 1.

NEWMAN, P. B. (1984) *Anal. Proc.* **21**, 496.

NEWTON, K G. and GILL, C. O. (1978) *Appl. Environ. Microbiol.* **36**, 375.

NICOL, D., SHAW, M. K and LEDWARD, D. A. (1970) *Biochem. J.* **105**, 127.

NICOL, T., MCKELVIE, P. and DRUCE, C. G. (1961) *Nature, Lond.* **190**, 418.

NICOL, T. and WARE, C. C (1960) *Nature, Lond.* **185**, 42.

NIELL, J. M. and HASTINGS, A. B. (1925) *J. Biol. Chem.* **63**, 479.

NIELSEN, G. P., PETERSEN, B. R. and MØLLER, A. J. (1995) *Meat Sci.* **41**, 293.

NISHIMURA, T., HATTORI, A. and TAKAHASHI, K. (1995) *Meat Sci.* **39**, 127.

NISHIMURA, T., HATTORI, A. and TAKAHASHI, K. (1996) *Meat Sci.* **42**, 251.

NIVEN, C. F., Jr. (1951) *Amer. Meat Inst. Found. Circ.*, No. 2.

NIVEN, C F., Jr. (1958) *Ann. Rev. Microbiol.* **12**, 507.

NIVEN, C. F., Jr. (1963) *Int. J. Appl. Radn. Isotopes* **14**, 26.

NIVEN, C. F., Jr. and CHESEBRO W. R. (1956) *Proc. 8th Res. Conf., Amer. Meat Inst. Found., Chicago*, p. 47.

NIVEN, C. F., Jr. DEIBEL, R. H. and WILSON, G. D. (1958) *Amer. Meat Inst. Found. Circ.*, No. 41.

NIVEN, C. F., Jr. and EVANS, J. B. (1957) *J. Ract.* **73**, 758.

NOBLE, I. (1965) *J. Amer. Diet. Ass.* **47**, 205.

NOCI, F., O'KIELY, P., MONAHAN, F. J., STANTON, C. and MOLONEY, A. D. (2005) *Meat Sci.* **69**, 509.

NORMAN, G. A. (1991) In *Developments in Meat Science*–5 (Ed. R. A. LAWRIE), p. 57 (Elsevier Appl. Sci.: Lond.).

NORTJÉ, G. L., NEL, L., JORDAAN, E., NAUDE, R. T., HOLZAPPEL, W. H. and GRIMBEEK, R. J. (1989) *Meat Sci.* **25**, 81, 99.

NOTTINGHAM, P. M. (1960) *J. Sci. Fd. Agric.* **11**, 436.

NOTTINGHAM. P. M. (1963) *Hygiene in Meat Works,* Meat Ind. Res. Inst. N.Z. Inc. Publ., No. 72.

NOTTINGHAM, P. M., PENNEY,.N. and HARRISON, J. C. L. (1973) *N.Z. J. Agric. Res.* **17**, 79.

NOTTINGHAM, P. M. and URSELMANN, A. J. (1961) *N.Z. J. Agric. Res.* **4**, 449

NUTSCH, A. L., PHEBUS, R. K., RIEMANN, M. J., KOTVOLA, J. S., WILSON, R. C., and BROWN, T. L. (1998) *J. Fd. Prot.* **61**, 571.

OBANU, Z. A., LEDWARD, D. A. and LAWRIE, R. A. (1975) *J. Fd. Technol.* **10**, 667.

O'DEA, P. and SINCLAIR, A. J. (1985) *J. Nutr. Sci. Vitaminol.* **31**, 441.

ODEBRA, B. R. and MILLWARD, D. J. (1982) *Biochem. J.* **204**, 663.

O'DONOGHUE, K. and FISK, N. M. (2004) *Biologist*, **51**(3), 125.

O'HALLORAN, G. R., TROY, D. J., BUCKLEY, D. J. and REVILLE, W. J. (1997) *Meat Sci.* **47**, 187.

OFFER, G. W. (1972) *Cold Spr. Harb. Sympos. Quant. Biol.* **87**, 87.

OFFER, G. W. (1991) *Meat Sci.* **30**, 157.

OFFER, G. W., JEACOCKE, R. E. and RESTALL, D. J. (1983–85a) Bienn. Rept. A.F.R.C. *Food Res. Inst., Bristol*, p. 136.

OFFER, G. W., KNIGHT, P. J. and VOYLE, C. A. (1983–86b) Bienn. Rept. A.F.R.C. *Food Res. Inst. Bristol*, p. 131.

OFFER, G. and TRINICK, J. (1983) *Meat Sci.* **8**, 245.

OGILVY, W. S. and AYRES, J. C. (1951) *Food Tech.* **5**, 97.

OGLE, C. and MILLS, C. A. (1933) *Amer. J. Physiol.* **103**, 606.

OHASHI, K. and MARUYAMA, K. (1979) *J. Biochem.* **85**, 1103.

OHIKUNI, K., MUROYA, S. and NAKAJIMA, I. (2004) *Meat Sci.* **67**, 87.

OHLOFF, G. and FLAMANT, I. (1978) *Heterocycles* **11**, 663.

O'KEEFE, M. and HOOD, D. E. (1982) *Meat Sci.*, **7**, 209.

OKITANI, A., MATSUKURA, V., KATO, S. and FUIIMAKI, M. (1980) *J. Biochem.* **87**, 1133.

OLLIVIER, L., SELLIER, P. and MONIN, G. (1975) *Ann. Génét. Sel. Anim.* **7**, 159.

OLLSON, V., ANDERSSON, I., RANSON, K., LUNDSTRÖM, K. (2003) *Meat Sci.* **64**, 287.

OLSMAN, W. J. and SLUMP, P. (1981) In *Developments in Meat Science* Vol. II (Ed. R. A. LAWRIE), p. 195. (Applied Sci.: Lond.).

OLSON, E. N. (1992) *Dev. Biol.* **154**, 262.

OLSON, E. N. and KLEIN, W. H. (1992) *Genes Develop.* **8**, 1.

O'NEILL, E., MULVIHILL, D. M. and MORRISEY, R. A. (1989a) *Meat Sci.* **25**, 1.

O NEILL, E., MULIHILL, D. M. and MORRISEY, R. A. (1989b) *Meat Sci.* **26**, 131.

ONYANGO, C. A., IZUMIMOTO, M. and KUTIMA, P. M. (1998) *Meat Sci.* **49**, 117.

OPIE, L. H. and NEWSHOLME, E. H. (1967) *Biochem. J.* **103**, 391.

ORLOWSKI, M. (1990) *Biochem.* **29**, 10289.

OSBORN, H. M., BROWN, H., ADAMS, J. B. and LEDWARD, D. A. (2003) *Meat Sci.* **65**, 631.

OSTOVAR, K, MACNELL, J. H. and O DONNELL, K. (1971) *J. Fd. Sci.* **36**, 1005.

OUALI, A. (1992) *Biochem.* **74**, 251.

OUALI, A., GARREL, N., OBLED, A., DEVAL, C., VALIN, C. and PENNY, I. F. (1987) *Meat Sci.* **19**, 83.

OUALI, A. and TALMANT, A. (1990) *Meat Sci.* **28**, 331.

OUALI, A., ZABARI, M., RENIOR, J. P., TOURAILLE, C., KOPP, J., BONNET, M. and VALIN, C. (1988) *Meat Sci.* **24**, 151.

OWEN, J. E. and LAWRIE, R. A. (1975) *J. Fd. Technol.* **10**, 169.

OWEN, J. E., LAWRIE, R. A. and HARDY, B. (1975) *J. Sci. Fd. Agric.* **26**, 31.

PAGE, S. (1968) *Brit. Med. Bull.* **24**, No. 2, 170,.

PAINE, F. A. and PAINE, H. Y. (1983) *A Handbook of Food Packaging* (Leonard Hill: Glasgow).

PALEARI, M. A., CAMISASCA, S., BARETTA, G., RENON, P., CORSICO, G., BERTOLO, G. and CRIVELLI, G. (1998) *Meat Sci.* **48**, 205.

PALKA, K. (1999) *Meat Sci.* **53**, 189.

PÁLLSON, H. (1939) *J. Agric. Sci.* **29**, 544.

PÁLLSON, H. (1940) *J. Agric. Sci.* **30**, 1.

PÁLLSON, H. (1955) *Progress in the Physiology of Farm Animals* (Ed. J. HAMMOND), vol. **2**, p. 430, Butterworths, London.

PÁLLSON, H. (1962) *J. Reprod Fertil.* **3**, 55.

PÁLLSON, H. and VERGES, I. B. (1952) *J. Agric. Sci.* **42**, 1.

PALMER, A. Z. (1963) *Proc. Mmeat Tenderness Symp.*, p. 161, Campbell Soup Co., Camden, N.J.

PALMER, A. Z., BRADY, D. E., NAUMANN, H. D. and TUCKER, L N. (1955) *Food Tech.* **7**, 90.

PALMER, R. M., REEDS, P. J. and SMITH, R. H. (1982) *J. Physiol. Lond.* **322**, 15P.

PALMITER, R. D., BRINSTER, R. L., HAMMER, R. E., TRUMBAUER, M. E., ROSENFELD, M. G., BRINBERG, N. C. and EVANS, R. M. (1982) *Nature, Lond.* **300**, 611.

PANIANGVAIT, P., KING, A. J., JONES, A. D. and GERMAN, B. G. (1995) *J. Fd. Sci.* **60**, 1159.

PARAKKAL, P. F. (1969) *J. Cell. Biol.* **41**, 345.

PARK, R. J. and MINSON, D. J. (1972) *J. Agric. Sci.* **79**, 473.

PARK, R. J. and MURRAY, K. E. (1975) *Meat Res. in C.S.I.R.O.* p. 22.

PARKER, C. J. JR. and BERGER, C. K. (1963) *Biochim. Biophys. Acta* **74**, 730.

PARR, T., BARDSLEY, R. G., GILMOUR, R. S. and BUTTERY, P. J. (1992) *Eur. J. Biochem.* **208**, 333.

PARRY, D. A. (1970) In *Proteins as Human Food* (ed. R. A. LAWRIE), Butterworths, London, p. 365.

PARRY, D. A. D., BARNES, G. R. C. and CRAIG, A. S. (1978) *Proc. Roy. Soc.* **B 203**, 305.

PARSONS, A. L., PARSONS, J. L., BLANSHARD, J. M. V. and LAWRIE, R. A. (1969) *Biochem. J.* **112**, 673.

PARSONS, S. E. and PATTERSON, R. L. S. (1986a) *J. Fd. Technol.* **21**, 117.

PARSONS, S. E. and PATTERSON, R. L S. (1986b) *J. Fd. Technol.* **21**, 123.

PARTIS, L., CROAN, D., GUO, Z., CLARK, R., COLDHAM, T. and MURBY, J. (2000) *Meat Sci.* **54**, 369.

PARTRIDGE, S. M. (1959) *Ann. Rept. Low Temp. Res. Sta.*, Cambridge, p. 19.

PARTRIDGE, S. M. (1962) *Adv. Prot. Chem.* **17**, 227.

PARTRIDGE, S. M. and DAVIS, H. F. (1955) *Biochem. J.* **61**, 21.

PASSBACH, F. L., JR., MULLINS, A. M., WIPF, V. K. and PAUL, B. A. (1969) *Proc. 15th Meeting European Meat Res. Workers, Helsinki*, p. 64.

PATERSON, J. L. (1975) In *Meat* (Eds. D. J. A. COLE and R. A. LAWRIE), Butterworths, London, p. 471.

PATTERSON, R. L. S. (1968a) *J. Sci. Fd. Agric.* **19**, 31.

PATTERSON, R. L. S. (1968b) *J. Sci. Fd. Agric.* **19**, 434.

PATTERSON, R. L S. (1975) In *Meat* (Eds. D. J. A. COLE and R. A. LAWRIE), Butterworths, London, p. 359.

PATTERSON, R. L. S. (1984) In *Recent Advances in the Chemistry of Meat* (Ed. A. J. BAILEY) p. 192 (Roy. Soc. Chem.: Lond.).

PATTERSON, R. L S. and MOTTRAM, D. S. (1974) *J. Sci. Fd. Agric.* **25**, 1419.

PATTERSON, R. L S. and SALTER, L. J. (1985) *Meat Sci.* **14**, 191.

PATTERSON, R. L. S. and STENSON, C. G. (1971) *Proc. 17th Meeting Meat Res. Inst. Bristol*, p. 148.

PAUL, A. A. and SOUTHGATE, D. A. T. (1978) In *The Composition of Foods* (Eds. R. A. MCCANCE and E. M. WIDDOWSON), 4th Ed. (H.M.S.O.: Lond.).

PAUL, M. H. and SPERLING, E (1951) Unpublished observations, cited by D. E. Green (1951) *Biol. Rev.* **26**, 445.

PAUL, M. H. and SPERLING, E. (1952) *Proc. Soc. Exp. Biol. Med.* **79**, 352.

PAUL, P. C. (1975) In *Meat* (Eds. D. J. A. COLE and R. A. LAWRIE), Butterworths, London, p. 403.

PAUL, P. and BRATZLER, L. J. (1955) *Food Res.* **20**, 626.

PAVEY, R. L (1972) Contract 17–70-C–001, U.S. Army, Nattick.

PAYNE, S. R., DURHAM, C. J., SCOTT, S. M. and DEVINE, C. E. (1998) *Meat Sci.* **49**, 277.

PAYNE, S. R., SANDFORD, D., HARRIS, A. and YOUNG, O. A. (1994) *Meat Sci.* **37**, 429.

PAYNE, S. R. and YOUNG, O. A. (1995) *Meat Sci.* **41**, 147.

PEARSE, A. S. (1939) *Animal Ecology,* 2nd ed., McGraw-Hill, New York.

PEARSON, A. M. (1981) In *Developments in Meat Science*, Vol. II. Ed. R. A. LAWRIE, p. 241. (Applied Sci.: Lond.)

PEARSON, A. M. and DUTSON, T. R. (1985) *Advances in Meat Research.* Vol. I. Electrical Stimulation (AVI: Westport, Conn.).

PEARSON, A. M. and DUTSON, T. R. (1987) *Advances in Meat Research* Vol. 3 (AVI: New York).

PEARSON, A. M. and DUTSON, T. R. (1995) *HACCP in Meat, Poultry & Fish Processing* (Blackie: Glasgow).

PEARSON, A. M., HARRINGTON, G., WEST, R. G. and SPOONER, M. (1962) *J. Food Sci.* **27**, 177.

PEARSON, A. M., WEST, R. G. and LEUCKE, R. W. (1959) *Food Res.* **24**, 515.

PEARSON, R. T., DUFF, I. D., DERBYSHIRE, W. and BLANSHARD, J. M. V. (1974) *Biochim. Biophys. Acta* **362**, 188.

PENNY, I, F. (1960a) *Food Process. Packag.* **29**, 363.

PENNY, I, F. (1960b) *Chem. Ind.* **11**, 288.

PENNY, I, F. (1967) *Biochem. J.* **104**, 609.

PENNY, I, F. (1968) *J. Sci. Fd. Agric.* **19**, 578.

PENNY, I, F. (1969) *J. Fd. Technol.* **4**, 269.

PENNY, I, F. (1972) *J. Sci. Fd. Agric.* **23**, 403.

PENNY, I, F. (1974) *J. Sci. Fd. Agric.* **25**, 1273.

PENNY, I, F. (1977) *J. Sci. Fd. Agric.* **28**, 329.

PENNY, L F. (1980) In *Developments in Meat Science*, Vol. 1. Ed. R. A. LAWRIE, p. 115. (Applied Sci. Lond.).

PENNY, I, F. and DRANSFIELD, E. (1979) *Meat Sci.* **3**, 135.

PENNY, I, F. and FERGUSON-PRICE, R. (1979) *Meat Sci.* **3**, 121.

PENNY, I. F. and LAWRIE, R. A. (1964) Unpublished data.

PENNY, I, F., VOYLE, C. A. and DRANSFIELD, E. (1974) *J. Sci. Fd. Agric.* **25**, 703

PENNY, I. F., VOYLE, C. A. and LAWRIE, R. A. (1963) *J. Sci. Fd. Agric.* **14**, 535.

PENNY, I. F., VOYLE, C. A. and LAWRIE, R. A. (1964) *J. Sci. Fd. Agric.* **15**, 559.

PEPE, F. A. (1982) In *Cell and Muscle Motility*, Vol. 12 (Eds. J. W. SHAY and R. M. DOWREN), p. 141 (Plenum Press: New York).

PEREZ, C. and DESOUBRY, S. (2002) *Int. Rev. Fd. Sci. Tech.* **1**, 101.

PEREZ-REYES, E., KIM, H. S., LACARDA, A. E., HORNE, W., WEI, X., RAMPE, D., CAMPBELL, K. P., BROWN, A. M. and BUMBAUMER, L. (1989) *Nature, Lond.* **340**, 233.

PERIGO, J. A., WHITING, E. and BASHFORD, T. E. (1967) *J. Fd. Technol.* **2**, 377.

PERNOTTY, J. M. (1963) *Nature, Lond.* **200**, 86.

PERRIE, W. T. and PERRY, S. V. (1970) *Biochem. J.* **119**, 31.

PERRON, R. B. and WRIGHT, B. A. (1950) *Nature, Lond.* **166**, 863.

PERRSON, T. and VAN SYDOW, E. (1974) *J. Fd. Sci.* **39**, 406.

PERRY, D. A. D., CRAIG, A. S. and BARNES, G. R. G. (1978) *Proc. Roy. Soc. B* **203**, 293.

PERRY, S. V. (1974) *Biochem. Soc. Sympos.* **39**, 115.

PERRY, S. V, COLE, H. A., MORGAN, M., MOIR, A. J. G. and PIRES, E (1975) *Abs. 9th Febs Meeting* **31**, 163.

PERRY, T. W., BEESON, W. M., ANDREWS, F. N. and STOTS, M. (1955) *J. Anim. Sci.* **14**, 329.

PERUTZ, M. F. (1962) *Proteins and Nucleic Aids: Structure and Function,* Elsevier, London.

PERYAM, D. R. (1958) *Food Tech.* **12**, 231.

PETAJA, E. (1983) *Proc. 29th Europ. Meeting Meat Res. Workers, Salsomaggiore*, p. 117.

PETERS, J. B., BARNARD, R. J., EDGERTON, V. R., GILLESPIE, C. A. and STEMPEI, K. E (1972) *Biochemistry* **11**, 2627.

PETERS, R. A. (1957) *Adv. Enzymol.* **18**, 113.

PETERSEN, A. C. and GUNDERSON, N. F. (1960) *Food Tech.* **14**, 413.

PETERSEN, J. S., HENCKEL, P., MARIBO, H., OKSBJERG, N. and SØRENSEN, M. T. (1997) *Meat Sci.* **46**, 259.

PETERSEN, O. H. (1999) *Biologist* **46**, 227.

PETERSEN, O. H., PETERSEN, C. C. H. and KASAI, H. (1994) *Ann. Rev. Physiol.* **56**, 297.

PETROVIC, L. (1982) Ph.D. Dissertation, Univ. Novi Sad, Yugoslavia.

PHILLIPS, B., RUTHERFORD, N., GORSUCH, T., MABEY, M., LOCKER, N. and BOGGIANO, R. (1995) *Food Technol. Today* **9**(1), 19.

PHILLIPS, G. O. (1954) *Nature, Lond.* **172**, 1044.

PHILLIPS, G. O. (1988) *S. Afr. Food Rev.*, Feb./Mar., p. 6.

PICARD, B., GAGNIÈRE, H., ROBELIN, J., PONS, F. and GEAY, Y. (1995) *Meat Sci.* **41**, 315.

PICARD, B., LEGER, J. and ROBELIN, J. (1994) *Meat Sci.* **36**, 333.

PICKEN, L (1960) *The Organization of Cells*, Clarendon Press, Oxford.

PICKERING, K., EVANS, C. L., HARGIN, K. D. and STEWART, C. A. (1995a) *Meat Sci.* **40**, 319.

PICKERING, K., GRIFFIN, M., SMETHURST, P., HARGIN, K. D. and STEWART, C. A. (1995b) *Meat Sci.* **40**, 327.

PIEZ, K. A. (1965) *Biochemistry, N.Y.* **4**, 2590.

PIEZ, K. A. (1968) *Ann. Rev. Biochem.* **26**, 305.

PIPER, L. R., BINDON, B. M. and DAVIS, G. H. (1985) In *Genetics of Reproduction in Sheep* (Eds. R. B. LAND and D. W. ROBSON) p. 115 (Butterworths: Lond.).

POLGE, C. (1965) *Vet. Rec.* **77**, 232.

POLIDORI, P., TRABALZA, M., MARINUCCI, M., FANTUZ, F., RENIERI, C. and POLIDORI, F. (2000) *Meat Sci.* **55**, 197.

POMEROY, R. W. (1941) *J. Agric. Sci.* **31**, 50.

POMEROY, R. W. (1960) *J. Agric. Sci.* **54**, 57.

POMEROY, R. W. (1971) *Ann. Rept. A.R.C. Meat Res. Inst.*, 1970–71, p. 26.

POMEROY, R. W. and WILLIAMS, D. (1962) *Anim. Prod.* **4**, 302.

PORTER, K. (1961) *The Sarcoplasmic Reticulum, Suppl. J. Biophys. Biochem. Cyt.* **10**, 219, Rockefeller Inst. Press, New York.

PORZIO, M. A., PEARSON, A. M. and CORNFORTH, D. P. (1979) *Meat Sci.* **3**, 31.

POSPIECH, E., GREASER, M., MIKOLAJCZAK, B., CHIANG, W. and KRZYWDZINSKA, M. (2002) *Meat Sci.* **62**, 187.

POTTHAST, K. (1975) *Fleischwirts.* **55**, 1492.

POWELL, V. H., BOUTON, P. E., HARRIS, P. V. and SHORTHOSE, W R. (1983) *Proc. 29th Europ. Meeting Meat Res. Workers, Salsomaggiore*, p. 76.

POWELL, V H. and GRIFFITHS, L (1988) *Proc. 34th Intl. Congr. Meat Sci. Technol., Brisbane*, p. 226.

PRATER, A. R. and COOTE, G. G. (1962) *C.S.I.R.O. Div. Food Pres. Tech. Paper*, No. 28.

PRATT, G. B. and EKLUND, O. F. (1954) *Quick Frozen Foods* **16**, 50.

PRESTON, T. R. (1962) *Antibiotics in Agriculture* (Ed. M. WOODBINE), p. 214, Butterworths, London.

PRESTON, T. R., WHITELAW, F. G., AITKEN, J. N., MCDIARMID, A. and CHARLESTON, E. B. (1963) *Anim. Prod.* **5**, 47.

PRUNIER, A., CARITEZ, J. C. and BONNEAU, M. (1987) *Ann. Zootech.* **36**, 49.

PRUSINER, S. B. (1995) *Sci. Amer.* **272**, 47.

PURCHAS, R. W. and AUNGSUPAKORN, R. (1993) *Meat Sci.*, **34**, 163.

PURCHAS, R. W., JOHNSON, C. B., BIRCH, E. J., WINGER, R. J., HAGYARD, C. J. and KEOGH, R. G. (1986) *Flavour Studies with Beef and Lamb*, Dept. Animal Sci. Massey Univ., N.Z.

PURCHAS, R. W., RUTHERFORD, S. M., PEARCE, P. D., VATHER, R. and WILKINSON, B. H. P. (2004a) *Meat Sci.* **66**, 629.

PURCHAS, R. W., RUTHERFORD, S. M., PEARCE, P. D., VATHER, R. and WILKINSON, B. H. P. (2004b) *Meat Sci.* **68**, 201.

PURSEL, V. G., BOLT, D. J., MILLER, K. F., PINKERT, C. A., HAMMER, R. E., PALMITER, R. D. and BRINSTER, R. L. (1990) *J. Reprod. Fert.*, Suppl., **40**, 235.

PURSLOW, P. P. (1985) *Meat Sci.* **12**, 39.

PUTNAM, F. W. (1953) *The Proteins*, vol. 1, Part B (Eds. II. NEURATH and K. BAILEY), p. 808, Academic Press, New York.

RACCACH, M. and HENRICKSON, R. L. (1980) *Proc. 26th Europ. Meeting Meat Res. Workers, Colorado Springs*, **2**, 70.

RADOUCO-THOMAS, C., LATASTE-DOROLLE, C., ZENDER, R., BUSSET, R, MEYER, H. M. and MOUTON, R. F. (1959) *Food Res.* **24**, 453.

RAHELIC, S., PUAC, S. and GAWWAD, A. H. (1985) *Meat Sci.* **14**, 63, 73.

RAHKIO, M., KERKEALA, H., SIPPOLA, I. and PELTONEN, M. (1992) *Meat Sci.* **32**, 173.

RALSTON, A. T. and DYER, I. A. (1959) *J. Anim. Sci.* **18**, 1181.

RAMIREZ-SUAREZ, J. C. and XIONG, Y. L. (2003) *Meat Sci.* **65**, 899.

RAMSBOTTOM, J. M. and KOONZ, G. H. (1939) *Food Res.* **4**, 425.

RAMSBOTTOM, J. M. and STRANDINE, E. I. (1948) *Food Res.* **13**, 315.

RAMSBOTTOM, J. M. and STRANDINE, E. J. (1949) *J. Anim. Sci.* **8**, 398.

RANDALL, C. J. and BRATZLER, L. J. (1970) *J. Fd. Sci.*, **35**, 245, 248.

RANDALL, C. J. and MACRAE, H. E. (1967) *J. Fd. Sci.*, **32**, 182.

RANDALL, C. J. and VOISEY, P. W. (1977) *J. Int. Canad. Sci. Technol. Aliment.* **10**, 88.

RANGELEY, W. R. and LAWRIE, R. A. (1976) *J. Fd. Technol.* **11**, 143.

RANGELEY, W. R. D. and LAWRIE, R. A. (1977) *J. Fd. Sci. Technol.* **12**, 9.

RANKEN, M. R. and EVANS, G. G. (1979) *BFMIRA Res. Dept.* No. 296, p. 18.

RAO, M. V. and GAULT, N. E. S. (1989) *Meat Sci.* **26**, 5.

RAO, M. V., GAULT, N. E. S. and KENNEDY, S. (1989) *Meat Sci.* **26**, 19.

RATCLIFF, D., BOUTON, P. E., FORD, A. L, HARRIS, P. V., MACFARLANE, J. J. and O'SHEA, J. J. (1977) *J. Fd. Sci.* **42**, 857.

RAYMOND, J. (1929) *The Frozen and Chilled Meat Trade*, vol. 2, p. 28, Gresham Publishing Co., London.

RAYMOND, J. M., STRANDINE, E. J. and KOONZH, G. H. (1945) *Food Res.* **10**, 497.

REAGAN, J. O. (1983) *Food Technol.* (May) p. 79.

REES, M. P., TROUT, G. R. and WARNER, R. D. (2002a) *Meat Sci.* **60**, 113.

REES, M. P., TROUT, G. R. and WARNER, R. D. (2002b) *Meat Sci.* **61**, 169.

REGIER, L W. and TAPPEL, A. L. (1956) *Food Res.* 21, 630.

REMANIN, G., GRÖSSWAGEN, P and SCHINDLER, H. (1991) *Pflüg. Arch.* **418**, 86.

RENERRE, M., TOURAILLE, C., BORDES, P., LABAS, R., BAYLE, M. C. and FOURNIER, R. (1989) *Meat Sci.* **26**, 233.

RESSLER, C. (1962) *J. Biol. Chem.* **237**, 733.

REUTER, B. J., WULF, D. M., SHANKS, B. C., BOK, J. M. and MADDOCK, R. J. (2001) *Proc. 54th Ann. Recip. Meat Conf.*, Indianapolis. p. 187 (AMSA: Chicago).

REVERTE, D., XOING, Y. L. and MOODY, W. G. (2003) *Meat Sci.* **65**, 533.

RHEE, K. S., ZIPRIN, Y. A., ORDONEZ, G. and BOHAC, C. E. (1988) *Meat Sci.* **23**, 293.

RHODES, D. N. (1966) *J. Sci. Fd. Agric.* **17**, 180.

RHODES, D. N. (1969) *Ann. Rept. ARC Meat Res. Inst.* 1968–69, p. 34 (H.M.S.O.: Lond.).

RHODES, D. N. (1972) *J. Sci. Fd. Agric.* **23**, 1483.

RHODES, D. N. and DRANSFIELD, E. (1973) *J. Sci. Fd. Agric.* **24**, 1583.

RHODES, D. N. and DRANSFIELD, E. (1974) *J. Sci. Fd. Agric.* **25**, 1163.

RHODES, D. N. and SHEPHERD, H. J. (1966) *J. Sci. Fd. Agric.* **17**, 287.

RICE, E. E. and BEUK, J. E. (1953) *Adv. Food Res.* **4**, 233.

RICE, E. E., BEUK, J. E., KAUFFMAN, E. L., SCHULTZ, H. W. and ROBINSON, H. E. (1944) *Food Res.* **9**, 491.

RICHARDSON, W. D. and SCHERUBEL, E. (1909) *J. Ind. Eng. Chem.* **1**, 95.

RICKENBACKER, J. R. (1959) *Farmers Cooperative Services,* U.S. Dept. Agric. Service Dept., No. 42.

RICKERT, J., BRESSLER, L., BALL, C. O. and STIER, E. E. (1957) *Food Tech.* **11**, 625.

RITCHEY, S. J., COVER, S. and HOSTETLER, R. L. (1963) *Food Tech.* **17**, 76.

RIXSON, D. (2000) *The History of Meat Trading* (Nottingham Univ. Press).

ROBERT, N., BRIAND, M., TAYLOR, R. and BRIAND, T. (1999) *Meat Sci.* **51**, 149.

ROBERTS, D. C. K. (1999) *BNF Nutr. Bull.* **23**, 165.

ROBERTS, P. C. B. and LAWRIE, R. A. (1974) *J. Fd. Technol.* **9**, 345.

ROBERTS, T. A. (1971) *Ann. Rept. Meat Res. Inst.,* Bristol, p. 42.

ROBERTS, T. A. and GIBSON, A. M. (1986) *Food Technol.* **40**, 176.

ROBERTS, T. A., GIBSON, A. M. and ROBINSON, A. (1981a) *J. Fd. Technol.* **16**, 239.

ROBERTS, T. A., GIBSON, A. M. and ROBINSON, A. (1981b) *J. Fd. Technol.* **16**, 267.

ROBERTS, T. A. and INGRAM, M. (1976) *Proc. 2nd Intl. Sympos. Nitrite Meat Prod,* Zeist, p. **29**, Pudoc, Wageningen.

ROBERTS, W. K and SELL, J. L. (1963) *J. Anim. Sci.* **22**, 1081.

ROBERTSON, J. D. (1957) *J. Physiol.* **140**, 58.

ROBERTSON, J., RATCLIFF, D., BOUTON, P. E., HARRIS, P. V. and SHORTHOSE, W. R. (1986) *J. Fd. Sci.* **51**, 47.

ROBINS, S. P. and BAILEY, A. J. (1972) *Biochem. Biophys. Res. Commun.* **48**, 76.

ROBINS, S. P., SHIMOKOMAKI, M. and BAILEY, A. J. (1973) *Biochem. J.* **131**, 771.

ROBINSON, D. (1939) *Science* **90**, 276.

ROBINSON, M. E. (1924) *Biochem. J.* **18**, 225.

ROBINSON, R. H. M., INGRAM, M., CASE, R. A. M. and BENSTEAD, J. G. (1953) *Spec. Rept. Fd. Invest. Bd., Lond.,* No. 59.

ROBINSON, T. J. (1951) *J. Agric. Sci.* **41**, 16.

ROBSON, R. M., GOLL, D. E., ARAKAWA, N. and STROMER, M. M. (1970) *Biochim, Biophys. Acta* **200**, 296.

ROBSON, R. M. and HUIATT, T. W. (1983) *Proc. 36th Ann. Recip. Meat Res. Conf.* p. 116.

ROBSON, R. M., TUBATABAI, L. B., DAYTON, W. R., ZEECE, M. G., GOLL, D. E. and STROMER, M. H. (1974) *Proc. 27th Ann. Recip. Meat Conf.* (Net. Livestock & Meat Bd.: Chicago), p. 199.

RØDBOTTEN, M., KUBBERØD, F., LEA, P. and OELAND, Ø (2004) *Meat Sci.* **68**, 137.

RODRIGUEZ, J. M., SOBRINO, O. J., MOREIRA, W. L., CINTAS, L. M., CASAUS, P., FERNANDEZ, M. F., SANZ, B. and HERNANDEZ, P. E. (1994) *Meat Sci.* **37**, 305.

ROGERS, P. J. (1969) *Proc. 22nd Ann. Recip. Meat Conf., Pomona, Calif.,* p. 166.

ROGERS, S. A., HOLLYWOOD, N. W. and MITCHELL, G. E. (1992) *Meat Sci.* **32**, 437.

ROGERS, S. A., TAN, L. T., BICANIC, J-A. and MITCHELL, G. E. (1993) *Meat Sci.* **33**, 51.

ROGOWSKI, B. (1980) *Wld. Rev. Nutr. Diet.* **34**, 46.

ROHNER, G. A., ALEXANDER, L. J., KEELE, J. W., SMITH, T. P. and BEATTIE, C. W. (1994) *Genetics* **136**, 231.

ROLFE, E. J. (1958) *Fundamental Aspects of the Dehydration of Foodstuffs,* p. 211, Society of Chemical Industry, London.

ROLLER, S. (2002) In *Chitosan in Pharmacy and Chemistry* (Eds. R. A. A. MUZZARELLI and E. MUZZARELLI), p. 177, Atec Edizioii, Italy.

ROLLER, S., SAGOO, S., BOARD, R., O'MAHONY, T., CAPLICE, E., FITZGERALD, G., FOGDEN, M., OWEN, M. and FLETCHER, H. (2002) *Meat Sci.* **62**, 165.

ROMANUL, F. C. A. (1964) *Nature, Lond.* **201**, 307.

ROMANUL, E. C. A. and VAN DER MEULEN, J. P. (1967) *Acta Neurol.* **17**, 387.

RONGEY, E. H., KAHLENBERG, O. J. and NAUMANN, H. D. (1959) *Food Tech.* **13**, 640.

RONZONI, E., WALD, S. M., BERG, L. and RAMSEY, R. (1958) *Neurology* **8**, 359.

ROSENVOLD, K. and ANDERSEN, H. J. (2003) *Meat Sci.* **64**, 219.

ROSS COCKRILL, W. (1975) In *Meat* (Eds. D. J. A. COLE and R. A. LAWRIE), Butterworths, London, p. 507.

ROUSSET-AKRIM, S., GOT, F., BAYLE, M.-C. and CULIOLI, J. (1996) *Int. J. Df. Sci. Technol.* **31**, 333.

ROWE, R. W. D. (1974) *J. Fd. Technol.* **9**, 501.

ROWE, R. W. D. (1978) *Meat Sci.* **2**, 275.

ROWE, R. W. D. (1986) *Meat Sci.* **17**, 293.

ROWE, R. W. D. (1989) *Meat Sci.* **26**, 271.

ROWSON, L. E. A., TERVIT, R. and BRAD, A. (1972) *J. Reprod. Fertil.* **29**, 145.

RUBASHKINA, S. S. (1953) *Coll. Works AR-Union Res. Inst. Meat Industry, U.S.S.R.* **5**, 91.

RUBENSTEIN, H. S. and SOLOMON, M. L (1941) *Endocrin.* **28**, 229.

RUDÉRHUS, H. (1980) *Proc. 26th Europ. Meeting Meat Res. Workers, Colorado Springs* **2**, 96.

RUMEN, N. M. (1959) *Acta Chem. Scand.* **13**, 1542.

RUSIG, O. (1979) *Meat Sci.* **3**, 295.

RUSSELL, F. C. and DUNCAN, D. L. (1956) *Minerals in Pasture: Deficiencies and Excess in Relation to Animal Health,* 2nd ed., Commonwealth Bureau of Animal Nutrition. Tech. Comm., No. 15, Commonwealth Agricultural Bureaux, Slough.

RUUSUNEN, M. and PUOLANNE, E. (2004) *Meat Sci.* **67**, 533.

SADLER, D. H. and YOUNG, O. A. (1993) *Meat Sci.* **35**, 259.

SAFFLE, R. L. and BRATZLER, L J. (1959) *Food Technol.* **13**, 236.

SAFFLE, R. L. and GALBREATH, J. W. (1964) *Food Technol.* **18**, 119.

SAIR, L. and COOK, W. H. (1938) *Can. J. Res.* **D 16**, 255.

SAIR, R. A., LISTER, D., MOODY, W. G., CASSENS, R. G., HOEKSTRA, W. G. and BRISKEY, E. J. (1970) *Amer. J. Physiol.* **218**, 108.

SALES, J. and MELLETT, F. D. (1996) *Meat Sci.* **42**, 235.

SALMINEN, P., DEIGHTON, M. A., BENNO, Y. and GORBACH, S. L. (1998) In *Lactic Acid Bacteria,* p. 211 (Eds S. Salminen & von Wrigh) (New York: Marcel Dekker).

SANDERSON, M. and VAIL, G. E. (1963) *J. Food Sci.* **28**, 590

SANGER, F. (1945) *Biochem. J.* **39**, 507.

SANUDO, C., ENSER, M., CAMPO, M. M., NUTE, G. R., MARIA, G. R., SIERRA, I. and WOOD, J. D. (2000) *Meat Sci.* **54**, 339.

SANZ, P. D., DE ELVIRA, C., MARTINO, M. N., ZARITZKY, N. E. and OTERO, L. (1999) *Meat Sci.* **52**, 275.

SARTORE, S., MASCARELLO, F., ROWLERSON, A., GORZA, L, AUSONI, S., VIANELLO, M. and SCHAFFIN S. (1987) *J. Muscle Res. Cell Motility* **8**, 161.

SATIN, M. (2002) *Meat Sci.* **62**, 277.

SATORIOUS, M. J. and CHILD, A. M. (1938) *Food Res.* **3**, 619.

SAVAGE, A. W. J., RICHARDSON, R. I., JOLLEY, P. D., HARGIN, K. D. and STEWART, C. A. (1995) *Meat Sci.* **40**, 303.

SAVAGE, A. W. J., WARRISS, P. D. and JOLLEY, P. D. (1990) *Meat Sci.* **27**, 289.

SAVAGE, W. G. (1918) *J. Hyg. Camb.* **17**, 34.

SAVARY, P. and DESNUELLE, P. (1959) *Biochim. Biophys. Acta* **31**, 26.

SAVELL, J. W., DUTSON, T. R, SMITH, G. C. and CARPENTER, Z. L. (1978) *J. Fd. Sci.* **43**, 1606.

SAVELL, J. W., SMITH, G. C. and CARPENTER, Z. L. (1977b) *J. Fd. Sci.* **42**, 866.

SAVELL, J. W., SMITH, G. C. and CARPENTER, Z. L. (1976) *J. Anim. Sci.* **46**, 1221.

SAVELL, J. W., SMITH, G. C., DUTSON, T. R, CARPENTER, Z. L. and SUTER, D. A. (1977a) *J. Fd. Sci.* **42**, 702.

SAVIC, I. and KARAN-DJURDJIC, S. (1958) *Proc. 4th Meeting European Meat Res. Workers, Cambridge,* Paper No. 36.

SAVIC, I. and SUVAKOV, M. (1963) *Proc. 9th Meeting European Meat Res. Workers, Budapest,* Paper No. 49.

SAYRE, R. N., BRISKEY, E. J. and HOEKSTRA, W. G. (1963a) *Proc. Soc. Exp. Biol. Med.* **122**, 223.

SAYRE, R. N., BRISKEY, E. J. and HOEKSTRA, W. G. (1963b) *J. Food Sci.* **28**, 292.

SAYRE, R. N., BRISKEY, E. J. and HOEKSTRA, W. G. (1963c) *J. Anim. Sci.* **22**, 1012.

SAYRE, R. N., KIERNAT, B. and BRISKEY, E. J. (1964) *J. Food Sci.* **29**, 175.

SAZILI, A. Q., PARR, T., SENSKY, P. L. and JONES, S. W. (2005) *Meat Sci.* **69**, 17.

SCARBOROUGH, D. A. and WATTS, B. M. (1949) *Food Tech.* **3**, 152.

SCHAUB, M. C. and PERRY, S. V. (1969) *Biochem. J.* **115**, 993.

SCHAUB, M. C., PERRY, S. V. and HÄCKER, W. (1972) *Biochem. J.* **126**, 237.

SCHILLER, S. P. (1966) *Ann. Rev. Physiol.* **28**, 137.

SCHILLING, M. W., MINK, L. E., GOCHENOUR, P. S., MARRIOTT, N. G. and ALVARADO, C. Z. (2003) *Meat Sci.* **65**, 547.

SCHMIDT, C. F. (1961) *Rept. Europ. Meeting. Microbiol. Irrad. Foods,* Append. II, E.A.O., Rome.

SCHMIDT, C. F., LECHOWICH, R. V. and FOLINAZZO, J. E. (1961) *J. Food Sci.* **26**, 626.

SCHMIDT, G. R. and GILBERT, K. V. (1970) *J. Fd. Technol.* **5**, 331.

SCHMIDT, G. R, GILBERT, K. V., DAVEY, C. L., NOTTINGHAM, P. M. and DE LAMBERT, M. (1970) *Ann. Rept. Meat Ind Res. Inst.,* No. 2, p. 28.

SCHMIDT, G. R., YEMM, R. S., CHILDS, K. D., O'CALLAGHAN, J. P. and HOSSNER, K. L. (2002) *Meat Sci.* **62**, 79.

SCHMIDT, G. R., ZUIDAN, L. and SYBESMA, W. (1971) In *2nd Symposium Condition and Meat Quality in Pigs* (Animal Res. Inst.: Zeist), p. 245.

SCHNELL, P. G., VADHERA, D. V., HOOD, L. R. and BAKER, R. C. (1974) *Poult. Sci.* **53**, 419.

SCHOENHEIMER, R. (1942) *The Dynamic State of Body Constituents,* Harvard Univ. Press, Cambridge, Mass.

SCHÖN, L. and STOSIEK, M. (1958a) *Fleischwirts.* **10**, 550.

SCHÖN, L. and STOSIEK, M. (1958b) *Fleischwirts.* **10**, 769.

SCHÖNFELDT, H. C., NAUDÉ, R., BOK, W., VAN HEERDEN, S. M., SMIT, R. and BOSHOFF, E. (1993) *Meat Sci.* **34**, 363.

SCHWARTZ, K. (1962) *Vits. Hormones* **20**, 463.

SCHWARTZ, W. N. and BIRD, J. W. C. (1977) *Biochem. J.* **167**, 811.

SCHWEIGERT, B. S. (1959) *Int. J. Appl. Radn. Isotopes* **6**, 76.

SCHWEIGERT, B. S., MCINTIRE, J. M. and ELVEHJEM, C. A. (1944) *J. Nutr.* **27**, 419.

SCHWEIGERT, B. S. and PAYNE, B. J. (1956) *Amer. Meat Inst. Bull*, No. 30.

SCHWIMMER, S. and FRIEDMAN, M. (1972) *Flav. Ind.* **3**, 137.

SCOPES, R. K (1964) *Biochem. J.* **91**, 201.

SCOPES, R. K. (1966) *Biochem. J.* **98**, 193.

SCOPES, R. K. (1970) In *The Physiology and Biochemistry of Muscle as a Food,* vol. II, p. 471 (Eds. E. J. BRISKEY, R. G. CASSENS and B. B. MARSH), Univ. Wisconsin Press: Madison.

SCOPES, R. K. (1974) *Biochem. J.* **142**, 79.

SCOPES, R. K. and LAWRIE R. A. (1963) *Nature, Lond.* **197**, 1202.

SCOPES, R. K. and NEWBOLD, R. P. (1968) *Biochem. J.* **109**, 197.

SCOTT, W. J. (1936) *J. Coun. Sci. Ind. Res. Aust.* **9**, 177.

SCOTT, W. J. (1953) *Aust. J. Biol. Sci.* **6**, 549.

SCOTT, W. J. (1957) *Adv. Food Res.* **7**, 84.

SCOTT, W. J. and VICKERY, J. R. (1939) *C.S.I.R.O. Bull.*, No. 129.

SEARCY, D. J., HARRISON, D. L and ANDERSON, L. L. (1969) *J. Food Sci.* **34**, 486.

SEAWARD, M. R. D., CROSS, T. and UNSWORTH, B. A. (1976) *Nature, Lond.* **261**, 407.

SEIFTER, S. and GALLOP, P. M. (1966) In *The Proteins: Composition, Structure and Function* (2nd Ed.) (Ed. H. NEURATH), Acad. Press, New York, p. 155.

SELF, H. L. (1957) *Proc. 9th Res. Conf., Amer. Meat Inst. Found.*, p. 53.

SELLIER, P. (1988) In *Qualita della Carcassa e della Carne Suina*, p. 163 (Univ. Bologna).

SELLIER, P. (1994) *Meat Sci.* **36**, 29.

SELYE, H. (1936) *Nature, Lond.* **138**, 32.

SELYE, H. (1944) *Canad. Med. Ass. J.* **50**, 426.

SELYE, H. (1946) *J. Clin. Endocrin.* **6**, 117.

SELYE, H. (1950) *The Physiology and Pathology of Exposure to Stress*, Acta Inc., Montreal.

SEYDL, K., KARLSSON, J-O. DOMINIK, A., GRUBER, H. and ROMANIN, C. (1995) *Pflüg. Arch. Eur. J. Physiol.* **429**, 503.

SHACKELFORD, S. D., KOOHMARAIE, M., CUNDIFF, I. V., GREGORY, K. E., ROHNER, G. A. and SAVELL, J. W. (1994) *J. Anim. Sci.* **72**, 857.

SHACKELFORD, S. D., KOOHMARAIE, M., MILLER, M. F., CROUSE, J. D. and REAGAN, J. D. (1991) *J. Anim. Sci.* **69**, 171.

SHACKLADY, C. A. (1970) In *Proteins as Human Food*, p. 317 (Ed. R. A. LAWRIE), Butterworths, London.

SHAHAMAT, M., SEAMAN, A. and WOODBINE, M. (1980) In *Microbial Growth and Survival in Extremes of Environment* (Ed. G. W. GOULD and J. E. L. CORRY), p. 227 (Acad. Press: New York).

SHAHIDI, F., BEGG, S. B. and SEN, N. P. (1994) *Meat Sci.* **37**, 327.

SHARMA, N. K., REES, C. E. D. and DODD, C. E. R. (2000) *Appl. Environ. Microbiol.* **66**(4), 1347.

SHARP, J. G. (1953) *Spec. Rept. Fd. Invest. Bd., Lond.*, No. 57.

SHARP, J. G. (1957) *J. Sci. Fd. Agric.* **8**, 14, 21.

SHARP, J. G. (1958) *Ann. Rept. Fd. Invest. Bd., Lond.*, p. 7.

SHARP, J. G. (1959) *Proc. 5th Meeting European Meat Res. Workers, Paris,* Paper No. 17.

SHARP, J. G. (1963) *J. Sci. Fd. Agric.* **14**, 468.

SHARP, J. G (1964) *Proc. 1st Intl. Congr. Food Sci. Tech., Lond.*

SHARP, J. G. and MARSH, B. B. (1953) *Spec. Rept. Fd. Invest. Bd., Lond.*, No. 58.

SHARP, J. G. and ROLFE, E. J. (1958) *Fundamental Aspects of the Dehydration of Foodstuffs*, p. 197, Society of Chemical Industry, London.

SHAW, B. G. and HARDING, C. D. (1984) *J. Appl. Racter.* **56**,

SHAW, B. G. and HARDING, C. D. (1986) *System. Appl. Racter.* **6**, 291.

SHAW, B. G. and LATTY, J. B. (1984) *J. Appl. Racter.* **57**, 59.

SHAW, C., STITT, J. M. and COWAN, S. T. (1951) *J. Gen. Physiol.* **5**, 1010.

SHAW, F. D. (1973) *Meat Res. in C.S.I.R.O.*, p. 16.

SHAW, F. D., HARRIS, P. V., BOUTON, P. E., WEST, R. R. and TURNER, R. H. (1976) *Meat Res. in C.S.I.R.O.*, p. 7.

SHAW, M. G. (1977) *J. Med. Microbiol.* **10**, 29.

SHAW, M. K. (1963) *Proc. 9th Meeting European Meat Res. Workers, Budapest,* Paper No. 61.

SHAY, B. J. and EGAN, A. F. (1976) *Meat Res. in C.S.I.R.O.*, p. 29.

SHAY, B. J. and EGAN, A. F. (1986) *Food Technol. in Aust.* **38**, 144.

SHEA, K. G. (1958) *Food Tech.* **12**, 6.

SHEARD, P. R., ENSER, M., WOOD, J. D., NUTE, G. R., GILL, B. P. and RICHARDSON, R. I. (2000) *Meat Sci.* **55**, 213.

SHEARD, P. R., TAYLOR, A. A., SAVAGE, A. W. J., ROBINSON, A. M., RICHARDSON, R. I. and NUTE, G. R. (2001) *Meat Sci.* **59**, 423.

SHELEF, L. A., NAGLIK, O. A. and BOGEN, D. N. (1980) *J. Fd. Sci.* **45**, 1042.

SHERIDAN, J. J. (1990) *Meat Sci.* **28**, 31.

SHERIDAN, J. J. and MCDOWALL, D. A. (1998) *Meat Sci.* **49**, S 151.

SHERIDAN, J. J. and SHERRINGTON, J. (1982) *Meat Sci.* **7**, 245.

SHERMAN, J. D. and DAMESHEK, W. (1963) *Nature, Lond.* **197**, 469.

SHERMAN, P. (1961) *Food Tech.* **15**, 79.

SHERMAN, W. C., HALE, W. H., REYNOLDS, W. M. and LUTHER, H. G. (1957) *J. Anim. Sci.* **16**, 1020.

SHERMAN, W. C., HALE, W. H., REYNOLDS, W. M. and LUTHER, H. G. (1959) *J. Anim. Sci.* **18**, 198.

SHESTAKOV, S. D. (1962) *Proc. 8th European Meeting Res. Workers, Moscow,* Paper No. 18.

SHIGEHISA, T., OHMORI, T., SAITO, A., TAJI, S. and HAYASHI, R. (1991) *Intern. J. Fd. Microbiol.* **12**, 207.

SHIKAMA, K. (1963) *Sci. Rept. Tohoku Univ.* **22**, 91.

SHIMADA, K., SAKUMA, Y., WAKAMATSU, J., FUKUSHIMA, M., SEKIKAWA, M., KUCHIDA, K. and MIKAMI, M. (2004) *Meat Sci.* **68**, 357.

SHIMOKOMAKI, M., ELSDEN, D. F. and BAILEY, A. J. (1972) *J. Fd. Sci.* **37**, 892.

SHORLAND, F. B. (1953) *J. Sci. Fd. Agric.* **4**, 497.

SHORTHOSE, W. R, HARRIS, P. V. and BOUTON, P. E. (1972) *Proc. Soc. Anim. Prod.* **9**, 387.

SHRIMPTON, D. H. and BARNES, E. M. (1960) *Chem. Ind.* 1492.

SIEGEL, D. G., CHURCH, K. E. and SCHMIDT, G. L. (1979) *J. Fd. Sci.* **44**, 1276.

SIEMERS, L. L and HANNING, F. (1953) *Food Res.* **18**, 113.

SIMÕES, J. A., MENDES, M. I. and LEMOS, J. P. F. (2005) *Meat Sci.* **69**, 617.

SIMON, R. A. (1998) *Allergy* **53**, 28.

SIMONE, H., CARROLL, F. and CHICHESTER, C. O. (1959) *Food Tech.* **13**, 337.

SIMPSON, R. K. and GILMORE, A. (1997) *J. Appl. Microbiol.* **83**, 181.

SIMS, T. J. and BAILEY, A. J. (1982) In *Developments in Meat Science*, Vol. II (Ed. R. A. LAWRIE) (Applied Sci.: Lond.).

SINCLAIR, A. J., JOHNSON, L., O'DEA, K. and HOLMAN, R. T. (1994) *Lipids* **29**, 337.

SINGH, O. H., HENNEMAN, H. A. and REINEKE, E. P. (1956) *J. Anim. Sci.* **15**, 624.

SISSON, S. and GROSSMAN, J. D. (1953) *The Anatomy of Domestic Animals,* Saunders, London.

SJOSTRÖN, L. B. (1963) *Food Tech.* **17**, 266.

SKARPEID, H.-J., MOE, R. E. and INDAHL, U. (2001) *Meat Sci.* **57**, 227.

SKOVGAARD, N. (1987) In *Elimination of Pathogenic Organisms from Meat and Poultry* (Ed. F. J. M. SMULDERS), p. 39 (Elsevier Appl. Sci.: Amsterdam).

SLINDE, E. and KRYVI, H. (1986) *Meat Sci.* **16**, 45.

SLIWINSKI, R. A., MARGOLIS, R., PIH, K., LANDMANN, W. A. and DOTY, D. M. (1961) *Amer. Meat Inst. Found. Bull.*, No. 45.

SMALL, A., REID, C. A., AVERY, S. M., KARABASIL, N., CROWLEY, C. and BUNSIC, S. (2002) *J. Fd. Protect.* **7**.

SMITH, A. U. (1958) *Biol. Rev.* **33**, 197.

SMITH, D. S. (1966) *Prog. Biophys. Molec. Biol.* **16**, 107.

SMITH, E. C. (1929) *Proc. Roy. Soc.* **B 105**, 198.

SMITH, E. L., PARKER, L. F. J. and FANTES, K. H. (1948) *Biochem. J.* **43**, proc. xxx.

SMITH, G. C., DUTSON, T. R, CARPENTER, Z. L. and HOSTETLER, R. L. (1977) *Proc. Meat Ind Res. Conf.* p. 47.

SMITH, G. C., PIKE, M. I. and CARPENTER, Z. L. (1974) *J. Fd. Sci.* **39**, 1145.

SMITH, G. C., SAVELL, J. W., DUTSON, T. R, HOSTETLER, R. L., TERRELL, R. N., MURPHY, C. E. and CARPENTER, Z. L. (1980) *Proc. 26th Europ. Meeting Meat Res. Workers, Colorado Springs*, **2**, 19.

SMITH, J. L. and PALUMBO, S. A. (1981) *J. Food Protect.* **44**, 936.

SMITH, M. G. (1973) *Meat Res. in C.S.I.R.O.*, p. 21.

SMITH, M. G. and GRAHAM, A. (1978) *Meat Sci.* **2**, 119.

SMITH, S. B., GARCIA, D. K. and ANDERSON, D. B. (1989) *J. Anim. Sci.* **67**, 3495.

SMOLENSKI, K. A., FALLON, A., LIGHT, N. D. and BAILEY, A. J. (1983) *Biosc. Rep.* **3**, 93.

SMORODINTZEV, I. A. (1934) *Z. Lebensmitt. Untersuch.* **67**, 429.

SNOKE, J. E. and NEURATH, H. (1950) *J. Biol. Chem.* **187**, 527.

SNOWDEN, L. MCK. and WEIDEMANN, J. F. (1978) *Meat Sci.* **2**, 1.

SOFOS, J. N., BELK, K. E. and SMITH, G. C. (1999) *Proc. 45th Int. Congr. Food Sci. Tech.*, Yokohama, Japan, p. 596.

SOLOMON, M. B., EASTRIDGE, J. S., ZUCKERMAN, R. and JOHNSON, W. (1997) *Proc. 43rd Int. Congr. Fd. Sci. Tech.*, p. 121.

SOLOV'VEV, V. I. (1952) *Myasnaya Industriya, U.S.S.R.* **23**, 43.

SOLOV'EV, V. I. and KARPA, I. N. (1967) *Proc. 13th Meeting European Meat Res. Workers, Rotterdam*, Paper No. D 11.

SOMLYO, A. P., DEVINE, C. E., SOMLYO, A. V. and NORTH, R. V. (1973) *Proc. Roy. Soc.* **B 262**, 333.

SOMLYO, A. V, BUTLER, T. M., BOND, M. and SOMLYO, A. P. (1981) *Nature, Lond.* **294**, 567.

SOMMERS, C. H., NIEMIRA, B. A., TUNICK, M. and BOYD, G. (2002) *Meat Sci.* **61**, 223.

SØRHEIM, O., NISSEN, H. and NESBAKKEN, T. (1999) *Meat Sci.* **52**, 157.

SORIMACHI, H., ISHIURA, S. and SUZUKI, K. (1997) *Biochem. J.* **328**, 721.

SORINMADE, S. O., CROSS, H. R, ONO, K. and WERGEN, W. P. (1981) *Meat Sci.* **6**, 71.

SOTELO, I., PÉREZ-MUNUERA, I., QUILES, A., HERNANDO, I., LARREA, V. and LLUCH, M. A. (2004) *Meat Sci.* **66**, 823.

SPANIER, A. M. and ROMAMOWSKI, R. D. (2000) *Meat Sci.* **56**, 193.

SPENCER, G. S. G. (1977) In *Scoliosis.* (Ed. P. A. ZORAB). (Acad. Press.: New York).

SPENCER, G. S. G. (1981) In *Hormones and Metabolism in Ruminants.* (Eds. J. M. FORBES and M. A. LOMAX), p. 80. (ARC: Lond.).

SPENCER, G. S. G., WILKINS, L. J. and LISTER, D. (1983) *Meat Sci.* **8**, 53.

SPENCER, G. S. G. and WILLIAMSON, E. D. (1981) *Anim. Prod.* **32**, 376.

SPENCER, R. (1971) *Proc. 17th Meeting Europ. Meat Res. Inst., Bristol,* p. 161.

SPERRING, D. D., PLATT, W. T. and HINER, R. L. (1959) *Food Tech.* **13**, 155.

SRÉTER, F. A. (1964) *Fed. Proc.* **23**, 930.

STANLEY, D. W. (1983) *Am. J. Clin. Nutr.* **26**, 1251.

STANTON, C. and LIGHT, N. (1988) *Meat Sci.* **23**, 179.

STANTON, C. and LIGHT, N. D. (1990) *Meat Sci.* **27**, 41.

STARLINGER, V. H. (1967) *Hoppe-Seyler. Z. Physiol.* **348**, 864.

STATON, R. S. and PETTE, D. (1990) In *The dynamic state of muscle fibres.* Cited by Y. GEAY and B. PICARD (1995) Proc. 48th Ann. Recip. Meat Conf., San Antonio, Texas, p. 51.

STEHLING, O., DORING, H., ERTL, J., PROBISCH, G. and SCHMIDT, I. (1996) *Am. J. Physiol.* **272**, R 1770.

STEINBERG, M. P., WINTER, J. D. and HUSTRALID, A. (1949) *Food Tech.* **3**, 367.

STENBOCK-FERMOR, COUNT (1915) *Anthr.* **26**, 298.

STEPHEN, P., CLARKE, F. M. and MORTON, D. J. (1980) *Biochem. Biophys. Acta* **873** 127.

STETTER, K. O. (1998) In *The Molecular Origin of Life* (Ed. A. BRACK), (Cambridge Univ. Press). Cited by HOWLAND (2001).

STEVENSON, M. H., CROSS, A. V. J., HAMILTON, J. T. G. and MCMURRAY, C. H. (1993) *Radiat. Phys. Chem.* **42**, 363.

STEWART, F., ALLEN, W. R. and MOOR, R. M. (1976) *J. Endocrin.* **71**, 371.

STEWART, M., MORTON, D. J. and CLARKE, F. M. (1980) *Biochem. J.* **186**, 99.

STOCKDALE, F. E. (1992) *Dev. Biol.* **154**, 284.

STOLL, N. R. (1947) *J. Parasit.* **33**, 1.

STONE, N. and MEISTER, A. (1962) *Nature, Lond.* **194**, 555.

STORM, E. and ORSKOV, E. R. (1984) *Brit. J. Nutr.* **52**, 63.

STRANDBERG, K., PARRISH, F. C., GOLL D. E. and JOSEPHSON, S. A. (1973) *J. Fd. Sci.* **38**, 69.

STRATTON, J. M. and BROWN, J. H. (1978) *Agricultural Records Ad220–1977* (Ed. R. WHITLOCK). Cited by Rixson (2000) *ioc. cit.*

STRAUB, F. B. (1942) *Stud. Inst. Med. Chem. Univ. Szeged* **2**, 3.

STREHLER, E. F., STREHLER PAGE, M., PERRIARD, J., PERIASAMY, M. and NADAL-GINARD, B. (1986) *J. Mol. Biol.* **190**, 291.

STRENSTROM, W. and LOHMANN, A. (1928) *J. Biol. Chem.* **79**, 673.

STROTHER, J. W. (1975) In *Meat* (Eds. D. J. A. COLE and R. A. LAWRIE), Butterworths, London, p. 183.

SULZBACHER, W. L. (1952) *Food Tech.* **6**, 341.

SULZBACHER, W. L and GADDIS, A. M. (1968) In *The Freezing Preservation of Foods,* 4th ed., vol. 2, 159, AVI: Westpoint, Conn.

SULZBACHER, W. L. and MCLEAN, R. A. (1951) *Food Tech. Champaign,* III. **5**, 7.

SUTHERLAND, E. W. and ROBINSON, G. A. (1966) *Pharmacol. Rev.* **18 (1)**, 145.

SUZUKI, A., HOMMA, Y., KIM, K., IKEUCHI, Y., SUGIYAMA, T. and SAITO, M. (2001) *Meat Sci.* **59**, 193.

SUZUKI, A., HOMMA, N., FUKODA, A., HIRAO, K. and URYU, T. (1994) *Meat Sci.* **37**, 369.

SUZUKI, A., KITAMURI, Y., INOUE, S., NONAMI, Y. and SAITO, M. (1985) *Meat Sci.* **14**, 243.

SUZUKI, A. and TAMATE, H. (1988) *Anat. Rec.* **42**, 39.

SUZUKI, A., WATANABE, M., IKEUCHI, Y., SAITO, M. and TAKAHASHI, K. (1993) *Meat Sci.* **35**, 17.

SUZUKI, T. and MACFARLANE, J. J. (1984) *Meat Sci.* **11**, 263.

SVEDBERG, T. and BROHULT, S. (1939) *Nature, Lond.* **143**, 938.

SWAIN, M. V. L., GIGIEL, A. J. and LIMPENS, G. (1999) *Meat Sci.* **51**, 363.

SWAN, H. and COLE, D. J. A. (1975) In *Meat* (Eds. D. J. A. COLE and R. A. LAWRIE), p. 71 (Butterworths: Lond.).

SWANSON, A. A., MARTIN, B. J. and SPICER, S. S. (1974) *Biochem. J.* **137**, 223.

SWARTZ, D. R. and GREASER, M. L. (1995) *Proc. 47th Ann. Recip. Meat Cont.* (Natl. Meat and Livestock Bd., Chicago).

SWARTZ, D. R., GREASER, M. L. and MARSH, B. B. (1993) *Meat Sci.* **33**, 139.157.

SWARTZ, D. R., LIM, S.-S., FASSEL, T. and GREASER, M. L. (1995) *Proc. 47th Recip. Meat Res. Conf.* p. 141 (Natl. Livestock & Meat Bd.: Chicago).

SWATLAND, H. (2002) *Meat Sci.* **62**, 225.

SWATLAND, H. (2003) *Meat Sci.* **63, 463.**

SWATLAND, H. (2004) *Meat Sci.* **67**, 371.

SWATLAND, H. J. (1973) *J. Anim. Sci.* **36**, 355.

SWATLAND, H. J. (1988) *J. Anim. Sci.* **66**, 379.

SWATLAND, H. J. (1994) *Structure and Development of Meat Animals and Poultry* (Technomic, Lancaster, Philadelphia).

SWIFT, C. E. and SULZBACHER, W. L. (1963) *Food Tech.* **17**, 106.

SWINGLER, G. R. and LAWRIE, R. A. (1979) *Meat Sci.* **3**, 63.

SWINGLER, G. R., NAYLOR, P. E. L. and LAWRE, R. A. (1979) *Meat Sci.* **3**, 83.

SWINGLER, G. R., NEALE, R. J. and LAWRIE, R. A. (1978) *Meat Sci.* **2**, 31.

SZENT-GYÖRGI, A. G. (1953) *Arch. Biochem.* **60**, 180.

TAGGERT, P., PARKINSON, P. and CARRUTHERS, M. (1972) *Brit. Med. J.* **iii**, 71.

TAKADA, F., VANDER WOUDE, D. L., TONG, H.-Q., THOMPSON, T. G., WATKINS, S. C., KUNKEL, L. M. and BEGGS, A. H. (2001) *J. Cell Sci.* **117**(10), 1971.

TAKAHASHI, G., LOCHNER, J. V. and MARSH, B. B. (1984) *Meat Sci.* **11**, 207.

TAKAHASHI, G., WANG, S.-M., LOCHNER, J. V. and MARSH, B. R. (1987) *Meat Sci.* **19**, 65.

TAKAHASHI, K. (1992) *Biochemie* **74**, 247.

TALLON, H. H., MOORE, S. and STEIN, W. H. (1954) *J. Biol. Chem.* **211**, 927.

TALMANT, A., MONIN, G., BRIAND, M., DADET, M. and BRIAND, Y. (1986) *Meat Sci.* **18**, 23.

TAM, L. G., BERG, E. P., GERRARD, D. E., SHEISS, E. B., TAN, F. J., OKOS, M. T. and FORREST, J. C. (1998) *Meat Sci.* **49**, 41.

TAMM, W. (1930) *Z. yes. Kält.*, Beih. R. **3**, H. 4, cited by J. KUPRIANOFF (1956).

TANNENBAUM, S. R., FETT, D., YOUNG, V. R., LANE, P. D. and BRUCE, W. R. (1978) *Science*, **200**, 1488.

TANNER, E. B. (1944) *The Microbiology of Foods*, 2nd ed., Garrard Press, Champaign, 111.

TAPPEL, A. L (1952) *Food Res.* **17**, 550.

TAPPEL, A. L. (1956) *Food Res.* **21**, 195.

TAPPEL, A. L. (1957a) *Food Res.* **22**, 408.

TAPPEL, A. L. (1957b) *Food Res.* **22**, 404.

TAPPEL, A. L, BROWN, W. D., ZALKIN, H. and MAIER, V. P. (1961) *J. Amer. Oil Chem. Soc.* **38**, 5.

TAPPEL, A. L., ZALKIN, H., CALDWELL, K. A., DESAI, I. D. and SHIBKO, S. (1962) *Arch. Biochem. Biophys.* **96**, 340.

TARLADGIS, B. G. (1962) *J. Sci. Fd. Agric.* **13**, 481.

TARR, H. L. A., SOUTHCOTT, B. H. and BISSETT, H. M. (1952) *Food Tech.* **6**, 363.

TARRANT, P. V. (1981) In *The Problem of Dark-Cutting in Beef* (Eds. D. E. HOOD and V. TARRANT), p. 3 (Martinus Nijhoff: The Hague).

TARRANT, P. V. (1998) *Meat Sci.* **49**, S 1.

TATSUMI, R. and TAKAHASHI, K. (1992) *J. Biochem.* **112**, 775.

TAUSIG, E. and DRAKE, M. P. (1959) *Food Res.* **24**, 224.

TAYLER, C. R. (1968) *Nature, Lond.* **219**, 181.

TAYLER, C. R. (1969) *Sci. Amer.* **220**, 86.

TAYLER, E. L. and PARFITT, J. W. (1959) *Int. J. Appl. Radn. Isotopes* **6**, 194.

TAYLOR, A. A. (1971) *Proc. 17th Meeting European Meat Res. Workers*, Bristol, p. 662.

TAYLOR, A. A. (1985) In *Developments in Meat Science* Vol. 3 (Ed. R. A. LAWRIE), p. 89 (Elsevier Appl. Sci.: Lond.).

TAYLOR, A. A. (1992) *Meat Sci.* **31**, 381.

TAYLOR, A. A. and DANT, S. J. (1971) *J. Food Technol.* **6**, 131.

TAYLOR, A. A., NUTE, G. R. and WARKUP, C. C. (1995) *Meat Sci.* **39**, 339.

TAYLOR, A. A., SHAW, B. G. and JOLLEY, P. D. (1980) *J. Fd. Technol.* **15**, 301.

TAYLOR, A. A., SHAW, B. G. and JOLLEY, P. D. (1982a) *J. Fd. Technol.* **17**, 339.

TAYLOR, A. A., SHAW, B. G., JOLLEY, P. D. and NUTE, G. R. (1982b) *J. Fd. Technol.* **17**, 339.

TAYLOR, A. A., SHAW, B. G. and MACDOUGALL, D. B. (1981) *Meat Sci.* **5**, 109.

TAYLOR, A. J., COLE, D. J. A. and LEWIS, D. (1973) *Proc. Brit. Soc. Anim. Prod.* **2**, 87.

TAYLOR, A. J., COLE, D. J. A. and LEWIS, D. (1974) *Proc. Brit. Soc. Anim. Prod.* **3**, 111.

TAYLOR, A. J., LINFORTH, R., WEIR, O., HUTTON, T. and GREEN, B. (1993a) *Meat Sci.* **33**, 75.

TAYLOR, A. J., PONCE-ALQUICIRA, E. and LINFORTH, R. (1993b) *Proc. IFST* **7**, 225.

TAYLOR, A. MCM. (1958) *Food Manuf.* **33**, 286.

TAYLOR, A. MCM. (1963) *Inst. Meat Bull.*, July, p. 4.

TAYLOR, C. B., PENG, S. K., WERTHESSEN, N. T., THAM, P. and LEE, K. T. (1979) *Am. J. Clin. Nutr.* **32**, 40.

TAYLOR, G. R. (1982) *The Great Evolution Mystery* (Seeker & Warburg: Lond.)

TAYLOR, M. A. J. and ETHERINGTON, D. J. (1991) *Meat Sci.* **29**, 211.

TERRELL, R. N., SUESS, G. G. and BRAY, R. W. (1969) *J. Anim. Sci.* **28**, 449, 454

TEUGEL, P. (1987) In *Elimination of Pathogenic Organisms from Meat and Poultry* (Ed. F. J. M. SMULDERS), p. 79 (Elsevier Appl. Sci.: Amsterdam).

THEORELL, H. (1932) *Biochem. Z.* **252**, 1.

THOMAS, J., ELSDEN, D. J. and PARTRIDGE, S. M. (1963) *Nature, Lond.* **200**, 651.

THOMAS, L. (1956) *J. Exp. Med.* **104**, 245.

THOMPSON, J. S., FOX. J. B., JR. and LANDMANN, W. A. (1962) *Food Tech.* **16**, 131.

THOMPSON, T. (2002) *Meat Sci.* **62**, 295.

THORNLEY, M. J. (1963) *J. Appl. Bact.* **26**, 334.

THORNLEY, M. J., INGRAM. M. and BARNES, E. M. (1960) *J. Appl. Bact.* **23**, 487.

THORNTON, H. (1973) *Textbook of Meat Inspection*, Baillière, Tindall & Cox, London.

THORNTON, R. A. and ROSS, D. J. (1959) *N.Z. J. Agric. Res.* **2**, 1002.

TIDBALL, J. G., O'HALLORAN, T. and BURRIDGE, K. (1986) *J. Cell Biol.* **103**, 1465.

TILGNER, D. J. (1958) *Fleischwirts*, **8**, 741.

TILGNER, D. J., MILER, K, PROMINSKI, J. and DARNOWSKI, G. (1962) *Tech. mesa. Beograd*, Paper No. 1, 12.

TIMS, M. J. and WATTS, B. M. (1958) *Food Technol.* **12**, 240.

TING, E. Y. (1999) *Proc. 52nd Recip. Meat Conf.*, Oklahoma. p. 61.

TISCHER, R. G., HURWICZ, H. and ZOELLNER, J. A. (1953) *Food Res.* **18**, 539.

TOLBERT, N. E. (1981) *Ann. Rev. Biochem.* **50**, 133.

TOLDRA, E., TORRERO, Y. and FLORES, J. (1991) *Meat Sci.* **29**, 177.

TOLMACHOFF, I. P. (1929) *Trans. Amer. Phil. Soc.* **23**, Part 1.

TOMBS, M. P. (1972) British Patent 1,265,661.

TOPEL, D. G. (1969) In *Points of View on the Condition and Meat Quality of Pigs for Slaughter*, p. 91 (Eds. W. SYBESMA, P. G. VAN DER WALS and P. WALSTRON), Res. Inst. Animal Husbandry, Zeist.

TOPEL, D. G., MARKEL, R. A. and WISMER PEDERSEN, J. (1967) *J. Anim. Sci.* **26**, 311.

TORLEY, P. J., REID, D. H., YOUNG, O. A. and ARCHIBALD, R. D. (1988) *Food Technol. N.Z.*, 51.

TORNBERG, E. and PERSSON, K (1988) *Proc. 34th Int. Congr. Meat Sci. Technol.*, Brisbane, p. 190.

TORRESCANO, G., SÁNCHEZ-ESCALANTE, A., GIMÉNEZ, B., RONCALES, P. and BELTRAN, J. A. (2003) *Meat Sci.* **64**, 85.

TOURAILLE, C., MONIN, G. and LEGAULT, C. (1989) *Meat Sci.* **25**, 177.

TOWER, S. S. (1937) *J. Comp. Neurol.* **67**, 241.

TOWER, S. S. (1939) *Physiol. Rev.* **19**, 1.

TOWNSEND, W. E. and BRATZLER, L. J. (1958) *Food Tech.* **12**, 663.

TOWNSEND, W. E., WITANAEUR, L. P., RELOFF, J. A. and SWIFT, C. E. (1968) *Food Tech. Champaign*, **22**, 71.

TRESSL, R., GRÜNEWALD, K. G., SILWAR, R. and BAHNI, D. (1979) In *Progress in Flavour Research*, (Eds. D. G. LAND and H. NURSTEN), p. 197. (Applied Sci.: Lond.).

TRESSLER, D. K., BIRSEYE, C. and MURRAY, W. T. (1932) *Ind. Eng. Chem.* **24**, 242.

TRESSLER, D. K. and EVERS, C F. (1947) *The Freezing Preservation of Foods*, AVI Publishing Co., New York.

TRINICK, J. A. (1981) *J. Mol. Biol.* **151**, 309.

TRINICK, J. A. and COOPER, J. (1982) *J. Muscl. Res. Cell. Motil.* **3**, 486.

TRINICK, J. A., COOPER, J. A. M. and WALKER, M. L. (1983–85) *Bienn. Rept. A.F.R.C. Food Res. Inst. Bristol*, p. 120.

TULLOH, N. M. (1964) *Aust. J. Agric. Res.* **15**, 333.

TULLOH, N. M. and ROMBERG, B. (1963) *Nature, Lond.* **200**, 438.

TUMA, H. L., HENRICKSON, R. L., STEPHENS, D. F. and MOORE, R. (1962) *J. Anim. Sci.* **21**, 848.

TUME, R. K. (1980) *Aust. J. Biol. Sci.* **33**, 43.

TURKKI, P. R. and CAMPBELL, A. M. (1967) *J. Food Sci.* **32**, 151.

TURMAN, E. J. and ANDREWS, F. N. (1955) *J. Anim. Sci.* **14**, 7.

TURNBULL, P. C. B. and ROSE, P. (1982) In *Campylobacter: Epidemiology Pathogenesis and Biochemistry* (Ed. D. G. NEWELL), p. 271 (MTP Press: Boston, Mass.).

TVERAAEN, T. (1935) *Hvalråd Skr.* No. 11, p. 5.

UDENFREND, S., SHORE, P. A., BOGDANSKI, D. F., WASSBACH, H. and BRODIE, B. B. (1957) *Rec. Progr. Hormone Res.* **13**, 1.

UENO, Y., IKEUICHI, Y. and SUZUKI, A. (1999) *Meat Sci.* **52**, 143.

UMENO, S. and NOBATA, R. (1938) *J. Jap. Soc. Vet. Sci.* **17**, 87.

UNDERWOOD, E. J. (1977) *Trace Elements in Human Nutrition* 4th edn. (Acad. Press: New York).

URBAIN, W. M. and JENSEN L. B. (1940) *Food Res.* **5**, 593.

URBIN, M. C. and WILSON, G. D. (1958) *Proc. 10th Res. Conf., Amer. Meat Inst. Found.*, p. 13.

URBIN, M. C., ZESSIN, D. A. and WILSON, G. D. (1962) *J. Anim. Sci.* **21**, 9.

US SURGEON-GENERAL (1965) 'Statement on the Wholesomeness of Irradiated Foods', June 10th.

USHERWOOD, E. J. (1979) *Phil. Trans. Roy. Soc., Lond.* **B 288**, 5.

VALIN, C. (1968) *J. Fd. Technol.* **3**, 171.

VALIN, C. (1970) *Ann. Biol. Anim. Bioch. Biophys.* **10**, 317.

VALLYATHAN, N. V., GRINYAR, I. and GEORGE, J. C. (1970) *Can. J. Zool.* **48**, 377.

VAN DEN OUWELAND, G. A. M., OLSMAN, H. and PEER, H. G. (1978) In *Agricultural and Food Chemistry: Past, Present and Future*. (Ed. R. TERANISHI), p. 292 (AVI: Westport, Conn.).

VAN DER FLIER, A. and SONNENBERG, A. (2001) *Cell. Tiss. Res.* **305**, 285.

VAN DER WAL, P. G., ENGEL, B., VAN BEEK, G. and VEERKAMP, C. H. (1995) *Meat Sci.* **40**, 193.

VAN LOON, G. R. and BROWN, G. M. (1975) *J. Clin. Endo. Metab.* **41**, 640.

VANICHSENI, S., HAUGHHEY, D. P. and NOTTINGHAM, P. M. (1972) *J. Fd. Technol.* **7**, 259.

VAN OECKEL, M. J. and WARRANTS, N. (2003) *Meat Sci.* **68**, 293.

VARNAM, A. H. and SUTHERLAND, J. P. (1995) *Meat and Meat Products* (Chapman & Hall: Lond.).

VEGA, G. L., GROSZEK, E., WOLF, R. and GRUNDY, S. M. (1982) *J. Lipid Res.* **23**, 811.

VENUGOPAL, B. and BAILEY, M. E. (1978) *Meat Sci.* **2**, 227.

VERATTI, E. (1902) *Mem. reale Inst. Lombardo* **19**, 87.

VERKAAR, E. L. C., NIJMAN, I. J., BANTAGA, K. and LENSTRA, J. A. (2002) *Meat Sci.* **60**, 365.

VETTERLAIN, R. and KIDNEY, A. J. (1965) *Proc. 11th Meeting Meat Res. Workers*, Belgrade.

VICKERY, J. R. (1932) *Spec. Rept. Fd. Invest. Bd., Lond.* No. 42.

VICKERY, J. R. (1953) *J. Aust. Inst. Agric. Sci.* **19**, 222.

VICKERY, J. R. (1968) *Meat Ind. Bull.*, *Aust. Jan.*, p. 31.

VICKERY, J. R. (1977) *C.S.I.R.O. Div. Fd. Res. Tech. Pap. No.* 42.

VIGNOS, P. J. and LEFKOWITZ, M. (1959) *J. Clin. Invest.* **38**, 873.

VILLEE, C. A. (1960) *Developing Cell Systems and their Control* (Ed. D. RUDNICK), p. 93, Ronald Press, New York.

VOLEKNER, H. H. and CASSON, J. L. (1951) *J. Anim. Sci.* **10**, 1065.

VOLODKEVITCH, N. N. (1938) *Food Res.* **3**, 221.

VON MICKWITZ, G. and LEACH, T. M. (1977) *Review of Preslaughter Stunning in E.C.*, C.E.C. Inform. Agnc. No. 30.

VOYLE, C. A. (1969) *J. Fd. Technol.* **4**, 275.

VOYLE, C. A. (1979) In *Food Microstructure* (Ed. J. G. VAUGHAN), p. 193 (Acad. Press: New York).

VOYLE, C. A. (1981) *Scanning Electron Micros.* **3**, 427.

VOYLE, C. A. and LAWRIE, R. A. (1964) *J. Roy. Microsc. Soc.* **81**, 173.

WAHLROOS, O. and NIINIVAARA, F. P. (1969) *Proc. 15th Meeting European Meat Res. Workers, Helsinki*, p. 226.

WAKAMATSU, J., NISHIMURA, T. and HATTORI, A. (2004) *Meat Sci.* **67**, 95.

WALLACE, L. R. (1945) *J. Physiol.* **104**, 33.

WALLACE, L R. (1948) *J. Agric. Sci.* **38**, 93.

WALLIMAN, I. and EPPENBERGER, H. M. (1985) In *Cell and Muscle Motility* (Ed. J. W. SHAY), p. 239 (Plenum: New York).

WALLS, E. W. (1960) *The Structure and Function of Muscle* (Ed. G. H. BOURNE), vol. 1, p. 21, Academic Press, New York.

WALTERS, C. L. (1973) *Proc. Inst. Fd. Sci. Technol.* **6**, 106.

WALTERS, C. L. (1983) *Brit. Nutr. Foundn. Nutr. Bull.* **8**, 164.

WALTERS, C. L., CASSELDEN, R. J. and TAYLOR, A. MCM. (1967) *Biochim. Biophys. Acta* **143**, 310.

WALTERS, C. L and TAYLOR, A. MCM. (1963) *Food Tech.* **17**, 119.

WALTERS, C. L. and TAYLOR, A. MCM. (1964) *Biochim. Biophys. Acta* **86**, 448.

WALTERS, C. L. and TAYLOR, A. MCM. (1965) *Biochim. Biophys. Acta* **96**, 522.

WANG, H. (1991) 'Causes and solutions of iridescence in precooked meat'. Ph.D. dissert., Kansas State Univ., Manhattan.

WANG, H., ANDREWS, F., RASCH, E., DOTY, D. M. and KRAYBILL, H. R. (1953) *Food Res.* **18**, 351.

WANG, H., AUERBACH, E., BATES, V., ANDREWS, F., DOTY, D. M. and KRAYBILL, H. R. (1954a) *Food Res.* **19**, 154.

WANG, H., AUERBACH, E., BATES, V., ANDREWS, F., DOTY, D. M. and KRAYBILL, H. R. (1954b) *Food Res.* **19**, 543.

WANG, H., DOTY, D. M., BEARD, F. J., PIERCE, J. C. and HANKINS, O. G. (1956) *J. Anim. Sci.* **15**, 97.

WANG, H. and MAYNARD, N. (1955) *Food Res.* **20**, 587.

WANG, H., WEIR, C. E., BIRKNER, M. and GINGER, B. (1957) *Proc. 9th Res. Conf. Amer. Meat Inst. Found.*, p. 65.

WANG, H., WEIR, C. E., BIRKNER, M. and GINGER, B. (1958) *Food Res.* **23**, 423.

WANG, I. H. and THOMPSON, J. M. (2001) *Meat Sci.* **58**, 137.

WANG, K. (1985) In *Cell and Muscle Motility*, Vol. 6 (Eds. J. W. SHAY and R. M. DOWBEN), p. 315 (Plenum Press: New York).

WANG, K. and WILLIAMSON, C. L. (1980) *Proc. Nat. Acad. Sci., USA*, **77**, 3254.

WANG, K. and WRIGHT, J. (1988) *J. Cell Biol.* **107**, 2199.

WARNER, K. F. (1928) *Proc. Amer. Soc. Anim. Prod.*, p. 114.

WARNER, R. D., KAUFFMAN, R. G. and RUSSELL, R. I. (1993) *Meat Sci.* **33**, 359.

WARREN, S. (1943) *Arch. Path.* **35**, 347.

WARRINGTON, R. (1974) *Vet. Bull.* **44**, 617.

WARRISS, P. D. (1978) *Meat Sci.* **2**, 155.

WARRISS, P. D. (2000) *Meat Science: An Introductory Text* (CAB International: Wallingford).

WARRISS, P. D., BEVIS, E. A., EDWARDS, J. E., BROWN, S. N. and KNOWLES, T. G. (1991a) *Vet. Rec.* **128**, 419.

WARRISS, P. D., BROWN, S. N., ADAMS, S. J. M. and CORLETT, L. K. (1994) *Meat Sci.* **38**, 329.

WARRISS, P. D., BROWN, S. N., ADAMS, S. J. M. and LOWE, D. B. (1990a) *Meat Sci.* **28**, 321.

WARRISS, P. D., BROWN, S. N., BARTON GADE, P., SANTOS, C., NANI COSTA, L., LAMBOOIJ, E. and GEERS, R. (1998b) *Meat Sci.* **49**, 137.

WARRISS, P. D., BROWN, S. N., FRANKLIN, J G. and KESTIN, S. C. (1990b) *Meat Sci.* **28**, 21.

WARRISS, P. D., BROWN, S. N., KNOWLES, T. G., EDWARDS, J. E., KETTLEWELL, P. J. and GUISE, H. J. (1998a) *Meat Sci.* **50**, 447.

WARRISS, P. D., KESTIN, S. C., BROWN, S. N. and NUTE, G. R. (1996) *Meat Focus Int.* **5**, 179.

WARRISS, P. D., KESTIN, S. C., BROWN, S. N. and WILKINS, L. J. (1984) *Meat Sci.* **10**, 53.

WARRISS, P. D., NUTE, G. R., ROLPH, T. P., BROWN, S. N. and KESTIN, S. C. (1991b) *Meat Sci.* **30**, 75.

WARRIS, P. D. and WOTTON, S. B. (1981) *Res. Vet. Sci.* **31**, 82.

WASSERMAN, A. E. (1966) *J. Food Sci.* **31**, 1005.

WASSERMAN, A. and SPINELLI, A. M. (1972) *J. Agric. Food Chem.* **20**, 171.

WATANABE, A. and DEVINE, C. E. (1996) *Meat Sci.* **42**, 407.

WATANABE, A., DALY, C. C. and DEVINE, C. E. (1996) *Meat Sci.* **42**, 67.

WATANABE, K. and SATO, Y. (1971) *Agric. Bio. Chem.* **35**, 756.

WATSON, J. D. and CRICK, F. H. C. (1953) *Nature, Lond.* **171**, 737, 964.

WATTS, B. M. (1954) *Adv. Food Res.* **5**, 1.

WATTS, B. M., KENDRICK, J., ZIPSTER, M., HUTCHINS, B. and SALEM (1966) *J. Food. Sci.* **31**, 855.

WATTS, B. M. and LEHMANN, B. T. (1952a) *Food Res.* **17**, 100.

WATTS, B. M. and LEHMANN, B. T. (1952b) *Food Tech.* **6**, 194.

WEBER, A., HERZ, R. and REISS, I. (1963) *J. Gen. Physiol.* **46**, 679.

WEBER, H. H. and MEYER, K. (1933) *Biochem. Z.* **266**, 137.

WEBSTER, A. J. F. (1974) In *Heat Loss from Animals and Man* (Eds. J. L. MONTEITH and L. E. MOUNT), p. 205, Butterworths, London.

WEBSTER, A. J. F. (1976) In *Principles of Cattle Production* (Eds. H. SWAN and W. H. BROSTER), p. 103, Butterworths, London.

WEBSTER, A. J. F. and YOUNG, B. A. (1970) *Univ. Alberta Feeders Day Rept. No.* 49, p. 34, Univ. Alberta, Edmonton.

WEBSTER, H. L. (1953) *Nature, Lond.* **172**, 453.

WEBSTER, J. D., LEDWARD, D. A. and LAWRIE, R. A. (1982) *Meat Sci.* **7**, 147.

WEINSTOCK, I. M., GOLDRICH, A. D. and MILHORAT, A. T. (1956) *Proc. Soc. Exp. Biol. Med.* **91**, 302.

WEIR, C. E. (1960) *The Science of Meat and Meat Products* (Ed. Amer. Meat Inst. Found.), p. 212, Reinhold Publishing Co., New York.

WEISER, H. H., KUNKLE, L. E. and DEATHERAGE, E. E. (1954) *Appl. Microbiol.* **2**, 88.

WEISS, A. and LEINWARD, L. A. (1996) *Ann. Rev. Cell Develop. Biol.* **12**, 417.

WEISS, J. (1952) *Nature, Lond.* **169**, 460.

WEISS, J. (1953) *Ciba Found. Colloq. Endocrin*, **7**, 142.

WELCH, G. R. (1977) *Prog. Biophys. Mol. Biol.* **32**, 103.

WEST, R. G., PEARSON, A. M. and MCARDLE, E. J. (1961) *J. Food Sci.* **26**, 79.

WEST, W. T. and MASON, K. E. (1958) *Amer. J. Anat.* **102**, 323.

WHARTON, D. (2002) *Chem. in Britain* Oct. p. 38.

WHIPPLE, G., KOOHMARAIE, M. and ARBONA, J. R. (1994) *Meat Sci.* **38**, 133.

WHITE, W. J. P. and LAWRIE, R. A. (1985) *Meat Sci.* **15**, 173.

WHITEHAIR, L. A., BRAY, R. W., WECKEL, K. G., EVANS, G. W. and HEILIGMAN, F. (1964) *Food Tech.* **18**, 114.

WHITEHEAD, W. L., GOODMAN, C. and BREGER, I. A. (1951) *J. Chim. Phys.* **48**, 184.

WHITEHOUSE, M. W. and LASH, J.W. (1961) *Nature, Lond.* **189**, 37.

WHITMANN, H. G. (1961) *Proc. 5th Intl. Congr. Biochem., Moscow*, Symp. No. 1, Pergamon Press, Oxford.

WHITTINGTON, F. M., NUTE, G. R., HUGHES, S. I., MCGIVAN, J. D., LEAN, I. J., WOOD, J. D. and DORAN, E. (2004) *Meat Sci.* **67**, 569.

WHO (1977) 'HEALTH HAZARDS FROM DRINKING WATER', REPT. WORKING GP., Lond. WHO REG. OFFICE EUROP ICP/PPE/005.

WIDDOWSON, E. M. (1970) In *The Physiology and Biochemistry of Muscle as a Food*, vol. II, p. 511 (Eds. E. J. BRISKEY, R. G. CASSENS and B. B. MARSH), Univ. Wisconsin Press, Madison.

WIDDOWSON, E. M. (1971) *Biol. Neonat.* **19**, 329.

WIDDOWSON, E. M., DICKERSON, J. W. T. and MCCANCE, R. A. (1960) *Brit. J. Nutr.* **14**, 457.

WIENER, C. (1972) U.S. Pat. No. 3,650,771.

WIENER, P. D., KROPF, D. H., MACKINTOSH, D. L. and KOCH, B. A. (1964) *J. Anim. Sci.* **23**, 864.

WIERBICKI, E. (1980) *Proc. 26th Europ. Meeting Meat Res. Workers, Colorado Springs*, **1**, 194.

WIERBICKI, E. and DEATHERAGE, E. E. (1958) *J. Agric. Food Chem.* **6**, 387.

WIERBICKI, E and HEILIGMAN, E. (1980) *Proc. 26th Europ. Meeting Meat Res. Workers, Colorado Springs*, **1**, 198.

WIERBICKI, E., KUNKLE, L. E., CAHILL, V. R. and DEATHERAGE, E. E. (1954) *Food Tech.* **8**, 506.

WIERBICKI, E., KUNKLE, L. E., CAHILL, V. R. and DEATHERAGE, E. E. (1956) *Food Tech.* **10**, 80.

WIERBICKI, E., TIEDE, M. G. and BURRELL, R. C. (1963) Fleischwirts. **15**, 404.

WIKLUND, E., STEVENSON-BARRY, J. M., DUNCAN, S. T. and LITTLEJOHN, R. P. (2001) *Meat Sci.* **59**, 211.

WILDE, W. S. and SHEPPARD, C. W. (1955) *Proc. Soc. Exp. Biol. Med.* **88**, 249.

WILDING, P., HEDGES, N. and LILLPORD, P. (1986) *Meat Sci.* **18**, 55.

WILEY, H. (1908) *U.S. Dept. Agric., Washington, D.C., Burl Chem. Bull.*, No. 84.

WILK, S. and ORLOWSKI, M. (1983) *J. Neurochem.* **40**, 842.

WILK, S., PEARCE, S. and ORLOWSKI, M. (1979) *Life Sci.* **24**, 457.

WILKINSON, J. M., PERRY, S. V., COLE, H. A. and TRAYER, I. P. (1972) *Biochem. J.* **127**, 215.

WILLIAMS, A. A. and ATKINS, R. K. (1983) *Sensory Quality in Foods and Beverages* (Ellis Horwood Ltd.: Chichester).

WILLIAMS, B. E. (1964a) Canad. Pat. No. 683,929.

WILLIAMS, B. E. (1964b) U.S. Pat. No. 3,128,191.

WILLIAMS, E. E. (1968) in *Quality Control in the Food Industry*, vol. II, p. 252 (Ed. S. M. HERSCHDOERPER), Acad. Press, London.

WILLIAMS, J. C., VIMINI, R. J., FIELD, R. A., RILEY, M. L. and KUNSMAN, J. E (1983) *Meat Sci.* **9**, 181.

WILLIAMS, J. D., ANSELL, B. M., REIFFEL, L. and KARK, R. M. (1957) *Lancet* **ii**, 464.

WILLIAMS, R. J. (1956) *Biochemical Individuality*, John Wiley, New York.

WILLIAMS, P., BAINTO, N. J., BAINTON, S., CHHABRA, S. R., WILSON, M. V., STEWART, G. S. A. B., SALMON, G. P. and BYCROFT, B. W. (1992) *FEMS Microbiol. Lett.*, 161.

WILLIAMSON, E. D., PATTERSON, R. L. S., BUXTON, E R., MITCHELL, K G., PARTRIDGE, Y. G. and WALKER, N. (1985) *Livest. Prod. Sci.* **12**, 251.

WILSON, G. D. (1960) *The Science of Meat and Meat Products* (Ed. Amer. Meat Inst. Found.), pp. 328, 349, Reinhold Publicating Co., New York.

WILSON, G. D. (1961) *Proc. 13th Res. Conf. Amer. Meat Inst. Found. Chicago*, p. 113.

WILSON, G. D., BRAY, R. W. and PHILLIPS, P. H. (1954) *J. Anim. Sci.* **13**, 826.

WILSON, G. D., BROWN, P. D., CHESBRO, W. R, GINGER, B. and WEIR, C. E. (1960) *Food Tech.* **14**, 143, 186.

WILSON, G. S. and MILES, A. A. (1955) *Topley and Wilson's Principles of Bacteriology and Immunity*, 4th ed., vol. 1, p. 977, Edward Arnold, London.

WINDRUM, G. M., KENT, P. W. and EASTOE, J. E. (1955) *Brit. J. Exp. Path.* **36**, 49.

WINEGARDEN, M. E., LOWE, B., KASTELLIC, J., KLINE, E. A., PLAGGE, A. R. and SHEARER, P. S. (1952) *Food Res.* **17**, 172.

WINKLER, C. A. (1939) *Canad. J. Res.* **D 17**, 8.

WINTERØ, A. K., THOMSEN, P. D. and DAVIES, W. (1990) *Meat Sci.* **27**, 75.

WISMER-PEDERSEN, J. (1959a) *Food Res.* **24**, 711.

WISMER-PEDERSEN, J. (1959b) *Acta Agric. Scand.* **9**, 69, 91.

WISMER-PEDERSEN, J. (1960) *Food Res.* **25**, 789.

WISMER-PEDERSEN, J. (1966) *J. Fd. Sci.* **31**, 980.

WISMER-PEDERSEN, J. (1969a) *Proc. 15th Meeting Meat Res. Workers, Helsinki*, p. 454.

WISMER-PEDERSEN, J. (1969b) In *Recent Points of View on the Condition and Meat Quality of Pigs for Slaughter* (Eds. W. SYBESMA, P. G. VAN DER WALS and P. WALSTRON), Res. Inst. Animal Husbandry, Zeist.

WISTREICH, H. E., MORSE, R. E and KENYON, L. J. (1959) *Food Tech.* **13**, 441.

WOESSNER, J. E., JR. and BREWER, T. H. (1963) *Biochem. J.* **89**, 75.

WONG, E., JOHNSON, C. B. and NIXON, L. N. (1975a) *N.Z. J. Agric. Res.* **18**, 261.

WONG, E., JOHNSON, C. B. and NIXON, L. N. (1975b) *Chem. Ind.*, p. 40.

WOOD, J. D. (1984a) *Res. Develop. Agric.* **1**, 129.

WOOD, J. D. (1984b) In *Fats in Animal Nutrition* (Ed. J. WISEMAN), p. 407 (Butterworths, Lond.).

WOOD, J. D., GREGORY, N. G., HALL, G. M. and LISTER, D. (1977) *Brit. J. Nutr.* **37**, 167.

WOOD, J. D. and LISTER, D. (1973) *J. Sci. Fd. Agric.* **24**, 1449.

WOOD, J. D., NUTE, G. R., FURSEY, G. A. J. and CUTHBERTSON, A. (1995) *Meat Sci.* **40**, 127.

WOOD, J. D., RICHARDSON, R. I., NUTE, G. R., FISHER, A. V., CAMPO, M. M., KASAPIDOU, E., SHEARD, P. R. and ENSER, M. (2003) *Meat Sci.* **66**, 21.

WOOD, J. M. and EVANS, G. G. (1973) *Proc. Inst. Fd. Sci. Technol.* **6**, 111.

WOOD, T. and BENDER, A. E (1957) *Biochem. J.* **57**, 366.

WRAY, C. and SOJKA, W. J. (1977) *J. Dairy Res.* **44**, 383.

WRENSHALL, C. L (1959) *Antibiotics: their Chemistry and Non-medical Uses* (Ed. H. S. GOLDBERG), p. 549, D. van Nostrand, New York.

WRIGHT, N. C. (1954) *Progress in the Physiology of Farm Animals* (Ed. J. HAMMOND), vol. 1, p. 191, Butterworths, London.

WRIGHT, N. C. (1960) *Hunger: Can it be averted?* (Ed. E. J. RUSSELL and N. C. WRIGHT), p. 1, Brit. Ass. Adv. Sci., London.

WRIGHT, R. H., HUGHES, J. R. and HENDRIX, D. E. (1967) *Nature, Lond.* **216**, 404.

WYCKOFF, R. W. G. (1930a) *J. Exp. Med.* **51**, 921.

WYCKOFF, R. W. G. (1930b) *J. Exp. Med.* **52**, 435.

WYTHES, J. R, SHORTHOSE, W. R., SCHMIDT, P. J. and DAVIS, C. B. (1980) *Aust. J. Agric. Res.* **31**, 849.

YANCEY, P. H. (2003) *Biologist* **50**, 126.

YAROM, R. and MEIRI, U. (1971) *Nature, Lond.* **234**, 254.

YEATES, N. T. M. (1949) *J. Agric. Sci.* **39**, 1.

YEH, E., ANDERSON, B., JONES, P. N. and SHAW, F. D. (1978) *Vet. Rec.* **103**, 117.

YEREX, D. and SPIERS I. (1987) *Modern Deer Farming Management* (Ampersand Assoc. Ltd.: Casterton, N.Z.).

YESAIR, J. (1930) *Canning Tr.* **52**, 112.

YOSHIDA, T. and KAGEYAMA, H. (1956) Jap. Pat. No. 732.

YOUNATHAN, M. T. and WATTS, B. M. (1960) *Food Res.* **25**, 538.

YOUNG, O. A. (1984) *Meat Sci.* **11**, 123.

YOUNG, O. A. and BASS, J. J. (1984) *Meat Sci.* **11**, 139.

YOUNG, O. A., BERDAGUE, J.-L., VIALLON, C., ROUSSET-AKRIM, S. and THERIEZ, M. (1997) *Meat Sci.* **45**, 183.

YOUNG, O. A. and BRAGGINS, T. J. (1993) *Meat Sci.* **35**, 213.

YOUNG, O. A. and CURSONS, R. T. M. (1988) *Proc. 34th Intl. Congr. Meat Sci. Technol., Brisbane*, p. 176.

YOUNG, O. A and DAVEY, C. L. (1981) *Biochem. J.* **195**, 317.

YOUNG, O. A., FOOTE, D. M. and BASS, J. J. (1986) *Meat Sci.* **16**, 189.

YOUNG, O. A, GRAAFHUIS, A. E and DAVEY, C. L. (1980) *Meat Sci.* **5**, 41.

YOUNG, O. A., TORLEY, P. J. and REID, D. H. (1992) *Meat Sci.* **32**, 45.

YOUNG, R. B. and ACHTYMUCHUK, G. W. (1982) *J. Ala. Acad. Sci.* **53**, 119.

YOUNG, R. B. and DENOME, R. M. (1984) *Biochem. J.* **218**, 871.

YOUNG, R. H. (1980) In *Developments in Meat Science*, Vol. 1 (Ed. R. A. LAWRIE), p. 145 (Applied Sci. *Lond.*).

YOUNG, R. H. and LAWRIE, R. A. (1974) *J. Fd. Technol.* **9**, 69.

YOUNG, R. H. and LAWRIE, R. A. (1975) *J. Fd. Technol.* **10**, 453.

YOUSIF, O. KH. and BABIKER, S. A. (1989) *Meat Sci.* **26**, 245.

YUDKIN, J. (1964) *Proc. Nutr. Soc.* **23**, 149.

YUDKIN, J. (1967) *Am. J. Clin. Bact.* **20**, 108.

YUEH, M. H. and STRONG, F. M. (1960) *J. Agric. Food Chem.* **8**, 491.

ZABALA, A., MARTIN, R., HAZA, A., FERNANDEZ, L., MORALES, P. and RODRÍGUEZ, J. J. (2001) *Meat Sci.* **59**, 79.

ZAKULA, R. (1969) *Proc. 15th Meeting European Meat Res. Workers, Helsinki*, p. 157.

ZANORA, F., CHAIB, F. and DRANSFIELD, E. (1998) *Meat Sci.* **49**, 127.

ZEBE, E. (1961) *Ergeb. Biol.* **24**, 247.

ZEMBAYASHI, M., LUNT, D. K. and SMITH, S. B. (1999) *Meat Sci.* **53**, 221.

ZENDER, R., LATASTE-DOROLLE, C., COLLET, R. A., ROWINSKI, P. and MOUTON, R. F. (1958) *Food Res.* **23**, 305.

ZENKER, F. A. (1860) *Firchows Arch. Path. Anat.* **18**, 561.

ZEROUALA, A. C. and STICKLAND, N. C. (1991) *Meat Sci.* **29**, 263.

ZEUNER, F. E. (1963) *A History of Domesticated Animals*, Hutchinson, London.

ZEUTHEN, P. (1995) In *Fermented Meats*, p. 53 (Blackie: Glasgow).

ZHANG, R. (2002) *Nature, Long.* **417**, 971.

ZHAO, T., DOYLE, M. P., HARMON, B. G., BROWN, C. A., MUELLER, P. O. E. and PARKS, A. H. (1998) *J. Clin. Microbiol.* **36**, 641.

ZHOU, G. H., YANG, A. and TUME, R. K. (1993) *Meat Sci.* **35**, 205.

ZIEMBA, Z and MALKKI, Y (1969) *Proc. 15th Meeting Meat Res. Workers, Helsinki*, p. 461.

ZINNERMANN, B K., TIMPL, R. and KUHN, Z. (1973) *Eur. J. Biochem.* **35**, 216.

ZIPRIN, Y. A., RHEE, K. S. and DAVIDSON, T. L. (1990) *Meat Sci.* **28**, 171.

ZIPSTER, M. W., KWON, T. W. and WATTS, B. M. (1964) *J. Agric. Fd. Chem.* **12**, 105.

ZÖLLNER, N. (1975) *Proc. 9th. Int. Congr. Nutr., Mexico*, **1**, 267.

ZUBER, F. (1993) *Indust. Aliment. Agri.* **110**, 431.

ZUBRZYCHA-GAARN, E. E., BULMAN, D. E., KARPATI, G., BURGHES, A. H. M., BELFALL, B., KLAMUT, H., TALBOT, J., HODGES, R. 1., RAY, P. N. and WORTON, R. G. (1988) *Nature, Lond.* **333**, 466.

ZUCKERMAN, H. and SOLOMON, M. B. (1998) *J. Muscle Fds.* **9**, 419.

Index